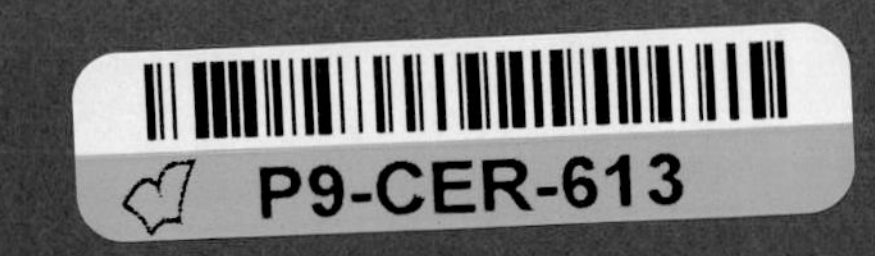

Where to find useful lists

NOTE: Other useful lists and tables, such as size-prediction tables and critical-temperature tables, may be found by referring to the **subject index** that begins on page 409.

TEMPERATE-ZONE POMOLOGY

TEMPERATE-ZONE POMOLOGY

Melvin N. Westwood
Oregon State University

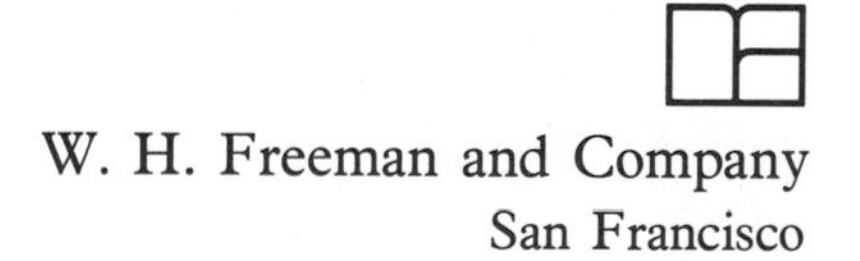

W. H. Freeman and Company
San Francisco

Cover: A flowering branch of peach (*Prunus persica*)
from a linoleum-block print by Henry Evans.

Library of Congress Cataloging in Publication Data

Westwood, Melvin Neil, 1923–
Temperate-zone pomology.

Bibliography: p.
Includes index.
1. Fruit-culture. 2. Nuts. I. Title.
SB355.W43 634′.0912 77–26330
ISBN 0-7167-0196-0

Printed in the United States of America

9 8 7 6 5 4 3 2 1

To the memory of
Neil Westwood
my first horticulture teacher

The greatest service which can be rendered any country is, to add an useful plant to its culture. . . . There is not a sprig of grass that shoots uninteresting to me.

THOMAS JEFFERSON

Contents

Preface

Pomology is a congenial blend of science and art. It is not, strictly speaking, the study of the physiological functioning of plants, yet an understanding of physiology is essential to the pomologist. This work attempts to convey the physiological concepts and principles that underlie sound horticultural practices. A mere enumeration of practices, complete in every detail, with no discussion of causes or interrelationships, would ill equip the pomologist to solve the myriad and complex problems to be found in the orchard or vineyard. An understanding of both theory and practice, together with specific information about the specific situation, are the ingredients necessary for problem solving.

In addition to its reliance on knowledge of plant physiology, pomology integrates specialized knowledge from the fields of geography, climatology, meteorology, soil science, chemistry, physics, nutrition, genetics, pathology, entomology, economics, business, and resource management.

Because of the nature of the subject matter, it seemed to me that the most useful format for the book would be to have the chapters treat topics like "flowering," "harvest," and "pests," rather than particular crops. Naturally, the unique characteristics of species and cultivars in relation to the topic under discussion are noted in the appropriate chapters. Because of this topical format, diligent use of the index will be essential when the book is used as a reference for a particular fruit or nut species.

The scope of the book is confined to the important deciduous crops of the temperate zone, including both orchard and small fruit crops. With the exceptions of the herbaceous strawberry and cranberry, which are evergreens, the species considered are deciduous, woody perennials.

As much as possible, the chapters are organized chronologically. The first few chapters deal with factors crucial to the proper siting and establishment of the planting. Decisions at this stage are critical because they are fixed—i.e., they cannot be readily changed once the planting is established. Chapters 7 through 13 move sequentially through the season from anthesis to harvest and post-harvest handling. Chapters 14 (on growth regulators), 15 (on dormancy and hardiness), and 16 (on pests) stand by themselves. Chapter 17 is a general integration and overview of all of the components of fruit growing. The material in the chapter on pests, insects, and diseases is not usually covered in pomology courses, but is included here because of the importance of pest management in fruit growing. This material should make the book more useful as a reference for fruit growers, fieldmen, and extension agents.

This book has been written primarily as a text for pomology classes whose students have already acquired a knowledge of general botany or biology and usually have taken at least one course in general horticulture. The instructor should use discretion in the use of certain tabular data presented primarily as reference material rather than for study. Judicious choice of textual materials, together with the use of a wealth of information available on local crops and problems, is a prerogative which every teacher should exercise. The Appendixes contain certain useful conversion tables and formulae, many of which will aid the student in the use of the metric units used throughout the book. The conversion to the metric system in the United States is to take place during the next ten years.

No specific chapters are designed as laboratory exercises, but both field and lab projects are readily apparent in Chapters 3 through 15. I have found that more than 70 percent of my students in recent years are not from farms; thus, it is essential to their training to include a good number of practical field activities to familiarize them with the techniques of orchard layout, planting, training, pruning, pollination, fruit set, and thinning.

The examples I have used to elucidate a principle may not always be appropriate to every fruit growing area. For example, fruit-thinning practice and size prediction for irrigated plants might not be appropriate to nonirrigated areas if levels of soil moisture there limit fruit growth.

The literature cited is by no means exhaustive, but rather gives the reader access to a number of supporting works, many of which are review articles that list many references on the topic. I probably have omitted some important references or have hidden them under cited authors or editors of books or review articles. A few general references are listed at the end of each chapter, with the complete listing of literature cited at the end of the book.

I am indebted to many people for their help in this work. Individual chapters or drafts of sections were reviewed by R. H. Groder, P. J. Breen, F. J. Lawrence, M. M. Thompson, M. W. Williams, A. N. Roberts, Mary Tingley Compton, O. C. Compton, R. L. Stebbins, M. J. Bukovac, D. G. Richardson, M. T. AliNiazee, P. B. Lombard, and C. J. Weiser. Valuable suggestions were made regarding the format and on early drafts of the manuscript by R. Garren, E. Bates, J. Janick, G. H. Oberly, E. L. Proebsting, Jr., and E. Hansen. I am especially grateful to F. G. Dennis and G. C. Martin, who reviewed the entire manuscript, and to P. H. Westigard and H. R. Cameron, who gave me extra help on the portions dealing with insects and diseases. H. O. Bjornstad has given me valuable technical assistance during the years of writing, and Janice Seibel provided a number of line drawings. None of them reviewed the final draft; thus, any errors or weaknesses are entirely my own. A number of people helped type and collate various drafts, especially Bonnie Dasenko, Dawn Nelson, Angie Clement, and Gene Brown.

I am indebted also to those who trained me for this task—to my father, who taught me more farming than I suspected; to S. W. Edgecombe, a gifted teacher and advisor; to R. M. Bullock, L. P. Batjer, and J. R. Magness, who were my early mentors in pomological research.

I have appreciated the considerable patience shown by my wife, Wanda Shields Westwood, and by my family during the long course of this effort.

MELVIN N. WESTWOOD

July 1978

TEMPERATE-ZONE POMOLOGY

1

Fruit and Nut Production Areas

WORLD PRODUCTION AREAS

Regardless of their place of origin, deciduous fruit and nut species are grown wherever climate, soil, and moisture conditions are suitable. European grape, for example, grows very well in the Napa Valley of California, and European pear is adapted to such diverse areas as South Africa, Australia, and the western United States (Fig. 1-1).

Climate determines the limits of distribution of species, and world climate is regulated by such things as wind direction in relation to bodies of water the angle of incoming solar rays, the length of day (latitude), and the amount of carbon dioxide (CO_2) in the atmosphere. Water vapor and carbon dioxide molecules trap heat waves, which are radiated outward from the sun-warmed surface of the earth. Figure 1-2 indicates the main factors regulating the concentration of carbon dioxide. In recent years carbon dioxide in the atmosphere has been increasing at the rate of about 1 part per million (ppm) per year due to increased burning of fossil fuels.

The world's natural vegetation zones are mapped in Figure 1-3, along with the length of day for four temperate-zone latitudes. Deciduous fruits and nuts generally are grown in areas where the species are climatically adapted and in desert areas where irrigation has been developed. Temperate deciduous species are confined mostly to the middle latitudes, from about 30° to about 50°. Production may extend to lower latitudes at high elevations and to higher latitudes in regions where large bodies of water have a moderating influence, as in western Europe, where the Atlantic Ocean has a warming effect. In Figure 1-4, the darkened areas on the world map indicate the principal areas of deciduous fruit and nut production.

The quantities of fruits and nuts produced commercially in nine major areas of the world are given

Figure 1-1 Vineyards of the world. **(A)** Rhine River Valley, West Germany. **(B)** Ontario, Canada. **(C)** North coastal region of California. **(D)** Barussa Valley, South Australia. Grapes are grown worldwide in temperate latitudes and in subtropical and tropical climates. In the tropics, the vines are defoliated and pruned after harvest, causing a new cycle of growth and a second crop the same year. [B: Courtesy of A. Hutchinson, Vineland Experiment Station; C: Courtesy of R. J. Weaver, University of California; D: Courtesy of R. Garren, Jr.]

in Table 1-1. Deciduous crops (grapes, pome fruits, stone fruits, figs, berries, and nuts) account for just under 100 million metric tons (m tons) per year. A substantial unreported tonnage of fruits is grown and consumed in local areas. This is because many fruits are soft and cannot be stored or shipped without cold storage.

Table 1-2 gives a breakdown of deciduous tree fruit and nut production by country. The greatest production occurs in Europe, but substantial production occurs also in North America, Asia Minor, Asia, South Africa, Australia, New Zealand, Argentina, and Chile. Unfortunately, production in Russia and some other eastern European countries is not known. The following list of deciduous tree crops shows the principal countries of production for each crop:

Crop	*Where Grown*
Apple	U.S., Italy, West Germany, France, Japan, and China
Pear	Italy, China, U.S., France, Japan, West Germany, and Spain
Peach	U.S., Italy, France, Japan, Spain, Argentina, and Greece
Cherry	West Germany, U.S., Italy, France, and Yugoslavia
Plum	Yugoslavia, U.S., West Germany, France, and Italy
Apricot	U.S., Spain, France, Italy, and Turkey
Almond	U.S., Spain, and Italy
Filbert	Turkey, Italy, Spain, and U.S.
Pecan	U.S., and Mexico
Pistachio	Iran, Turkey, and Syria
Walnut	U.S., France, Italy, India, and Turkey
Chestnut	Italy, Spain, and Portugal

Production data from China are incomplete, but Pieniazek (1973) reported a double or triple increase between 1957 and 1973 to a total of 6 to 9 million metric tons. The most important cultivated deciduous species in China are apple, pear, peach, apricot, grape, persimmon, jujube, cherry, plum, walnut, and chest-

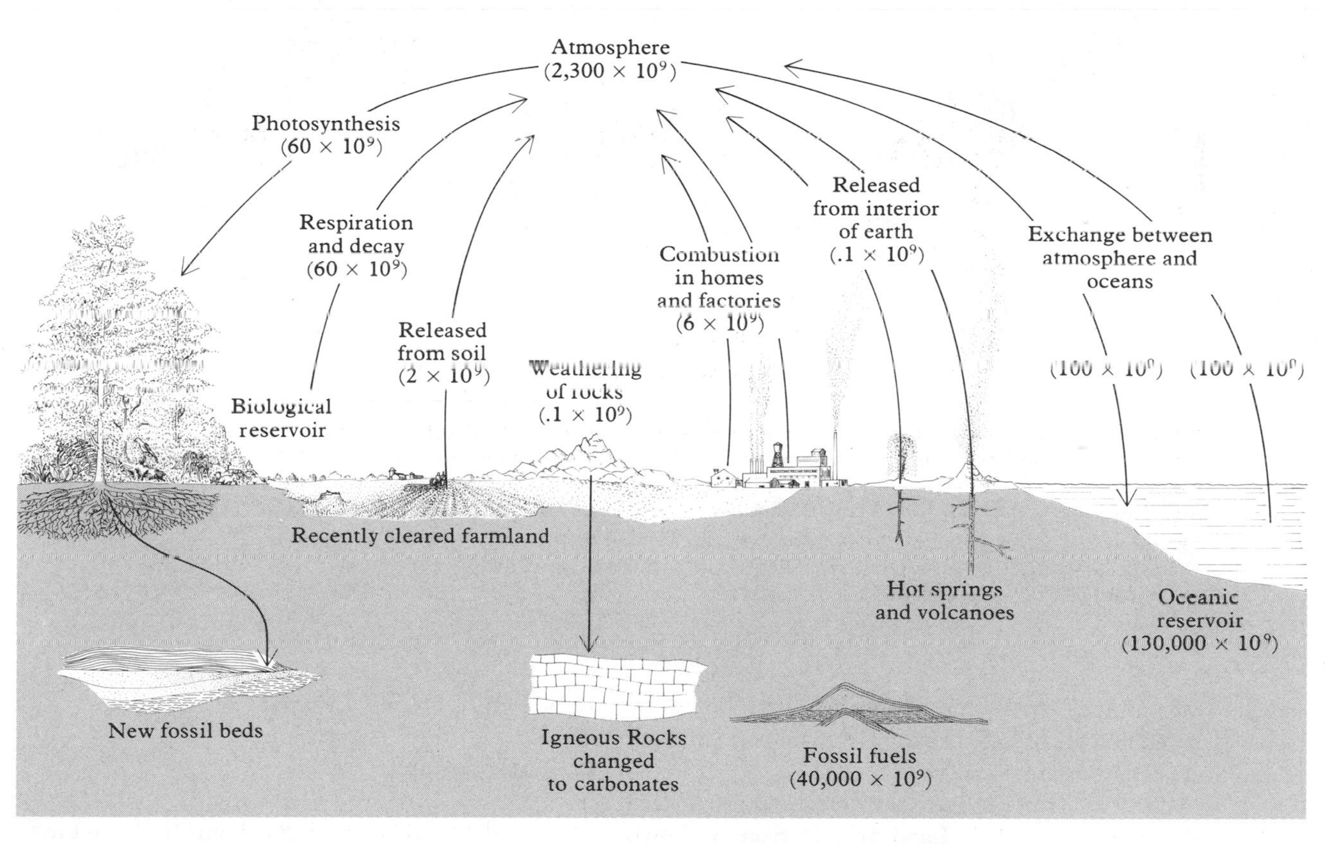

Figure 1-2 Processes that influence the concentration of carbon dioxide in the atmosphere. The numbers in parentheses indicate the number of tons of carbon dioxide being used in a process each year. Because carbon dioxide molecules trap escaping heat rays, its concentration is important in determining world temperature. Recent increases resulting from burning of fossil fuels could cause a rise in world mean temperature. [After G. N. Plass, "Carbon Dioxide and Climate." Copyright © 1959 by Scientific American, Inc. All rights reserved.]

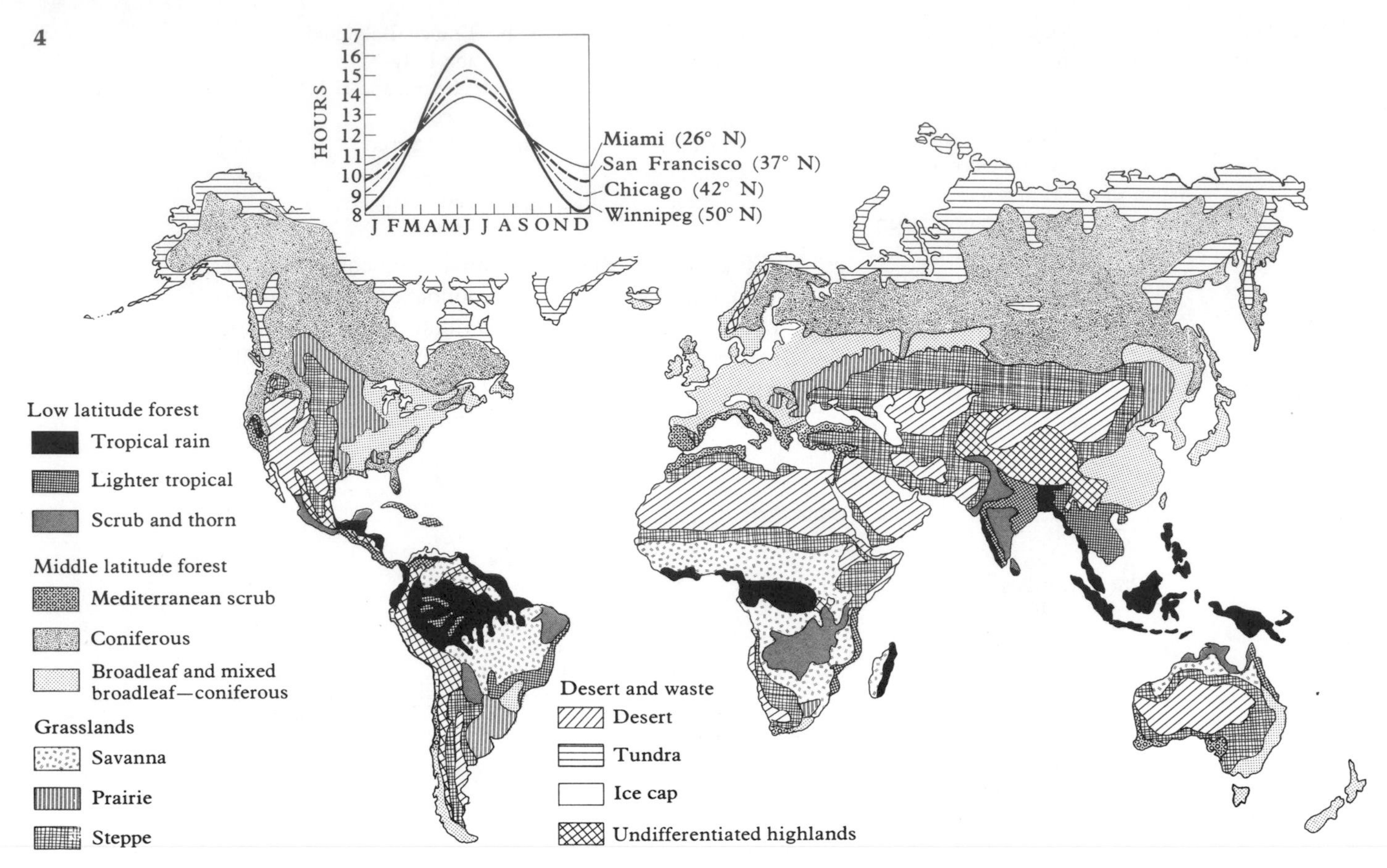

Figure 1-3 The world's natural vegetation zones are created by the effects of latitude, rainfall, elevation, and nearness to large bodies of water. The inset graph shows that in going from Miami north to Winnipeg one finds that climate and seasonal daylength change significantly—the plant species that can be cultivated within this temperate zone vary widely, as do the natural species. [After Jules Janick, *Horticultural Science.* W. H. Freeman and Company. Copyright © 1963, 1972; Inset graph after F. W. Went, "Climate and Agriculture." Copyright © 1957 by Scientific American, Inc. All rights reserved.]

nut. The principal apple cultivar is Ralls, followed by several others of U.S. origin. The main pear cultivars are from Oriental species.

Canadian fruit production by crop and province is shown in Table 1-3. Land area in fruit is shown graphically in Figure 1-5. The fruit-growing provinces are Ontario, British Columbia, Quebec, Nova Scotia, and New Brunswick. The most important crops in Canada are apple, grape, peach, cherry, and strawberry.

U.S. PRODUCTION AREAS

Climatic Areas

Most of the production of deciduous fruits and nuts in the United States is concentrated in a few areas. The agricultural areas of California alone account for half the U.S. annual production of about 10 million metric tons. Other important fruit-growing areas are located in the Pacific Northwest, the eastern shores

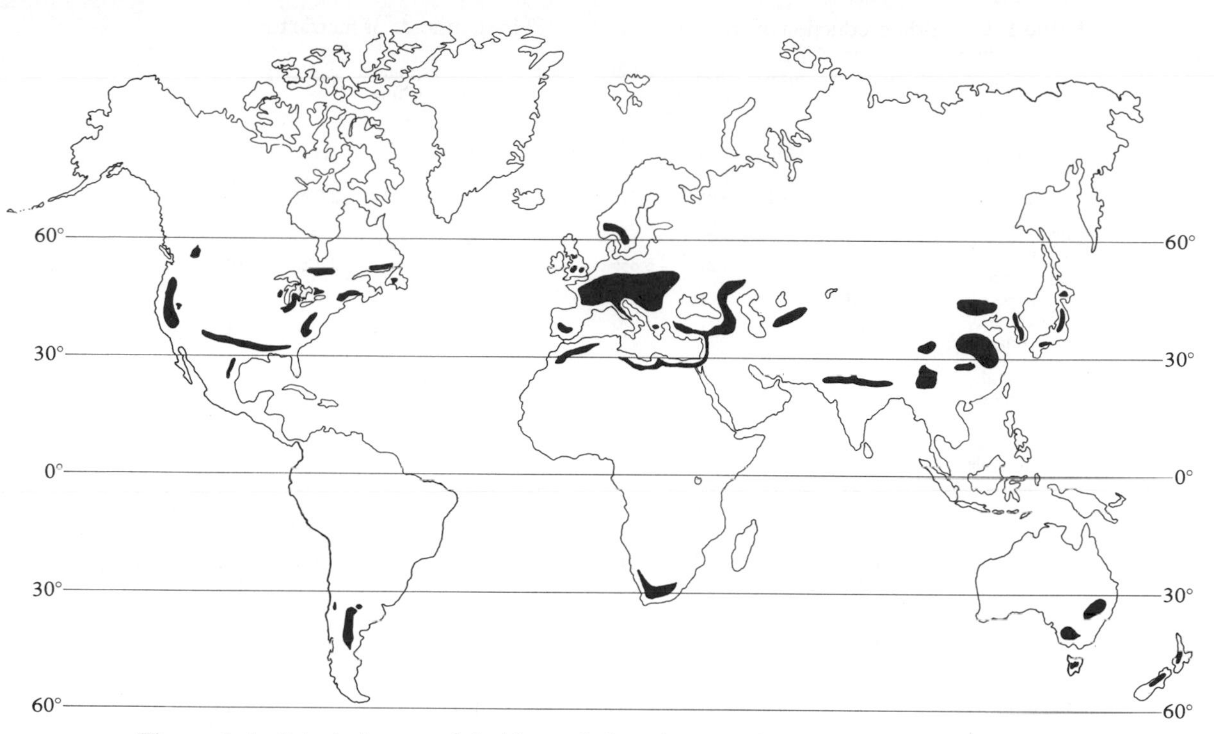

Figure 1-4 Principal areas of deciduous fruit and nut production. In both hemispheres, they lie mostly within latitudes 30° to 50° but are extended into higher latitudes by the moderating influence of nearby bodies of water, and into lower latitudes by the cooling influence of higher elevations.

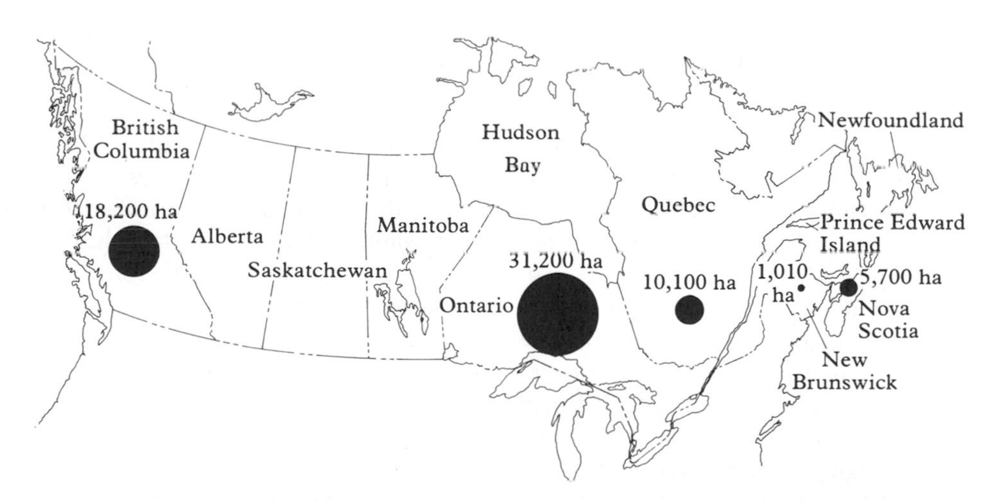

Figure 1-5 Fruit-growing areas of Canada, showing the moderating influence of large bodies of water on the climate. The circles represent the number of hectares being used for fruit production. Very little fruit is produced in the central provinces because of the short summers and cold, harsh winters. [After Krueger, 1972]

Table 1-1 World production of fruits and nuts in 1967 in thousands of metric tons.

Crop	Europe	USSR	North America	Latin America	Near East	Far East	Mainland China	Africa	Oceania	World Total
Grapes										
Wine	29,553	2,340	1,651	3,896	189	25	—	1,823	251	39,728
Raisins	1,540	—[1]	1,642	98	3,335	[2]	—	99	1,084	7,798
Total	32,755	3,441	3,847	4,473	4,727	314	130	2,059	706	52,452
Citrus										
Oranges and tangerines	3,876	35	5,273	5,856	2,221	3,314	650	2,011	198	23,434
Grapefruit	10	—	1,616	212	268	67	—	103	10	2,286
Lemons, limes, etc.	940	—	666	475	390	609	—	79	23	3,182
Tropical and subtropical fruits										
Bananas	409		5	16,077	171	7,864	—	1,137	158	25,991
Pineapples	3	—	884	912		1,404	—	345	130	3,678
Dates	21	—	19	7	1,416	150	—	352		1,965
Figs	897	—	38	37	336	2	—	121	1	1,432
Pome fruits										
Apples	14,406	—	2,561	746	912	1,333	360	225	481	21,424
Pears	3,754	—	460	226	221	536	880	87	153	6,297
Stone fruits										
Plums, prunes	2,957	—	540	161	138	88	—	39	35	3,958
Peaches	2,233	—	1,294	464	147	359	—	183	147	4,827
Cherries	1,193	—	203	6	87	9	—		9	1,507
Apricots	623	—	137	30	202	109	—		44	1,145
Berries										
Strawberries	553	—	236	122	2	103	—		6	1,022
Raspberries	94	—	44				—		3	142
Other berry[3] fruits	340	—	85		100	4	—		5	534
Nuts (incl. coconuts)	1,132	—	254	2,463	214	22,547	—	1,642	1,664	30,016

SOURCE: *Production Yearbook, 1968*, F.A.O. 1969.
[1]Dash signifies that figures are not available.
[2]Absence of data signifies insignificant crop.
[3]Gooseberries, currants, cranberries, blueberries, others.

of the Great Lakes, a narrow belt along the Appalachian Mountains, the Northeast, and the southeastern states. Some fruits or nuts are grown in every state, but most of the production is located in ten or twelve states (Table 1-4; Figs. 1-6, 1-7, 1-8, 1-9, and 1-10). The distribution pattern was not by design, because commercial fruit growing was tried in nearly every agricultural region of the country. The overriding determinant of success was climate, and man recognized this by expanding production in the favorable areas and changing to other crops in those unsuited to fruit. Although success is not solely determined by climate, it is climate that sets the limits within which other factors can operate.

The northern limit of deciduous fruit growing is found where the hardiest species, such as apple, will survive the coldest winters and where there is a sufficient number of frost-free days during summer to mature the crop. The northernmost orchards are found near large bodies of water, which mitigate the effects of cold air masses during winter and prolong the frost-free growing season. The southern limit for deciduous orchards is determined by the least amount of winter chilling required to break rest in a given cultivar (cultivated variety). In warmer zones, subtropical fruits, such as citrus, are grown. In border areas, where both citrus and deciduous fruit (peaches, for example) are grown, the citrus suffer

Table 1-2 Deciduous tree fruit annual production by country, 1968–1974, in thousands of metric tons (USDA Production Survey).

Country	Apple	Pear	Peach	Cherry	Plum	Apricot	Almond	Filbert	Pecan	Pistachio	Walnut
North America											
Canada	434	39	57	18	8	3					
Mexico	157	35							22		
United States	2797	602	1400	234	530	156	65	9	100		113
Europe											
Austria	172	49	7	28	86	14					
Belgium-Luxembourg	257	64		18	10						
Denmark	79	8		3							
France	1812	476	559	112	160	97					26
Germany—West	1838	450	38	274	503	5					
Greece	206	104	194	28	18	38					
Italy	1901	1629	1166	196	138	96	28	83			20
Netherlands	438	84		6	10						
Norway	51	10		5	14						
Portugal							5				
Spain	535	320	246	52	70	141	38	20			
Sweden	42	6			2						
Switzerland	104	22		37	42	10					
United Kingdom	414	60		8	54						
Yugoslavia	358	111	49	93	892	24					
Asia Minor											
Iran						47	8			14	4
Lebanon	147	6									
Turkey	760	175	103	88	109	90		194		9	10
Syria										1	
Asia											
Japan	1028	470	273	11	53						
India											14
South America											
Argentina	424	90	226	2	62	16					
Chile	88	20	37	4	22	4					
Africa											
Morocco							4				
South Africa	243	97	144		8	20					
Oceania											
Australia	407	171	115	9		39					
New Zealand	128	18	24		4	5					
Crop Totals	14826	5146	4652	1223	2796	803	148	306	122	24	187

Table 1-3 Canadian deciduous fruit production and land planted in fruit.

		Annual Production			
		By Province (as percent of total)			
Crop	Total (in thousands of metric tons)	Maritime Provinces[1]	Quebec	Ontario	British Columbia
Apple	434.5	17	26	28	29
Pear	39.0	4	0	58	38
Peach	57.0	0	0	88	12
Cherry	18.3	8	0	78	22
Plum	8.5	0	0	57	43
Apricot	3.0	0	0	0	100
Strawberry	15.0	16	21	34	29
Raspberry	6.6	0	5	6	89
Grape	75.5	0	0	84	16
Blueberry	11.5	58	24	0	18
Cranberry	5.0	2	0	0	98
		Land Area in Fruit			
Total		By Province (as percent of total)			
60,520 hectares		2	17	52	30

[1]Nova Scotia and New Brunswick.

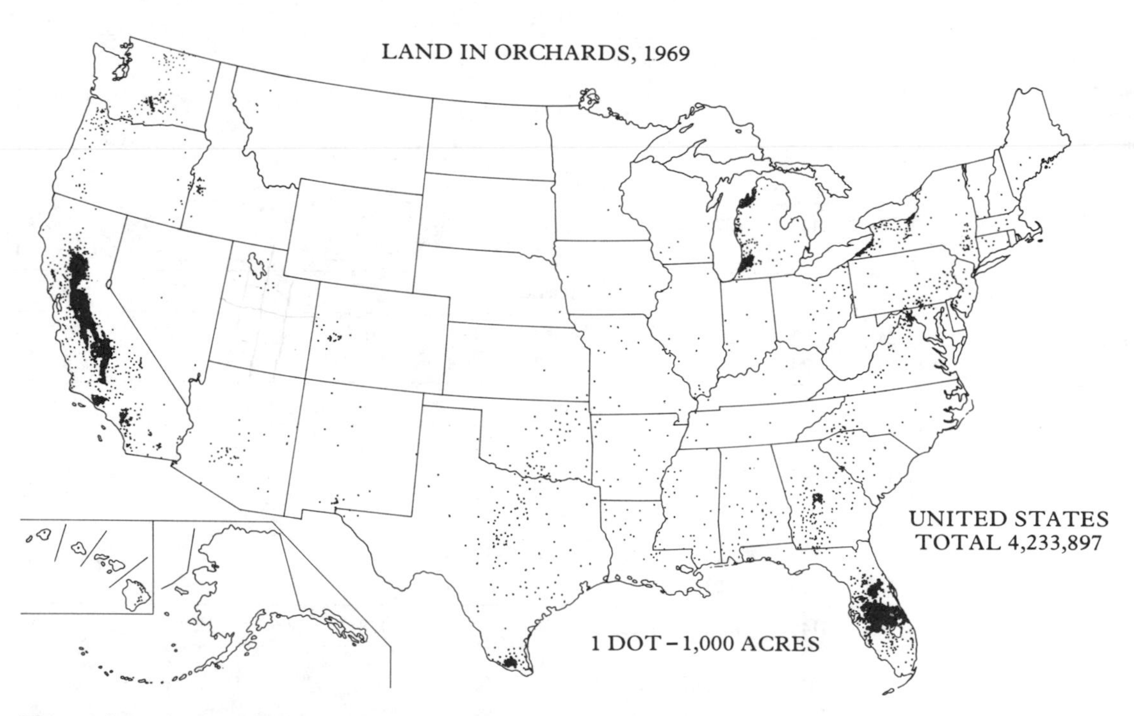

Figure 1-6 Areas of fruit and nut production in the U.S. Those in the most-southern regions produce mostly citrus and other subtropical species. The north central region has very few orchards because of frequent damaging winter freezes. [After U.S. Department of Commerce]

Table 1-4 Average annual production of deciduous fruit and nut crops in the U.S., 1969–1973, by crops and principal states, expressed in thousands of metric tons and each crop as percent of total.

Crop	Average Annual Production of Crop, 1969–1973 (in thousands of metric tons)	Percent of Total Average Annual Production, 1969–1973	Principal States Where Grown
		Stone Fruit	
Apricot	160	1.7	Cal., Wash., Utah
Peach, Freestone	648	6.9	Cal., S.C., Ga., Mich., N.J., Pa. (34 states produce peaches)
Peach, Cling	678	7.2	Cal.
Nectarine	51	0.5	Cal.
Plum	92	1.0	Cal.
Prune (fresh basis)	388	4.1	Cal., Ore., Wash., Mich., Id.
Sweet Cherry	116	1.2	Ore., Wash., Cal., Mich., N.Y.
Sour Cherry	119	1.3	Mich., N.Y., Pa., Utah, Ore.
Total	2251	24.0	
		Pome Fruit	
Apple	2364	25.2	Wash., N.Y., Mich., Cal., Pa.
Pear	604	6.4	Cal., Wash., Ore., Mich., N.Y.
Total	2968	31.7	
		Tree Nuts	
Almond	116	1.2	Cal., Ariz.
Walnut	116	1.2	Cal., Ore.
Pecan	93	1.0	Ga., Tex., Ala., La., Okla., Miss., Ark., N. Mex.
Filbert	9	0.1	Ore., Wash.
Tung	54	0.6	Miss., Fla., La., Ala.
Total	388	4.1	
		Other Fruits	
Fig	42	0.4	Cal., Tex.
Persimmon	2	0.02	Cal.
Pomegranate	3	0.03	Cal.
Total	46	0.5	
Subtotal (tree fruits and nuts)	**5654**	**60.3**	
		Small Fruit	
Grape	3281	35.0	Cal., N.Y., Mich., Wash., Pa., Ohio, Ariz., Ark.
Strawberry	253	2.7	Cal., Ore., Wash., Mich., Fla.
Blackberry	35	0.4	Ore., Tex., Wash., Mich., Cal., Ark.
Raspberry	30	0.3	Wash., Ore., Mich., N.Y., Ohio, Pa., Minn.
Blueberry	30	0.3	Maine, N.J., Mich., N.C., N.H., Mass., Wash.
Cranberry	92	1.0	Mass., Wisc., N.J., Wash., Ore.
Currant	1	0.01	N.Y., Wash., Mich., Ore., Cal., Pa.
Gooseberry	1	0.01	Ore., Mich., Wash., Cal., N.Y., Mo., Minn.
Other	0.3	0.003	
Total	**3724**	**39.7**	
Total U.S. fruits and nuts	**9378**	**100**	

SOURCE: USDA

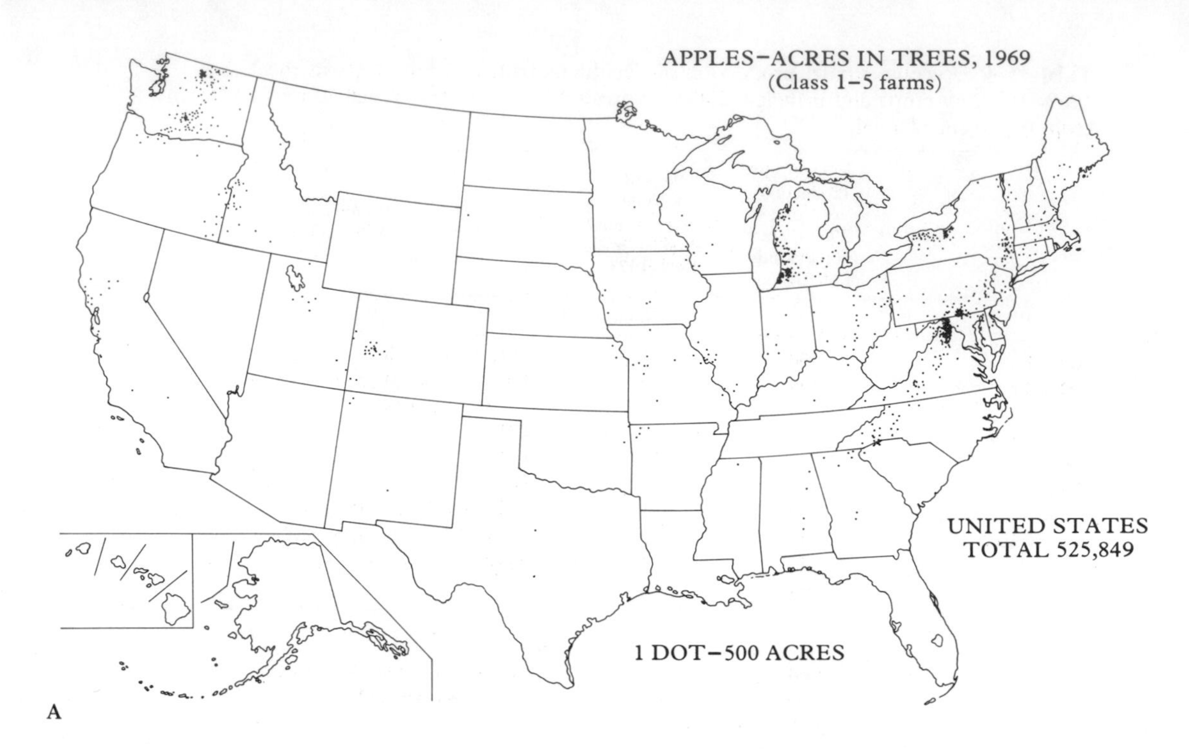
APPLES–ACRES IN TREES, 1969
(Class 1–5 farms)
UNITED STATES
TOTAL 525,849
1 DOT–500 ACRES

A

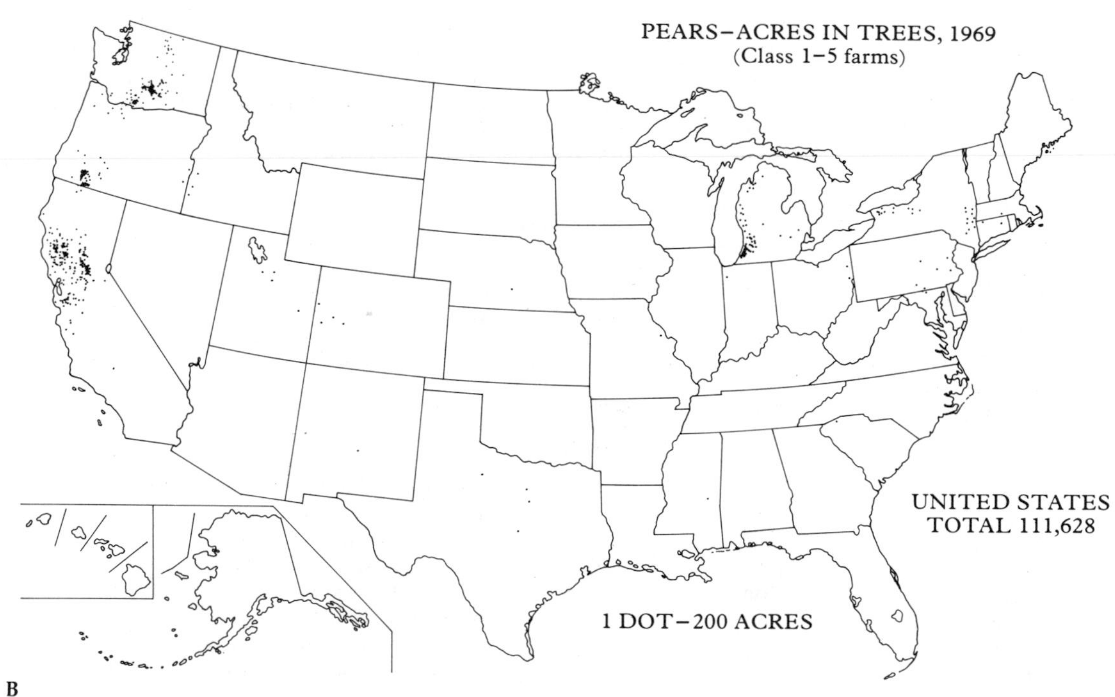
PEARS–ACRES IN TREES, 1969
(Class 1–5 farms)
UNITED STATES
TOTAL 111,628
1 DOT–200 ACRES

B

Figure 1-7 Areas of pome-fruit production in the U.S. Apples are grown mainly in the eastern and Great Lakes states and in the western states. Because of the prevalence of fire blight in the humid eastern states, most of the pear production is in the arid regions of the Pacific slope. [After U.S. Department of Commerce]

cold injury in some years and the deciduous fruit suffer from inadequate chilling in others.

Climate not only sets the limits for a given crop but also determines to a great degree the consistency of annual cropping and the quality to be expected. Thus a would-be grower should not merely seek a climate within the limits of survival for a cultivar, he should seek one that is optimum for fruit set and quality. Such a climate produces fruit more economically than others, and thus, other factors being equal, it places the grower in an advantageous economic position.

As more is learned about the climates of specific areas, and as more is learned about the physiologic responses of plants to their environments, the better the two can be matched. Recent studies have given us better knowledge of how freezing kills plants and of how winter chilling affects dormant plants. This new information may lead to more efficient fruit growing by better use of genetic material in breeding and selection, chemical treatments to increase hardiness, better physical protection of tender parts, and more effective modification of microclimates in orchards or fields at critical times.

The suitability of a plant species depends in part upon its adaptation to the climate in which man places it. Many kinds of trees are used in areas to which they are not native. Thus, an understanding of the interrelationship of climate and plant function is necessary to matching them properly. The general climatic requirements for temperate-zone trees and shrubs are as follows:

1. Winter temperatures must not be so cold that they kill the plants (see Fig. 1-11).

2. Winters must be cold enough to give buds adequate chilling to break winter rest.

3. The growing season (number of frost-free days) must be long enough to mature the crop.

4. Temperature and light during the growing season must be adequate for the species in question to develop fruit of good quality.

Besides temperature, many other factors differ between the subtropical latitudes and those of the temperate regions. Summer days are longer and nights shorter in high latitudes; the length of the frost-free growing season is shorter. Also, there are differences in elevation and nearness to large bodies of water that can alter not only the range of temperatures but also can alter the light intensity, light quality, and the diurnal temperature pattern. Species that evolved in a particular climate are, by virtue of built-in control mechanisms, closely synchronized with their environments, not only during the four seasons of the year but during the important transitions between seasons.

Because of the many different climatic factors a plant must be adapted to in order to function optimally, it is generally unwise to plant a cultivar or species at the very limit (north or south) of its natural range. Such plants must compete in productivity and quality with others more ideally located.

The concentration of the commercial production of each species in relatively few areas has occurred because the species are best adapted to the climates of those areas: apples are best adapted to the coldest commercial areas; pecans and almonds do best in the warmest areas; other nut crops and peaches are grown in latitudes just north of the pecan belt; pears, cherries, and plums are grown in latitudes just south of the apple belt. The many exceptions to these generalizations are due to influences of elevation, the proximity of large bodies of water, and to varietal differences within a species.

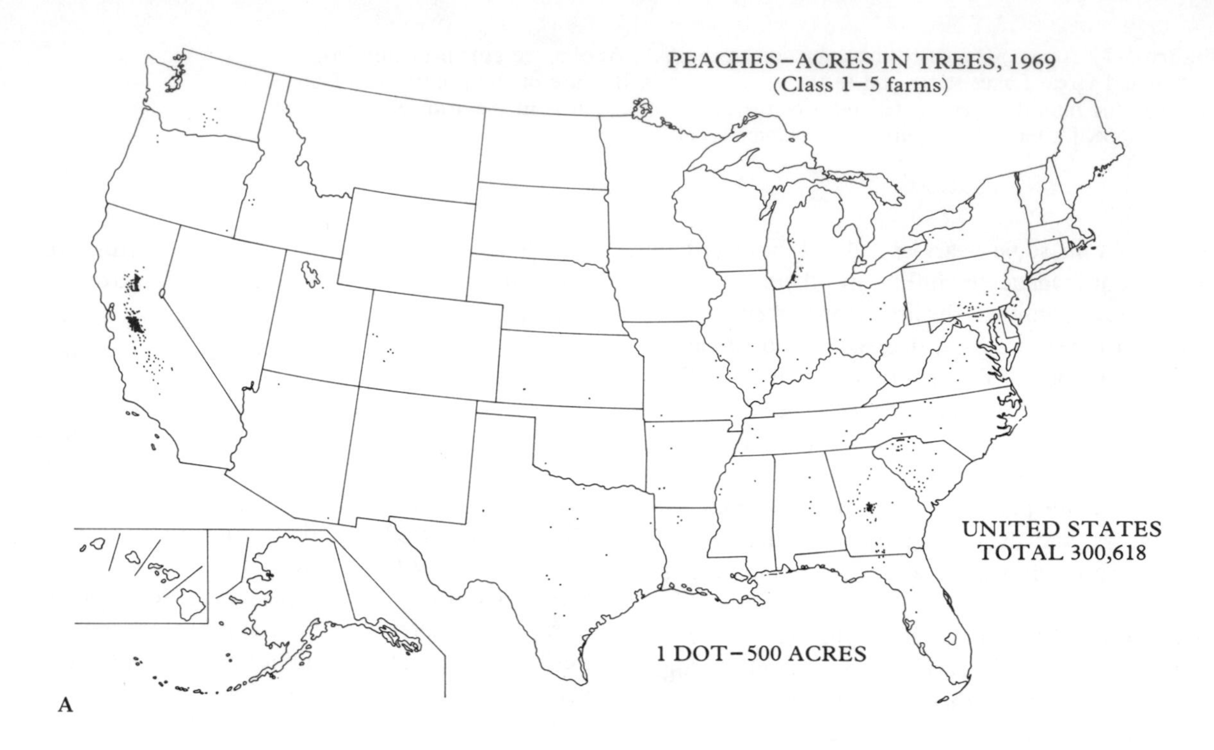
PEACHES–ACRES IN TREES, 1969
(Class 1–5 farms)
UNITED STATES TOTAL 300,618
1 DOT–500 ACRES

A

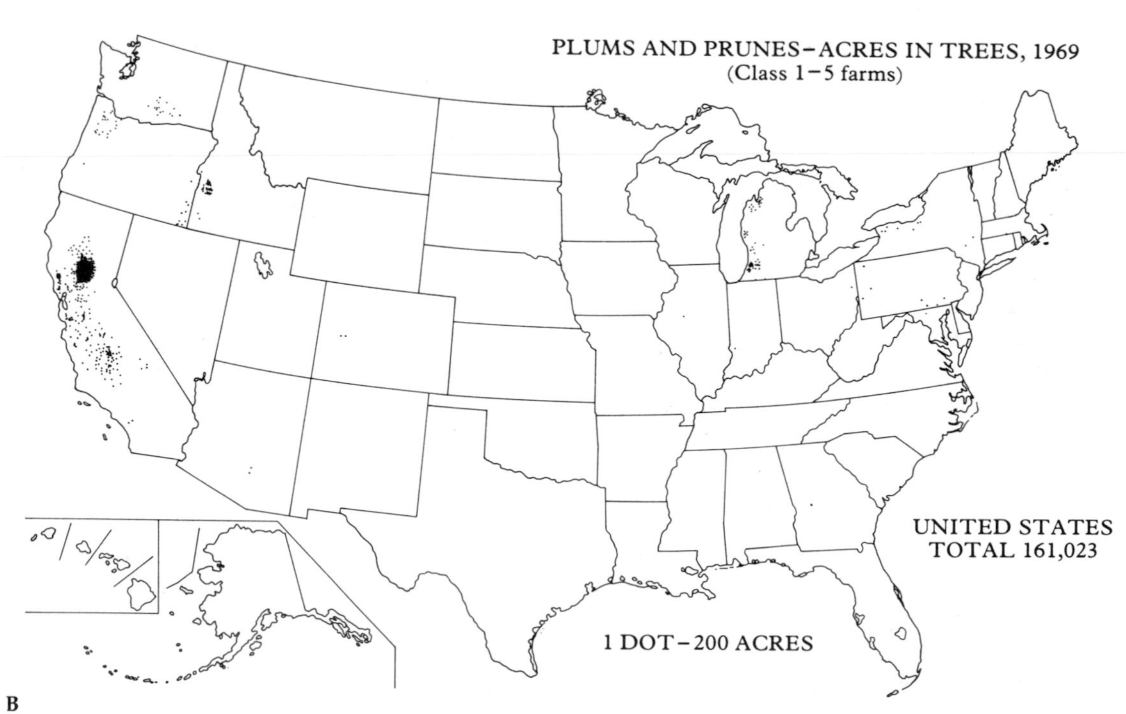
PLUMS AND PRUNES–ACRES IN TREES, 1969
(Class 1–5 farms)
UNITED STATES TOTAL 161,023
1 DOT–200 ACRES

B

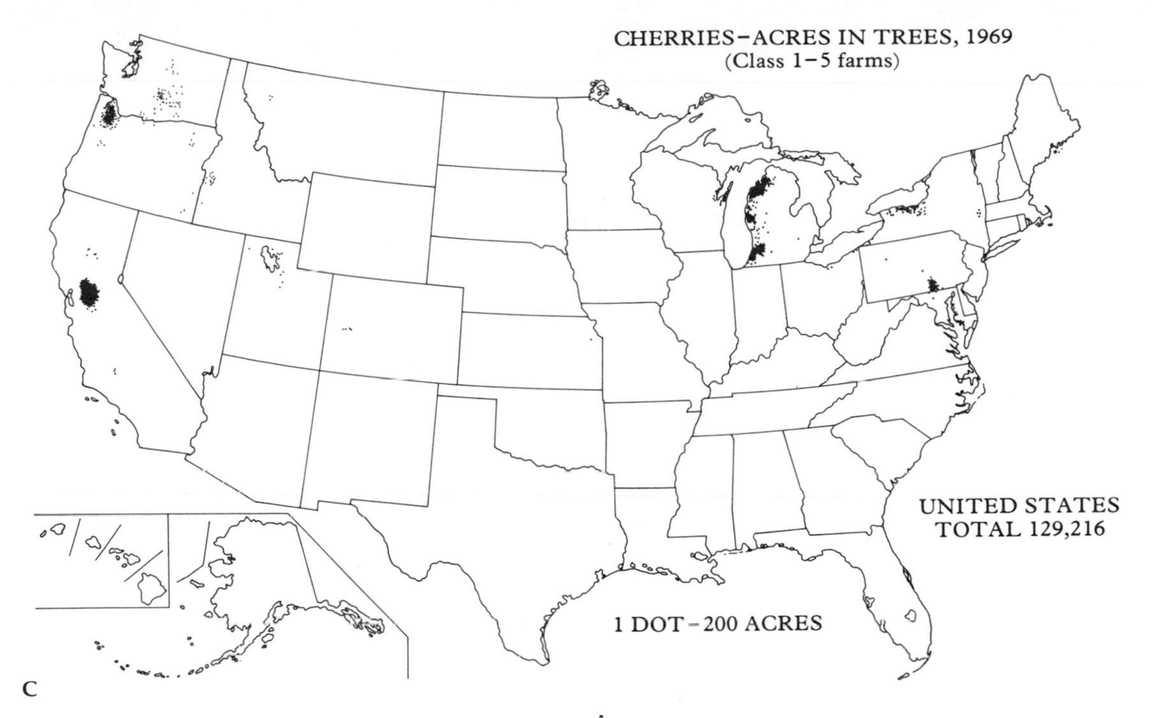

Figure 1-8 Areas of stone-fruit production in the U.S. Peaches are grown in most states except the cold central-plains states. Cherries and plums are grown mostly in the Great Lakes states and in the western states. Apricot and almond production (not shown) is located in the states of the Pacific slope. [After U.S. Department of Commerce]

Orchard Site

The site of an orchard is very important and can determine the success or failure of the venture. Upland, rolling, or benchland sites are good, provided the soil is suitable. The choice should take into account the local effects of temperature inversions in the spring, average spring air drainage, and wind velocity. Because the cold air from radiation cooling at night moves down a slope to the lowest part of a valley, the orchard should not be planted less than about 15 meters up from the base of a slope (see Chapter 15). Steep slopes may prevent efficient orchard operations, but some slopes can be improved by contouring.

The summer and winter temperatures at the site are important, but spring temperatures may be the critical factor determining suitability. In some valleys, each 30-meter rise in elevation results in higher nighttime minimums in the spring of three to six degrees Celsius. Such an advantage can save the grower thousands of dollars in frost protection costs over the years. Level land also can be a suitable site if local conditions, such as the effects of a nearby lake or river, naturally prevent frost there, or if there is a good means of air drainage from the area.

Local obstructions should be considered in choosing sites. Timbered lands are reservoirs of cold air and may also compete for soil nutrients, water, and sunlight. Also, woods or windbreaks below orchards prevent good air drainage and may create frost pockets. For these reasons, plantings should not be closer than about 25 meters to timber.

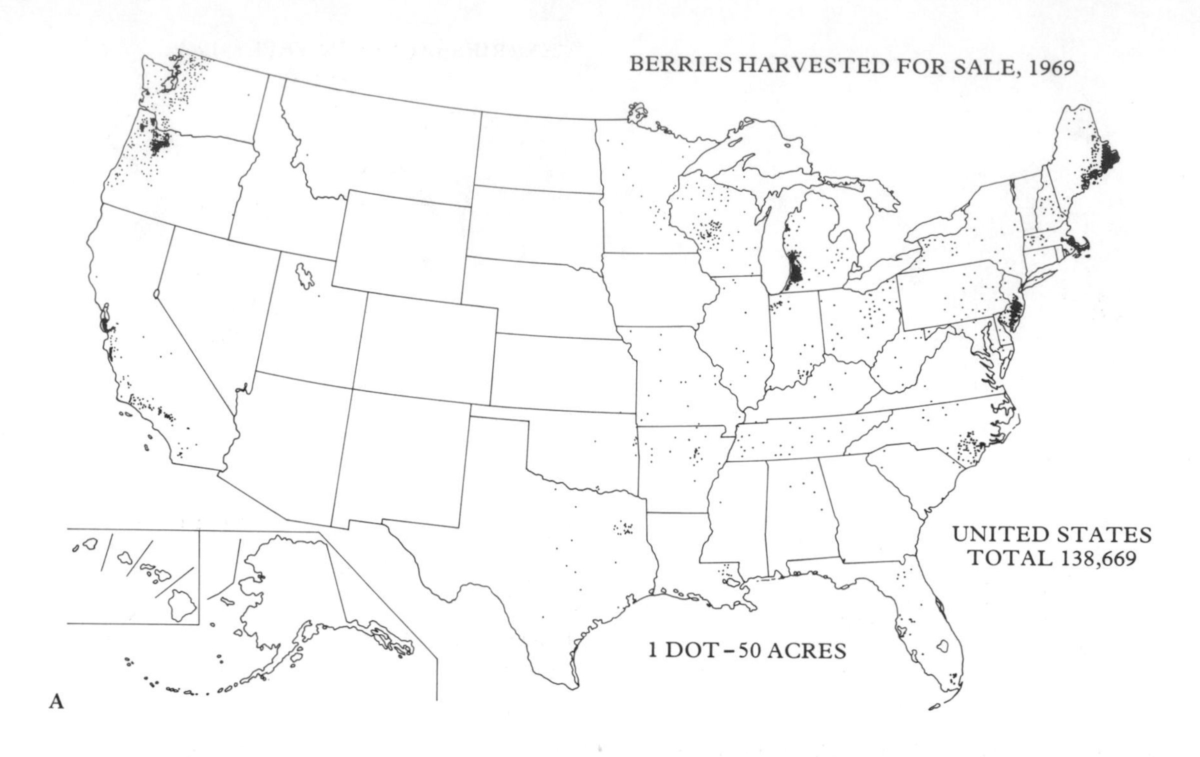
BERRIES HARVESTED FOR SALE, 1969
UNITED STATES
TOTAL 138,669
1 DOT – 50 ACRES

A

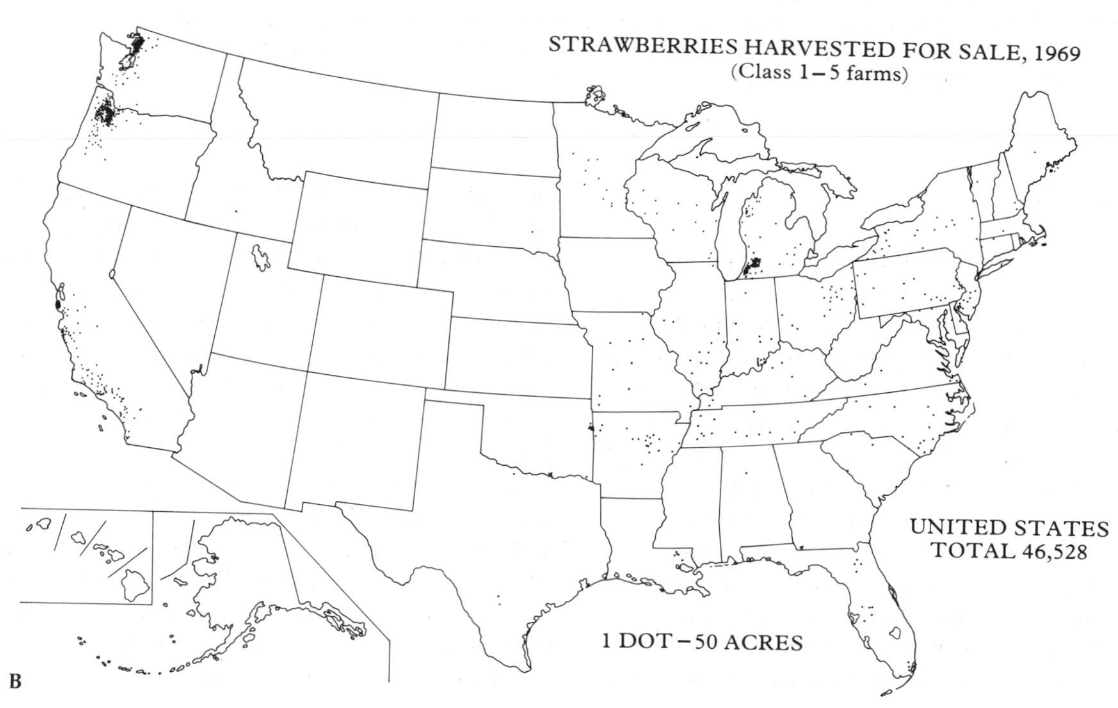
STRAWBERRIES HARVESTED FOR SALE, 1969
(Class 1 – 5 farms)
UNITED STATES
TOTAL 46,528
1 DOT – 50 ACRES

B

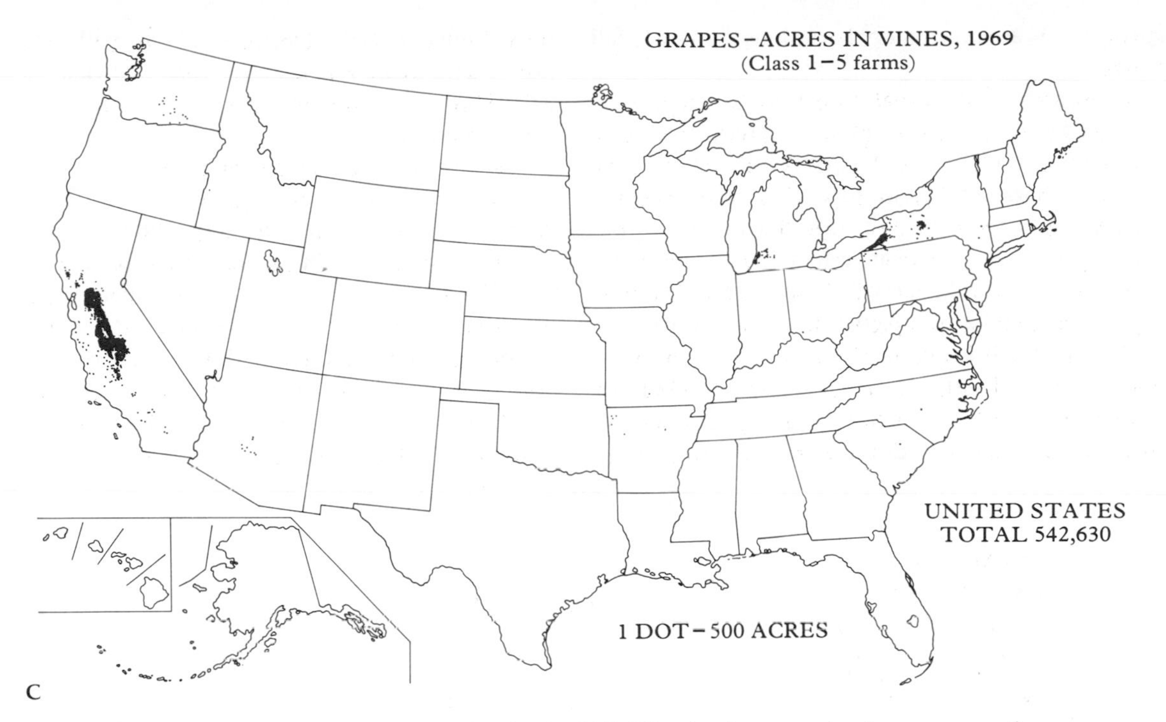

Figure 1-9 Areas of small-fruit production in the U.S. The dominant production areas are the coastal states, the Great Lakes area, and the southern states. [After USDA]

The direction of a slope is important and should be considered in choosing a site. North-facing slopes retard bud development in the spring; south-facing slopes accelerate development. The delay of bloom on north facing slopes can be of value in preventing frost damage. If early maturity of the fruit is of economic importance, however, then south-facing slopes should be considered. On cold nights in spring, the warmest slope is one against which a slight wind is blowing. The layer of cold air generated by radiation inversion at the earth's surface will move down a slope, unless a gentle breeze mixes it with the warmer air aloft. In North America, in places where spring radiation frosts rather than advective freezes occur, slight winds are usually from the north or northwest. Thus slopes in those directions are sometimes planted with windbreaks on one or more sides of an orchard to protect it from strong winds or to deflect cool ocean winds upward if a warmer growing temperature is desired for the specific kind of fruit being grown. Hilltops or ridges usually are not good sites for orchards because of both wind and eroded soil. In some places, however, appropriate windbreaks can be used to take advantage of a windy but otherwise favorable site. Windbreaks must be used with care because the hazard of frost can be increased if they prevent the flow of cold air from an orchard or if they prevent air mixing that would bring warm air to the trees from an inversion layer.

As previously mentioned, it is not desirable to choose a site near the climatic limit (with regard to winter lows, summer temperature, length of growing season, and so forth) for that particular crop. Flowers and young fruit are more tender than mature fruit, so spring frosts are much more hazardous than early fall frosts.

Trees and shrubs usually do best on deep, well-drained sandy or silty loam soils. Trees need deep soil (with a depth of 1 to 2 meters) and do not thrive in soils with water tables or impervious layers near the surface. Most trees do best in well-aerated soil, but some species, such as pear, tolerate wet soils better than other species. About half the average soil volume is pore space and is occupied either by air or water. A well-aerated soil is one with about 50 percent of its pore space filled with air. Roots require a good supply of oxygen as well as water for normal functioning. In some plantings poor drainage rather than low fertility limits growth. Dormant plants will tolerate excess soil water much better than growing plants will. Thus winter flooding may not be serious unless it is prolonged.

Both water table and drainage characteristics of a soil can be observed by digging a trench 2 meters deep. After heavy rains the water table might rise nearly to the surface, but it should recede in a day or two. The color of the soil at a given depth indicates how well drained it is. Well-drained sandy loam is a uniform brown; imperfectly drained soil will have a gray layer plus mottled subsoil; poorly drained soil will be gray to black. These dark colors result from anaerobic activity where there is too much water and

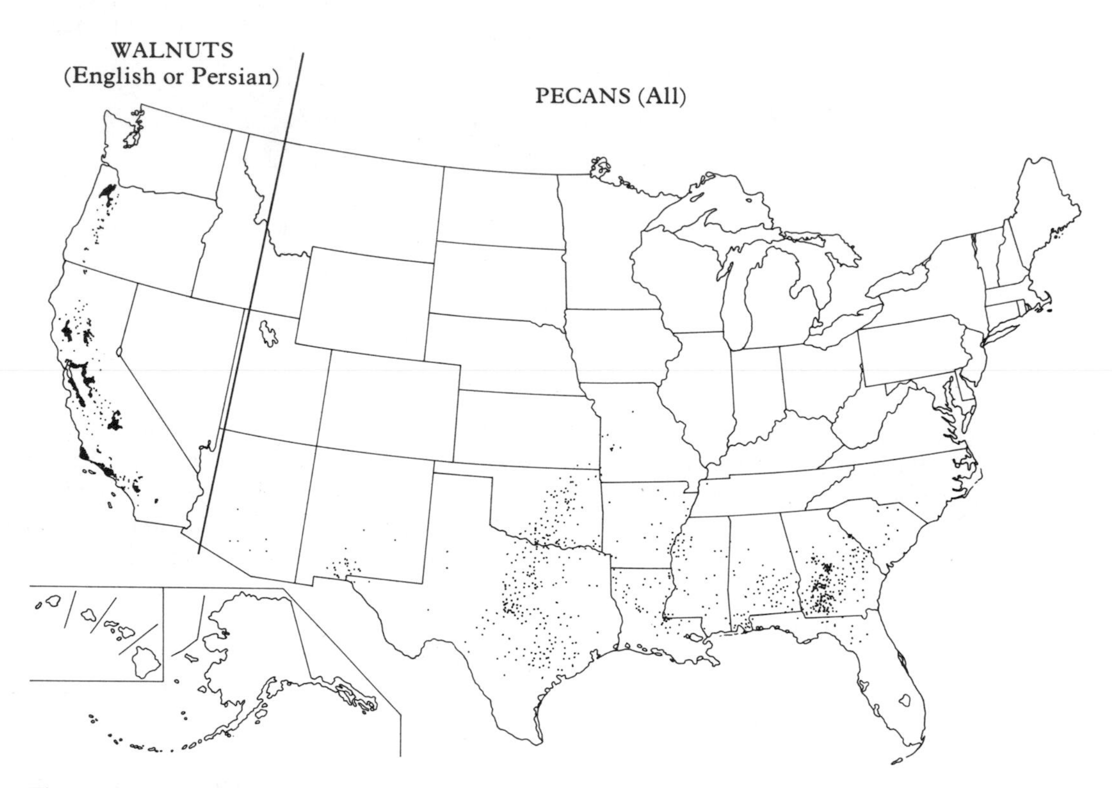

Figure 1-10 Areas of walnut and pecan production in the U.S. Pecans are produced in the southern states, while regia walnuts are produced on the West Coast. Black walnuts (not shown) are produced in several midwestern states for lumber and shells as well as for nuts. For walnuts, one dot represents 10,000 trees of all ages. For pecans, one dot equals 500 acres. [Walnut map: After U.S. Bureau of Census; Pecan map: After USDA]

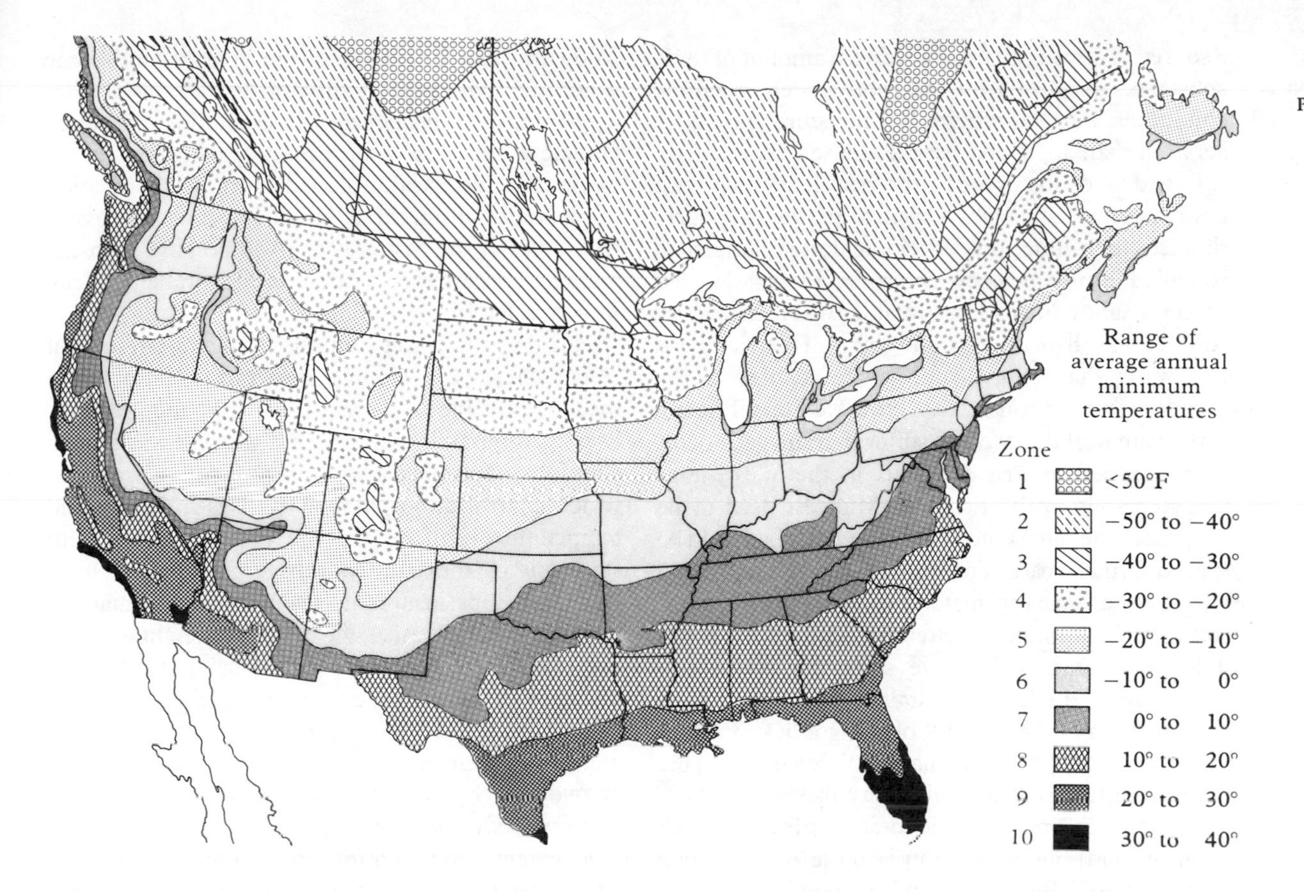

Figure 1-11 Hardiness zones for the U.S. and Canada. Note the influence of the oceans in moderating winter temperatures in the higher latitudes. The zones are defined by decimals of the Fahrenheit scale:

−50°F = −45°C
−40°F = −40°C
−30°F = −34.44°C
−20°F = −28.89°C
−10°F = −23.33°C
0°F = −17°C
10°F = −12.2°C
20°F = −6.67°C
30°F = −1.11°C
40°F = 4.44°C

[After USDA]

too little air (oxygen) in the soil. The best-aerated soils are gravelly, and the poorest are clay.

The following guidelines indicate some general soil requirements for several fruits: Apples require a large moisture supply and are adapted to open- or medium-textured soils 2 to 2.5 meters deep; very dry or very wet soils should be avoided. Pear requirements are similar to apple, but pears will tolerate greater extremes of both moisture and soil texture. Greenhouse tests with potted pear and apple seedlings indicate that pear is much more tolerant of wet soil than apple. Potted apples submerged in water to above the soil line die in about six days. After six weeks, similarly treated pears stop growing but remain alive. When removed from the water, the pear seedlings' feeder roots in the soil are dead and black, but new functional roots grow out of the stems just below the water line and above the soil line in the pot. Thus the tolerance of pear to wet situations appear to result either from its ability to regenerate roots near a source of oxygen or from a reduced amount of toxic substances produced by the roots under anaerobic conditions. Peaches require less moisture and better aeration than apple and pear. Because of this, sandy soils and sandy loams are better than clayey soils with poor internal drainage. Peaches grown in poorly drained, high-lime soils develop chlorosis similar to so-called iron-deficiency chlorosis. Plums do well in good sandy loams but are able to tolerate heavy soils better than other stone fruits. Cherries need less water than apples, and so are adapted to the coarser-textured soils. They grow well on silt loams if they are well drained but suffer if drainage is poor.

In contrast to irrigated soils in the West, non-irrigated soils in the East and Midwest need to be deeper for the same kind of fruit, and plants should be spaced farther apart. The ground water table should not be closer than one meter during the growing season, because it limits the effective depth of the roots. Tile drainage may improve soils, but adobe-clay soils cannot be effectively drained by tiling. Excess moisture can be used up by planting a heavy winter and spring cover crop that may be disced up when the excess moisture is gone. Soil acidity also is important. Orchard trees appear to do best at pH 5.5 to 6.5 (slightly acid) but soils with pH up to 8.5 (alkaline) have been used. Blueberries and cranberries, however, grow best in soils of pH 3 to 5, too acid for other fruits. Shallow soils are to be avoided, but cool, humid climates permit trees to survive with less soil depth. Under adverse conditions (very hot, very dry, or very wet) trees on shallow soils may die. Trees with extensive root systems are better able to withstand unfavorable conditions.

The nearness of the orchard site to suburban areas and housing developments is of economic importance. Growers must try to plan far enough ahead to keep out of the way of rapidly spreading suburbs.

Location Related to Markets

It was previously shown that climate has greatly influenced the distribution of fruit and nut plantings. For example, large acreages of apples planted in the Midwest were killed by severe freezes, and some peach orchards in the southern states were unsatisfactory because of inadequate winter chilling. Although climate sets the limits for these crops, the nearness to market has, historically, often determined the location of plantings within these limits. The most successful fruit areas in the past have thrived because both climate and access to markets were favorable and not merely tolerable.

Nearness to market continues to be important, but recent changes in storage and transportation have made it possible to market fresh fruit months after harvest, thousands of miles from where it was grown. Carefully controlled cold-storage facilities, some of which control atmosphere as well as humidity and temperature, prolong the life of fruit; subsequent shipment of the fruit in refrigerated trucks, train cars, and ships insures its reaching distant markets in good condition. Also, the use of air freight is growing rapidly in transporting perishable fruits long distances in a short time. For example, producers of sweet cherries in South America (during our winter season) are delivering this fruit to our cities in one or two days by air, when previously it took two or more weeks to deliver by surface vessel. Thus, to a great extent, modern transportation has made it possible to grow crops in the areas best suited to the crops regardless of the distance to market. But the price for this progress is higher costs and a greater use of energy than when crops are produced near their markets.

Concentrated Versus Isolated Plantings

Growing fruit in an area where there is a concentration of acreage has a number of advantages over growing fruit in isolated locations.

1. There are substantial savings in the purchase of supplies and equipment by cooperative organizations.

2. Better service and repair of equipment is available.

3. Growers have more opportunity to exchange progressive ideas than does the isolated grower.

4. Cooperative marketing can be done, permitting better service to the grower and more uniform grading and packaging of the product.

5. There is more opportunity to develop stable processing outlets for the crop.

6. More and better service from state experiment stations, extension specialists, county extension agents, and commercial fieldmen is possible.

7. There is better use of labor.

8. There may be access to an irrigation project.

9. There is more opportunity for growers to act in concert on legislation and on other matters of importance to some aspect on fruit growing.

GENERAL REFERENCES

Janick, J.; R. W. Schery; F. W. Woods; and V. W. Ruttan. 1974. *Plant science: An introduction to world crops* 2nd ed. W. H. Freeman and Company, San Francisco.

Krueger, R. R. 1972. The geography of the orchard industry in Canada. In *readings in Canadian geography,* edited by R. M. Irving, Holt, Rinehart, & Winston, Ltd., Toronto. Pp. 216–241.

Olmstead, C. W. 1956. American orchard and vineyard regions. *Econ. Geograph.* 32:189–236.

Rehder, A. 1947. *Manual of cultivated trees and shrubs.* (Excludes nonhardy subtropical and tropical plants.) The Macmillan Company, New York.

Upshall, W. H., Ed. 1976. *History of fruit growing and handling in the United States of America and Canada: 1860–1972.* The American Pomological Society, University Park, Pennsylvania.

USDA. 1974. *Crop production, 1974 annual summary.* Stat. Reporting Service.

2

The General Plant Environment

This chapter is an overview of the general environment of the above-ground portion of the plant, including seasonal changes in climate, the climatic requirements of fruit species, and the nature and effects of light, water, and temperature. Details of microclimate (that is, the climate around individual plants and plant parts), and details of the environment of underground plant parts will be discussed in other chapters.

World climates and their geographic distribution are shown in Figure 2-1. Note that the latitudes between 30° and 50° encompass most of Europe, the U.S., parts of Canada, Japan, northern China, and Manchuria in the Northern Hemisphere, and southern South America, the tip of South Africa, southern Australia, Tasmania, and New Zealand in the Southern Hemisphere. Fruit areas nearer the equator than about 35° latitude must be of low-chilling species, or they must be grown at higher elevations in order to obtain the necessary winter chilling to break rest. Such areas include Australia, South Africa, North Africa, southern Asia minor, southern China, India, Mexico, the southern U.S., and northern South America. Compare Figure 2-1 with Figure 1-3, which shows natural vegetation zones. Deciduous fruits and nuts are grown best in the temperate zones, but whether an arid or humid climate is suitable depends upon the specific requirements of a species and the degree to which the grower can alter a natural environment to the needs of that species.

SEASONAL CHANGES

The temperate deciduous tree or shrub responds to seasonal changes in a number of ways, depending upon the extent to which its internal physiology is affected by external environment. If the plant is ideally suited to the climate, then each seasonal change evokes such physiological changes in the plant as are necessary to survive that season and anticipate the next.

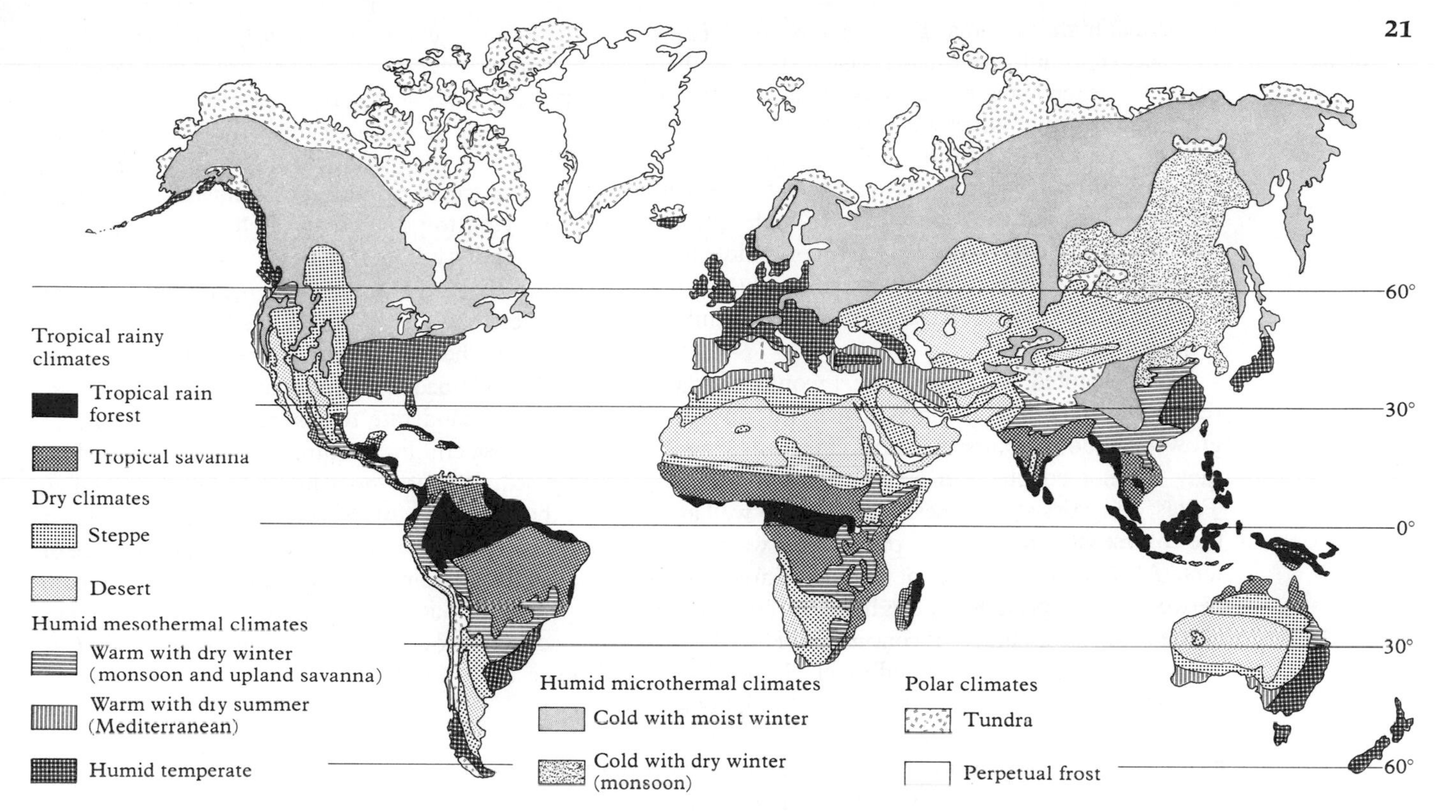

Figure 2-1 Major climatic regions of the world. The main areas of deciduous fruit and nut production lie between latitudes 30° and 50° in both hemispheres. [After USDA]

Fall

Beginning during the later summer season, the deciduous tree must stop growing, drop its leaves, and acquire winter hardiness. The details by which this is accomplished are not all known, but it seems clear that growth inhibitors and growth promoters play important roles. Recent studies indicate that abscisic acid (ABA, an isoprenoid related to vitamin A) is produced in rather large amounts by leaves as the day length begins to shorten in late summer. This substance, being a growth inhibitor, can cause terminal buds to form, if they have not formed earlier; after a short time, enough of the inhibitor will accumulate in the buds to prevent further growth. There is recent evidence that winter rest is regulated by a balance of inhibitors and promoters rather than simply by the inhibitor level.

Winter

The state of dormancy just described, in which growth will not occur even when the plant is put in a favorable environment for growth, is termed rest. Rest is naturally broken by winter chilling, the required amount of which depends upon the species or clone. Many temperate-zone trees have optimum chilling temperatures near 5°C (see the appendix for equivalent Fahrenheit temperatures), but a few have somewhat higher optima. Temperatures much below 0°C seem to be inffective in breaking rest. Thus, there seem to be enzymatic reactions at above-freezing temperatures that alter the growth promoter/inhibitor balance. The winter-chilling requirement is relatively short for plants native to low latitudes with warm winters and for those native to high latitudes with long, very cold winters. Natives of middle temperate regions with cold, but at times fluctuating, winter temperatures have the longest chilling requirements.

Plants adapted to warm winters would naturally require short chilling. At times, man has attempted to grow deciduous plants in warm climates to which they are not adapted. In areas such as California, Texas, and Mississippi, these trees may do well after the coldest winters but grow poorly after warm winters. Also, with intermittent chilling (interspersed between warm periods), more hours of chilling are needed than if the chilling is continuous. Thus a warm period tends to reverse the effect of chilling. This situation often occurs in areas with cold nights and sunny winter days. For this reason, winter fogs are of great value to growers in these areas because they prevent direct solar radiation from warming the buds.

Plants adapted to long and continuously cold winters, such as are found in the high latitudes away from water bodies, also have short or moderate chilling requirements, but for quite a different reason. Because much of the winter remains continuously below freezing, the chilling requirement is apparently not satisfied until late winter and early spring, when the temperature again rises above freezing but remains cool for a time. One pear species, *Pyrus ussuriensis,* is adapted to such a climate. It is native to Manchuria and parts of Siberia where winter temperatures drop to minus 45°C. Not only do these trees have a short chilling requirement (20 to 30 days) but the optimum chilling temperature is 7° to 10°C, somewhat higher than the 3° to 5°C optimum for many temperate-zone species (Westwood and Bjornstad, 1968). Although *P. ussuriensis* is hardy in the very cold winters of its native habitat, it is often injured in a mild marine climate, because it is not adapted to mild winters. Its short chilling requirement is satisfied during early winter, after which it begins to grow whenever the temperature rises above freezing. Flowers and young shoots develop in February and early March and often are killed by spring frosts only slightly below freezing. This is an extreme example, but it illustrates the problem of trying to grow crops in climates with which their physiology is out of tune.

Plants from the middle-temperate region have the longest chilling requirements. This is adaptive because, in the fluctuating winters of this region, warm periods occur in midwinter and at times are followed by moderate to severe freezes. Species or clones whose chilling requirement is satisfied before such a sequence are more likely to be winter killed than those still in rest. Recent work indicates that a plant in rest does not always have the same degree of hardiness, depending upon the preconditioning cool temperatures, but that buds in rest are always hardier than actively growing tissue of the same species (see Chapter 15).

Spring

Plants adapted to a specific spring climate respond to their environment in such a way that flowering and growth proceed only after the danger of late freezes or killing frosts are over. Some trees whose native habitat is at high elevations or at high latitudes respond unfavorably to the average temperate spring. Often the growing season is short in high elevations and latitudes, and in that circumstance it is adaptive to begin growth at lower temperatures in the spring, so that the crop can mature before the first killing frosts of autumn. But when trees native to that climate are placed in a climate having several frosts during the spring season, both the blossoms and young shoots often are killed because growth starts too early. Examples of such species are the apricot and the regia (Persian) walnut. Some selections of regia walnut *(Juglans regia)* from the Carpathian Mountains are very hardy during the winter season but begin growth too early in the spring and are injured by moderate frosts. As was mentioned in Chapter 1, orchard site is important in preventing damage from spring frosts, but a correct choice of species and cultivar also is important.

With the coming of spring, the rest period of the dormant tree should be broken so that active growth can occur when spring weather permits. As rest ends in response to adequate chilling, levels of growth inhibitors in the buds may fall or may remain the same as levels of growth promoters increase. The ability of a bud to grow in the spring seems to depend upon the proper *balance* of both inhibitors and promoters, rather than on an absolute level of either.

Summer

Young vigorous trees under favorable conditions will grow during the entire summer, stopping only when temperatures drop in early winter. But mature bearing trees usually grow most in early summer, after which terminal buds form and the major growth for the rest of the season is root and fruit growth. The buds, however, can be forced into growth by summer pruning or by the stimulus of fertilizer and irrigation, until the production of inhibitors and reduction of promoters in the buds, as earlier described, results in the onset of rest.

Figure 2-2 accounts for the dispersion of solar radiation (heat and light) on the surface of the earth at noon on a summer day. As the lower graph shows, the shift from incoming radiation during the day to outgoing radiation at night produces a characteristic diurnal temperature pattern. This diurnal pattern varies between summer and winter and between different climates.

As trees approach the fall and winter season, current summer cropping also influences their leaf-fall and readiness for winter. During an unusual early freeze in November, 1955 (minus 20°C) in the Pacific Northwest, apple trees that had been harvested the previous week sustained much less damage than did those still holding their crop. Soon after harvest the leaves turned yellow and showed obvious signs

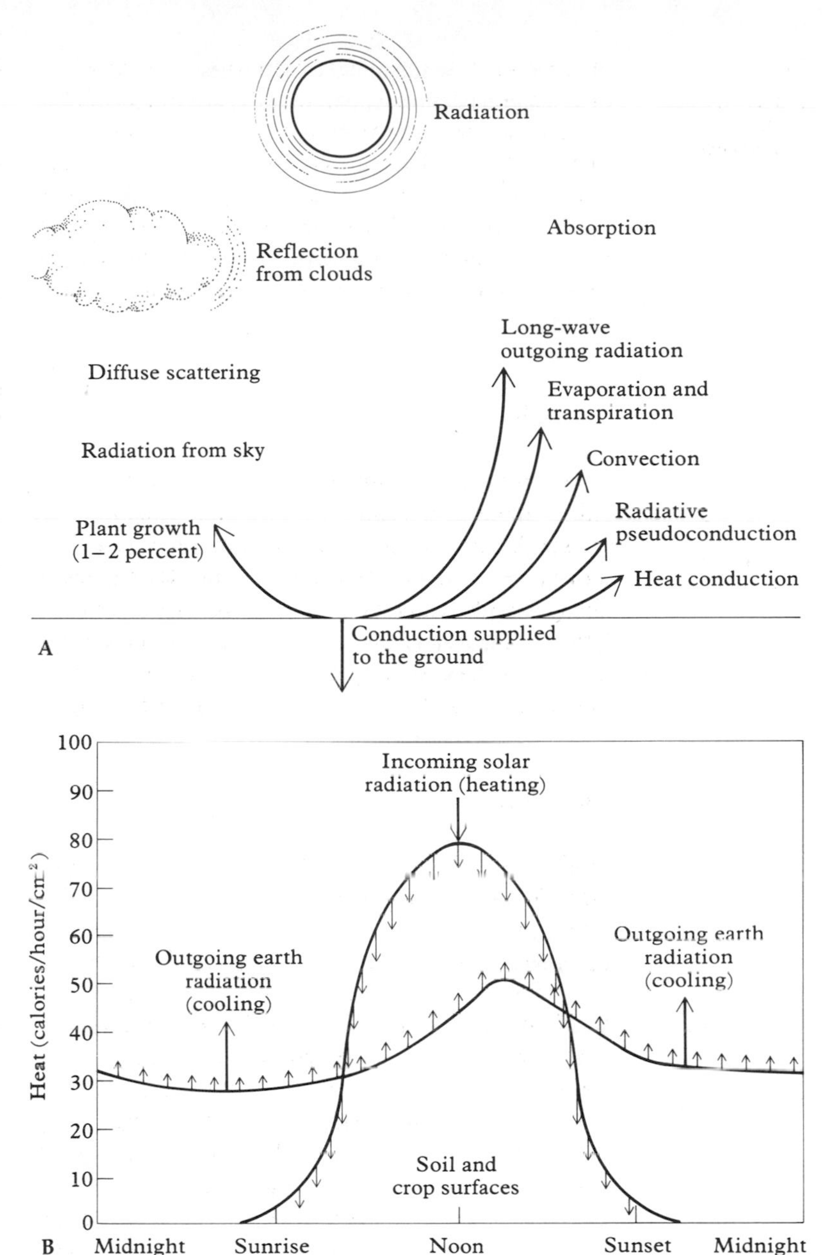

Figure 2-2 **(A)** The energy balance at midday in summer and **(B)** a graph of diurnal changes in outgoing and incoming radiant energy. A small but important amount of solar energy is fixed in plant growth. [A: After Geiger, *The Climate Near the Ground.* Cambridge: Harvard Univ. Press, 1950. © 1965 by the President and Fellows of Harvard College; B: After Newman and Blair, *Crops and Soils Magazine,* June–July 1964.]

of senescence, while unharvested trees retained leaves of a rich green color. Apparently the promoter/inhibitor balance is quite different after the fruit is removed. These observations suggest that inhibitors that cause leaf-fall, rest, and readiness for winter predominate only after the fruit is removed. The interaction of cropping with decreasing day length in autumn in bringing on rest is incompletely understood.

Latitude

Other influences remaining constant, winters are colder and longer as one goes from the equator north (or south), that is, as one goes from a low to a high latitude. Also, the relative length of the day and night changes with latitude (see Fig. 1-3). The higher the latitude, the longer the summer days and the shorter the winter days. Woody plants must, in order to survive, adapt to both the temperature and the day length.

Elevation

High elevations are colder than low ones at the same latitude, except when the effect of elevation is offset by temporary temperature inversions and influences of bodies of water, both of which will be discussed later. Some temperate-zone fruits can be grown at low latitudes only at higher elevations. However, this situation, which provides enough cool temperature during the winter to satisfy the bud-chilling requirement, may not be satisfactory if the change in day length between summer and winter is not great enough. For example, some plants require the shortening of days to a critical length in late summer to trigger the internal mechanism that brings about leaf-fall, rest, and winter hardiness. Because they contain no internal mechanism to anticipate the onset of winter and to prepare for it, such plants might not survive a moderate freeze in early winter (see Chapter 15). Some deciduous species at low elevations in the tropics tend to be semi-evergreen because they do not receive the proper environmental cues to drop their leaves and go dormant.

Bodies of Water

As we know, at a given latitude and elevation, nearness to a large body of water will moderate extremes in temperature. That this effect is important to fruit growing can be seen from the fact that the danger of winter injury at 35°N latitude in the central United States is greater than at 45° or 50° N latitude in the Northwest or Northeast near the ocean. The influence of a body of water may be quite local where the land is relatively flat—as it is around the Great Lakes—or it may extend some distance inland—as it does on the Pacific slope. There, westerly winds from the ocean keep winter temperatures moderate, and high mountain barriers (the Cascade and Sierra Nevada ranges) prevent polar continental air masses from moving in. Wind currents moving across an unfrozen body of water modify temperatures at all seasons. Winter temperature is warmer than in a continental climate, spring is cooler (thus bloom is later), summer is cooler, and fall is warmer and extends longer. One of the striking features of this marine influence is the extension of the frost-free growing season. For example, Astoria, on Oregon's northwest coast, has an average frost-free growing season of 290 days. Corvallis, in the Willamette Valley east of the low Coast Range Mountains but well within the coastal influence west of the Cascade Mountains, has a 213-day season. Bend, Oregon, east of the Cascades and in a continental climate, has only 90 days of frost-free growing weather. Both Astoria and Bend are within 150 miles of Corvallis and all three locations are between 43° and 46° N latitude. The tremendous influence of water on climate results from its extremely high specific heat (thirty-three times that of lead) and its high latent heat of fusion and vaporization. These properties will be discussed when frost protection is considered.

CLIMATIC REQUIREMENTS OF FRUIT AND NUT SPECIES

The domestic apple is thought to have originated in Western Asia, Eastern Europe, and southwestern Siberia. Several crab apples and other *Malus* species are native also to China, India, and North America. Although some species are not hardy, the domestic apple *(Malus pumila)* is one of the hardiest of temperate zone fruits. Its winter chilling requirement to break rest (1000 to 1600 hours) is high enough that it does poorly in latitudes with warm winters.

European cultivated pears probably developed through selection from the wild pears of southeastern Europe. The present population of wild trees in the Caucasus Mountains have botanical characters similar to the cultivars, although the fruit is much smaller and possesses much more stony tissue in the flesh. Two oriental species, *Pyrus pyrifolia* and *P. ussuriensis,* also are cultivated, but only to a limited extent in the U.S. *P. pyrifolia* is native to central and eastern China, Korea, and Japan. *P. ussuriensis* is native to the colder regions of northeastern China, Korea, and Manchuria. Cultivated *P. communis* and *P. ussuriensis* are generally hardier than *P. pyrifolia* because they originated in colder climates. A number of other pear species grow in southern Europe, North Africa, Asia Minor, and Asia, but they are used as rootstocks rather than as edible cultivars. No pears are native to the Southern Hemisphere or the New World.

Pears grow well in an arid climate that is moderate to hot. Dry summers are essential in controlling the bacterial disease fire blight, which limits pear growing in North America more than any other factor. Bartlett usually sets good seedless crops in hot interior valleys but requires cross pollination for fruit set in the cooler coastal valleys of California and Oregon. Cool (10°C) nights during the month before Bartlett harvest causes softening and premature ripening, in which both fruit size and quality are impaired.

Except for a few hardy *P. ussuriensis* cultivars, pears are generally less hardy than apples and thus are more restricted to climates without such cold winters. Anjou, Clapp Favorite, and Flemish Beauty are hardier than other common pears. Asian cultivars vary a great deal in hardiness. In general, Ussuri types are very resistant to early or mid-winter freezes. *P. pyrifolia* cultivars resist late winter freezes by remaining dormant during late winter (see Chapter 15). Pears vary in winter chilling requirements as much or more than do apples. Barlett has a long chilling requirement (about 1500 hours); some *P. pyrifolia* hybrids, such as Kieffer, require much less chilling and thus can be grown at lower latitudes. However, Twentieth Century, the principal *P. pyrifolia* cultivar of Japan, has a much longer chilling requirement than many *P. communis* cultivars.

The quince *(Cydonia oblonga)* is native to southern Europe and Asia Minor. Quince trees require less winter chilling than apple or pear, but they flower later in the spring because some vegetative growth must occur before the flowers appear. Quince flowers appear to be self-fertile and set well in both cool and hot climates. Quince trees are less hardy than apples and pears, and they are about as susceptible to fire blight as pears. The roots are shallow and seem to require more oxygen than other pome types. Pear roots are the most tolerant of poorly drained soils, apples are intermediate, and quince are least tolerant. (See Chapters 4, 6, and 15.)

The peach *(Prunus persica)* is native to the warm areas of China. It does best in a hot summer climate and is only moderately winter hardy. The peento peach from south China is almost evergreen and has a very short chilling requirement. Peaches are not only less hardy than apples and pears but also generally require less winter chilling to break rest (400 to 1000) hours. Thus they are grown at lower latitudes than apple and pear. Peaches thrive and produce high-quality fruit in very hot, arid climates. Diseases, such as brown rot and leaf curl, are difficult to control under humid conditions. Because peaches bloom 20 to 30 days earlier than apples and are tender to frost, it is important to plant them in frost-free sites. (Further details are found in Chapters 6, 15, and 16.)

The sweet cherry *(Prunus avium)* appears to have originated in the Caucasus, between the Black and Caspian Seas. Sweet cherries are susceptible to brown rot and thus should be grown where it is too cool or too dry for the disease to develop. It can be partially controlled by sprays during the bloom and preharvest period. This fruit develops good quality in climates too cool for peaches or apricots. Thus areas with good winter rains and dry, cool summers are ideal. The crop matures early, so irrigation is not needed in some areas. The better-quality, firm cherries tend to be more susceptible to rain cracking than lower-quality, soft cherries. The selection of sites with dry summers to prevent cracking is the principal reason that the major U.S. acreage is in the Pacific-slope states. Sweet cherry trees and buds are somewhat more hardy to cold than are peaches, but less so than pears and common plums *(Prunus domestica)*. Windsor, Governor Wood, and Lyon seem to be hardier than other older cultivars. Sweet cherry buds require somewhat longer winter chilling than Elberta peach (about 1000 hours). Bing, Lambert, and Napoleon have longer chilling requirements than Black Tartarian, Chapman, or Black Republican.

The sour cherry *(Prunus cerasus)* also originated in southeastern Europe. It has a somewhat longer winter chilling requirement than sweet cherry. It is also more hardy, most cultivars being about as hardy as Northern Spy apple. Russian cultivars are about as hardy as McIntosh apple. Trees in low vigor and low-nitrogen status have flower buds that are often killed in winter when no wood injury is apparent.

The common plum *(Prunus domestica)* of Europe seems to have originated in southeastern Europe and has been cultivated by man for centuries. Damson plums *(Prunus insititia)* from Europe are somewhat hardier than *Prunus domestica* cultivars but vary greatly in their requirement for chilling. The so-called Japanese plum *(Prunus salicina)* is a native of China. It is considered to be less hardy than European plums, but its cultivars vary greatly. A number of native American plum species have been studied but few have been selected as cultivars. These species have rather wide distribution and thus vary widely in their hardiness and climatic requirements. Many cultivars of *Prunus americana* are about as hardy as the average apple cultivar. *P. domestica* cultivars are hardy in the same areas of Canada and the Pacific slope where pears are grown. *P. salicina* trees tend to be less hardy than *P. domestica,* but they vary a great deal. Burbank, Abundance, and First are as hardy as domestic plums, but Kelsey is less hardy than peach. Climates favorable to the development of brown-rot fungus are not good areas for plums. Brown rot can be serious at both the blossom and mature-fruit stages for North American species and *P. domestica,* but it is mostly a mature-fruit problem with *P. salicina.* Warm (not hot) dry climates tend to keep brown rot under control. Chilling requirements for domestica cultivars is somewhat less than for apples, except for President, which requires as much chilling as apples. Japanese plums generally need somewhat less chilling than European types. *P. salicina* blossoms are subject to spring frosts, not because they are less hardy than other plums but because they bloom very early in the spring.

The apricot *(Prunus armeniaca)* seems to have originated in western China and Siberia. It generally requires less winter chilling than the average peach, but is subject to bud drop following warm winters in California, especially if the December and January weather is warm. Apricots bloom very early in the spring and are thus subject to spring frosts. They are grown commercially only in the warm, arid areas of the Pacific slope. Most of the U.S. crop is produced in northern California, the remainder being grown in Washington, Utah, and Colorado. Some Russian cultivars are hardier than peach. Apricots are more susceptible to late winter freezes than most other fruits because their buds swell and deharden early. Riland and Perfection are hardier than other cultivars in Washington, and they transmit this hardiness to seedlings when crossed with other cultivars. Apricots are subject to brown rot in moist climates.

The almond *(Prunus amygdalus)* is native to western Asia and is adapted to the warmer areas of southern Europe. It is less hardy and requires less winter chilling than does the peach. It also blooms earlier in the spring and is thus subject to diseases that are associated with a cool, wet bloom period and that may reduce set. It requires 6 to 8 months from full bloom to mature the nuts. The warmest winters in southern California tend to delay bud opening and may also reduce yield.

The Persian or English walnut *(Juglans regia)* seems to have a wide natural distribution from the Carpathian Mountains of central Europe to the Caucasus, northern Asia Minor, and Manchuria. Both hardy and nonhardy cultivars are grown. Many cultivars are fairly hardy when fully dormant, but early growth in the spring and late growth in the autumn are susceptible to frost. For this reason, and because the crop needs a long season to mature, these walnuts are not usually grown north of 35° N latitude in this country, except in the marine climate of Oregon's Willamette Valley. Various black walnuts *(Juglans nigra, J. hindsii,* and *J. californica)* are native to the United States. The nuts are highly flavored, but their very thick shells and dirty black hulls have prevented these species from being widely cultivated. They are generally hardier than English walnut, but this would vary with the cultivars being compared.

Filberts are commonly classified as *Corylus maxima* and cobnuts as *C. avellana,* but crossing and selection has made this distinction questionable. These types are said to have originated in southern Europe and Asia Minor. A few selections of a native hazelnut *(C. americana)* have been made for adaptation to the climate of the eastern United States. The wood of the filbert is quite hardy, but because it blooms during the winter months, good cropping is confined to areas where a marine influence prevents winter temperatures that would kill the open flowers. Female flowers with open pistils may be killed at −10°C. The leaf buds of filbert require about as much chilling as do apple buds, but flower buds of filbert require much less chilling.

The pecan *(Carya illinoensis)* is the most important native nut produced in the U.S. Pecans require a long growing season to mature and are thus confined to the southern states. Many cultivars have a short-to-moderate chilling requirement and the trees are less than hardy. A few selections have been made, however, which are hardy in some of the northern states. Several other species of hickory are native to North America, but none except the pecan is important commercially. Some selections of these species are hardy enough to survive New England winters, but most of them grow in warmer regions.

The fig *(Ficus carica)* is a borderline temperate-zone species, and most cultivars are about as tender to cold as the hardiest orange trees. Chilling to break rest is so slight that a few weeks of cool weather are sufficient. Early winter temperatures of −3° to −6°C may kill figs to the ground. Thus, California and Texas are the only states to produce figs commercially. But home-garden trees of hardier cultivars are grown against walls in such places as Virginia, western Oregon, western Washington, and in parts of England. The fig has been grown in the warm areas around the Mediterranean Sea for thousands of years.

Chilling requirements of the papaw *(Asimina triloba)* range from slight in the southern U.S. to as long as apple for some strains in northern California. These northern strains seem also to be quite cold hardy. The dull purple flowers arise from nodes on shoots of the previous season. They appear to be self-sterile. The resinous aromatic flavor of the fruit is especially liked by some.

The pomegranate *(Punica granatum)* requires much summer heat to mature properly. The tree has a very slight chilling requirement, but will withstand −9° to −12°C cold.

The jujube tree *(Zizyphus jujuba)* is quite resistant to cold (it withstood −21°C at Corvallis, Oregon) but has only a short chilling requirement. Its fruit does not mature in cool or short summers. It is best adapted to hot interior valleys at low latitudes.

Pistachio nut trees *(Pistacia vera)* are as winter hardy as almond but have longer chilling requirements. There is some delay in bud opening after warm winters in southern California. Recent studies indicate that, although pistachio trees flower annually, many flower buds abscise the year following a heavy crop, resulting in an alternate-year cropping tendency (Crane, 1971).

Small fruits, like tree species, are very diverse and have widely differing climatic requirements depending upon the species or cultivar. Being smaller in size than tree species, many small fruits survive adverse winter climates when they are covered with a protective layer of straw, snow, or even—as is true of cranberry—water.

Grapes range from hardy *(Vitis labrusca),* to partly hardy *(V. vinifera),* to tender *(V. rotundifolia).* They all require a relatively long hot growing season to mature the fruit. Spring frosts are often avoided by their blooming very late in spring. Because they will grow and perform reasonably well with suboptimal winter chilling, grapes are grown in a wider range of latitudes than other fruits. *V. vinifera* are even grown in the tropics by pruning and defoliating the vines just after harvest; such practices prevent dormancy and induce another cycle of growth and fruiting.

Strawberries are widely adapted because of their low chilling requirements and the small size of the plants, which permits covering them to protect against cold. In subtropical climates, such as those of Florida and California, the strawberry is grown as an annual and will crop from early spring until fall. Because of the short period from bloom to harvest strawberries can be grown in higher latitudes than many species.

Red raspberries are as hardy as apples and pears. Because of their relatively long chilling requirements and the short season required to mature the crop, they are grown mostly in the northern part of the fruit regions of North America and Europe. Black raspberry is somewhat less hardy than red.

Blackberries are generally less hardy and have lower chilling requirements than raspberries and are thus usually grown at lower latitudes than raspberries. The erect forms of blackberry are considerably hardier than the western trailing forms and can be grown in more northern areas. The trailing blackberries are mainly confined to the moderate climates of western coastal valleys and to the south central states.

Blueberries grow in a wide range of climates, from Maine and eastern Canada (lowbush types), Michigan and eastern states (highbush types), to the southern states (rabbiteye types). The northern species have quite long chilling requirements and thus do not grow as far south as does the rabbiteye blueberry. Lowbush blueberry is about as hardy as apple; highbush blueberry is less hardy than pome fruits and sour cherry. Blueberries do not require a long season to mature the fruit.

Cranberries are grown in northern latitudes where summers are long enough to mature the crop. They avoid winter injury when they are covered with water or sprinkled to prevent the plants from reaching low temperatures. They bloom quite late (June in Oregon) and fruit are not mature until September or October.

Currants and gooseberries are quite as hardy as pome fruits and have relatively long chilling requirements. The fruit matures relatively early in the season so they do not require a long or hot summer season.

SUMMARY From the foregoing brief discussion, it is obvious that each species, having evolved and become adapted to a given climatic sequence, has particular limits within which it can be grown. But with careful selection for a given trait (such as hardiness) the culture of a type can be extended to colder areas. Likewise, selection within a hardy northern species can provide types with shorter winter-chilling requirements, permitting them to be grown in warm southern areas. An example of the first type is the selection of hardy pecans for the north central states. An example of the latter is the selection of the Beverly Hills apple, whose low chilling requirement permits one to grow it in southern California. However, survival in a given climate should not be taken to mean that the climate is optimal for a species.

LIGHT

General Effects

Green plants require light for photosynthesis. In addition to this familiar effect of light on plants, there are a number of others. The bending of shoots toward light (phototropism) profoundly affects plant form and function. Also, intense light in which there is a high proprotion of ultraviolet wavelengths tends to cause plants to be dwarfed, while low-intensity light, in which there is mostly red wavelengths, results in long, spindly growth. Absence of light causes etiolation. Length of day (photoperiod) regulates both flower initiation and cessation of vegetative growth in some species. For example, the strawberry initiates flowers in short days (when days are shorter than some critical value) but begins vegetative growth in long days (when days are longer than the critical value). Plants that respond to day length have an internal pigment system (phytochrome) for measuring the length of the dark period. During the day, red light induces one form of phytochrome (P_{fr}), and at night, it slowly shifts to another form (P_r) (Fig. 2-3). For other non-photoperiodic plants, day length is not important, yet light must shine on green leaves for the buds near these leaves to initiate flowers. The "flowering factor" produced by light and leaves has not yet been identified. Direct light to fruit skin is required for the synthesis of red and blue anthocyanin pigment in some species. In others, light may be required for

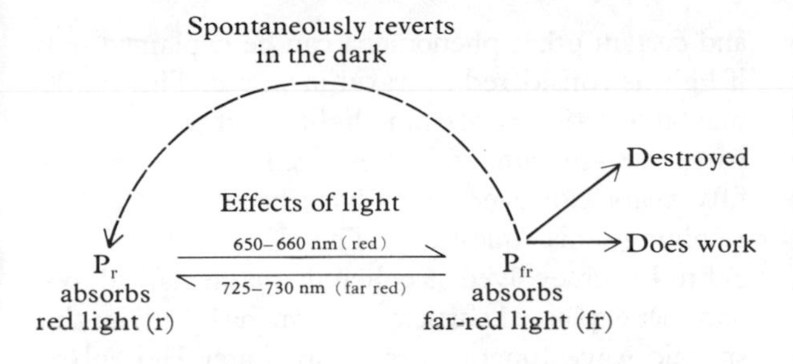

Figure 2-3 Phytochrome is a plant pigment with two forms, in which the P_r form is converted by red light (which is a constituent of sunlight) to the P_{fr} form. P_{fr} is used in biological responses, destroyed, or spontaneously reverts back to the P_r form in the dark. Thus phytochrome is used in photoperiodic phenomena to measure the length of the night.

seed germination. In the field, the effects of light and temperature are not easily separated because temperature usually increases with an increase in light intensity. Results of controlled tests that vary light and temperature independently have, however, shown which responses are due to light alone. Details of these responses are given in other sections of the book.

Historical Background

In classical times, much of the speculation about the nature light was concerned with the phenomenon of vision. Pythagoras (5th century B.C.) taught that vision results from particles continually streaming from the surface of objects into the eye. Aristotle maintained that light is not a material emission but a mere quality or action of a medium called the *pellucid.* From these beginnings came the two major theories on light. Pythagoras pointed the way for the corpuscular theory and Aristotle's surmise anticipated the wave theory.

In 1678 the Dutch astronomer Christian Huygens formulated a wave theory of light based on mathematical considerations. However, most scientists of that day rejected this theory in favor of the corpuscular theory proposed by the renowned physicist Sir Isaac Newton. He considered light rays to be very small bodies emitted from shining substances. The high velocity of these bodies seemed to furnish the kinetic energy required to account for the observed energy of light. Not until a century later did Thomas Young revive the wave theory by discovering the interference of light, a phenomenon best explained if light is considered as overlapping wavelets rather than as an homogeneous stream. In the early 19th century, the first polarizers were constructed. If the elements of a polarizer are aligned with their optical axes parallel (Fig. 2-4A), then light passes through. But if one of the elements is turned 90°, then extinction occurs (Fig. 2-4B). This phenomenon can be explained only if light is considered to be undulating waves (Fig. 2-4C). If light were merely a stream of particles, it should as easily penetrate the elements in B as in A.

From the early 19th century up to the beginning of the 20th century, the wave theory was held in high repute. To scientists of the time it seemed obvious that if light were indeed a wave, then it must be supported in space by some medium. It was known that sound waves were propagated through the air, and it

Figure 2-4 The elements of a polarizer **(A)** in parallel, with light passing through both elements, and **(B)** with the elements at right angles, causing extinction. **(C)** The drawing of the man with the rope shows the effect of a polarizer on waves.

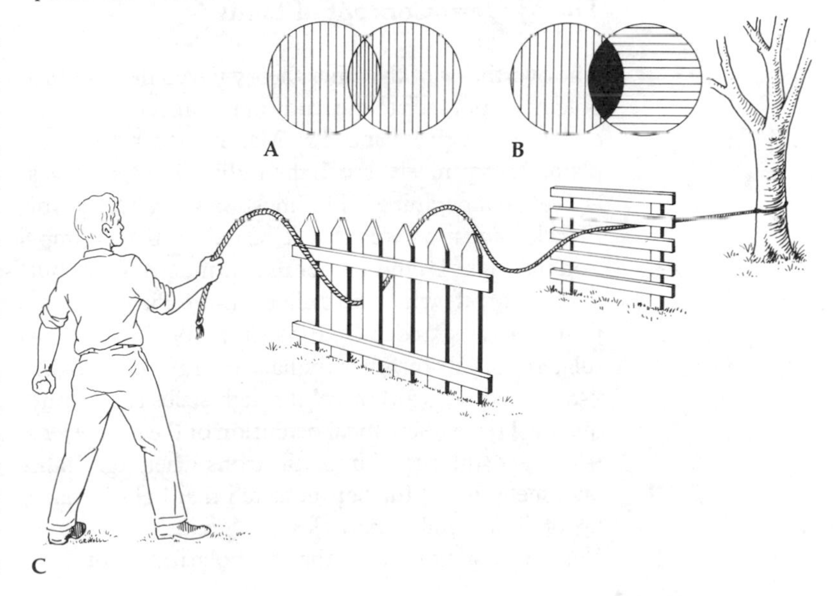

seemed logical that something provided the necessary medium to propagate light waves. Hence, when experiments showed that light travels in an airless vacuum, scientists invented the ether, an invisible medium that exists even in vacuums, such as that of outer space.

However, the ether theory posed some perplexing problems, one of which was that its actual existence had never been proved. And so, in 1881, two American physicists, A. A. Michelson and E. W. Morley, performed an experiment to determine if there was indeed an ether. They reasoned that the velocity of light *against* the ether stream (in the direction of the earth's orbital motion) should be reduced by the velocity of the earth in its orbit, which is 32.2 kilometers per second, and that the velocity of light measured in the opposite direction should be 32.2 kilometers per second faster. The classic experiment, when completed, indicated conclusively that there was in fact no difference in velocity whether it was measured with or against the ether stream. This meant either that there is no such thing as ether or that the earth's velocity is zero. It also meant that the velocity of light is a universal constant and is independent of the velocity of its source.

The Modern Concept of Light

The overthrow of the ether theory paved the way to a bold new approach to the study of radiant energy. The quantum theory advanced by Max Planck in 1900 explained very nicely the light radiated from bodies at high temperatures. The most striking feature of Planck's equation was that it rested on the assumption that light is emitted not in continuous waves but in discrete bits which he called *quanta.* Planck's conjecture was expanded by Einstein in 1905, who postulated that all forms of radiant energy (heat, light, x-rays, and so forth) travel through space as separate quanta. His mathematical definition of the photoelectric effect (the emission of electrons when light falls on a metal plate) further indicated that light is made up of innumerable particles or "grains" of energy. Yet the paradox remains that the polarization of light and certain other phenomena can be explained only if light is considered to travel in waves. This fundamental question concerning light—Is it waves or is it particles?—remains to be answered. During the past fifty years, however, gains have been made toward resolving this question. The Viennese scientist Schrödinger evolved a coherent mathematical system that explained particle phenomena by attributing specific wave functions to them. Later Heisenberg and Born developed a mathematical scheme that described quantum phenomena either as waves *or* particles as one wished. A simple analogy is the wave created by dropping a stone into a lake. The perception of a wave is clear since the component parts (crest and trough) are visible. But we all know that whatever it looks like, the wave is actually composed of billions of individual water molecules. The wave structure is there simply because certain areas, which we call a crest, contains many more water "particles" than the trough. Likewise, light appears to be made up of particles of energy, the whole exhibiting certain wave phenomena.

The Radiant Energy Spectrum

Visible light is only one type of electromagnetic radiation. As a matter of fact, it is but one seventieth of the electromagnetic spectrum. Waves of different lengths vary considerably in the amounts of energy they carry: the high-frequency, short-wavelength rays have much more energy than the low-frequency, long-wavelength rays. All wavelengths impinge upon plants, and the invisible (e.g., ultraviolet, infrared) may affect them as much as the visible.

Some of the common characteristics of all types of radiant energy should be mentioned. In each type, the product of the wavelength and the frequency is always equal to the velocity of light. All types are apparently massless but can be converted into physical particles. For example, a gamma ray under the influence of an electric field can disappear and an electron and a positron are created. The reverse occurs when an atom bomb is exploded, and mass is converted into energy. All types of energy radiations have the same velocity

(299,517 kilometers per second), can move through a vacuum, and their velocities are *the same* regardless of the motion of the source or the receiver.

Figure 2-5 shows the positions of the major types of waves in the electromagnetic spectrum. There are no sharp dividing lines between them since their distinct properties depend upon wavelength and wave frequency, both of which change gradually from one end of the spectrum to the other. A particular property, such as penetrating power, will change in proportion to the wavelength change. For example, cosmic and gamma rays have the greatest penetrating power and cannot be stopped completely by several feet of lead shielding. X-rays are next in penetrating power, followed by ultraviolet rays, visible light, and infrared rays. Heat waves can be stopped by a thin piece of foil. In terms of plant function, gamma and x-rays are potentially damaging to plants because of their high energy and penetrating power. It is these rays that strike nucleic acid molecules (in which is stored the genetic code) causing mutations. Visible light (blue and red) is used in photosynthesis, while longer wavelengths (far-red and Hertzian waves) provide heat to maintain biological temperatures.

PHOTOSYNTHESIS

Historical Background

The life-giving power of the sun has been celebrated throughout human history, but only in the modern era have men come to understand how solar energy is imparted to the earth's green plants. The Greeks of Aristotle's day held that water, air, fire, and earth were basic and immutable elements and that plants

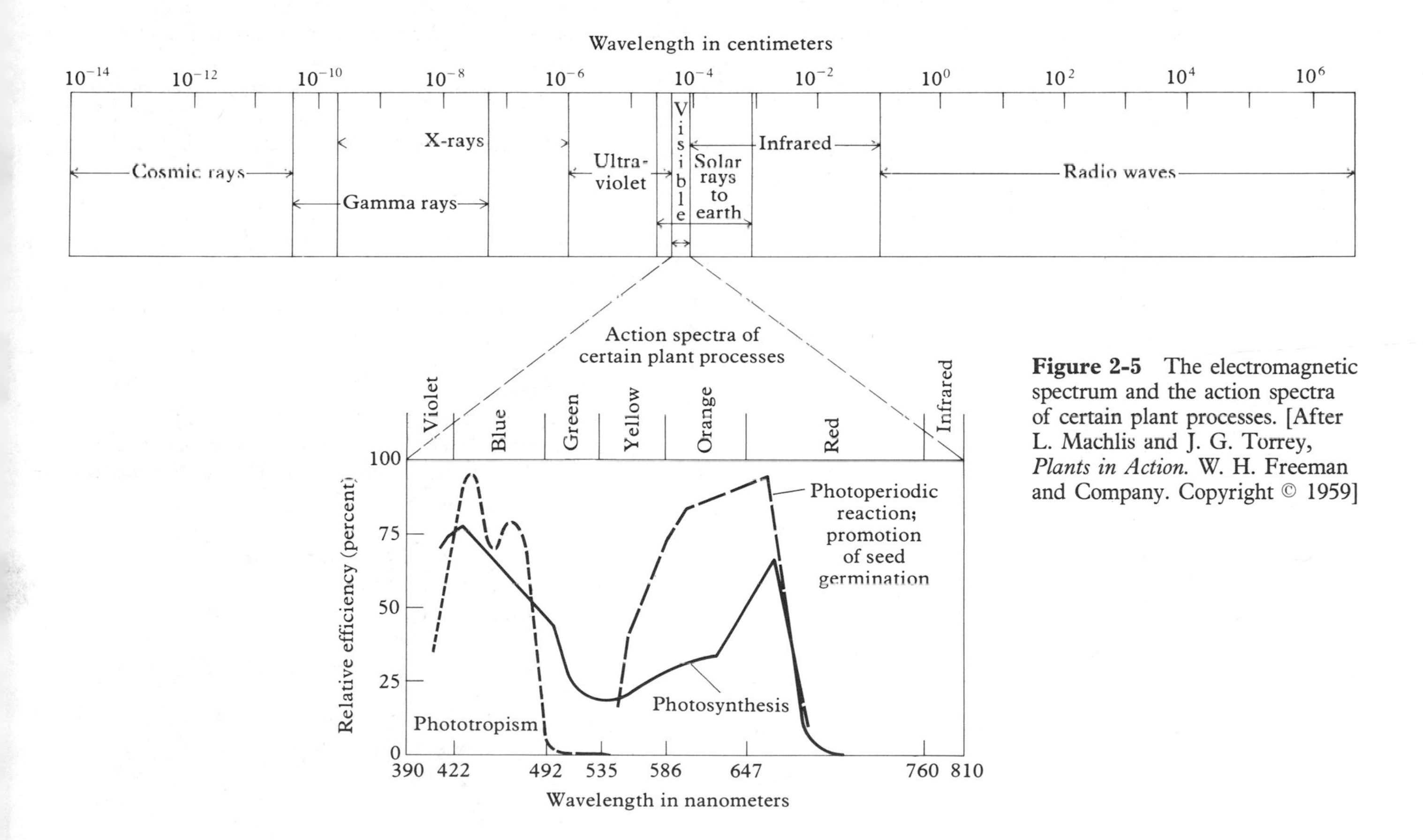

Figure 2-5 The electromagnetic spectrum and the action spectra of certain plant processes. [After L. Machlis and J. G. Torrey, *Plants in Action.* W. H. Freeman and Company. Copyright © 1959]

received their substance already elaborated from the earth. This theory went unchallenged until the 17th century, when Van Helmont, a Belgian chemist, grew a willow tree in 91 kilograms of soil for five years, only to find that the tree gained 74 kilograms, while the soil lost only 57 grams. He concluded that the weight gained by the tree had come almost entirely from the water he had applied, not yet realizing the role of carbon dioxide (CO_2) in the air.

Then in 1772, the English scientist Joseph Priestly discovered that green plants affect air in a manner reverse to that of animal respiration. Shortly afterward, the Dutch physician Ingen-Housz showed that the ability of plants to synthesize carbohydrates from carbon dioxide is dependent upon light. About that time Henry Cavendish found that pure water could be produced by igniting inflammable air (hydrogen) in air. It remained for the great French chemist Lavoisier to see the full significance of the work of Priestly and Cavendish, when he correctly concluded that water is a combination of two gases, which he named hydrogen and oxygen. The discovery of the correct composition of water was one of the most important discoveries of the century because it supplied the keystone for the building of modern chemistry, necessary to an understanding of how green plants function.

In the closing years of the 18th century the basic facts of photosynthesis were uncovered. Its raw materials and end products were identified. Its dependence on light and green plants was known. Ingen-Housz proposed that light splits apart the CO_2 absorbed by the plant, accounting for the release of O_2. Others proposed that water was added chemically to carbon, forming the basic unit of carbohydrate (CH_2O). And so, for more than 100 years the role of light was regarded as settled.

Then in the 1930's, Van Niel of the United States showed that bacteria can carry on photosynthesis without evolving oxygen. He correctly concluded that light splits *water* rather than CO_2 as previously supposed. In 1937, Hill of England showed that photosynthesis is a two-step process, only the first of which requires light. The second step (the reduction of CO_2 to carbohydrate) was found not unique to the photosynthetic process (Fig. 2-6). Here are some important terms used to describe the photosynthetic processes:

$$2\,H_2O^* + CO_2 \xrightarrow{\text{Photosynthesis}} \underbrace{\left[H{-}\overset{|}{\underset{|}{C}}{-}OH \right]}_{\text{Sugar}} + O_2^* + H_2O$$

Figure 2-6 Carbon dioxide enters the leaf through the stomata, and water through the xylem vessels from the roots. Note that the oxygen released in the reaction (*see asterisks*) comes from the water rather than from the carbon dioxide, showing that it is water that is photochemically split. The oxygen released by photosynthesis is either used in cell respiration or diffuses into the atmosphere from the stomata.

ADP = Adensosine diphosphate.
ATP = Adensosine triphosphate.
Carotenoids = Pigments containing a carotene molecule.
Decarboxylation = Chemical alteration of an organic acid resulting in the loss of CO_2.
NAD = Nicotinamide adenine dinucleotide (oxidized form).
$NADH_2$ = Reduced form of NAD.
e^- = Electron.
Glycolysis = The breakdown of fructose diphosphate to pyruvic acid to yield energy (ATP).
Grana = Ellipsoid bodies that contain the chlorophyll within the chloroplast.
H^+ = Proton (hydrogen ion).
H_2 = Hydrogen gas.
Hexose = A sugar with six carbons.
Krebs cycle = The stepwise decarboxylation of organic acids with the release of H (and energy) to DPN molecules.
O_2 = Oxygen gas.
Oxidation = The addition of O_2, the removal of H, or the removal of electrons.
Phospholipids = Fatty substances containing phosphorus.
Phosphorylation = The chemical addition of phosphate (PO_4).
Reduction = The removal of O_2, the addition of H, or the addition of electrons.

Pigment Systems

Chlorophyll is the principal photosynthetic pigment. It is green because relatively more of the blue and red rays of light are absorbed by it while green rays are transmitted or reflected. The light energy that is absorbed by it is transformed into chemical energy that can be stored and utilized later. Chlorophyll in the higher plants is found in grana within the chloroplasts. Intimately associated with chlorophyll in the grana are proteins, phospholipids, carotenoids, and water of hydration. Grana are embedded in a complex matrix called the stroma. Essential reactions not requiring light occur in the stroma. Figure 2-7 shows the main structure of the leaf in which photosynthesis takes place. The essential steps in photosynthesis are:

1. Light energy is intercepted by chloroplasts.
2. This energy is used to split water into H and O.
3. The H attaches to molecules of NAD, and some of the generated energy is used to form ATP molecules.
4. $NADH_2$ and ATP act upon CO_2 to form carbohydrate.

Limiting Factors

The main factors limiting photosynthesis are CO_2, light, heat, minerals, and water. Of these, CO_2 most often limits the process under normal field conditions. (The rapid burning of fossil fuels thus could increase worldwide photosynthesis by increasing the CO_2 content of the air.) In trees and dense shrubs, light may be limiting in the shaded portions.

The reverse of photosynthesis is the breakdown of carbohydrates to yield energy (in the form of ATP) for general life processes; this occurs in small bodies (mitochondria) within the protoplasm. When ATP

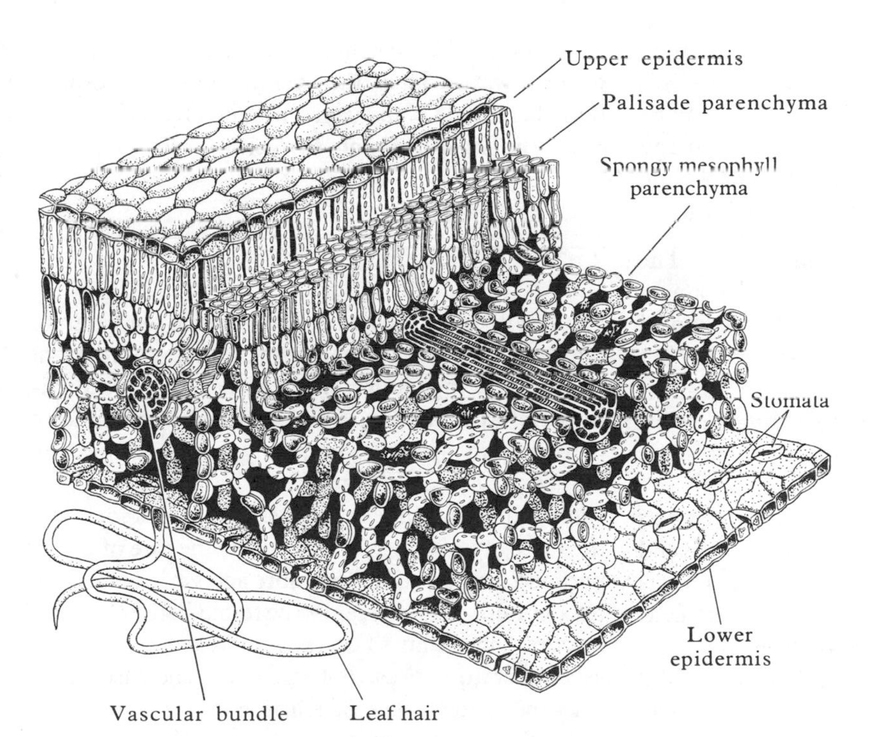

Figure 2-7 Diagrammatic section of an apple leaf showing the stomata on the lower surface for carbon dioxide entry, conducting tissue, and the parenchyma tissue containing chloroplasts for photosynthesis.

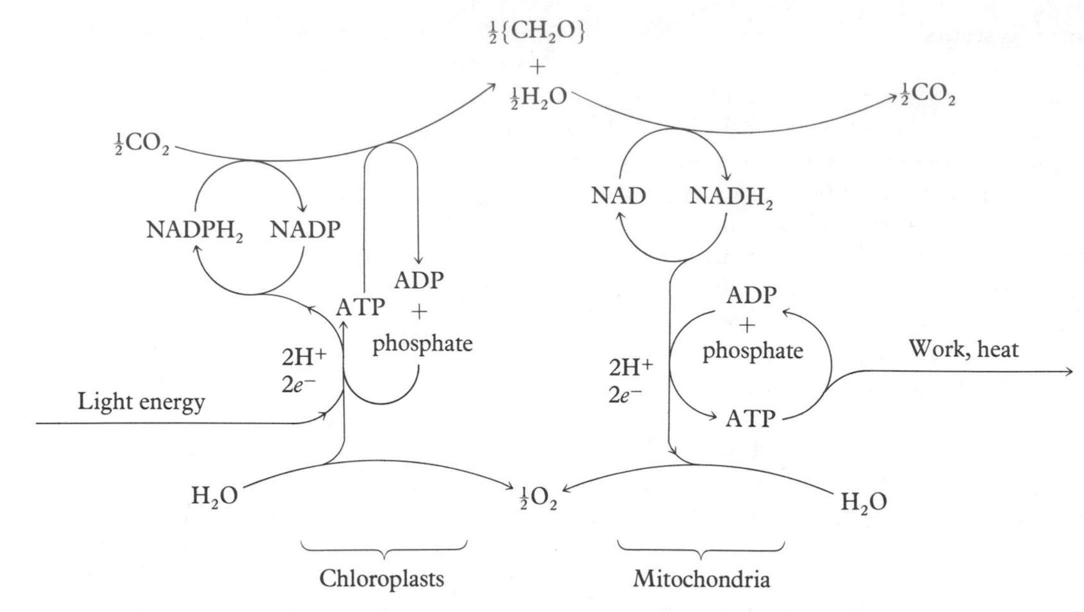

Figure 2-8 The bioenergetics of photosynthesis and the breakdown of carbohydrates. [From Park S. Nobel, *Biophysical Plant Physiology.* W. H. Freeman and Company Copyright © 1970]

gives up energy it becomes ADP, and the ADP is then regenerated into ATP. As depicted in Figure 2-8, the process is cyclic, and both ADP and ATP can be used over and over, since they merely transfer energy and are not "used up" in metabolism. Here are the component processes of carbohydrate breakdown.

1. Carbohydrate-phosphate compounds. The first step in sugar breakdown is the conversion of sucrose to glucose, followed by phosphorylation of glucose and fructose. This step releases no energy but uses ATP.

2. Glycolysis. In this process hexose sugar is converted stepwise into pyruvic acid, yielding some energy (as ATP).

3. Krebs cycle. This cycle is fed by a two-carbon fragment derived from pyruvic acid (which is the end product of glycolysis). In a series of reactions, CO_2 is released by decarboxylation, ATP molecules are generated, and H atoms are transferred to NAD.

4. Cytochrome respiration. The H is carried by $NADH_2$ to the complex cytochrome system, where more ATP is made by stepwise energy transfer.

5. O_2 is added chemically to the residual H atoms to form water. Thus, the overall process of respiration liberates CO_2 and water, uses up O_2, and releases energy in small amounts for cell processes.

TEMPERATURE

Plants will die above and below a certain temperature range. Within that range, there are many important effects of temperature, such as autumn preconditioning, breaking rest, and determining the time of flowering in spring. Many physical processes are influenced by temperature, such as the diffusion of gases and liquids in plants, the solubilities of ions, and the viscosity of water, which affects the rate of transport and transpiration. Far more and varied effects of temperature, however, are to be found in the chemical reactions in plants. The rates of reactions increase as temperature rises, but such rate increases vary with the specific type of reaction.

A useful relative measurement, the Q_{10}, (or "temperature coefficient") is used to indicate the relationship between temperature and a given reaction rate. Specifically, Q_{10} is the rate increase of a reaction or process with a 10°C rise in temperature. Growth, for example, may have a high Q_{10} over a given temperature range. Roots may have a high Q_{10} for excreting exogenous chemicals (e.g., herbicides); plants adapted to high-temperature photosynthesis may have a low Q_{10} for respiration. The Q_{10} of uncatalyzed chemical reactions in a physical system is about 2.4, while that relating to the overall plant growth process, with its interactive complexities, is about 1.2. Enzyme-catalyzed reactions have Q_{10} values somewhat higher, between 1.3 and 5.

Figures 2-9 and 2-10 show effects on plants of light and temperature. There is a dynamic balance between incoming and outgoing radiation and energy, to which the plant must adapt by rapid energy exchange to avoid lethal temperatures.

Because the processes of plant growth are so varied and complex, optimum temperatures vary for each process and tissue. For example, the optimum temperature for root growth is lower than that for shoot growth. Cell respiration, which supplies biological energy for vital processes, increases with a rise in temperature. This may or may not be of benefit to the fruit grower, depending on the situation. Moderately high temperatures may aid in fruit growth and matu-

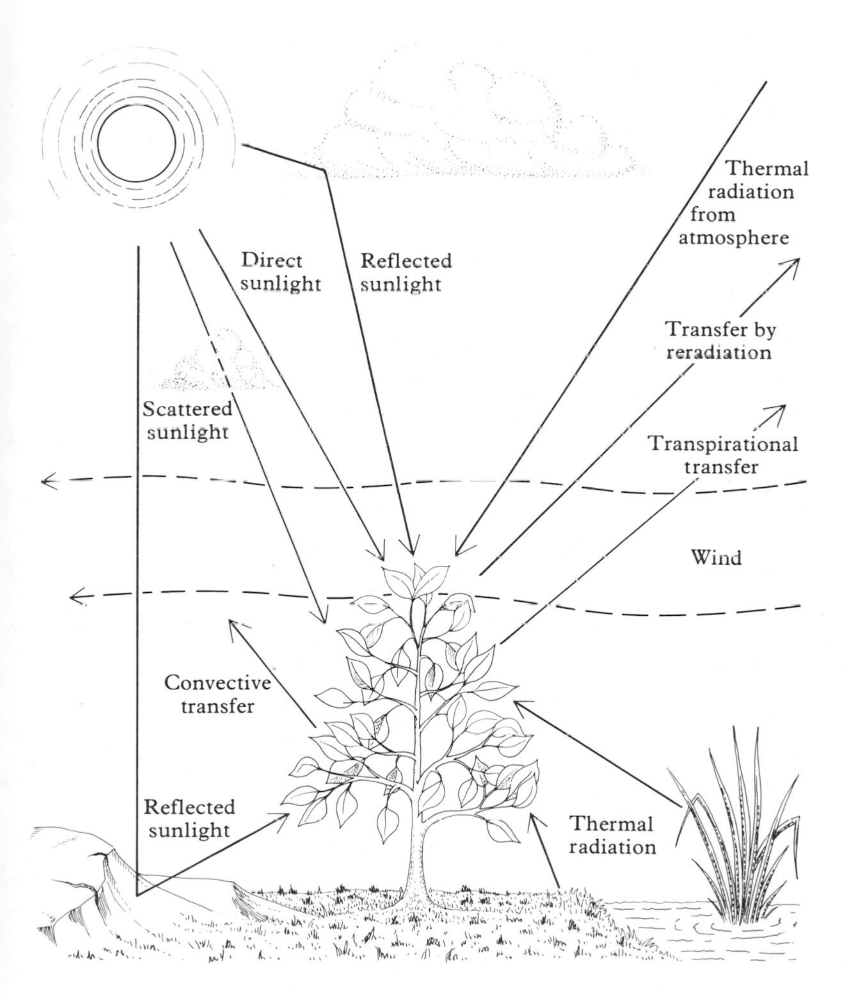

Figure 2-9 The exchange of energy between plant and environment. Plants that are not adapted to a particular environment are often killed by stresses imposed by such energy exchanges. [After D. M. Gates, "Heat Transfer in Plants." Copyright © 1965 by Scientific American, Inc. All rights reserved.]

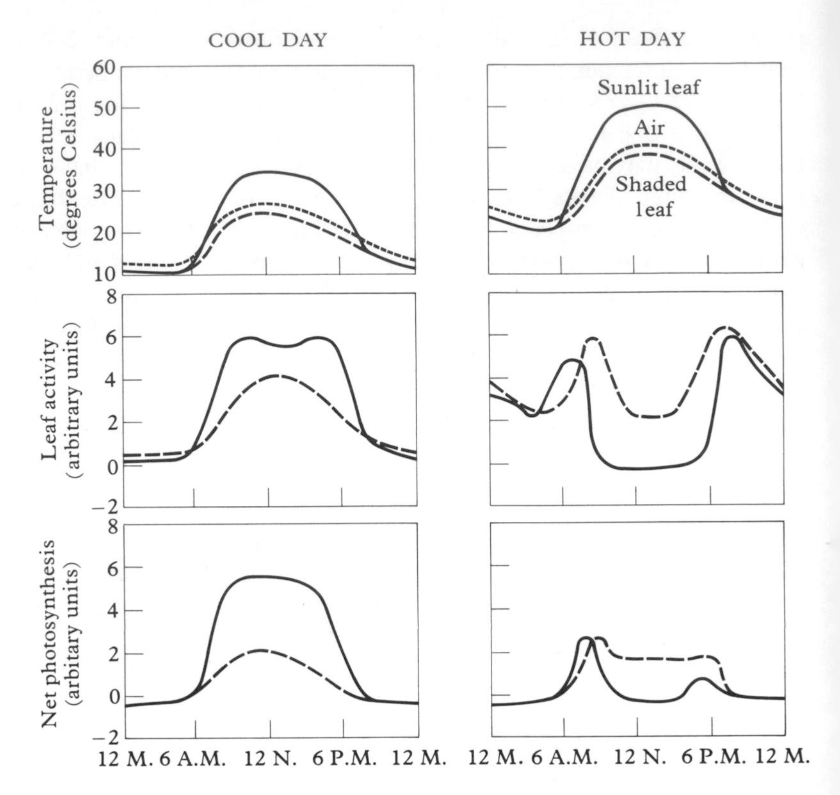

Figure 2-10 Differences in temperature, leaf activity, and net photosynthesis between a cool and a hot summer day. Note that the shaded leaf on a hot day has greater net photosynthesis than the sunlit leaf, although it is low for both. [After D. M. Gates, "Heat Transfer in Plants." Copyright © 1965 by Scientific American, Inc. All rights reserved.]

ration, but after harvest the fruit may be placed in cold storage to reduce respiration and thus prolong storage life. Whole plants are subjected to a variety of temperatures during the seasonal and diurnal cycles and each organ may be at a different temperature at a given time. In summer, the root temperature is lower than other organs during the day, and usually the temperature at the soil surface is higher than that of branches and leaves, unless the soil is shaded. At night, the tissues at the soil surface may be the coolest part of the plant. The important effects of temperature on plant performance and survival are covered in detail in other chapters.

WATER

The early Greeks thought that water was an immutable element, equal in rank to air, fire, and earth. Water's vital role in supporting life has been recognized throughout the ages, but little was learned about its nature until the 18th century.

The discovery that water is made up of the two gases, H_2 and O_2, permitted a firmer understanding of all kinds of combustion, including biological oxidation, and of photosynthesis and chemistry in general. Although the determination of the simple formula "H_2O" was a scientific triumph, the truly unique properties of water remained a mystery for many years after the formula was known.

The properties of water are of prime interest to the biologist, soil scientist, soil conservationist, farmer, and meteorologist. In fact, water is the most important compound in the biological world. Most living things are by weight 70 to 90 percent water. Water takes part in the fundamental energy-storing reaction of photosynthesis, the most important of all biochemical reactions. More than three-quarters of the earth's surface is covered with water, the total amount of which is estimated at 1.4×10^{18} tons. This

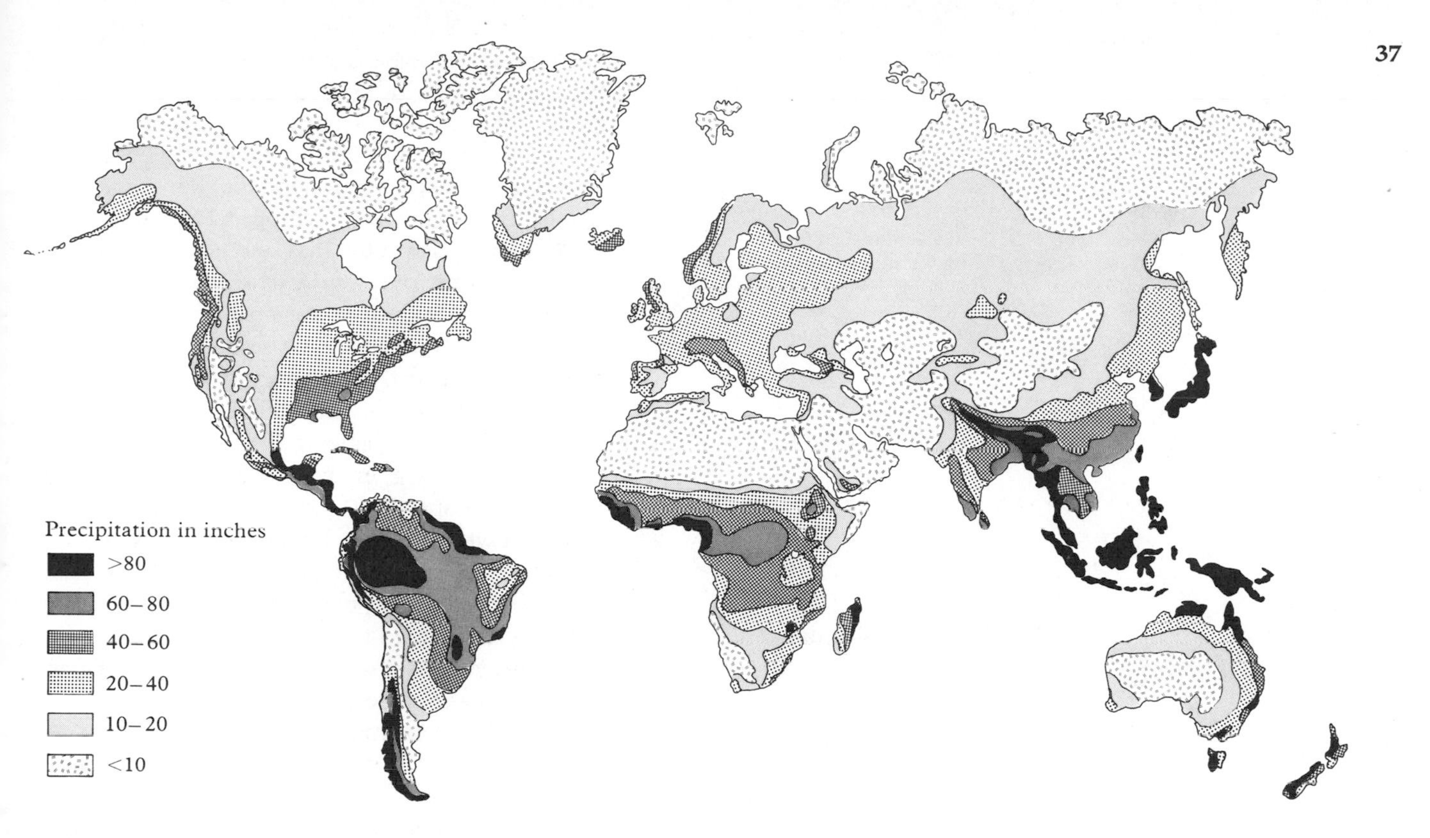

Figure 2-11 The world's annual distribution of precipitated moisture. Note that some of the desert areas have been made into productive fruit regions by the development of irrigation. The zones are defined by precipitation in inches: 80 inches = 203.2 cm; 60 inches = 152.4 cm; 40 inches = 101.6 cm; 20 inches = 50.8 cm; 10 inches = 25.4 cm [After USDA]

large amount of water, plus its great energy-storing capacity makes it the greatest single influence on climate. The worldwide distribution of precipitation is shown in Figure 2-11. In order to grow intensive fruit crops in arid areas of the temperate zone, irrigation must be provided. In most areas of the world, new irrigation projects are essential in developing new croplands.

Water is the only common liquid in nature that becomes less dense when it freezes, and thus ice floats on water. This characteristic permits the survival of marine life during extremely cold weather; it also is the cause of freeze damage to plants. Chemists have devoted much time to the study of the water molecule in an effort to learn why it has such unique properties.

The Hydrogen Bond

The concept of the hydrogen bond has become one of the most important concepts in modern biochemistry, and it has done a great deal toward clarifying the properties of water. Indeed, Linus Pauling was awarded the Nobel prize for his elucidation of the nature of the hydrogen bond.

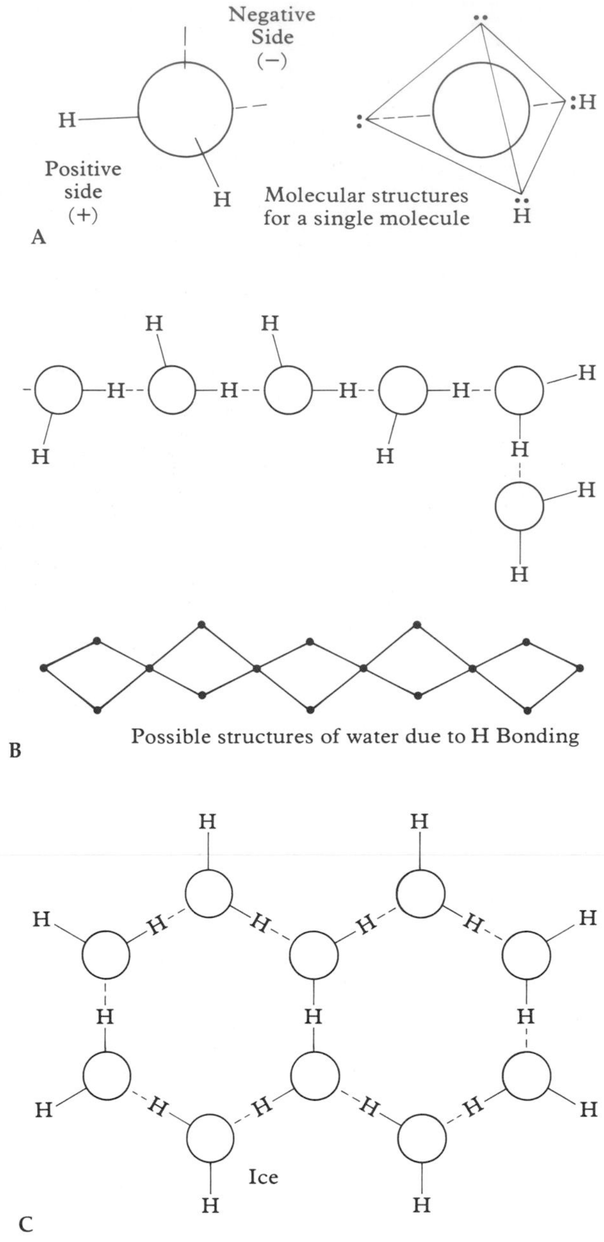

Figure 2-12 Diagrammatic models for water, showing (A) the molecular structure and spatial arrangement of the atoms that result in a strongly polar molecule; (B) possible polymolecular structures of water due to hydrogen-bond associations; (C) the hexagonal crystal structure of ice; the relatively large open spaces show why ice is less dense than liquid water.

Other studies have shown water to be a strongly polar molecule with the H's attached to the O at an angle of about 105° from each other. This corresponds closely to the 109° angles formed by the lines drawn from the center through the four corners of a tetrahedron (Fig. 2-12). Hydrogen, consisting of one electron and one proton, has only one stable orbital and thus forms only one true covalent bond with another atom. But due to the small size of H, and its positive nature, it can form a secondary bond when between two strongly electronegative atoms, such as O and N. The dotted lines (Fig. 2-11) indicate the secondary H-bonds between O atoms in water. Extensive H-bonding in water explains most of its anomalous properties. A single H-bond is relatively weak, but when it occurs by the millions, as in water, it changes completely the properties of the compound. Because of H-bonding, liquid water is now generally considered to consist of large, loose molecules with various possibilities as to the packing and linkage of basic molecules (Fig. 2-11).

The molecular grouping of liquid water is in a constant state of flux and gives no distinct x-ray diffraction pattern. But ice gives a definite diffraction pattern that is consistent with a hexagonal arrangement of molecules (Fig. 2-11). This structure explains why ice is less dense than liquid water (the specific gravity of ice = .917; that of water at 4°C = 1.00). The long linear and/or diagonally linked molecules in the liquid state can pack closer together than those in the hexagonal ice crystal, which has much larger empty spaces in it than does the liquid.

Heat Absorption

The extremely high specific heat, heat of vaporization, and heat of fusion of water make it the greatest temperature-stabilizing influence in the world. One gram of water must absorb 720 calories of heat in going from ice to liquid to vapor. The Atlantic gulf stream carries tremendous amounts of heat northward and prevents Europe from being as extremely cold as other lands that lie in the same northern latitudes. Thus ocean waters serve as a vast storehouse for solar energy. In addition to warming the oceans, the sun's rays evaporate an estimated 417,000 cubic kilometers of water per year. This water must absorb about 2.1 $\times 10^{23}$ calories (or 8.3 $\times 10^{20}$ BTU) before it can vaporize. To help visualize the amount of energy required, let us imagine a furnace set to burn one million BTU's per hour. Such a furnace would have to burn for 100 million years to use that much energy.

Water in Living Things

Water's very high specific heat and high thermal conductivity permit good heat distribution and dissipation in living things and tend to prevent rapid temperature changes that might destroy protoplasm.

The viscosity (resistance to flow) of water varies considerably with temperature. Thus its viscosity is twice as great at 0° than at 25°C. During the springtime, when plants take up cold water from the soil, the water moves more slowly in the conducting system than it does later in the year when it is warmer. For this reason, transpiration "leaf scorch" may occur in early spring, when the soil is cold and the evaporating power of the atmosphere is high (because of warm, dry air).

The water-conducting system of a tree is shown in Figure 2-13. Water taken up by the roots is transported across the cortex and into the xylem vessels, where it is raised to the leaves by the suction created by evapo-transpiration. The high internal pressure

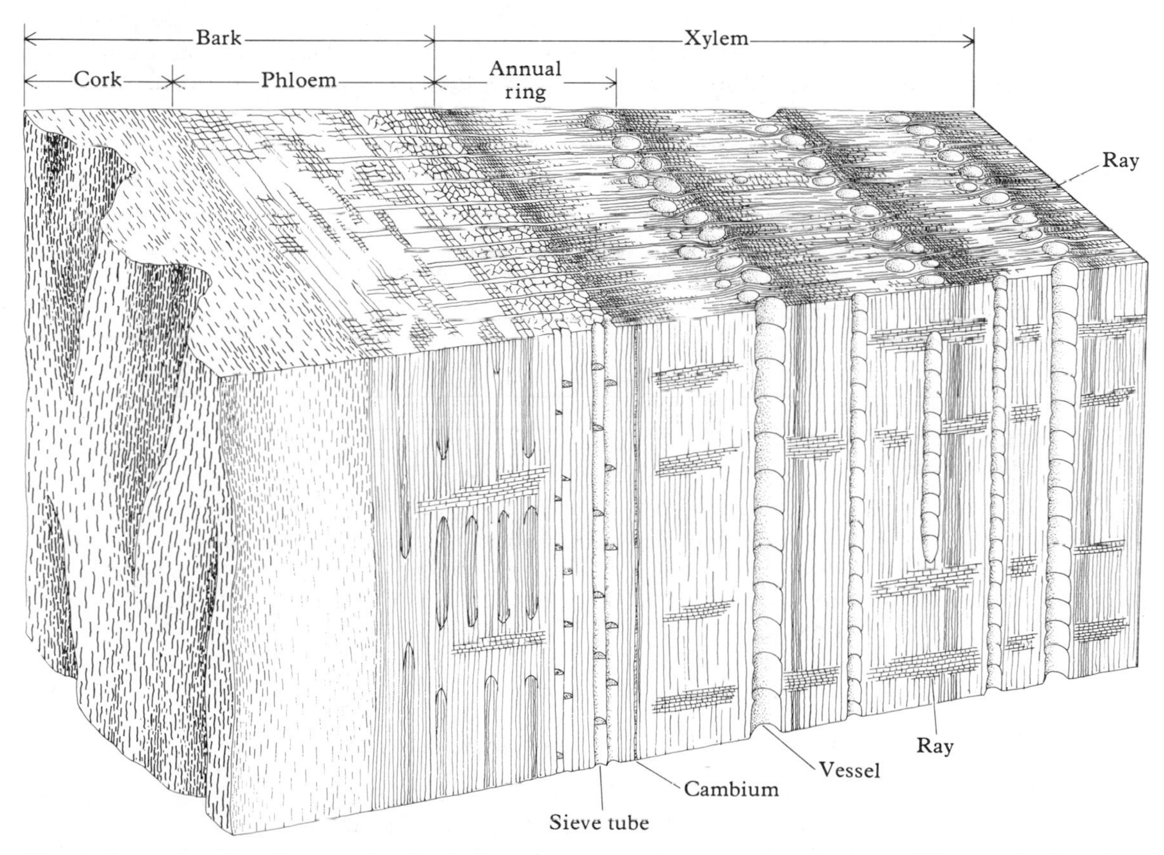

Figure 2-13 Section of the trunk of a deciduous tree showing the transport system. Water and minerals ascend from the roots through dead xylem elements (vessels). Photosynthate and elaborated substances, such as hormones, move downward from leaves and growing points through the living phloem (sieve tubes). [After M. H. Zimmermann, "How Sap Moves in Trees." Copyright © 1963 by Scientific American, Inc. All rights reserved.]

Table 2-1 Some important properties of water.

Specific gravity at 4°C = 1.00	(The standard for measuring density.)
Specific gravity of ice = 0.917	
Specific heat at 15°C = 1.00	(The specific heat of lead at 20°C = 0.03; water requires thirty-three times as much energy to increase 1°C as does lead.)
Heat of vaporization at 100°C = 540 cal/g	(Important in world energy storage and in evaporative cooling of plants.)
Heat of fusion at 0°C = 80 cal/g	(Important release of energy on freezing. Used to protect flowers from frost by over-tree sprinkling.)
Surface tension at 20°C 72.5 dynes/cm	(CH_3OH (molecular weight 32) = 22.6 dynes/cm.)
Specific viscosity at 0°C = 1.000	(The standard of viscosity measurement.)
Specific viscosity at 25°C = .499	(Water at 25° flows much faster than at 0°.)
Dipole moment = $1.85 \div 10^{18}$ e.s.u.	(H_2S = $0.95 \div 10^{18}$ electrostatic units; the strong dipole moment of water makes it a polar liquid (separated + and − charges) which permits H-bonding and makes it a liquid at biological temperatures.)

(cohesion) of water is very important in water translocation within plants. It is estimated that about 50 atmospheres of pressure are needed to raise water to the tops of tall trees. The suction created by water vapor loss from the leaves is adequate to supply the pull required, but this places considerable stress (negative pressure) on the water columns in the conducting system. It is only by virtue of the strong cohesion of water molecules (which has a force of approximately 10,000 atmospheres) that these columns are not broken. This high internal pressure is due mainly to H-bonding.

Water in plants and animals contributes as much to the essential properties of life as do fats, proteins, carbohydrates, and minerals. It is the principal medium in which all metabolic processes occur. Water not only serves as a vehicle for the transport of ions to and from cells but also participates in acid-base equilibria and provides the cell turgor so necessary to fruit growth and to leaf and stem orientation and support in plants. It is not surprising that half of all life (by weight) is found in water. The well-ordered arrangement of water in the hydrated protoplast has led one scientist to conclude that its spatial and energy relationships are more aptly described by solid-state physics than by classical biochemistry.

In conclusion, we shall enumerate (in Table 2-1) the specific energy-storing and energy-releasing properties of water. As we have seen, these properties make it important in determining climate, in protecting plants from frost and heat, and in regulating many biological processes. (We shall discuss these properties in more detail in Chapters 6 and 15.)

GENERAL REFERENCES

Crafts, A. S.; H. B. Currier; and C. R. Stocking. 1949. *Water in the physiology of plants. Chronica Botanica,* Waltham, Mass.

Devlin, R. M. 1967. *Plant physiology.* Reinhold Pub. Corp., New York.

Geiger, R. 1957. *The climate near the ground.* Harvard Univ. Press, Cambridge, Massachusetts.

Hodgeman, C. D. 1947. *Handbook of chemistry and physics.* Chemical Rubber Co. Press, Cleveland, Ohio.

Leopold, A. C., and P. E. Kriedemann. 1975. *Plant growth and development.* 2nd ed. McGraw-Hill Inc., New York.

Nobel, P. S. 1974. *Introduction to biophysical plant physiology.* W. H. Freeman and Company, San Francisco.

Platt, R. B., and J. F. Griffiths. 1965. *Environmental measurement and interpretation,* Reinhold Pub. Corp., New York.

Went, F. W. 1957. Climate and agriculture. *Scientific American,* June, pp. 82–94.

3

Fruit and Nut Species

To use the principles set forth in other sections properly, one should know the general characteristics of each crop species as well as the specific idiosyncrasies of important cultivars. This chapter will discuss groups of related species and cultivars in terms of their botanical characteristics and their importance and distribution as crop plants. To aid in understanding the descriptive terms used and to acquaint the reader with the important horticultural facts, Figure 3-1 illustrates many types of leaves, inflorescences, flowers, buds, and fruit. Some terms that are not found in this figure may be found in the Glossary.

In the following discussion of fruit and nut species, the chromosome numbers for each group (as given by Bolkhovskikh et al., 1969) appear following the names of the groups. These numbers are important in breeding and cross-pollination for fruit set. The basic chromosome number for each group (that is, the number of chromosomes in each genome) is designated by the symbol x. The number of chromosomes in the somatic cells of the various members of a group (their sporophytic chromosome numbers) are referred to as *somatic numbers* in this book.

POME FRUITS

The most important genera of pome fruits (Family Rosaceae, Subfamily Pomoideae, $x = 17$) are *Malus* (apple), *Pyrus* (pear), and *Cydonia* (quince). Other genera of minor importance for their edible fruit but

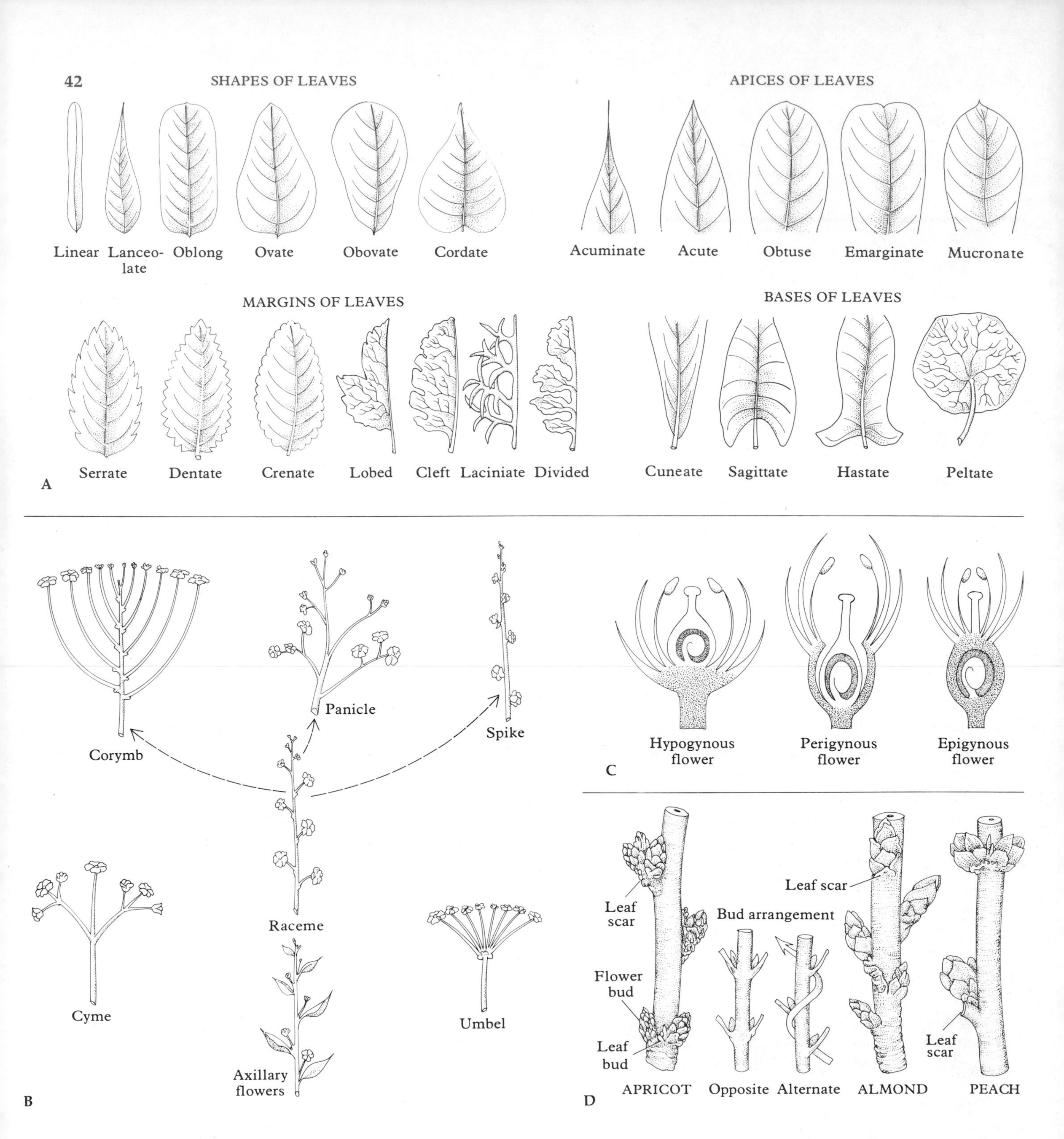
SHAPES OF LEAVES
Linear
Lanceo-late
Oblong
Ovate
Obovate
Cordate
APICES OF LEAVES
Acuminate
Acute
Obtuse
Emarginate
Mucronate
MARGINS OF LEAVES
Serrate
Dentate
Crenate
Lobed
Cleft
Laciniate
Divided
BASES OF LEAVES
Cuneate
Sagittate
Hastate
Peltate
A
Corymb
Panicle
Spike
Raceme
Cyme
Axillary flowers
Umbel
B
Hypogynous flower
Perigynous flower
Epigynous flower
C
Leaf scar
Flower bud
Leaf bud
Bud arrangement
Leaf scar
Leaf scar
APRICOT
Opposite
Alternate
ALMOND
PEACH
D

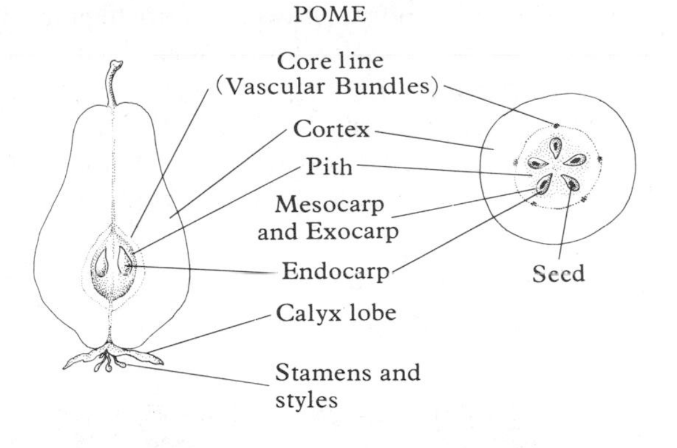

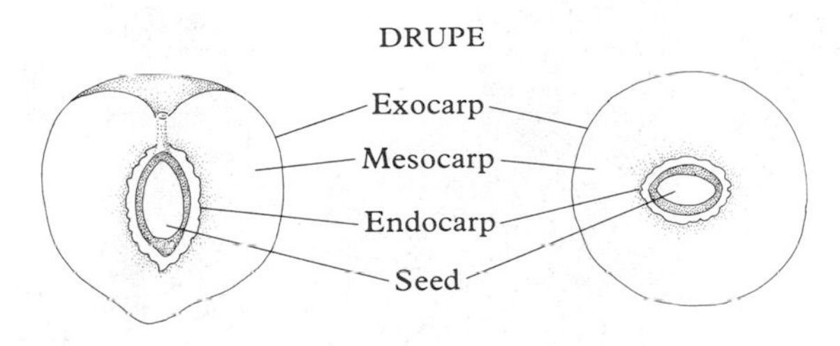

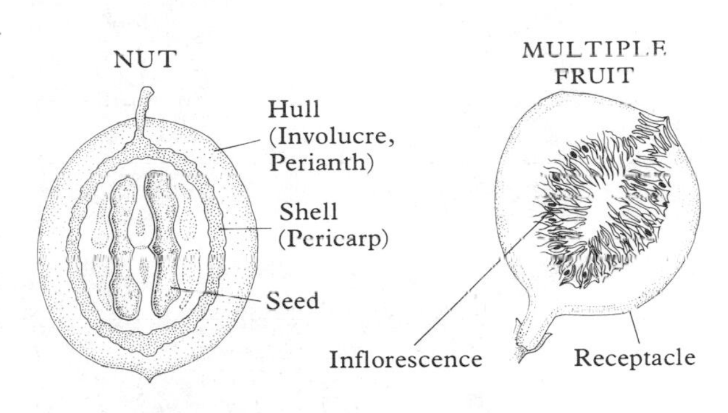

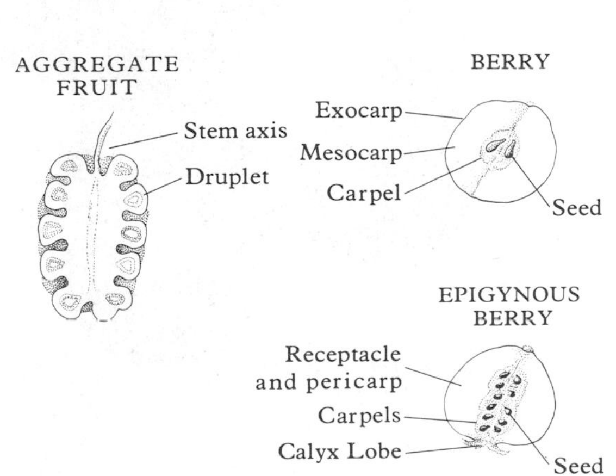

Figure 3-1 Useful botanical terms. **(A)** Specific types and margins of leaves. **(B)** Inflorescence types: the raceme, corymb, panicle, and umbel are indeterminate, with basal flowers opening first; the cyme is *determinate.* **(C)** The basic flower types: the superior ovary with *hypogynous* insertion of stamens, petals, and sepals; the *perigynous* insertion on the hypanthium; the inferior ovary with *epigynous* insertion of the corolla above the ovary on receptacle tissue. **(D)** Bud arrangement: *opposite* and *alternate* types; also, leaf and flower buds of some stone fruits. **(E)** The seven main fruit types: A *pome,* such as the pear, is derived from the fusion of ovary and receptacle. A *drupe,* such as the peach, is a one-seeded fruit derived entirely from an ovary. A *nut,* such as the walnut, is derived from the fusion of ovary and perianth. A *multiple fruit,* such as the syconium of the fig, is derived from the fusion of many ovaries and receptacles of many flowers. An *aggregate fruit,* such as the blackberry, is derived from many ovaries of a single flower. A *berry,* such as a grape, is a multiseeded fruit derived from a single ovary. An *epigynous* or false *berry,* such as a blueberry, is a multiseeded fruit derived from the fusion of ovary and receptacle. [A, B, C: After Swingle, 1946; D: After Zielinski, *Modern Systematic Pomology,* Brown, Dubuque, 1955.]

of possible use as rootstocks are *Chaenomeles* (Chinese quince), *Mespilus* (medlar), *Crataegus* (hawthorn), *Sorbus* (mountain ash), *Amelanchier* (service berry). The degree of graft compatibility among these and other genera of the subfamily Pomoideae is such that some are being developed as rootstocks with special features, such as dwarfing, tolerance to adverse soil conditions, and resistance to diseases or insects. These are discussed in detail in Chapters 4, 6, and 16.

The basic chromosome number of all genera of the Pomoideae is 17. This group probably originated by spontaneous doubling in a sterile hybrid between primitive species of the rosaceous subfamilies Prunoideae ($x = 8$) and Spiroideae ($x = 9$), resulting in allopolyploids that function as normal sexual diploids. Some species, for example, apple, however, at present are polyploids of $3x$, $4x$, $5x$, etc., most of which are apomictic.

The plants that constitute the subfamily Pomoideae are trees or shrubs generally described as follows: Leaves simple or pinnate, stipulate; flowers solitary or in umbels, racemes, cymes, panicles or corymbs; 2–5 carpels, usually two-ovuled, more or less united and adnate to the cup-shaped calyx tube, forming an inferior ovary; fruit a fleshy pome.

The fleshy part of the pome fruit is nonovarian calyx and receptacle tissue, so at times fruit may be produced without fertilization.

Apple (*Malus* Mill.)

DESCRIPTION: $x = 17$, somatic nos. $= 34, 51, 68, 85$; deciduous, rarely evergreen trees or shrubs, rarely with spiny branches; **buds:** ovoid, with several imbricate scales; **leaves:** serrate or lobed, folded or convolute in bud, stipulate; **flowers:** white to pink or carmine, in cymes; **petals:** usually suborbicular or obovate; **stamens:** 15–50, with usually yellow anthers; **ovary:** inferior, 3–5 cells; **styles:** 2–5, connate at base; **fruit:** a pome, usually without grit cells, persistent or deciduous calyx (Fig. 3-2).

Figure 3-2 Buds, flowers, and fruit of **(A, B, C)** apple and **(D, E, F)** pear. The apple inflorescence is a cyme; that of the pear is a raceme. In both species, the fruit type is a pome.

The genus consists of about fifteen primary species, including two from Europe, four from North America and the rest from Asia. Most of our domestic cultivars derive from *M. pumila* Mill., the common apple of Europe. These large-fruited derivatives have been under selection and improvement by man for thousands of years, so it is not possible to know their origin with certainty. Domestic crab apples are hybrids between *M. pumila* and one of the primitive species such as *M. baccata* (L.) Borkh. Recent apple breeding has used primitive *Malus* species as a source of disease or insect resistance.

When Columbus discovered the New World there existed several hundred named cultivars of apple. Today 90 percent of U.S. production is accounted for by thirteen cultivars (Table 3-1), of which five account for more than two-thirds of the production. A relatively small number of cultivars accounts for most of the world production of apples. (Table 3-2). Its great diversity of climatic adaptation has made the apple the most widely planted fruit of the temperate zone. Some cultivars mature in as few as 70 days; others take 180 or more days. Some are very hardy; others are relatively tender. Some have a long winter-chilling requirement; others require a very short period. Such genetic diversity makes it hard to generalize about specific climatic requirements of the apple.

Table 3-1 Apple production in the U.S., 1970–1972.

Cultivar	Rank	Millions of Bushels	Percent of Total	Main Region
Delicious (all strains)	1	43.6	28.8	West
Golden Delicious	2	20.0	13.2	West
McIntosh	3	17.6	11.6	East
Rome Beauty	4	12.0	7.9	East
Jonathan	5	9.3	6.1	Central
York	6	8.2	5.4	East
Stayman	7	6.5	4.3	East
Winesap	8	4.3	2.8	West
Cortland	9	4.0	2.6	East
Newtown	10	3.9	2.6	West
Rhode Island Greening	11	3.8	2.5	East
Northern Spy	12	3.4	2.2	East
Gravenstein	13	2.3	1.5	West
All Others		12.6		
Total		151.5		

SOURCE: *American Fruit Grower*, October 1972, p. 8.
NOTES: (a) The first six cultivars account for 73 percent of the crop; the next six account for 16 percent. (2) The eastern states account for 46 percent, the central states 19 percent, and the western states 34 percent of the total crop.

Pear (*Pyrus* L.)

DESCRIPTION: $x = 17$; somatic nos. = 34, rarely 51, 68; deciduous, rarely evergreen trees or shrubs, sometimes thorny; **leaves:** serrate, crenate, or entire, rarely lobed, involute in bud, petiolate stipulate; **buds:** with imbricate scales; **flowers:** with or before the leaves, in umbel-like racemes, white, rarely pinkish; **sepals:** reflexed or spreading; **petals:** clawed, suborbicular to broad-oblong; **stamens:** 20–30; anthers red or purple; **styles:** 2–5, free, closely constricted at base by the disk; **ovules:** 2 per locule; **fruit:** a globose or pyriform pome, calyx persistent or deciduous, flesh usually with grit cells; **walls of locules:** cartilaginous; **seeds:** black or nearly so (Fig. 3-2).

The genus *Pyrus* consists of about twenty species, of which about half are found in Europe and in North Africa and Asia Minor around the Mediterranean Sea. All others are native to Asia. The domestic pear of Europe, like the apple, has been selected, improved and cultured since prehistoric times. *Pyrus communis* L. is an entirely cultivated form but possibly derives from the relatively large-fruited wild species *P. caucasica* Fed. and *P. nivalis* Jacq. The domestic pears of Asia derive mostly from *P. pyrifolia* (Burm.) Nak., known as the Japanese sand pear, or from selections of Ussuri pear, *P. ussuriensis* Maxim. Written records exist from 1100 B.C. describing selection and culture of these oriental pears. Present cultivar characteristics indicate that many are complex hybrids between the two species listed, or with other Asian species. All species are self-sterile, cross-fertile, sexual diploids ($2n = 34$). Even though some cultivars in some climates set seedless fruit, best yields are obtained where pollinizers are planted. All species are graft-compatible with each other, so that several special rootstock species have been developed that are resistant to pests and tolerant of adverse soils (see Chapter 4).

As is true of apple, a relatively few cultivars account for most of the world production (Table 3-3).

Table 3-2 Apple cultivars of the world and their important characteristics.

Cultivar	Somatic Chromosome Number	Tree Size	Bloom to Harvest (days)	Fruit Size	Fruit[1] Color	Use[2]	Maximum Storage Life (days)	Precocity[3]	Alternate Bearing	Overall Productivity[3]	Self-fruitful
Yellow Transparent	34	S–M	70–100	S	Y	C	80	Good	Yes	Mod	Partly
Gravenstein	51	L	110–130	L	SR	F,C	90	Poor	Yes	Mod	No
James Grieve	—	M	110–130	L	YRS	F	100	Good	No	Mod	—
Antonovka	—	M	110–130	M	Y	C	100	Good	Yes	Mod	—
Wealthy	34	M	120–135	M	R	F,C	80	Good	Yes	Good	Partly
Winter Banana	34	M	150–165	L	YB	C,F	150	Good	No	Good	No
Cortland	34	M	125–140	S	R	F,C	150	Good	No	Good	Slight
McIntosh	34	M–L	125–145	M	R	F,C	130	Good	No	Good	Slight
Cox's Orange Pippin	34	M	130–160	M	BY	F	130	Good	Mod	Fair	No
Rhode Island Greening	51	L	130–155	M–L	G	C	180	Poor	Mod	Fair	No
Bramley	51	L	135–155	L	G	C	190	Fair	Yes	Mod	Partly
Ralls	—	M	—	M	R	F	185	Poor	Yes	Fair	—
Jonathan	34	M	135–150	S	R	C,F	120	Good	No	Good	Slight
Grimes Golden	34	M	140–150	M	Y	F,C	120	Good	Mod	Good	Partly
Golden Delicious	34	M	140–160	M–L	Y	F,C	160	V.Good	Yes	V.Good	No
Delicious (all strains)	34	M–L	140–160	M–L	R	F,C	180	Fair	Mod	Fair	No
Spur Delicious	34	M	145–165	M–L	R	F,C	180	Good	Mod	Good	No
Boskoop	51	L	145–165	L	GYR	F,C	160	Poor	Mod	Fair	
Northern Spy	34	L	145–170	L	B	F,C	180	V.Poor	Mod	Poor	No
Mutsu	51	L	145–170	L	GY	F,C	190	Good	No	Good	No
York Imperial	34	L	155–175	M–L	R	C	180	Poor	Yes	Fair	Partly
Rome Beauty	34	M–S	160–175	L	R	C	240	V.Good	No	V.Good	Slight
Newtown	34	M	160–175	M	GY	C,F	200	Poor	Yes	Fair	Partly
Winesap	34	M	160–180	S–M	R	C,F	240	Fair	Mod	Mod	No
Stayman	51	L	160–175	M–L	R	F,C	180	Good	No	Good	Slight
Sturmer Pippin	—	M	160–175	M–L	GY	F	210	—	—	—	—
Granny Smith	34	L	180–210	M–L	G	F,C	210	Fair	No	Good	—

NOTE: Dashes indicate missing data.
[1]Y = Yellow, G = Green, R = Red, B = Blush, S = Striped.
[2]F = eaten fresh, C = culinary.
[3]V. = Very.

The oriental pears grown in Japan, China, and Korea have been erroneously described as being inferior to European cultivars because they were judged by European standards of texture and flavor. Asian pears are crisp and juicy with some apple-like flavors, in contrast to the "pear-flavored," very soft, melting texture of European pears.

The principal cultivars grown in the U.S. are susceptible to bacterial fire blight, which is caused by the bacterium *Erwinia amylovora* (Burr.) Winslow et al. Thus the main acreage is centered in the arid regions of the Pacific Coast, where the disease is most easily controlled. The main cultivars of the world are described in Table 3-4.

Quince (*Cydonia*)

DESCRIPTION: $x = 17$; somatic no. $= 34$; deciduous thornless shrub or small tree; **buds:** small, pubescent, with few scales; **leaves:** petioled, entire stipulate; **flowers:** white or pink, terminal, solitary at the end of leafy shoots; **sepals:** 5, entire reflexed; **petals:** 5, obovate; **stamens:** 20; **styles:** 5, free, pubescent below; **ovary:** inferior, 5-celled, each with numerous ovules; **fruit:** a many-seeded pome, pyriform.

The genus consists of a single species, *Cydonia oblonga* Mill, native to the warmer regions of south-

Table 3-3 Pear production in the U.S. (1966–1970).

Cultivar	Yield Rank	Annual Average Production (1000 m tons)	Production (percent of total)	Area Planted Hectares (1962–1964)	Hectares (as % of total)	Ripening Season	Fire Blight[1]
Bartlett (Williams)	1	408.0	75	31,889	76	Mid	Sus.
Anjou	2	88.4	16	4,897	12	Late	Sl. Res.
Bosc	3	22.7	4	1,578	4	Late	Sus.
Hardy	4	9.1	1.7	647	2	Mid	Sus.
Comice	5	2.7	0.5	850	2	Late	Sus.
Winter Nelis	6	2.4	0.4	202	0.5	Late	Sus.
Seckel	7	1.8 (est.)	0.3	445	1	Mid	Mod. Res.
Kieffer	8	0.9 (est.)	0.2	162 (est.)	0.4	Late	Mod. Res.
Clapp Favorite	9	.7 (est.)	0.1	162	0.4	Early	V. Sus.
All other	—	10.2	1.9	971	3		
Total		546.9		41,803			

NOTES: (1) Bartlett accounts for three-quarters of the production and acreage (one hectare = 2.47 acres). (2) The three West Coast states produce 90 percent of the total pear crop on about 87 percent of the acreage. (3) About 95 percent of the total tonnage comes from the top three cultivars (Bartlett, Anjou, Bosc).
[1]Sl. = Slightly, V. = Very, Sus. = Susceptible, Res. = Resistant.

Table 3-4 Pear cultivars of the world and their important characteristics.

Cultivar	Tree Size	Bloom to Harvest (days)	Bloom Date	Fruit Size	Fruit Color	Use	Cold Storage Life[1] (days)	Needs Pollinizer	General Productivity
Giffard	M	100–120	M	M–Lg	Y–Bl	F	(short)	yes	G
Precoce Morettini		100–125	E			F		yes	G
Dr. J. Guyot	M	105–125	M	L	Y	F	(short)	yes	G
Clapp Favorite	Lg	105–130	M	Lg	Y–Bl	F	50–70	yes	VG
Bartlett (Williams)	M	110–135	M	M	Y–Bl	F,C	70–85	yes	VG
Seckel	M–Lg	120–140	M	S	R–Bl	F	90–100	yes	VG
Hardy	M	130–150	M	M	R	F,C	75–140	yes	G
Eldorado	M	140–160	M	M	Gr–Y	F	180–220	yes	G
Anjou	Lg	140–165	E–M	M–Lg	Gr–Bl	F	175–185	yes	P–G
Bosc	M	150–165	L	M–Lg	R	F,B	90–100	yes	G
Packham's Triumph	M	150–165	M	M–Lg	Gr	F	170–190	yes	G
Comice	Lg	150–170	L	Lg	Gr–Bl	F	90–105	yes	P–G
Angoulême	M	150–170	E	Lg	Y	F		yes	G
Flemish Beauty	M	160–180	L	M	Y–Bl	F		yes	G
Conference	M	160–180		M–Lg	Gr	F		yes	VG
Easter	M	160–185		M	Y–Bl		90–100	yes	
Winter Nelis	M	160–185	L	S	Gr	F	175–230	yes	VG
Forelle	M	160–190	E	S	Gr–Bl	F		yes	VG
Kieffer	M	170–190	E	Lg	Y–Bl	F,C	90–120	yes	VG
Glou Morceau	M	170–200		Lg	Gr–Y	F	(long)	yes	G
Clairgeau	M	170–200	M	Lg	Y–Bl	F		yes	G
Passe Crassane	M	180–210	M	M–Lg	Gr–R	F	(long)	yes	
Nijiseike	M		L	M–L	Y	F	(long)	yes	VG
Ya Li	M		E	L	Y–Gr	F	(long)	yes	

KEY: S = Small, M = Medium, Lg = Large.
E = Early, L = Late.
Y = Yellow, Gr = Green, R = Russet, Bl = Blush.
F = Fresh, B = Baking, C = Canning.
VG = Very Good, G = Good, P = Poor.
[1]Storage life is extended several weeks with controlled atmosphere storage.

A
B
C
D
E
F
G
H
I

eastern Europe and Asia Minor. Few studies have been made of this minor crop. It is used mostly as a dwarfing rootstock for pear but is also sometimes used for preserves, jams, and jellies.

The main cultivars of quince are Orange, Champion, Pineapple, Smyrna, Van Deman, Rea, and Meech. There are something less than 200 acres grown in the U.S., but in Argentina it is an important crop of 20,000 tons annual production.

STONE FRUITS

DESCRIPTION: Family Rosaceae, Subfamily Prunoideae, Genus *Prunus;* $x = 8$; deciduous or evergreen trees or shrubs; **winter buds:** with many imbricate scales; **leaves:** alternate, serrate, rarely entire, stipulate; **flowers:** perfect, solitary or in fascicles or racemes; **sepals:** 5, **petals:** 5, usually white, sometimes pink or red; **stamens:** numerous, perigynous; **pistil:** 1, with elongated style, 2-ovuled; **fruit:** a drupe, usually 1-seeded. Nearly 200 species, mostly in the temperate zone, a few in the Andes of South America.

The large genus *Prunus* includes peach, nectarine, plum and prune, cherry, apricot, almond, and many species used only as rootstocks or ornamentals. The basic chromosome number for *Prunus* is 8, but the somatic chromosome numbers of the various species range from 16 to as high as 176 (in *P. laurocerasus*). For convenience, the species will be grouped into three subgenera for discussion. Graft compatibility varies considerably among the many *Prunus* species (see Chapter 4). In the three subgenera, *Amygdalus, Prunophora,* and *Cerasus,* flower buds do not contain leaves but only flowers, and they are always lateral buds, never terminal (Fig. 3-3). Because the drupe consists entirely of a developed ovary, both pollination and fertilization must take place to give the necessary stimulus for the fruit to develop. Unlike the members of the subfamily Pomoideae, stone fruits do not set parthenocarpically.

Subgenus *Amygdalus* (L.) Focke. (Peaches and Almonds)

DESCRIPTION: $x = 8$; **flowers:** sessile or short-stalked, open before the leaves; **fruit;** tomentose (glabrous in nectarine), usually dehiscent; **stone:** pitted or smooth; **leaves:** conduplicate in bud; **buds:** 3 in each axil, the lateral ones flower buds. Seven primary species.

Both peach (*Prunus persica* [L.] Batsch.) and almond (*P. amygdalus* Batsch.) have somatic numbers of 16. They have a number of common characteristics, including both graft and pollen compatibility.

PEACH (*PRUNUS PERSICA*) The peach ($x = 8$; somatic no. $= 16$) is native to China, not Persia, but it seems to have been cultivated in Persia for some time before being introduced into Europe, from whence it was brought to North America. Except for the apple, it is the most widely planted deciduous tree fruit in the U.S. It is a relatively short-lived tree of medium to small size (Fig. 3-3).

Genetically, peaches are much less variable than apples or pears. The fruit of every seedling grown is at least edible, and peach is naturally precocious, thus it is relatively easy to breed new cultivars. Most of the main cultivars of apple and pear were selected from chance seedlings, whereas many of the peach cultivars in use today are products of experimental-breeding programs. Simple dominant genes are responsible for the freestone condition, white flesh, and hairy fruit surface. The nectarine is merely a peach with recessive genes that result in fuzzless fruit. Successful breeding programs in different areas

Figure 3-3 Buds, flowers, and fruit of representative stone fruits are: **(A, B, C)** sweet cherry; **(D, E, F)** plum; and **(G, H, I)** peach. Flower buds are lateral, unmixed. Cherry and plum flowers are in fascicles; the peach flower is solitary. In all three species, the fruit type is a drupe. [G, H: Courtesy of E. L. Proebsting, Jr.]

Table 3-5 Characteristics of peach cultivars in order of ripening (rated on a scale from 0 to 10, where 10 = best).

Cultivar	Ripening Days Before or After Elberta	Flesh Color[1]	Score (from 0 to 10)							Chilling Hours Below 7°C	Main Value[2]
			Fruit Size	Stone Freeness	Attrac-tiveness	Flesh Firmness	Dessert Quality	Canning Quality	Bacteria Spot Resistance		
Marcus	−63	Y	5	4	6	4	6	4	5	800	HLC
Mayflower	−58	W	3	3	6	3	5	3	7	1200	HLC
Earlired	−50	Y	6	5	8	6	7	4	7	850	HLC
Collins	−49	Y	5	5	8	6	7	4	—	—	HLC
Cardinal	−46	Y	6	4	8	6	7	4	7	950	LC
Early Redhaven	−44	Y	6	5	8	7	7	—	—	—	LC
Dixired	−42	Y	6	4	7	6	7	4	8	1000	LC
Redcap	−42	Y	6	4	8	6	8	4	7	750	LC
Erly-Red-Fre	−40	W	7	6	7	6	8	4	8	900	HL
Sunhaven	−38	Y	7	6	7	7	8	5	8	900	HLC
Merrill Gemfree	−38	Y	7	6	6	7	7	5	6	—	LC
Early East	−37	Y	7	3	6	5	5	3	—	—	HL
Jerseyland	−33	Y	8	8	6	8	8	7	4	850	LC
Dixigem	−32	Y	7	7	7	7	9	7	7	850	LC
Arp Beauty	−32	Y	8	4	8	7	7	3	—	—	HL
Prairie Dawn	−32	Y	7	5	8	7	7	6	—	—	HL
Redhaven	−30	Y	7	7	8	8	8	7	8	950	HLC
Raritan Rose	−27	W	8	8	8	6	9	7	9	950	HL
Golden Jubilee	−25	Y	8	8	7	6	8	8	8	850	HL
Prairie Daybreak	−25	Y	9	6	6	6	5	—	—	—	HL
Ranger	−25	Y	8	8	7	8	8	8	9	950	LC
Newday	−25	Y	8	6	6	6	6	6	—	—	HL
Washington	−24	Y	8	8	8	8	8	8	—	—	LC
Triogem	−22	Y	8	8	7	8	8	7	6	850	HLC
Fairhaven	−19	Y	8	8	7	7	8	8	7	850	L
Glohaven	−19	Y	8	8	7	8	8	6	—	—	LC
Western Pride	−19	Y	8	6	6	7	6	—	—	—	HL
Sunhigh	−17	Y	9	8	8	8	9	9	4	750	C
Vedette	−17	Y	8	5	6	6	6	3	—	—	HL
Richhaven	−16	Y	9	9	9	9	9	9	4	1000	LC
July Elberta	−15	Y	8	9	7	8	9	9	5	750	HL
Southland	−14	Y	9	9	9	9	9	9	6	750	HLC
Halehaven	−14	Y	9	9	6	7	9	8	7	850	LC
Redglove	−14	Y	9	9	10	10	9	9	7	850	HLC
Loring	−11	Y	9	9	9	9	9	9	8	800	LC
Veteran	−11	Y	8	8	8	7	9	8	8	1100	HL
Slappey	−11	Y	8	9	6	6	6	7	—	—	HL
Delight	−11	Y	7	9	7	6	7	—	—	—	HL
Gene Elberta	−11	Y	7	8	8	7	9	—	—	—	HLC
Goldeneast	−10	Y	8	6	6	6	6	—	—	—	HLC
Belle	−8	W	8	9	6	7	9	8	9	850	HL
Redelberta	−8	Y	9	9	7	9	9	—	—	900	LC
Suncrest	−8	Y	9	9	9	10	9	9	4	850	LC
Sullivans Early Elberta	−4	Y	9	9	7	8	8	8	6	900	C
Early Elberta	−3	Y	9	9	8	9	9	9	8	850	LC
Merrill 49'er	−3	Y	8	7	7	7	6	6	—	—	LC
Elberta	0	Y	9	9	8	9	8	8	7	900	C

Table 3-5 (continued)

Cultivar	Ripening (days before or after Elberta)	Flesh Color[1]	Score (from 0 to 10) Fruit Size	Stone Freeness	Attrac-tiveness	Flesh Firmness	Dessert Quality	Canning Quality	Bacteria Spot Resistance	Chilling (hours below 7°C)	Main Value[2]
Redskin	0	Y	9	9	9	9	9	8	9	650	HLC
H. H. Brilliant	0	Y	9	9	9	9	9	8	5	750	LC
Dixiland	0	Y	9	9	9	9	9	—	7	750	C
Madison	0	Y	9	9	8	8	9	—	—	—	HL
J. H. Hale	+1	Y	10	9	9	10	9	7	6	900	C
Halberta Giant	+2	Y	10	7	7	8	6	—	—	—	HL
Gold Medal	+2	Y	7	6	7	7	6	—	—	—	HLC
Afterglow	+5	Y	9	9	7	9	8	7	7	750	C
Alamar	+5	Y	10	9	9	8	7	—	—	—	HLC
Rio Oso Gem	+6	Y	10	9	8	10	9	8	7	900	C
Constitution	+10	Y	9	9	7	8	8	7	7	750	C
Autumn	+12	Y	9	9	7	8	7	7	8	850	C
Late Elberta	+12	Y	8	6	5	7	6	6	—	—	—
Krummel	+27	Y	9	9	7	9	7	7	7	900	C

SOURCE: USDA Handbook 280 by H. W. Fogle et al., 1965
[1]Y = Yellow, W = White.
[2]H = Home, L = Local, C = Commercial.

of the world have resulted in many more commercial cultivars being grown than are found for apple and pear. Some of these cultivars are grown only in one area because of their special adaptation. Characteristics of a number of cultivars are given in Tables 3-5 and 3-6.

Most peach cultivars are self-fertile. Some self-sterile ones are J. H. Hale, Halberta, Candoka, Mikado, and Alamar. Except when damaged by spring frosts, peaches usually set an excess of fruit and must be thinned to get good size and quality.

ALMOND (*PRUNUS AMYGDALUS*) The almond ($x = 8$; somatic no. $= 16$) is native to the hot arid regions of western Asia and probably was carried to Greece and North Africa prior to written history. Current production is from Italy, Spain, Iran, Morocco, and Portugal. U.S. production is limited to the hot valleys of California, whose annual production is about 116,000 metric tons. From 1962 to 1971, almond acreage in California doubled (from 50,000 to 103,000 hectares).

The leading California cultivar is Nonpareil, followed by Texas, Ne Plus Ultra, Drake, IXL, Peerless, Jordanolo, and Eureka. New cultivars with some promise are Kaparcil, Davey, Thompson, and Merced. Almost all cultivars of almond are self-sterile and thus require a pollinizer. Also a few cultivars, for example, IXL and Nonpareil, are cross-sterile as well. Because almond blooms very early during cool and moist weather, it is important to have plenty of bees in the orchard for good cropping.

Subgenus *Prunophora* Focke. (Plums and Apricots)

DESCRIPTION: $x = 8$; somatic no. $= 16, 32, 48$; **fruits:** sulcate, glabrous, usually with bloom on epidermis; **stone:** compressed, usually longer than broad and smooth or nearly so; **flowers:** solitary or in umbel-like clusters, appearing before, or rarely with, the leaves; **pedicel:** usually remaining with the fruit. There are nineteen species of plum, six of apricot.

PLUMS Section 1 (Euprunus) has one or two flowers, rarely three, while section 2 (Prunocerasus) has flowers in clusters of two to five. Section 3 (Armeniaca) comprises the apricots.

Table 3-6 Clingstone canning peaches grown in California.

Cultivar	Ripening Time[1]	Hectares Planted 1967	Origin	Tree Vigor[2]	Yield[3]	Fruit Quality	Remarks
			Extra Early				
Fortuna	0	1499	USDA 1941	Vig.	Av.	F–G	Set heavy, needs heavy thinning
Carson	0	495	Carson Orchard, Stanislaus, Co. 1943	Vig.	Av.	F	Average, sizes well
Loadel	0	1558	Chance seedling, Harter Packing Co. 1947	Vig.	Heavy	F	Good color, size
Vivian	5	1298	USDA 1950	Vig.	Av.	F	Needs heavy thinning
Dixon	10	2968	Canners League of California 1952	V.Vig.	Heavy	G	Some split pit, red flesh
			Early				
Cortez	15	1624	USDA 1944	Vig.	Av.	F	Erratic set
Jungerman	15	172	Univ. of Calif. 1964	Vig.	Heavy	G	Red pit, flesh
Andross	18	100	Univ. of Calif. 1964	Vig.	Heavy	G	Pink pit, promising variety
Paloro	18	1728	Chance seedling 1914	Weak	Av.	F	Peach blight, rust, mildew
Johnson	21	378	Chance seedling 1914	Vig.	Heavy	F	Needs early heavy thinning
Klamt	21	227	Univ. of Calif. 1964	V.Vig.	Heavy	G	Good color, pit brown
Peak	21	1339	Chance seedling 1914	Vig.	Av.	F	Similar to Paloro
Andora	23	372	USDA 1941	Vig.	Av.	Poor	Requires long chilling
			Late				
Gaume	26	2160	Chance seedling 1913	Vig.	Irreg.	G	Split pit, fruit drops
Carolyn	28	2515	USDA 1941	V.Vig.	Av.–Heavy	Poor	Good size, poor color
Everts	36	85	Univ. of Calif. 1962	Av.	Heavy	G	Pink pit, good size
Halford	36	6994	Chance seedling 1901	Vig.	Heavy	G	Good size, best quality
Stanford	37	179	USDA 1935	Vig.	Heavy	F	Good conventional variety
			Extra Late				
Starn	38	1388	Chance seedling, Starn Orchard, Modesto 1935	Vig.	Av.	F–G	May set light, may split
Wiser	39	1102	4-H Club Project, Ray Wiser 1918	Vig.	Heavy	G	A satisfactory variety
McKune	40	12	Univ. of Calif. 1964	Vig.	Heavy	G	Pink pit, promising variety
Sullivan #4	41	229	Chance seedling, Sullivan Ranch, Sutter Co. 1936	Vig.	Av.	F	Satisfactory for reason
Stuart	43	1239	Chance seedling 1923	Vig.	Av.	F	Average late variety
Carona	44	778	USDA 1942	V.Vig.	Heavy	Poor	Poor color, tends to drop

[1]Days after July 20 to ripen in California.
[2]Vig. = Vigorous, V. = Very.
[3]Av. = Average, Irreg. = Irregular.

NORTH AMERICAN PLUMS Of the hundreds of cultivars of North American plums, the vast majority are of limited local importance. Most of them appear to be self-unfruitful, so pollinizers should be placed in the orchards. Some of the species and clones will be discussed briefly. Cultivars of *P. americana* are

Table 3-7 Plum species.

Species	Origin	Somatic Chromosome Number	Uses
	Section One: Euprunus		
P. insititia (L.) Bullace	Europe, W. Asia	48	Fresh, jelly—rootstock
P. cerasifera Ehrh.	E. Europe, W. Asia	16	Jelly—rootstock
P. spinosa L.	Europe, W. Asia	32	Jelly? —rootstock
P. domestica L.	Europe	48	Fresh, canned, drying
P. salicina Lindl.	China	16	Fresh
	Section Two: Prunocerasus		
P. americana Marsh.	North America	16	Fresh—rootstock
P. nigra Ait.	North America	16	Fresh—rootstock
P. hortulana Bailey	North America	16	Fresh, jam—rootstock
P. munsoniana Wight & Hedr.	North America	16	Fresh, jam—rootstock
P. maritima Marsh.	North America	16	Jelly, jam—rootstock
P. subcordata Benth.	North America	16	Jelly, jam—rootstock
P. besseyi Bailey*	North America	16	Fresh, jelly—rootstock

*This species is a true plum, although it is often misplaced in the subgenus *Cerasus* and bears the common name "sand cherry."

grown in the upper Mississippi Valley and in Canada and are hardy to the region. Fruits are yellow or orange with golden yellow flesh. The skin of this species tends to be tough, reducing its culinary value. Leading cultivars are Desoto, Hawkeye, Wyant, Weaver, and Terry. Cheney is from a related species *P. nigra* and is hardy in eastern Canada. *P. hortulana* is grown in the middle Mississippi Valley. It is less hardy but more resistant to brown rot, and its fruit less flavorful than *P. americana.* It is used for jams and marmalades. Two cultivars are Wayland and Golden Beauty. *P. munsoniana,* Wild Goose, is a bright red plum grown in the southern Mississippi Valley because of its resistance to spring frost and to brown rot of the fruit. *P. besseyi* is a hardy shrub of the high central plains. It is used both for hybridizing and as a dwarfing rootstock for other plums. *P. maritima* grows along the sandy beaches from Virginia to New Brunswick. Wild fruit, mostly blue to dark purple in color, are made into jams and jellies. Fruit is usually astringent. *P. subcordata* is a shrubby species native only to south central Oregon and northeast California. The red fruit are small and astringent but have been used commercially for preserves and jellies. They are locally used as a sauce or relish for meat in much the same way as cranberries. Some clones have been selected for improved fruit types. but all seem highly susceptible to brown rot of the blossoms. The tree is hardy and is being tested as a dwarfing rootstock for peach and plum.

EUROPEAN PLUMS *P. domestica* is by far the most important plum species, world wide. Relatively minor use is made of the fruit of *P. cerasifera, P. spinosa,* and *P. insititia. P. domestica* seems to have been cultivated in Europe for at least 2000 years, but there is no distinctly wild form known. The fact that its seeds were not found under the ash of Pompeii indicates that it is of relatively recent origin. Evidence suggests that it arose by spontaneous doubling of a sterile triploid hybrid of *P. cerasifera* ($2n = 16$) and *P. spinosa* ($2n = 32$), which would result in the known chromosome number ($2n = 48$) for *P. domestica* (Fig. 3-3).

Commercially important cultivars of *P. domestica* are shown in Table 3-8. They can be divided into four classes:

1. Prunes: Purplish blue, oval plums with high-enough sugar content to permit drying without removing the seed. Important cultivars are Agen (French), Stanley, Sugar, Imperial Epineuse, Italian

Table 3-8 European plums and prunes of commercial importance.

Cultivar	Ripening Season	Needs Pollinizer	Fruit Shape	Fruit Size	Free-Stone	Skin Color[1]	Flesh Color[1]	Use[2]
Tragedy	Early	Yes	Oval	L	Yes	B–P	G–Y	F
California Blue	Early	No	Oval	L	Yes	B	Y	FC
Iroquois	Early	No	Oval	M	Yes	B	G–Y	FC
Early Italian	Early	No	Oval	M	Yes	B	Y	FC
Lombard	Early	No	Oval	L	Yes	R	Y	F
Queenston	Early	—	Oval	M	Yes	B	Y	F
Parsons	Early	Yes	Oval	M	Yes	R–B	A–Y	FD
Stanley	Mid	No	Obovate	M	Yes	B	Y	FC
Bluefre	Mid	No[3]	Oval	L	Yes	B	Y	F
Damson	Mid	No	Oval	S	Yes	B	Y	FC
Bluebell	Mid	Yes	Oval	L	Yes	B	Y	FC
Grand Duke	Mid	Yes	Oval	L	No	B	Y	F
Agen (French)	Mid	No	Obovate	M	Yes	B	Y	FCD
German	Mid	Yes	Oval	M	Yes	B	Y	FCD
Italian (Fellenberg)	Mid	No	Oval	M	Yes	B	A–Y	FCD
Imperial	Mid	Yes	Oval	M	Yes	B	Y	FD
Brooks	Mid	No	Oval	VL	Yes	B	Y	FD
Sugar	Mid	No	Oval	M	Yes	B	Y	FD
Sargeant	Mid	—	Oval	M	Yes	B	Y	FD
Reine Claude	Late	No	Oval	M	Yes	G–Y	G–Y	FC
President	Late	Yes	Oval	L	Yes	P	Y	F
Pozegaća	Late	—	Oval	S	—	B	—	Br
Standard	Late	Yes	Oval	L	Yes	B–P	G–Y	F
Vision	Late	Yes	Oval	L	Yes	B	Y	F
Victoria	Late	No	Oval	M	Yes	R	Y	F
Moyer Perfecto	Late	No	Oval	L	Yes	B	Y	FD

[1]B = Blue, P = Purple, G = Green, Y = Yellow, R = Red, A = Amber.
[2]F = Fresh, C = Canned, D = Dried, Br = Brandy.
[3]Cross-pollination may improve fruit set in some areas.

(Fellenberg), German, Giant, and Tragedy. Used for drying, canning, and fresh market. Pozegaća in Yugoslavia is used for brandy.

2. Reine Claude (Green Gage): Roundish, sweet, green-yellow or golden plums of high quality. Cultivars include Reine Claude, Bavay, Jefferson, Yellow Gage, Washington, Imperial Gage, and Hand. Used for canning and fresh market.

3. Yellow Egg: A small group of plums used for canning. Cultivars include Yellow Egg, Red Magnum Bonum, and Golden Drop (Silver Prune).

4. Lombard: A group of large oval red or pink plums of somewhat lower quality than group 1 and 2, used mostly for fresh market. The most important fresh market variety in western Europe is Victoria. Other cultivars are Pond (incorrectly called Hungarian), Bradshaw, and Lombard.

ORIENTAL PLUMS *P. salicina,* the so-called Japanese plum, probably originated in China. It is the only oriental species important in developing high-quality plum cultivars. It is readily distinguished from *P. domestica* plums by its rough bark, more numerous presistent fruiting spurs throughout the tree, and by the mostly conic or heart-shaped fruits with a more pointed apex than other species. In 1870 Kelsey was the first *P. salicina* cultivar brought to the U.S. from Japan. Later *P. salicina* cultivars were introduced to Europe from America. Leading cultivars in California are Santa Rosa, Burmosa, Duarte, Casselman, Laroda, Late Santa Rosa, El Dorado, Queen Ann, Nubiana, Kelsey, and Beauty. The early cultivar, First, shows promise in Norway; in Canada and the northeastern U.S. there is limited production of Burbank, Crystal Red, Methley, and Shiro. Many cultivars and hybrids of *P. salicina* plums (Table 3-9) were introduced by Luther Burbank.

Table 3-9 Japanese and Japanese hybrid plums as grown in California. All are primarily consumed fresh.

Cultivar	Ripening Season	Needs Pollinizer	Fruit Shape	Fruit Size	Skin Color[1]	Flesh Color[1]
Red Beaut	VE	Yes	Conic	L	R	Y
Beauty	VE	No[2]	Heart	L	GY–R	A–R
Burmosa	VE	Yes	Semi-H. to round	L	Y–R	LA
Santa Rosa	VE	No[2]	Conic	L	P–R	A–R
Early Golden	VE	Yes	Round	M	Y–R	Y
Queen Rosa	E	Yes	Round	L	R–P	Y–R
Wickson	E	Yes	Heart	L	GY–R	Y
Mariposa	E	Yes	Heart	L	G–YR	R
July Santa Rosa	E	No	Conic	L	P–R	Y
Frontier	E	Yes	Heart	L	P–B	R
El Dorado	E	Yes	Flat-oval	L	R P	Y
Duarte	E	Yes	Heart	M	Dull R	R
Late Santa Rosa	E	No[2]	Conic	L	P–R	Y–R
Redroy	E	No[2]	Conic	L	R	Y
Methley	E	No[2]	Round	S	P	R
Shiro	E	Yes	Round	M	Y	Y
Ace	Mid	Yes	Heart	L	GY–R	R
Elephant Heart	Mid	Yes	Heart	L	GY–R	R
Kelsey	Mid	Yes	Heart	L	G–Y	G–Y
Laroda	Mid	Yes	Conic	L	R–P	Y
Nubiana	Mid	Yes	Flat-oval	L	P–B	Y
Queen Ann	Mid	Yes	Heart	L	P–B	LY
Simka	Mid	No	Heart	L	R	Y
Ozark Premier	Mid	Yes	Round	L	R	Y
Burbank	Mid	Yes	Round	M	R	Y–R
Friar	Late	Yes	Flat-oval	L	Black	A
Grand Rosa	Late	No	Conic	L	R–P	Y
Casselman	Late	No[2]	Conic	L	P–R	Y
Roysum	Late	No	Conic	L	P	Y–R

[1]R = Red, G = Green, Y = Yellow, P = Pink, B = Blue.
[2]Cross-pollination may increase fruit set in some areas.

APRICOT (*PRUNUS ARMENIACA* L.)

DESCRIPTION: $x = 8$; somatic no. = 16; **flowers:** solitary, white or pinkish, about 2.5 cm across; **fruit:** round, 3 cm across or more, yellowish with red cheek, nearly glabrous; **stone:** smooth with a thickened furrowed edge.

The apricot is native to China and Siberia. This delightful fruit was brought to Italy about 100 B.C., to England in the 13th century, and to North America by 1720.

Most of the apricots in the U.S. are grown in California; the next-largest amounts are grown in Washington and Utah. Commercial acreage is confined to frost-free sites of the Pacific slope. Blenheim and Royal are very similar types; both are important California cultivars. Others grown there are Tilton, Derby-Royal, Stewart, Wiggins, and Newcastle. Wenatchee Moorpark is the main Washington cultivar, followed by Tilton, Royal, Perfection, and Riland. Large Early Montgamet (Chinese) is the main cultivar of Utah and Colorado, followed by Perfection and Royal. Large Early Montgamet and some other cultivars have sweet edible kernels and can be used in the same ways as almonds. Recent introductions from New York are Alfred and Farmingdale, both of which ripen ahead of Large Early Montgamet but tend to be small in size. Two other new cultivars, Goldcot from Michigan and Goldrich from Washington, show some promise. Both are early maturing. Morden 604 from Manitoba is of poor quality, but its seeds are used to grow hardy rootstocks. Veecot is a good canning cultivar, introduced in 1964 at Vineland, Ontario. It ripens in midseason.

Subgenus *Cerasus* Pers. (Cherries)

DESCRIPTION: **Leaves:** conduplicate in buds; **flowers:** white or rosecolored, usually with pedicel, solitary or fascicled or racemose; **calyx-tube:** cup-shaped or tubular and flowers sessile; **style:** furrowed; **stigma:** emarginate; **ovary:** usually glabrous; **stone:** smooth or furrowed and pitted.

Two species are important for their fruit: sweet cherry, *Prunus avium,* L. and sour cherry, *Prunus cerasus* L. Both are native to southeast Europe and

Table 3-10 Sweet cherries and their characteristics.

Cultivar	Bloom Date[2]	Needs Pollinizer	Intra-incompatible Group[3]	Fruit Size	Fruit Color[4]	Rain Cracking	Flesh Firmness[5]	Flavor[5]	Use[6]
Very Early									
Seneca	M	Yes	10	S	B	Low	P	P	F
Vista	E	Yes	11	M	B	Low	G	G	F
Early Burlat[1]	M	Yes		L	BR	High	G	G	FC
Early Purple	M	Yes		M	B	High	F	F	FCB
Bigarreau de Schrecken	L	Yes		M	B	High	G	G	FCB
Early									
Black Tartarian	M	Yes	1	S–M	B	Low	P	F	F
Viva		Yes	4	M	RB	V. Low	F	G	F
Vega	M	Yes	12	L	W	Med	VG	F	BC
Venus	E	Yes	2	M	B	High	G	G	FC
Chinook	E	Yes	9	L	B	High	G	G	FC
Corum[1]	E	Yes		L	W	Med	F–G	G	BFC
Mona[1]	M	Yes		L	B	—			
Macmar	M	Yes		L	W	High	G	G	BFC
Knight's Early Black	L	Yes	1	M	B	High	G	F	CBF
Bada[1]	M–L	Yes		L	W		G	F	BC
Larian[1]	M	Yes		M	B		P	F	F
Mid-Season									
Merton Bigarreau	M	Yes	2	L	B	Low	G	G	FCB
Bing	M	Yes	2	L	B	High	VG	VG	FC
Rainier	M	Yes	9	L	W	Med	G	G	FC
Napoleon (Royal Ann)	M	Yes	3	L	W	Med	G	G	BCF
Sam[1]	L	Yes		M	B	Low		G	FC
Sue	M	Yes	4	L	W	High	VG	G	BCF
Schmidt	M	Yes	8	M	B	Low	G	G	FC
Vernon	L	Yes	3	L	B	Low	G	G	BCF
Vic	M	Yes	13	M	B	—	G	G	CB
Stella[1]	M	No	—	L	B	Med	F	F	F
Emperor Francis		Yes	3	B M	W–R	Low	G	G	
Berryessa	M	Yes		L	W		G	F	BC
Late									
Windsor	L	Yes	2	M	BR	Low	G	G	B
Gold		Yes	6	S	Y	V. Low	G	F	B
Lamida	L	Yes		L	B	Med	G	F	FCB
Spalding	L	Yes		M	B	Med	G	G	F
Van	M	Yes	2	L	B	Med	VG	G	FCB
Jubilee[1]	M	Yes		L	B	Low	F	G	FC
Hudson	M	Yes	9	M	RB	Low	VG	G	FB
Ulster	M	Yes	13	M	B	Low	G	G	FC
Hedelfingen	M	Yes	7	L	B	Low	G	G	FCB
Lambert	L	Yes	3	L	B	High	G	G	FC
Black Republican	E	Yes		M	B	High	VG	G	FC

[1] Cultivar is pollen-compatible with Bing, Lambert, and Napoleon.
[2] M = Midseason, E = Early, L = Late.
[3] Cultivars with the same number are intra-incompatible and will not cross pollinate each other.
[4] B = Black, R = Red, W = White, Y = Yellow.
[5] P = Poor, F = Fair, G = Good, VG = Very Good.
[6] F = Fresh market B = Brining C = Canning

western Asia. They are graft compatible and will cross genetically to form hybrid (Duke) cultivars.

SWEET CHERRY (*PRUNUS AVIUM*) The sweet cherry ($x = 8$, somatic no. $= 16$) probably originated between the Black Sea and the Caspian Sea, but birds seem to have carried it into Europe in ancient times. Heart (Gean) cherries are ovoid or heart-shaped, and the flesh is soft. Bigarreau cherries are firm or hard-fleshed. A sweet cherry is shown in Figure 3-3; Table 3-10 lists the cultivars and their characteristics.

Many cultivars grown throughout the world originated in Europe, but a number of important ones were selected or bred in local cherry districts. European cultivars still grown in the U.S. are Napoleon (Royal Ann), Black Tartarian, Eagle, Early Purple, Early Rivers, Elkhorn, Hedelfingen, Knight's Early Black, Lyon, and Schmidt. The cultivars Windsor, Van, Sam, Vista, Victor, Sue, Vega, Summit, and Stella were developed in Canada. Lamida, Ebony, and Spalding are Idaho introductions. Chinook and Rainier were developed in Washington. Bing, Lambert, Black Republican, Corum, and Hoskins were selected and developed in Oregon. Chapman, Burbank, Bush Tartarian, and the new cultivars Mona, Larian, Jubilee, Berryessa, and Bada originated in California. New York has introduced Ulster and Hudson, and Utah recently introduced Angela.

The most important cultivars in the western U.S., where about 80 percent of the U.S. crop is produced, are Napoleon, Bing, and Lambert. Because these three cultivars are intrasterile and will not cross-pollinate each other, pollinizers such as Van, Black Republican, Corum, and Black Tartarian are used. The main cultivars in Michigan and western New York are Windsor, Black Tartarian, Napoleon, Schmidt, Emperor Francis, and Hedelfingen. Light-fleshed sorts such as Napoleon, Corum, and Emperor Francis are best for making into maraschino cherries. Napoleon also is used for canning. Bing is mainly a fresh market cultivar, and Lambert is used both for canning and fresh market. Black Republican and other very firm, dark cherries are good for freezing. Dark cultivars are picked immature for "brining" in sulfur dioxide solution (preparatory for maraschino process) because pigment is undesirable.

SOUR CHERRY (*PRUNUS CERASUS*) The sour cherry ($x = 16$, somatic no. $= 32$) is native to the same area of eastern Europe as sweet cherry. In fact, there is good evidence that the sour cherry arose from an unreduced pollen grain of *P. avium* ($2n = 16$) crossed with *P. fruticosa* Pall. ($2n = 32$).

Fruit of sour cherry are soft, juicy, and depressed-globose in shape. The three cultivars important in North America are Early Richmond (a light red early cultivar), Montmorency (medium red), and English Morello (a late dark red cultivar). Two new hardy cultivars from Minnesota are North Star and Meteor. All are self-fertile and do not require pollinizers.

DUKE CHERRIES A few cultivars of Duke cherries ($x = 16$, somatic no. $=32$), such as May Duke, Royal Duke, Late Duke, and Reine Hortense are tetraploid hybrids of *P. avium* × *P. cerasus*. They are only partly self-fruitful and may require a suitable pollinizer. The Duke cherries as a whole resemble the sweet cherry more than the sour cherry.

MULBERRY AND FIG, FAMILY MORACEAE

Mulberry (*Morus* L.)

DESCRIPTION: $x = 14$; somatic nos. $= 28, 42, 56, 84, 112, 308$; deciduous thornless trees or shrubs, **bark:** usually scaly; **buds:** with 3-6 imbricate scales; **leaves:** undivided or lobed, serrate or dentate, with 3-5 nerves at base, stipules lanceolate, deciduous; **flowers:** monoecious or dioecious, both sexes in stalked axillary pendulous catkins; **calyx:** 4-parted, the filaments inflexed in the bud, later partly inclosed by the involute sepals; **stigmas:** 2; **fruit:** an ovoid compressed achene, covered by the succulent white to black calyx, aggregating into an ovoid to cylindric syncarp superficially resembling a blackberry; **seed:**

albuminous, cotyledons oblong, equal. Flowers with the leaves (Fig. 3-4).

There are about twelve species in temperate and subtropical regions of the northern hemisphere. They are trees or shrubs grown for their edible fruits or for their leaves, which are fed to silkworms or, in the Orient, made into paper. Fruit of the mulberry may be black, red, pink, or white. Many are sweet and delicious, but all are too tender to be handled for market. The fruit have weak abscission zones and often drop to the touch even when immature. Most mulberries are dioecious, so male trees may be grown as ornamentals where the fruit is not wanted. *Morus nigra* L. is reported to have 308 chromosomes in its somatic cells, the largest number for any plant.

The Fig (*Ficus carica* L.)

DESCRIPTION: $x = 13$; somatic no. $= 26$; deciduous tree to 10 m; **branches:** stout, glabrous; **leaves:** 3–5 lobed, rarely undivided, 10–20 cm long and about as broad, usually cordate at base, palmately nerved, the lobes usually obovate and obtuse at base and irregularly dentate, scabrous above and below with stout, stiff hairs; **petioles:** 2–5 cm long; **receptacle:** axillary and solitary, pear-shaped at maturity, 5–8 cm long, greenish or brownish violet (Fig. 3-4).

The fig comes from western Asia and has been cultivated since early times. It ranks with the grape, the date, and the olive as an important crop in the

Figure 3-4 Buds, flowers, and fruit of **(A, B, C)** fig and **(D, E)** mulberry. Flowers are borne laterally at the nodes. Fig bears its flowers inside a pyriform receptacle; mulberry produces flowers in a stalked catkin (flowers and fruit are seen in E). Both are multiple fruits. [D, E: Courtesy of D. K. Ourecky, New York Agricultural Experiment Station]

early Mediterranean civilizations. The fig is distantly related to the mulberry, but it will neither cross with the mulberry nor can it be grafted to it.

Common fig flowers produce no pollen, but many cultivars will set seedless, parthenocarpic fruits. Such cultivars include Celeste and Brown Turkey, the best known cultivars of the gulf states, Brunswick (Magnolia) grown in Texas, the first crop of some cultivars and the second crop of Calimyrna (Lob Injir), Adriatic, Mission, and Kadota grown in California. Kadota, properly called Dottato, is an important cultivar for canning. Its brebas (first crop figs) are large, pyriform, with yellow-green skin and violet-tinted pulp. Second-crop fruit have similar skin but amber pulp in the hot areas; in cooler districts its skin is green, with violet pulp. Brunswick is grown for canning in Texas, where the trees are more vigorous and productive than in California. Calimyrna is the only Smyrna-type fig of importance in California. It produces only a few golden brebas, which are of excellent quality when dried or fresh. However, second-crop figs will set with pollination by a type called caprifigs, pollen of which is only transferred by the wasp *Blastophaga psenes*. Caprifigs are grown apart from the orchard. When the edible figs are ready for pollination, caprifigs containing *Blastophaga psenes* ready to emerge are picked, carried in containers, and hung in the Calimyrna trees.

TREE NUTS

Walnut (*Juglans* sp., Family Juglandaceae)

DESCRIPTION: Deciduous trees; **branches:** with lamellate pith; **trunk:** smooth or with scaly, furrowed bark; **buds:** with few scales, sessile, rarely short-stalked; **leaves:** alternate, odd pinnate, large, aromatic, estipulate, leaflets opposite, serrate or entire; **flowers:** monoecious, the staminate flowers on last year's branchlets in lateral pendulous catkins, each consisting of a bract bearing 8–40 stamens, 2 bractlets, and 1–4 calyx lobes; pistillate flowers in few-to-many-flowered terminal racemes, with 4 calyx lobes, and a 3-lobed involucre consisting of a bract and 2 bractlets; **style:** divided into 2 plumose stigmas; **fruit:** a large indehiscent drupe; **nut:** thick-walled, incompletely 2–4 celled, indehiscent or finally separating into 2 values; **seed:** 2–4 lobed, remaining within the shell in germination. Flowers appear before or after the leaves. (Latin *Jovis glans,* Jupiter's Acorn.)

There are about fifteen species of walnut native to North and South America, southeastern Europe, to east Asia. *Juglans* species appear to hybridize when grown in one locale. In the western U.S., *J. hindsii* Jeps. and its hybrids have been used as a rootstock for *J. regia* L. All *Juglans* are edible, but *J. regia* (English walnut) is the most important, and *J. nigra* L., the eastern American black walnut, is somewhat important. The latter is of value also for its wood. The butternut, *J. cinera* L., and the Japanese walnut, *J. sieboldiana* Maxim., are also planted for their nuts.

BLACK WALNUT (*J. NIGRA* L.) The eastern black walnut ($x = 16$, somatic no. $= 32$) is the most important North American *Juglans* species. Its fine hardwood is its principal value. The nuts, although of the finest flavor, are usually relatively small and have very thick shells that make it hard to crack out kernel halves. Also, the powdery black outer hull that adheres to the shell presents a cleaning problem. Most of the marketed nuts are from wild trees. A number of clones have been selected and propagated from superior trees. Among them are Thomas, Ohio, Clark, Rowher, Stabler, Creitz, Ketler, Cresco, Allen, Tasteright, Wiard, Hines, Snyder, Ten Eyck, Bowser, Peanuts, Thorp, Seward, Berhow, Mintle, Scringer, Vanderstoot, Somers, NCL, Fately, Harrison, McDermid, Breslan, Jacobs, and Watts.

Production of black walnuts in a good year is 23,000 metric tons, from which 2.7 million kilograms of kernels are obtained (Forde, 1975). At Stockton, Missouri one of the very large processors of black walnuts shells up to 4,500 metric tons of nuts from Arkansas, Tennessee, Kentucky, Ohio, West Virginia, and Missouri. Because about 88 percent of the nut is shell, the problem of shell disposal until recently was a serious one. Now, however, the powdered or granular shells are used successfully in plastic mould-

A
B
C
D
E
F
G
H
I

Figure 3-5 Buds, flowers, and nuts of **(A, B, C)** walnut, **(D, E, F)** pistachio, and **(G, H, I)** pecan. The walnut and pecan are monoecious and the pistachio is dioecious. Walnut male flowers are borne laterally on one-year wood; female flowers (shown in B) are in racemes, usually terminal. Pistachio bears both types of flowers in panicles from one-year lateral buds. (In E, male flowers are on the left, female flowers are on the right.) The fruit type for both species is a drupaceous nut. [C: Courtesy of G. C. Martin; D, E, F: Courtesy of J. C. Crane, University of California; G, H, I: Courtesy of Darrell Sparks, University of Georgia]

ing, as a glue extender, as a "sand blast" cleaner for jet engines, and by car makers for polishing metal parts.

ENGLISH WALNUT (*J. REGIA* L.) The English or Persian walnut ($x = 16$, somatic no. $= 32$) is neither English nor Persian but is native to the Caucasus, the Carpathian Mountains, and the region lying eastward to Manchuria and Korea. It should more properly be called the "regia" nut to identify it with the species (Fig. 3-5). In North America, regia nuts are grown commercially only in California and Oregon. Table 3-11 summarizes the characteristics of cultivars as grown in California. More than 95 percent of the annual production of 136,000 metric tons is produced in California. The most important cultivars are Hartley, Payne, Franquette, Eureka, Ashley, and Placentia.

In Oregon, the most common cultivars are the Franquette (Vrooman strain), Hartley, Spurgeon, and Adams. In 1956 the Oregon and Washington Walnut Variety Committee stated that Franquette was not the cultivar needed in the Northwest. They further stated that, "the ideal variety needed is one that is a fast grower, less susceptible to winter injury

Table 3-11 Walnut cultivar characteristics, as found in California.

Cultivars	Time of Leafing (days after Payne)		Est. Percent of Lateral Buds Forming Female Flowers	Estimated Potential Yield[2]			Avg. Nut Size	Avg. Kernel Weight (gm)	Avg. Percent Kernel	Kernel Color (% light)	Shell Seal[3]	Tree Size	Harvest Period[4]
	Days	Class[1]		3–7 Yrs.	8–15 Yrs.	Mature							
Old Cultivars													
Placentia	−3	E	0	L	M	M	M	5.6	48	30	G	L	E
Payne	0	E	80	H	H	H	M	6.2	52	60	G	S–M	E
Eureka	10	M	0	L	M	M	M	7.7	50	40	VG	L	M
Hartley	13	M	10	M	H	H	L	6.5	47	70	F	M	M–L
Franquette	28	L	0	L	L	L	S–M	6.0	48	90	G	L	L
New Cultivars													
Ashley	4	E	90	H	H	H	M–L	7.0	56	90	F–G	S	E–M
Marchetti	6	E	90	H	H	H	M	6.4	53	60	G	S–M	E–M
Lompoc	5	E	50	—	—	—	L	7.7	54	70	G	—	M–L
Serr	5	E	50	—	—	—	L	8.1	59	96	F–G	—	E–M
Gustine	6	E	80	—	—	—	M	6.1	53	80	G	—	M
Chico	6	E	80	—	—	—	M	5.3	49	90	F–G	—	E–M
Vina	8	M	80	—	—	—	M	4.8	49	70	G	—	E–M
Midland	12	M	50	—	—	—	M	6.3	50	70	G	—	M
Amigo	14	M	80	—	—	—	M–L	6.3	54	80	P–F	—	E–M
Pioneer	14	M	40	—	—	—	M	6.0	48	70	F–G	—	M
Tehama	18	L	80	—	—	—	M–L	7.0	53	70	G	—	M
Pedro	18	L	80	—	—	—	L	7.5	50	70	F	—	M–L

[1]E = Early, M = Midseason, L = Late.
[2]L = Low, M = Medium, H = High.
[3]P = Poor, F = Fair, G = Good, VG = Very Good.
[4]E = Early, M = Midseason, L = Late.

than Franquette, and an early producer that will bear on lateral buds as well as terminal buds. It should leaf out about the same time as Franquette but mature two weeks earlier. It should be a consistently heavy producer of large size nuts of good quality. By good quality, we mean hard shell, well sealed, and with a good fill of light-colored kernels." With these criteria in mind, the newer cultivars UC 49–46, Adams No. 10, Chambers No. 9, and Wepster No. 2 were suggested by Stebbins et al. (1967). None are as late leafing out as Franquette but not as early as Payne, and all mature their nuts earlier in the fall.

Two events in recent years have decimated the walnut acreage in Oregon, which in 1960 stood at 4,850 hectares. A hurricane on October 12, 1962 blew down about half the trees. Then on December 8, 1972 a severe early freeze killed half the remaining trees, leaving about 1,200 hectares in 1976.

Most regia walnuts are self-fertile, but many cultivars shed their pollen ahead of the opening of the female flowers, which makes it necessary for pollinizer cultivars to be planted in the orchard. This will be discussed in Chapter 8.

Hickory (*Carya* sp.)

DESCRIPTION: $x = 16$; somatic nos. $= 32, 64$; deciduous trees; **branches:** with solid pith; **buds:** scaly; **leaves:** alternate, odd-pinnate, estipulate, leaflets 3–17, opposite, serrate; **flowers:** monoecious, with the leaves; **staminate flowers:** in axillary, usually ternate pendulous catkins, each flower in the axil of a 3-lobed bract; **stamens:** 3–10; **pistillate flowers:** sessile in terminal 2–10-flowered spikes; **ovary:** 1-celled, enclosed by a 4-lobed involucre; **stigmas:** 2, short; **fruit:** globose to oblong, with a husk separating more-or-less completely into 4 valves; **nut:** smooth or slightly rugose, often angled, 4-celled at base, 2-celled at apex, cotyledons remaining enclosed in the shell. There are about twenty species in North America, two in East Asia.

To the hickory genus belong the bitternut, *C. cordiformis* (Wang.) K. Koch; the pignut, *C. glabra* (Mill.) Sweet; the shellbark hickory, *C. laciniosa* (Michx. f.) Loud.; the sweet pignut, *C. ovalis* (Wang.) Sarg.; the shagbark hickory, *C. ovata* (Mill.) K. Koch; and pecan, *C. illinoensis* (Wang.) K. Koch, the most important native American orchard species (Fig. 3-5).

PECAN *(C. ILLINOENSIS)* Recent annual production of pecan ($x = 16$; somatic no. $= 32$) in the U.S. was 93,000 metric tons, with an in-shell value of more than $80 million. The total acreage harvested is uncertain because of the many wild groves that are harvested. Important producing states are Georgia, Texas, Alabama, Louisiana, Oklahoma, Mississippi, Arkansas, New Mexico, South Carolina, Florida, and North Carolina. Recent plantings in Arizona and California may also develop into major production areas. The pecan is largely self-fertile but dichogamous; in some cultivars there is no overlap of pollen shedding and stigma receptivity. Table 3-12 gives

Table 3-12 Characteristics of some pecan cultivars.

Cultivar	Kernel (as percent of Whole Nut)	No. of Nuts per kg	Pollen Shedding	Stigma Receptivity	Season of Maturity
		Western			
Burkett	56	110	Late	Early	Mid
Clark	54	154	Early	Late	Past mid
Halbert	58	154	Early	Late	Early
Ideal	56	143	Late	Early	Mid
Nugget	58	198	Late	Early	Mid
Onliwon	60	143	Early	Late	Mid
San Saba Improved	60	—	Early	Late	Early
Squirrel Delight	54	132	Early	Late	Early
Sovereign	54	132	Early	Late	Mid
Western Schley	58	143	Early	Late	Before mid
		Eastern			
Desirable	56	110	Early	Late	Mid
Mahan	55	110	Late	Early	Late
Moore	48	154	Early	Late	Early
Odom	56	110	Late	Early	Mid
Schley	58	132	Late	Early	Mid
Success	52	110	Early	Late	Mid
Stuart	49	116	Early	Late	Mid

SOURCE: Texas Agr. Exp. Sta. Bull. 162.

some data on kernel and nut size and on relative time of pollen shedding and nut maturity. Most cultivars require at least 180 days and up to perhaps 220 days for nuts to mature.

Filbert (Hazel; *Corylus* sp.), Family Betulaceae

DESCRIPTION: $x = 14$; somatic no. $= 28$; deciduous shrubs, rarely trees; **buds:** obtuse, rarely acute, with many imbricate scales; **leaves:** generally ovate, usually doubly serrate and more or less pubescent, conduplicate in bud; **staminate catkins:** cylindric, pendulous, naked during winter; **flowers:** without perianth, each bract with 4–8 stamens, filaments bifid, anthers pilose at apex; **pistillate inflorescence:** headlike, enclosed in a small scaly bud, with only the red styles protruding; **ovaries:** with 1, rarely 2 ovules per cell; **style:** bifid to the base; **fruit:** a sub-globose or ovoid nut, with ligneous pericarp, included or surrounded by a large leafy variously toothed, or dissected involucre, often tubular, in clusters at the ends of branchlets; **cotyledons:** thick, fleshy, remaining enclosed in the nut. Flowers appear before the leaves in winter and early spring; nuts mature in autumn. There are about fifteen species in North America, Europe, and Asia (Fig. 3-6).

Two species of filbert, *C. americana* and *C. cornuta*, are native to North America, but domestic cultivars are derived from the European hazel, *C. avellana*.

Filbert production of 9000 metric tons in North America is confined to western Oregon (97 percent) and western Washington (3 percent). Most of the 9300 hectares are of the cultivar Barcelona. Since filberts are self-sterile, pollinizers such as Daviana, DuChilly, Nooksack, Halls Giant, Butler, and Gem are planted at a suggested ratio of one pollinizer to fourteen Barcelona trees. Table 3-13 gives the nut characteristics of the important cultivars in Oregon. Daviana is the main pollinizer used for Barcelona.

Turkey is the principal filbert-producing country, accounting for about 60 percent of world production. The main cultivar of Turkey is Tombul. Italy produces 24 percent of the world crop; its main cultivar is Tonda Gentile della Langhe. Spain produces 10 percent of the world crop on about 30,000 hectares, the main cultivar being Negreta. A few hundred tons of filberts are produced in the southeast of France, and a new program of breeding and research has been started there (Lagerstedt, 1975).

Pistachio (*Pistacia vera* L.), Family Anacardaceae

DESCRIPTION: $x = 15$; somatic no. $= 30$; trees or shrubs; **buds:** with several outer scales; leaves alternate, simple, ternate or pinnate; **flowers:** dioecious, apetalous or naked, in lateral panicles; **staminate flowers:** with 2 bractlets at base and 1–2 sepals, 3–5 stamens with short filaments; **pistillate flowers:** with 2 bractlets, 2–5 sepals, ovary superior subglobose or ovoid, with 1 ovule; **style:** short, 3-parted; **fruit:** a dry, obliquely ovoid drupe; **seed:** compressed, embryo with plan-convex cotyledons (Fig. 3-5).

Pistachio is a member of the cashew family to which also belong sumac, poison oak, and mango. There are eight species in the genus found in Asia Minor, eastern Asia, and southern North America. *P. vera* comes from Syria and has been grown in Italy and Sicily for centuries. Until recently only a few trees were grown in California, but now at least 12,000 hectares have been planted in the San Joaquin Valley.

Because the trees are dioecious, staminate trees must be planted in the orchard at a ratio of about one to six bearing trees. The species is wind pollinated and large quantities of pollen are produced by the male trees. The tendency for pollen to be shed before the pistils are receptive has led to the practice of sometimes collecting and storing pollen to be dispersed when the female flowers are receptive.

In California, the male cultivar Peters is being used. In southern Europe, *P. terebinthus* L. has been used as a source of pollen for *P. vera* orchards. In California, Kerman is the best female cultivar and has the lightest

A
B
C
D
E
G
H

Figure 3-6 Buds, flowers, and nuts of **(A, B, C, D)** filbert and **(E, F, G, H)** chestnut. Both are monoecious. The filbert bears male flowers in lateral catkins (shown in B) and female flowers in lateral buds of one-year wood (shown in C). Chestnut flowers (shown in F and G) are borne laterally on new spring growth. (A single female flower is shown in G.) [A, B, C, D: Courtesy of H. B. Lagerstedt, USDA; E, F, G, H: Courtesy of M. M. Thompson, Oregon State University]

kernels with only a tinge of green. Red Aleppo is intermediate; Bronte and Trabonella produce greenish kernels.

Chestnut (*Castanea* sp.) Family Fagaceae

DESCRIPTION: $x = 12$; somatic no. $= 24$; deciduous trees, rarely shrubs; **bark:** furrowed, **buds:** with 3–4 scales; **branchlets:** without terminal bud; **leaves:** 2-ranked, serrate, with numerous parallel veins; **staminate flowers:** in erect cylindric catkins; **calyx:** 6-parted; **stamens:** 10–20; **pistillate flowers:** on the lower part of the upper staminate catkins, rarely on separate catkins, usually 3 in a prickly symmetrical involucre; **styles:** 7–9; **ovary:** 6-celled; **nuts:** large, brown, with a large pale scar at the base, 1–3, rarely 5 or 7 in a prickly involucre, splitting at maturity into 2–4 valves. There are about ten species of the temperate regions of the Northern Hemisphere (Fig. 3-6).

C. dentata Borkh. (American), *C. mollissima* Bl. (Chinese), *C. sativa* Mill. (European), and *C. crenata* Sieb. and Zucc. (Japanese) have in the past been the main cultivated species of chestnut. However, during the early 1900's the American chestnut was wiped out by chestnut blight, a disease caused by the fungus *Endothia parasitica.* The European chestnut is also susceptible to blight, and this disease is spreading rapidly through native forests. Japanese chestnuts are also moderately susceptible to blight. Only the Chinese chestnut is resistant to blight and can be grown where the disease occurs.

The principal Chinese cultivars are Abundance, Crane, Kuling, Meiling, Nanking, Orrin, and Hemming. They appear to have relatively short chilling requirements and are a bit hardier than most peach varieties. Fresh chestnuts contain about 50 percent moisture and may be stored at 4.4°C for eight weeks in that state, provided they do not mold. One hour in a 68°C water bath will suppress the mold without injuring the nuts. Chestnuts may be stored for a year at 4.5°C if they are dried to 10 percent moisture. The hard dry nuts are rehydrated by soaking or steaming for a half hour before use. Unlike most other nuts, the chestnut is low in oil and high in starch. It is boiled or roasted to make it more digestible.

Table 3-13 Varietal characteristics of filbert nuts.

Cultivar	Kernel (as Percent of Whole Nut)	Shape	Size	As Pollinizer for Barcelona	Productivity	Kernel
Barcelona	43	Round	M–L	—	Mod	Smooth
Daviana	52	Oval	M	VG	Light	Smooth
DuChilly	44	Long	L	G	High[1]	Shriveled
White Aveline	50	Flat-oval	S	VG	Mod	Smooth
Montebello	42	Round	M	None	Mod	Smooth
Brixnut	42	Round	L–M	None	High[1]	Some shrivel
Halls Giant	46	Round	M	G	Mod	Some roughness
Royal	Low	Oval	L	P	Mod	Some roughness
Nooksack	43	Long	M	P	Mod	Rough

SOURCE: Baron and Stebbins 1972.
[1]Alternate bearing.

BERRIES

Strawberry
(*Fragaria* L. spp.), Family Rosaceae

DESCRIPTION: $x = 7$; acaulescent, more or less hairy, perennial herbs with basal leaves and long filiform runners from the axils, which root and form new plants; **petioles:** mostly long and channeled above; **stipules:** adnate at the base of the petiole, large, mostly scarious and brown, persistent and covering the rootstock; **leaves:** 3-foliate, or sometimes unequally imparipinnate; i.e., with a pair of much smaller lateral leaflets below the normal ones; **leaflets:** sharply dentate, but entire at the more or less wedge-shaped base, the lateral ones oblique, the inner half usually smaller; **Scape:** mostly about as long as the petioles, cymosely branched, the lowest bracts with stipules and a more-or-less developed blade; **pedicels:** slender, erect when in flower, curved when in fruit; **flowers:** polygamo-dioecious, rarely hermaphrodite, the male flowers larger and showier, all 5-parted, central flowers open first, often 6- to 8-parted and larger than the later ones; **calyx-lobes:** form a flat hypanthium, augmented by as many shorter and mostly narrower outer calyx lobes or bractlets; **stamens:** about 20 or less, or abortive; **filaments:** mostly shorter than the receptacle; **anthers:** oblong; **receptacle:** roundish or conic, bearing numerous pistils with lateral styles; at maturity the receptacle becoming enlarged and juicy, popularly known as "strawberry" (Fig. 3-7).

CULTIVATED STRAWBERRY (*F.* × *ANANASSA* DUCH.) Present-day cultivated strawberries (somatic no. = 56) appear to be hybrids between the North American *F. virginiana* Duch. and the Chilean *F. chiloensis* (L.) Duch.; *F. virginiana* was imported to Europe prior to 1600 and began immediately to replace the small-fruited inferior species of Europe. Then in 1714, the French explorer A. G. Frézier brought plants of *F chiloensis* from the west coast of South America to Paris. Soon they were being grown in many of the private gardens of Europe, although few of them set berries. Later it was discovered that flowers of the Chilean plants lacked pollen, but if they were planted in alternate rows with the perfect-flowered *F virginiana,* they produced excellent crops of large berries. From these plantings came the hybrids of superior size, quality, and yield that were the forerunners of our present domestic strawberries (Wilhelm, 1974).

Figure 3-7 Plants, flowers, and fruit of strawberry. **(A)** The plant is an evergreen herb that bears its flower buds from the crown at the axils of new leaves. **(B)** The inflorescence is cymose; the flowers are perfect; the later-formed flowers are much smaller than earlier ones. **(C)** The fruit is an aggregate type. [Courtesy of F. J. Lawrence, USDA, and M. M. Thompson]

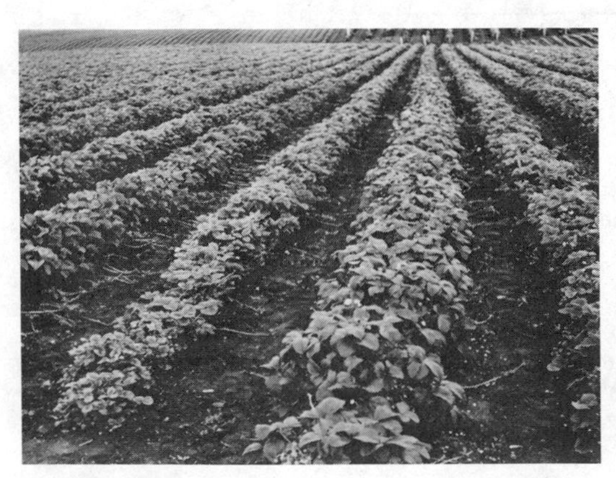
A

B

C

Table 3-14 Leading strawberry cultivars of the U.S.

Cultivar	Origin	U.S. Area Grown	Leaf Spot	Leaf Scorch	Red Stele	Verticillium Wilt	Virus Tolerance	Ripening (days after Midland)	Size	Flesh Firmness	Skin Firmness	Dessert Quality	Process Quality (Freezing)
Albritton	NC	SE–SC	R	VR	S	S	S	12	L	VF	F	E	G
Blakemore	MD	SE–MW	S	VR	S	R	T	3	Sm	F	F	M	G
Catskill	NY	E–NC	S	R	S	VR	VS	7	VL	So	So	G	M–G
Dabreak	LA	S	VR	R	S	U	T	0	M	M	M	G	G
Erlidawn	MD	SE–MW	S	M	S	S	S	0	L	M	M	M	VG
Florida 90	FL	SE	VS	VS	S	S	U	5	VL	So	So	VG	M
Fresno	CA	W	M	U	S	S	M	7	VL	F	F	M	M
Headliner	LA	S	R	U	S	U	U	7	L	M	M	G	G
Hood	OR	NW	R	R	R	R	S	14	L	M	M	VG	G
Howard 17	MA	NE	R	R	S	R	T	3	M	So	So	G	P
Jerseybelle	NJ	NE	VS	S	S	S	S	14	VL	So	F	M	P
Marshall	MA	NW	S	S	R	U	VS	7	L	So	So	E	VG
Midland	MD	SE–MW	R	R	S	S	S	0	L	F	So	E	VG
Midway	MD	NE–E	VS	S	R	M	U	10	L	F	F	G	VG
Northwest	WA	NW	R	U	S	M	T	14	M	M	M	G	VG
Pocahontas	MD	E–MW	R	M	S	S	U	7	L	M	M	G	VG
Robinson	MI	E–NC	M	S	S	R	T	10	L	So	So	M	P
Shasta	CA	W	S	U	S	S	T	7	L	M	M	G	G
Siletz	OR	NW	R	R	R	R	T	7	M	M	So	VG	VG
Sparkle	NJ	NE–NC	S	M	R	S	S	12	Sm	So	So	VG	VG
Surecrop	MD	E–MW	R	R	R	VR	T	5	L	F	M	G	G
Tennessee Beauty	TN	SE–SC	R	R	S	U	T	12	Sm	F	F	G	G
Tioga	CA	W	S	U	S	S	T	10	VL	F	F	G	G

SOURCE: Scott et al., 1972.
KEY: E = Excellent, F = Fair, G = Good, L = Large, M = Moderate, Medium, P = Poor, R = Resistant, S = Susceptible, Sm = Small, So = Soft, T = Tolerant, U = Unknown, V = Very.

The two main classes of cultivars, everbearing and June bearing, are grown in a wide variety of climatic situations. Strawberry cultivars are, perhaps more than any other fruit, bred and selected especially for specific climatic areas. They respond differently to specific combinations of temperature and day length. Long days and warm temperatures result in production of leaves and runners while short days and cool temperatures cause flower initiation of June bearers. Everbearing types initiate flowers under both long and short days. Southern cultivars require little winter chilling to break rest and are not hardy to northern climates. Northern cultivars have long chilling requirements and are very slow to flower when planted in the South. Under the cool climate of the coastal regions of California, cultivars such as Shasta and Tioga bloom continuously from early spring to December and act as everbearers. Yet in other areas they bear but one crop in the spring. True everbearers are not usually grown commercially but are used in home gardens. Principal everbearers are Gem (Brilliant, Superfection) and Ozark Beauty. Others are Geneva, Ogallala, Radiance, Red Rich, Rockhill, Streamliner, and Twentieth Century. Table 3-14 lists the characteristics of the most important U.S. cultivars and the regions where they are usually crop is produced in the West Coast states, mainly from the cultivars Tioga, Northwest, Hood, Shasta, and Fresno. A rather large proportion of this total is produced by the relatively new California cultivar Tioga. In 1974 in the coastal valley areas, this cultivar produced 45 metric tons per hectare.

Brambles (*Rubus* spp. [Tourn.] L.), Family Rosaceae

DESCRIPTION: $x = 7$; somatic nos. = 21, 28, 35, 42, 49, 56, 63, 70, 77, 84; deciduous or evergreen shrubs or suffruticose or herbaceous plants; **stems:** erect to trailing, mostly prickly and usually short-lived **leaves:** alternate, simple, 3-foliolate or pinnately or pedately compound, stipulate; **flowers:** perfect, rarely dioecious, white to pink, in racemes, panicles or corymbs, or solitary, terminal or rarely axillary; **calyx:** 5-parted, rarely 3–7 parted, with persistent lobes; **petals:** 5, sometimes lacking; **stamens:** many; **pistils:** many, sometimes few, on a convex torus; **styles:** nearly terminal; **mature carpels:** usually drupelets, sometimes dry. More than 400 species mainly in the colder and temperate regions of the Northern Hemisphere (Fig. 3-8).

Most raspberries, *R. idaeus* L. (red) and *R. occidentalis* L. (black), are $2n = 14$, while blackberries, *R. ursinus* Cham. & Schlecht., have ploidy from $2n = 14$ to $2n = 84$. There are many species and hybrid forms of blackberry, including the cultivated boysenberry, loganberry, youngberry, and dewberry (Fig. 3-8).

Raspberries are grown principally in Washington, Oregon, Michigan, and New York, with nearly 85 percent of the U.S. commercial acreage in the Northwest. Major early red cultivars are Chief, June, Madawaska, September, Hilton, Sunrise, and Viking for the East and Midwest. Midseason cultivars are Canby, Fairview, Newburgh, Willamette, and Meeker for the West. Late cultivars include Fallred, Latham, Milton, and Taylor for the East, and Puyallup and Sumner for the West. Basic fall-bearing cultivars include September, Fallred, and Heritage. Important black raspberries are Allen, Black Hawk, Black Pearl, Bristol, Cumberland, Dundee, Huron, New Logan, Morrison, Munger, and Plum Farmer. Purple cultivars (red × black hybrids) are Clyde, Marion, and Sodus.

The term "blackberry" is here used to mean all closely related types, including the Boysen, Logan, Young, and so forth. The most widely planted cultivars are Boysen, Cascade, Chehalem, Thornless Evergreen, Logan, Marion, Aurora, Darrow, Olallie, and Himalaya. Oregon, California, Arkansas, and Washington are the most important producing states, with the Pacific Coast states accounting for about 80 percent of the U.S. total.

Cranberry and Blueberry (*Vaccinium* spp.), Family Ericaceae

DESCRIPTION: $x = 12$; deciduous or evergreen shrubs, rarely trees; **winter buds:** small, ovoid, with 2 or several outer scales; **leaves:** alternate, short-petioled, entire or serrate; **flowers:** axillary or terminal, solitary or in racemes; **calyx lobes:** 4–5, rarely obsolete, corolla cylindric, urceolate or campanulate, 4-5 lobed or sometimes 4-parted; **stamens:** 8 or 10, **anthers:** awned or awnless, prolonged into terminal tubes, with an opening at the apex; **ovary:** inferior 4–10 celled; **fruit:** a many-seeded berry crowned by the persistent calyx lobes. About 130 species in the Northern Hemisphere, Alaska to the tropics.

CRANBERRY (*V. MACROCARPON* AIT.)

DESCRIPTION: somatic no. = 24; creeping, with slender stems to 1 meter, upright at ends; **leaves:** elliptic-oblong about 12 mm long, flat or revolute, lighter beneath; **pedicels:** with 2 leaf-like bracts near apex; **petals:** 8 mm; **filaments:** about $\frac{1}{3}$ the length of the anthers; **fruit:** 1–2 cm, red (Fig. 3-9).

Leading cultivars of cranberry are Early Black, Howes, Jersey, McFarlin, Searles, and Stevens. Principal producing states in the U.S. are Massachusetts, New Jersey, Wisconsin, Washington, and Oregon, with a total of about 8700 hectares which produced 90,700 metric tons in 1970. Canadian production (Quebec, Nova Scotia, and British Columbia) is about 1300 metric tons (Galletta, 1975).

HIGHBUSH BLUEBERRY (*V. CORYMBOSUM* L., *V. AUSTRALE* SMALL.)

DESCRIPTION: Shrub to 4 meters; **branches:** spreading; **leaves:** ovate-elliptic lanceolate, 5 cm long, acute, entire, half-grown at anthesis; **flowers:**

Figure 3-8 Buds, flowers, and fruit of *Rubus:* **(A, B, C)** blackberry, and **(D, E, F)** red raspberry. Flowers are borne from mixed buds on one-year canes, in an indeterminate inflorescence. Aggregate fruit formed by many drupelets. [B and E: Courtesy of M. M. Thompson; C: Courtesy of F. J. Lawrence, USDA; F: Courtesy of D. K. Ourecky, New York Agricultural Experiment Station]

Figure 3-9 Buds, flowers, and fruit of **(A, B, C)** blueberry and **(D, E, F)** cranberry. Flowers arise from one-year buds. Blueberry inflorescence is a raceme; cranberry is solitary in the axils of evergreen leaves. Fruit type is an epigynous (false) berry in both species. [C, D: Courtesy of R. Garren, Jr.; E: Courtesy of M. M. Thompson; F: Courtesy of R. W. Henderson]

in dense clusters; **calyx:** glaucous; **corolla:** cylindric-urceolate to narrow-ovoid, 8 mm long, white or pink; **fruit:** blue-black, 1 cm across (Fig. 3-9).

Leading highbush-blueberry cultivars (somatic no. = 48) are Jersey, Bluecrop, Weymouth, Wolcott, Coville, Earliblue, Berkeley, Blueray, Rubel, Stanley, and Dixi. Leading states in highbush-blueberry production are New Jersey, Michigan, North Carolina, and Washington.

LOWBUSH BLUEBERRY Maine is the principal producer of lowbush blueberries from three species: *V. myrtilloides* Michx. (somatic no. = 24); *V. angustifolium* Ait. (somatic no. = 48); and *V. brittonii* Porter ex Brickn. (somatic no. = 48). Wild berries are harvested from several thousand hectares in the northeastern U.S. and in southeastern Canada.

RABBITEYE BLUEBERRY *V. ashei* Reade (somatic No. = 72) is native to several southeastern states. It is more tolerant of heat and drought than other types, but the rabbiteye fruit is generally inferior to that of highbush cultivars.

Currant (*Ribes* spp.), Family Saxifragaceae

DESCRIPTION: $x = 8$; somatic no. $= 16$; usually unarmed shrubs, but sometimes with stipular spines and bristles; **leaves:** palmately veined and more or less deeply lobed and serrate; **flowers:** in racemes, rarely clustered, hermaphrodite or unisexual and then dioecious; **pedicels:** jointed below the ovary, often with 2 minute bractlets; **ovary:** glandular or smooth, never spiny; **receptacle:** from shallowly saucer-shaped or rotate to tubular, the top disc-like, often thickened or with knobs or rings; **fruit:** disarticulating from the pedicel, red, white, yellow, or black, often with a bloom, glabrous, glandular, or glandular-hairy.

More than 100 species of currant and gooseberry grow in the colder regions of the northern hemisphere and into the Andes of South America. The principal cultivated species are *Ribes sativum* Syme (red or white), *R. rubrum* L. (red), and *R. nigrum* L. (black). Black currant is not usually grown in the U.S. because it is susceptible to mildew and to white-pine blister rust. This is unfortunate because it is much higher in vitamin C than most other fruits and is used as a source of this vitamin in Europe (Fig. 3-10).

The main cultivars of red currant are Red Lake, Minnesota 71, Stephens No. 9, Wilder, Perfection, Rondom, and Rote Spätlese. The best white cultivar is White Imperial. Most of the currants are produced in Europe. In North America they are grown in New York, Michigan, Ontario, and to a lesser extent in Pennsylvania, Ohio, Minnesota, Colorado, British Columbia, Washington, and Oregon.

Gooseberry (*Ribes grossularia* L.), Family Saxifragaceae

DESCRIPTION: $x = 8$; somatic no. $= 16$; shrub to 1 m; **branches:** sometimes bristly with stout spines about 1 cm long and mostly 3-parted; **leaves:** cordate to broad-cuneate, 2–6 cm wide, 3–5 lobed with crenulate-dentate obtusish lobes, glabrous or pubescent, of firm texture; **flowers:** 1 or 2, greenish; **ovary:** pubescent and often glandular-villose; **sepals:** usually pubescent, about as long as the short-campanulate receptacle; **stamens:** shorter than sepals; **fruit:** globose to ovoid, pubescent, and glandular-bristly or smooth; red, yellow or green (Fig. 3-10).

The main cultivars grown in the U.S. are Oregon Champion, Chautauqua, and Downing (green), and Fredonia, Poorman, and Welcome (red). Most of the commercial berries of the U.S. are grown in Oregon, with a few being grown in Michigan and Washington. Because of their susceptibility to mildew and rust, the large-fruited English cultivars are not grown in the U.S.

Grape (*Vitis* spp.), Family Vitaceae

DESCRIPTION: $x = 19, 20$; somatic nos. $= 38, 40$; deciduous, rarely evergreen shrubs, climbing by tendrils; **pith:** brown, interrupted at the nodes by diaphragms; **leaves:** simple, dentate, usually lobed, rarely palmately compound; **flowers:** polygamo-dioecious, 5-merous, in panicles opposite the leaf; **sepals:** minute or obsolete; **petals:** cohering at the apex and falling as a whole at anthesis; **disk:** hypogynous, consisting of 5 nectariferous glands; **ovary:** 2-celled, cells 2-ovuled; **style:** conic, short; **fruit:** a pulpy 2- to 4-seeded berry; **seeds:** usually pyriform, with a contracted beak-like base, with 2 grooves on the ventral side. There are about sixty species in the Northern Hemisphere, chiefly in temperate regions (Fig. 3-11).

The grape is the most important and most widely grown deciduous fruit and is a major crop on every

Figure 3-10 Buds, flowers, and fruit of *Ribes:* **(A, B, C)** currant and **(D, E, F)** gooseberry. Currant flowers are borne in racemes from twig buds on two- and three-year wood. Gooseberry flowers are from similar buds, usually solitary. Fruit type is a many-seeded epigynous berry in both species. [B: Courtesy of M. M. Thompson; C and F: Courtesy of D. K. Ourecky, New York Agricultural Experiment Station]

continent. Two-thirds of the annual world production of 51,452,000 metric tons is produced in Europe. Other important production areas are Asia Minor, Latin America, the USSR, North America (mainly California), and Africa. The main species is *Vitis vinifera* L. (2n = 38), followed by the native North American *V. labrusca* L. (2n = 38) and *V. rotundifolia* Michx. (2n = 40). Other species, such as *V. riparia* Michx., *V. rupestris* Scheele, and *V. lincecumii* Buckl. have been used with *V. vinifera* in developing hybrid cultivars and rootstocks. Table 3-15 lists characteristics of grape cultivars grown in the U.S. and Canada.

In North America (besides California), grapes are

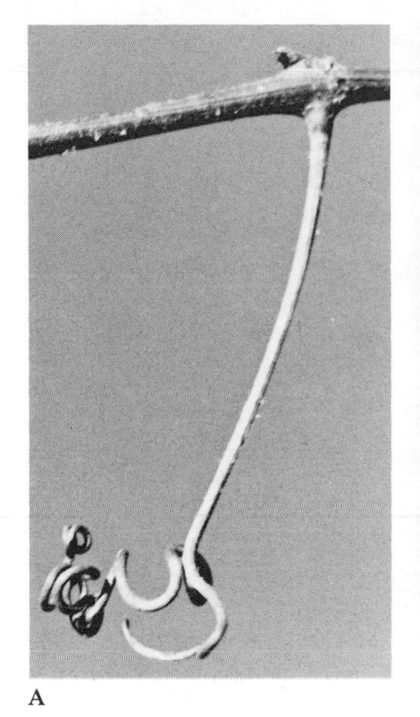
A

B

C

Figure 3-11 **(A, B, C)** Buds, flowers, and fruit of grape. Flowers are borne opposite the leaves on new shoots arising from buds on one-year canes. The inflorescence is a racemose panicle, and the fruit a two- to four-seeded true berry. [C: Courtesy of R. W. Henderson]

an important crop in New York, Ohio, Michigan, Ontario, Washington, and to a lesser degree, in the southeastern states. Because of the importance of climate to quality and production, cultivars must be chosen with great care for a given site.

MINOR SPECIES

Persimmon (*Diospyros* spp.), Family Ebenaceae

DESCRIPTION: $x = 15$; somatic nos. = 60, 90; deciduous or evergreen trees and shrubs; **winter buds:** ovoid, with about 3 outer scales, the terminal bud lacking; **flowers:** dioecious, whitish, the pistillate usually solitary, the staminate ones cymose; **calyx and corolla:** usually 4-lobed, rarely 3–7 lobed; **stamens:** usually 8–16; **ovary:** 4–12 celled; **styles:** 2–6; **fruit:** a large juicy berry, 1- to 10-seeded, usually with an enlarged calyx at base; **seeds:** large, flattened. There are about 200 species, mostly tropical and subtropical (Fig. 3-12).

Persimmons belong to the ebony family from which comes the ebony wood of commerce. Species important for their fruit are *Diospyros virginiana* of North America, *D. lotus* L., the Asian date plum, and *D. kaki* L. of China, the best of the edible species. *D. lotus* is used more as a rootstock than for its fruit.

D. virginiana ranges from Florida and Texas up to northern latitudes of 38° to 40°. It blooms late and is not often injured by spring frost. Even the northern types require less chilling than apples. Three selected cultivars are Early Golden, Ruby, and Miller. Pollination is by insects, which may carry pollen from a staminate tree several hundred feet away. Some cultivars will set fruit parthenocarpically if there is no

Table 3-15 Characteristics of grape cultivars grown in the U.S. and Canada.

Species and Cultivar	Fruit Color	Fruit Size	Seedless	Use[1]	Origin
V. vinifera					
Thompson's Seedless (Sultanina)	White	M	Yes	T,R,W	Persia?
Emperor	Rose–Purple	L	No	T	Unkn.
Flame Tokay	Red	L	No	T	Algeria
Ribier				T	
Cardinal	Red			T	U.S.
Perlette	White	M	Yes	T	U.S.
Red Malaga	Red		No	T	Spain
Carignane	Black		No	W	Spain
Zinfandel	Red		No	W	Unkn.
Muscat of Alexandria	White	L	No	R,W	N. Africa
Grenache	Purple		No	W	Spain
Alicante Bouschet	Black		No	W	France
Mission	Black	S–M	No	W	Spain
Palomino	White		No	W	Spain
Malaga	White	M	No	T,W	Spain
Petite Sirah	Black		No	W	France
Burger	White		No	W	
Mataro	Red		No	W	
French Colombard	White		No	W	
Salvador	Red		No	W	
Cabernet Sauvignon	Black		No	W	France
Corinthe Noir	Black	VS	Yes	R	Greece
V. labrusca					
Concord	Black	M	No	P,J,T	U.S.
Niagara	White		No	T	U.S.
Fredonia	Black	L	No	T	U.S.
Delaware	Red		No	T	U.S.
Catawba	Red		No	T	U.S.
V. vinifera hybrids					
Seibel 10878	Black	M	No	W	France
Foch	Black	M	No	W	France
Seibel 9549	Black	M	No	W	France
V. rotundifolia					
Scuppernong	White	L	No	T	U.S.
Flowers	Black	L	No	T	U.S.
James	Black	VL	No	T	U.S.
Thomas	Black	L	No	J,P	U.S.

[1]T = Table, R = Raisins, W = Wine, J = Juice, P = Processing.

local source of pollen. The fruit is 2.5–3.8 cm in diameter, rather seedy and very astringent until fully soft and ripe. The assertion that frost is required to remove the astringency is unfounded.

D. kaki is one of the most important orchard fruits of Japan and China. Many of the cultivars grown are centuries old. Until recently, most cultivars were of the astringent type. However, they can be treated with ethylene and other chemicals to remove the astringency while the fruit is still firm. The leading Japanese cultivar of California is Hachiya and that of Florida is Tenenashi. Both are conic, have orange skin and flesh. As with *D. virginiana,* many cultivars set fruit both parthenocarpically and with pollination. Parthenocarpic fruits tend to have lighter flesh and are more astringent than seeded ones, but this varies greatly with cultivar.

Northern papaw (*Asimina triloba* [L.] Dun.), Family Annonaceae

DESCRIPTION: $x = 9$; somatic no. $= 18$; small deciduous tree; **branchlets:** fulvous-pubescent while young, later glabrous and brown; **leaves:** obovate-oblong, 15–30 cm, short-acuminate, usually

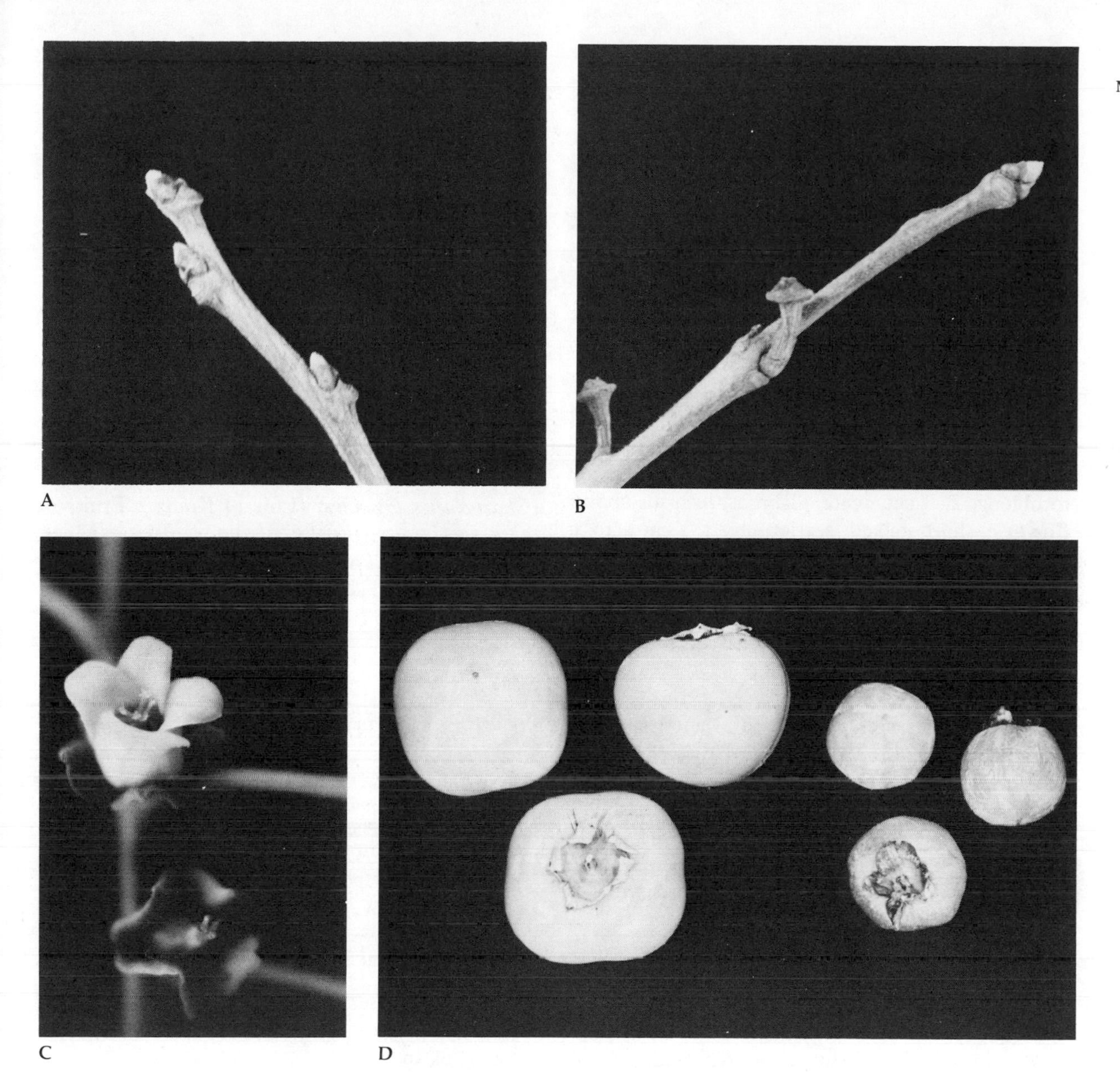

Figure 3-12 Persimmon. **(A, B)** Buds and old fruit stems, **(C)** flowers, and **(D)** fruit of persimmon. The dioecious flowers are borne at the axils of new leaves. The male flowers are in cymes; the female ones are solitary. The fruit is a large, several-seeded berry. [C: Courtesy of M. M. Thompson, O.S.U.; D: Courtesy of D. K. Ourecky, New York Agricultural Experiment Station]

gradually narrowed into a petiole 5–10 mm; **flowers:** 4–5 cm across, lurid purple, on short fulvous hairy stalks about 1 cm; **sepals:** ovate, acute, greenish pubescent outside; **outer petals:** broad-ovate, rounded, and later reflexed at apex; **inner petals:** upright, smaller, pointed; **carpels:** ellipsoid to oblong, 5–7 cm long, greenish-yellow, turning brown; **seeds:** 2–2.5 cm long.

The northern papaw, *A. triloba,* is the only hardy species of the tropical so-called custard apple family that consists of more than 600 species. In the U.S., it grows from New York South to Florida and west to Nebraska and Texas. The fruit vary considerably in size and color from tree to tree. They are usually green until ripe, turning yellow, then bronze, with yellow flesh when soft enough to eat.

Pomegranate (*Punica granatum* L.), Family Punicaceae

DESCRIPTION: $x = 8$; somatic no. $= 16$; small deciduous tree 5–10 m, with spiny branches, branchlets angled, glabrous; **leaves:** short petioled, obovate to oblong, 2–8 cm, acute, glabrous, lustrous above; **flowers:** short-stalked, scarlet, about 3 cm across; perfect, perigynous, **stamens:** numerous, **ovary:** inferior or nearly so, with the carpels in one or two superimposed series, **petals:** crumpled; **calyx:** purple; **fruit:** subglobose, 6–8 cm diameter, yellow, red, pinkish, persistent calyx.

The pomegranate is often used as an ornamental yard tree. It has been cultivated in Asia Minor for thousands of years. The apple-sized fruit are covered with a leathery skin that encloses many seeds. The edible portion consists of fleshy outer seedcoats (aril). The cultivar Wonderful produces the best fruit in California among the few commercial orchards.

Jujube (*Zizyphus jujuba* Mill.), Family Rhamnaceae

DESCRIPTION: $x = 12$; somatic no. $= 24$; deciduous shrub or tree, glabrous; **winter buds:** small; **stipules:** spiny, one spine to 3 cm, the other decurved; **branchlets:** often fascicled, resembling pinnate leaves; **leaves:** oblong-ovate to ovate-lanceolate, rarely ovate, 2–6 cm long, acute or obtuse, oblique and 3-nerved at base, crenate-serrulate, glabrous, firm; **petiole:** 1–5 mm; **flowers:** yellow, 2–3 axillary, short-stalked; **fruit:** a drupe, ovoid to oblong, dark red, turning black, 1.5–2.5 cm long, short stalked.

The jujube is a member of the Buckthorn family and is one of forty species. The fruit is mild and sweet and reportedly contains a high concentration of vitamin C. It is one of the principal fruits of China.

Other Species

A few other species of interest are elderberry (*Sambucus canadensis* L., *S. coerulea* Raf.), serviceberry (*Amelanchier alnifolia* Nut.), highbush cranberry (*Viburnum trilobum* Marsh.), cornelian cherry (*Cornus mas* L.), buffaloberry (*Shepardia argentea* Nut.), medlar (*Mespilus germanica* L.), sand cherry (*Prunus besseyi* Bailey), Chinese gooseberry or kiwi (*Actinidia chinensis,* Planch.) and oriental quince (*Chaenomeles lagenaria* [Loisel.] Koidz.). Fruits of these species are usually used for making jellies, sauces or wines. The kiwi is grown commercially in New Zealand and is being planted to a limited extent in California. Its fruit are eaten fresh.

GENERAL REFERENCES

Bolkhovskikh, Z.; V. Grif; T. Matvejeva; and O. Zakharyeva. 1969. *Chromosome numbers of flowering plants.* Academy of Science, U.S.S.R.

Chandler, W. H. 1957. *Deciduous orchards.* 3rd ed. Lea & Febiger, Philadelphia.

Crane, M. B., and W. J. C. Lawrence. 1938. *The genetics of garden plants.* Macmillan, London.

Esau, K. 1953. *Plant anatomy,* Chapters 19 and 20. Wiley, New York.

Forde, H. I. 1975. Walnuts. In *Advances in fruit breeding,* edited by J. Janick and J. Moore. Purdue Univ. Press, Lafayette, Indiana.

Hedrick, U. P. 1925. *The small fruits of New York.* N.Y. Agr. Exp. Sta., Geneva, N.Y.

Janick, J., and J. N. Moore. 1975. *Advances in fruit breeding.* Purdue University Press, Lafayette.

Madden, G. D., and H. L. Malstrom. 1975. Pecans and hickories. In *Advances in fruit breeding,* edited by J. Janick and J. N. Moore. Purdue Univ. Press, Lafayette, Indiana.

Serr, E. F., and H. I. Forde. 1968. Ten new walnut varieties released. *Calif. Agr.* 22:8–10.

4

Rootstocks: Their Propagation, Function, and Performance

Because neither rootstock nor scion cultivars propagate true from seed, they are propagated vegatatively (asexually) by budding, grafting, cuttage, and so forth. Sexual propagation by seedage is, however, a common practice for obtaining seedling rootstocks on which to graft or bud desirable types. The various kinds of asexual propagation are by apomictic embryos, by runners, by suckers, layerage, crown division, stem cuttings, root cuttings, grafting, and budding. Various kinds of physical and chemical treatments are often required to increase some species by these asexual methods. Also some degree of skill and knowledge of plant function are needed for successful propagation by any method. General methods of rootstock propagation are shown in Table 4-1. Once the root is produced, it must then be budded or grafted with the scion cultivar (unless scion rooting is done).

SEED PROPAGATION

Dry seeds of most deciduous trees and shrubs, even though mature, will not germinate and grow until they are chilled above freezing under moist conditions (stratification). This is often erroneously called afterripening, a term inferring an internal process rather than an external treatment. As Figure 4-1 shows, such chilling releases embryos from rest (physiologic dormancy). The optimum temperature and duration

Table 4-1 Methods of rootstock (or plant) propagation for temperate fruits and nuts.

Kind of Plant	Seedage		Cuttage		Layerage		Runners	Crown Division	Root Suckers
	Sexual	Apomict	Stems	Roots	Tip	Mound			
Apple	X	(X)	(X)			X			(X)
Pear	X		X	(X)					(X)
Quince	(X)		X	(X)		X			(X)
Plum	X		X	(X)		X			(X)
Sweet Cherry	X		X			X			(X)
Sour Cherry	X		X			X			(X)
Peach	X		(X)			(X)			
Apricot	X		(X)			(X)			
Almond	X		(X)			(X)			
Persimmon	X			(X)					
Fig	(X)		X						
Mulberry	(X)		X						
Jujube	X		(X)	(X)					
Walnut	X								
Pecan	X		(X)	(X)					
Filbert	(X)				X	(X)			(X)
Pistachio	X								
Chestnut	X								
Papaw	X								
Strawberry							X	X	
Raspberry			(X)		X				X
Blackberry			(X)		X				X
Blueberry	(X)		X						(X)
Cranberry			X						
Gooseberry	(X)		X			X			
Currant	(X)		X			(X)			
Grape	(X)		X			(X)			
Kiwi	X		X						
Pomegranate			X						

KEY: X = Common method, (X) = Uncommon method.

of chilling vary widely with species. Chilling optima usually fall within the range of 4° to 10°C, and the time required ranges from 0 to 160 days (Table 4-2). Temperature optima are usually higher for those species requiring fewer days of chilling (Westwood and Bjornstad, 1968a). Removal of the stony endocarp or pericarp of seeds often reduces the number of days of chilling for germination. Also removal of pome-fruit seedcoats results in complete germination without chilling. However, seedlings obtained in this way are stunted and their epicotyls will not grow until they are chilled or treated with GA (Fig. 4-1). Thus seedcoat inhibitors as well as embryo inhibitors are present. Both appear to be ABA, also found to be a principal rest-period inhibitor of buds. Seedcoat inhibitors are washed out by repeated washing with water, but embryo inhibitors seem only to be removed by the physiologic action of chilling. Very hard seed coverings may require special treatments to soften them sufficiently for germination. They may be scarified, treated with a strong acid or subjected to alternate freezing and thawing, or the covering may be removed, as from nuts and stone fruits, to facilitate germination. Drying fully chilled, imbibed seed may cause secondary dormancy, and cold-moist treatment is again required before germination will occur.

VEGETATIVE PROPAGATION

Apomictic Seeds

Apomixis is a form of vegetative reproduction because the seeds are formed entirely from diploid ovarian tissue rather than from pollination and fertilization of egg cells. Apomictic seeds produce juvenile seedlings, as do sexual seeds, but the former are identical genetically to the cultivar from which they came. Some Asian and North American *Malus* and some tung trees, along with some tropical species such as citrus and mango, exhibit apomixis. Because apomictic seedlings are often virus free, they may be used to clean up an infected cultivar; however, such trees come into bearing later than grafted stock because apomicts must outgrow the juvenile condition before flowering is possible.

Cuttings

Roots may form on both leafy and dormant cuttings under controlled environments and with chemical treatment. As early as 1880 the German scientist J. Sachs postulated that a root-forming substance is produced in leaves and is translocated down the stem to promote rooting. Not until the 1930's, however, was progress made in identifying chemical factors that promote rooting. As cited in a recent review by Westwood (1972), Zimmerman et al. in 1933 reported that ethylene gas would stimulate adventitious roots. In 1934, Thimann et al. established that auxin (IAA) is a principal rooting hormone. Then in 1935 Zimmerman et al. found that the synthetic auxins IBA and NAA promoted rooting of cuttings. Dr. Fritz Went in 1938 proposed an added cofactor, rhizocaline, which is produced in the leaves and moves down the stem, where, in combination with auxin, it causes roots to form. More recently, Hess (1961) found a number of co-factors that aid in rooting.

In today's practice, cuttings are usually treated with the auxin IBA (soaked in a solution of 25–200 ppm or dipped for five seconds in a solution of

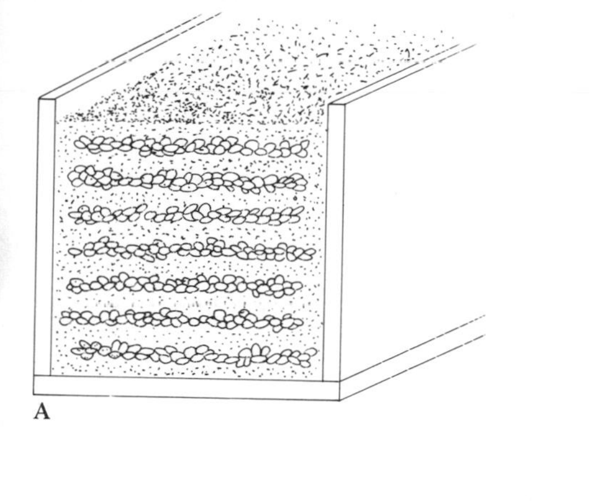

Figure 4-1 Propagation by seeds. **(A)** A cross-section of a stratification box filled with moist sand, perlite, or peat between the layers of seed, which must be kept moist during chilling. **(B)** Both chilling and washing of pear seed are necessary to good germination. The seed at left were chilled but not washed; those at right were chilled and washed. **(C)** Seedcoat removal only permits germination, but without chilling there is no epicotyl elongation. Plant at right had complete chilling prior to germination. [A: After USDA]

Table 4-2 Size, chilling requirement, and other characteristics of seeds of some fruit and nut species, including some species used only as rootstocks.

Species	Seed Size (no. per g)	Chilling Requirement (days)[1]	Optimal Chilling Temp. (°C)[2]	Speed of Germination (days)	Seed Viability (years)
Pome Fruits					
Apple (Malus)					
M. baccata	48–185	30	5	8	
M. coronaria	31	120	5	19	
M. ioensis	66	60	5	4	
M. pumila	15–60	75	5		
M. sylvestris	21–35	60–100	4	30	2–3
Christmas berry					
Heteromeles arbutifolia	53	90			
Hawthorn (Crataegus)					
C. mollis		75–90	5		
Mountain ash (Sorbus)					
S. americana	185–520	90	5	5	
S. aucuparia	230–375	60–90	1		
S. decora	280	90	5	5	
Pear (Pyrus)					
P. amygdaliformis	24	15–25	7		
P. betulaefolia	90	10–55	4		
P. calleryana	55	10–30	7		
P. communis (domestic)	22	60–90	4		2–3
P. communis (wild)					
P. cordata	86	40–100	4		
P. dimorphophylla	77	30–65	4		
P. elaeagrifolia	22	50–90	4		
P. fauriei	57	10–35	7		
P. nivalis	18	60–110	4		
P. pashia	55	0–15	10		<1
Serviceberry (Amelanchier)					
A. alnifolia	112–250	180	2.5	30	
Stone Fruits (Prunus)					
Apricot—P. armeniaca	.6–.8	60	5	15	5
Cherry					
P. avium	4.8–6.6	90–120	4	15	1–2
P. cerasus	7.0–8.8	90–120	1–4	18	1–2
P. cerasus cracked		60	1		
P. mahaleb	10–12	100		15	1–3
P. padus	14.6–15.5	60–90	5		
P. pennsylvanica	26–46	90–120	5	10	
P. serotina	7–18	90–120	5	10	
P. tomentosa		75	4		
P. tomentosa cracked		42	4		
P. virginiana	7–18	90–160	5	15	
Peach					
P. davidiana	.35–.49	100		15	5
P. persica	.25–.36	45–100	4–7	15	5
P. persica cracked		35–60	4		
Plum					
P. alleghaniensis	6.5	90–120	5	20	
P. americana	1.2–3.0	150	5	15	
P. americana	1.8–1.9	150	4	30	4–6
P. angustifolia	1.7–3.4	120–160	5	25	
P. augustifolia, var. Watsoni	2.2	120–150	5		
P. besseyi	3.3–5.3	90–120	5		
P. besseyi	5.6–6.0	80–100		15	4–6

Table 4-2 (continued)

Species	Seed Size (no. per g)	Chilling Requirement (days)[1]	Optimal Chilling Temp. (°C)[2]	Speed of Germination (days)	Seed Viability (years)
		Stone Fruits (Prunus) (continued)			
P. cerasifera	2.1–2.5	80–100		30	4–6
P. domestica	.9–1.0	120		30	4–6
P. insititia	3.5–4.2	100–120	4	30	4–6
P. insititia cracked		84	4		
P. (marianna) hybrid	1.8–2.5	100		30	4–6
P. munsoniana	4.2–4.9	80–100		15	4–6
P. pumila	5.5–6.0	90–120	5		
P. pumila, var. susquehanae	8–13	60–120	5	15	
P. salicina	.7–1.4	60–100		15	4–6
P. simonii		45	10		
P. simonii cracked		25	5		
		Nuts			
Almond, Prunus amygdalus	.28	20–30	7		
Chestnut, Castanea dentata	.2–.4		1		
Hazel (Corylus)					
C. americana	.4–1.6	60–77	5		
C. avellana	.17–1.2	30–150		10–30	
C. californica	.8–.9	90–100	5		
C. cornuta	.9–1.5	60–90	5		
Hickory (Carya)					
C. aquatica, water hickory	.4	90–150	1		
C. cordiformis, bitternut	.3–.4	90–120	4	30	
C. glabra, pignut	.4–.5	90–120	4		
C. illinoensis, pecan	.1–.4	30–90	5	20	1–3
C. laciniosis, shellbark	.06–.08	120–150	4		
C. ovata, shagbark	.2–.3	90–150	5	40	
C. tomentosa, mockernut	.07–.25	90–150	1		
Tung, Aleurites fordii	.35–.50	30–60		10	1–3
Walnut (Juglans)					
J. californica	.16				
J. cinerea	.03–.09	90–120	5	45	
J. hindsii	.07–.18	60–120		30	3–5
J. nigra	.04–.22	60–120	5	25	
J. nigra	.11	60–120		30	3–5
J. (Paradox) hybrid	.11–.14	60–80		25	3–5
J. (Royal) hybrid	.11–.18	60–100		25	3–5
J. regia	.07	30–60		20	1–3
J. regia	.07–.11	30–60	5		
J. rupestris	.17–.24	90	7	14	
		Small Fruits			
Brambles (Rubus):					
R. idaeus, Red raspberry	670–800	90	5	10	
R. occidentalis, Black raspberry	630–850	90	5	8	
Currant and Gooseberry (Ribes):					
R. americana, currant	545–735	90–120	5		
R. aureum, currant	440–509	90	5		
R. cynosbati, gooseberry	415–485	90–120	5	20	
R. missouriense, gooseberry	345–370	90	5	15	
R. odoratum, currant	230–342	60–90	5	12	
R. rotundifolia, gooseberry		0		60	
Grape, Vitis riparia	25–38	60–120	5	15	

(continued)

Table 4-2 (continued)

Species	Seed Size (no. per g)	Chilling Requirement (days)[1]	Optimal Chilling Temp. (°C)[2]	Speed of Germination (days)	Seed Viability (years)
Minor Fruits					
Buffaloberry, Shepardia argentea	40–148	60–90	5	15	
Cornelian cherry, Cornus mas		120	10		
Elder, Sambucus canadensis	385–715	90–150	5	15	
S. glauca	255–280	90	5	25	
S. pubescens	420–830	90–150	5	25	
Highbush cranberry, Viburnum trilobum	21–39	60–90	5		
Jujube, Ziziphus jujube	1.3–2.5	60	5		
Kiwi, Actinidia chinensis	v. small	14			
Mulberry, Morus alba	285–770	60–90	5	10	
M. alba, var. tatarica	540–815	30–60	5	15	
M. rubra	440–1100	90–120	5		
Pawpaw, Asimina triloba	.6–2.6	100	7–10	10	
Persimmon, Diospyros virginiana	1.5–3.9	60–90	10	20	

SOURCES: USDA Agr. Handbook 450 (1974) and USDA Agr. Yearbook (1961).
[1]Some seeds with hard impervious seed coverings usually require a period of 20°–25°C stratification, scarification, or sulfuric acid treatment prior to chilling at 1° to 5°C.
[2]In some cases the optimal temperature was not indicated.

1000–5000 ppm) and are sometimes treated with other chemicals, such as boric acid, adenine sulphate, and B vitamins. The condition of the cutting (whether dormant, leafy, succulent, or mature) and the environmental factors of temperature and light also affect rooting. Leafy cuttings of many species root well when treated with IBA and placed under intermittent water mist with 25°C bottom heat, using sand or perlite as a rooting medium. In some species tip cuttings root best; in others the basal portion of a shoot roots best. Rooting of hardwood dormant cuttings is sometimes best when buds are not in rest, but many species root better if buds are in rest. Typically, the cuttings are taken in late fall or early winter, treated with IBA, and callused for 2–3 weeks at 15°–20°C. When roots begin to show in the basal callus, they are placed in cool storage at 3°C to await spring planting. Depending upon the species, bud removal may enhance or hinder rooting of dormant cuttings. Enhancement of rooting by bud removal often seems to be due to the wounding action rather than to bud removal *per se*. Wounding may cause ethylene synthesis, which is known to stimulate rooting.

In some species, such as quince, currant, and some species and cultivars of apple, preformed root initials occur at the nodes. In these plants the problem is to induce root elongation rather than initiation. Many plants that root readily from cuttings, such as grape, do not have preformed root initials but form initials readily under proper conditions.

Root cuttings may be induced to form buds, which in turn form sprouts as a means of vegetatively propagating some species. Both blackberry and raspberry are multiplied by this means.

In summary, the balance between auxin and other plant constituents controls organ formation and is the basis both for rooting of stem and for budding of root cuttings. This balance may be achieved by manipulating genetic, environmental, and chemical factors. The following facts are important to an understanding of the processes of root and bud generation:

1. Budding and rooting are strongly polar (buds form at the top of roots and roots form at the base of stems). The movement of auxin and rooting cofactors is polar also, moving toward the base, while

movement of cytokinins is acropetal, from base to top. Auxin stimulates rooting and cytokinins stimulate budding. Thus they must be in proper balance for best results.

2. Nutritional deficiencies usually hinder rooting.

3. Juvenile tissues contain more rooting promoters than adult tissues. Juvenile cuttings also lack flower buds, which are known to inhibit rooting in some species (Ali and Westwood, 1968).

4. Whether active buds or auxin treatment aids in rooting depends upon genetics and the season in which cuttings are taken.

5. GA and cytokinins often inhibit rooting, while ethylene, ABA, and morphactins may improve rooting. Growth retardants such as Alar, CCC, and TIBA give variable responses.

6. Both environment and genetic controls affect the kind and amount of rooting co-factors. These cofactors include phenolic compounds, which interact with auxin to stimulate rooting.

7. Other factors, such as girdling of the stock plant, photoperiod, etiolation, orientation of the cuttings during callusing, pre-severence position, and tissue maturity may all affect rooting.

Layering

Layering is the process of developing new plantlets by rooting stems that are attached to a mother plant. Ultimately, the rooted stems (called "layers") are detached, forming a new plant on its own roots. Some plants, such as blackberry and black raspberry, naturally form layers. Others, such as filbert and apple, are manipulated to obtain layers. Root initiation during layering may be induced by the exclusion of light (etiolation) and by various mechanical treatments (notching, girdling) to slow or block the downward movement of carbohydrates, hormones, and other organic translocates. Rooting occurs just above the point of blockage (Fig. 4-2).

The two main types are (*a*) tip and (*b*) mound, or stool, layering. Tip layering is done in late summer on trailing caneberries for best results, at a time when the layered tips will soon set terminal buds (Fig. 4-2). Mound layering, the common method of obtaining clonal rootstocks for apple, quince, and sometimes cherry, is established as indicated in Figure 4-3. When available, sawdust is an excellent material to use for mounding the layers. It is easy to apply, is

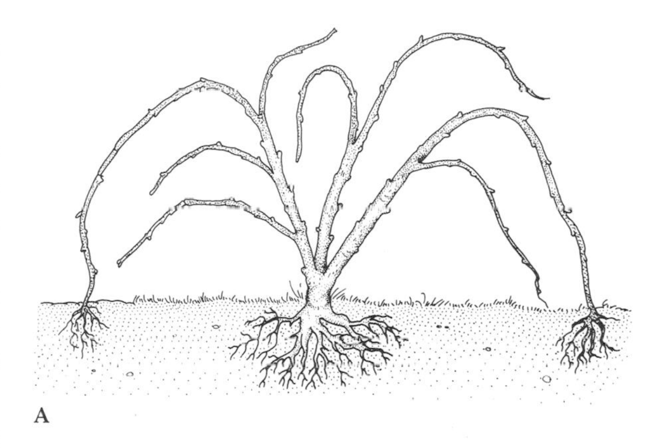

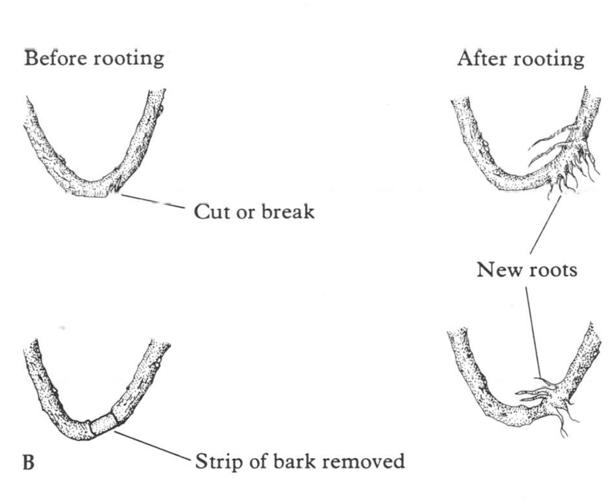

Figure 4-2 **(A)** Natural tip layering of black raspberry. **(B)** The beneficial effect of wounding or girdling on rooting. The latter treatments are used when tip layering difficult-to-root plants. [A: After Verner, 1946; B: After Hudson T. Hartmann and Dale E. Kester, *Plant Propagation: Principles and Practices*, 3rd ed. © 1975, p. 456. Reprinted by permission of Prentice-Hall, Inc., Englewood Cliffs, New Jersey]

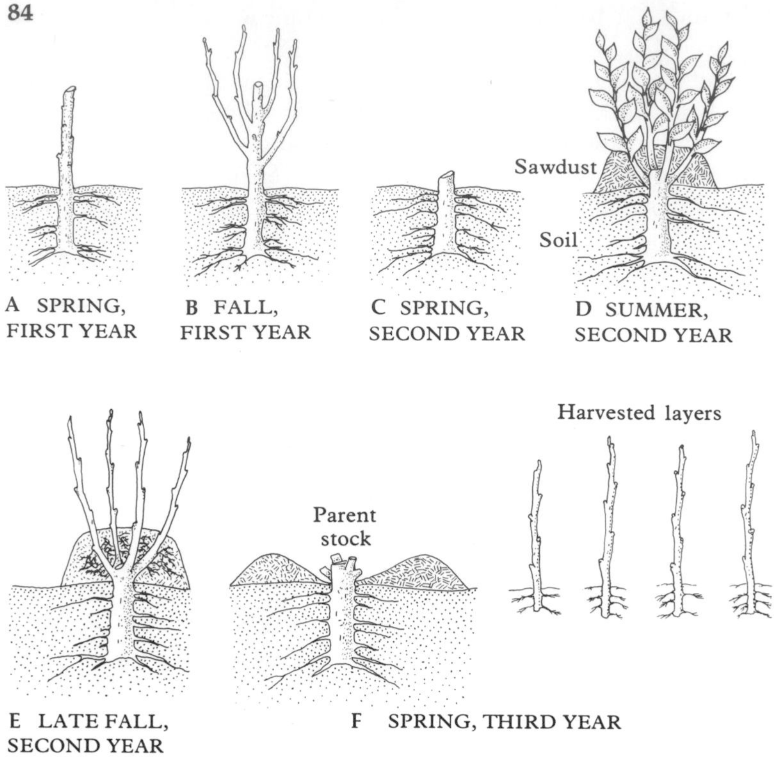

Sawdust
Soil
A SPRING, FIRST YEAR
B FALL, FIRST YEAR
C SPRING, SECOND YEAR
D SUMMER, SECOND YEAR
Harvested layers
Parent stock
E LATE FALL, SECOND YEAR
F SPRING, THIRD YEAR

Figure 4-3 (A–F) Mound layering, showing the two-year sequence from the time of planting the original stock (shown in A) through the removal of rooted layers in the early spring of the third year (shown in F). Sawdust is the preferred material for mounding because it is much easier to work with than soil. [After Roberts and Mellenthin, 1957]

better aerated than soil, and permits harvesting the rooted layers by a mechanical mower that can operate directly through the sawdust mound.

Budding and Grafting

Both budding and grafting are similar in that a portion of one plant is joined to another to form a compound system. With budding, a single bud is placed on a stock plant; whereas with grafting a larger portion of the stem is used as the scion. Budding is usually done in late summer but may be done at any time the scion buds are mature and dormant and when the bark on the stock is slipping and active (e.g., early summer) to save a year (June budding). The most common type is the T-bud or shield bud (Fig. 4-4). Others used are inverted T, patch bud, ring bud, plate bud, and chip bud. It is now known that when a compatibility bridge (interstem) is required, as with pear on quince, double-budding can be done in a single operation (Fig 4-5), saving a full season in getting the double-worked tree.

The most common types of grafts are the whip (tongue) graft and the cleft graft (Fig. 4-4). Other types are side, bark, notch, and approach graft. Specialized grafts used to repair trees are the inarch and the bridge graft.

For successful budding or grafting, follow these simple rules:

1. The knife should be razor sharp to permit smooth cuts.
2. The bud or scion should be dormant, but the understock may be growing it if is established in the orchard.
3. The cambium zones of the scion and stock must make intimate contact at one point at least. (With cleft grafting, do not try to line the cambiums up precisely but slant the scion into the cleft so that the cambiums cross at one point.)
4. Wrap or secure the union to prevent movement and by wrapping, waxing, or both wrapping and waxing, prevent any drying of the scion and the cambial zones.

MATERIALS AND TOOLS The usual pruning equipment can be supplemented with a number of specialized tools that have been developed for grafting. What is called the "grafting tool" is used for cleft grafting. The curved blade makes a clean cut in the stock stub when driven in with a mallet. The wedged tip of the grafting tool is used to keep the cleft open while inserting the tapered scion. The budding knife has a convexly curved cutting edge, and that of the whip-grafting knife is concave (Fig. 4-6).

Budding and grafting wraps are of various materials, but one of the most common is the rubber band.

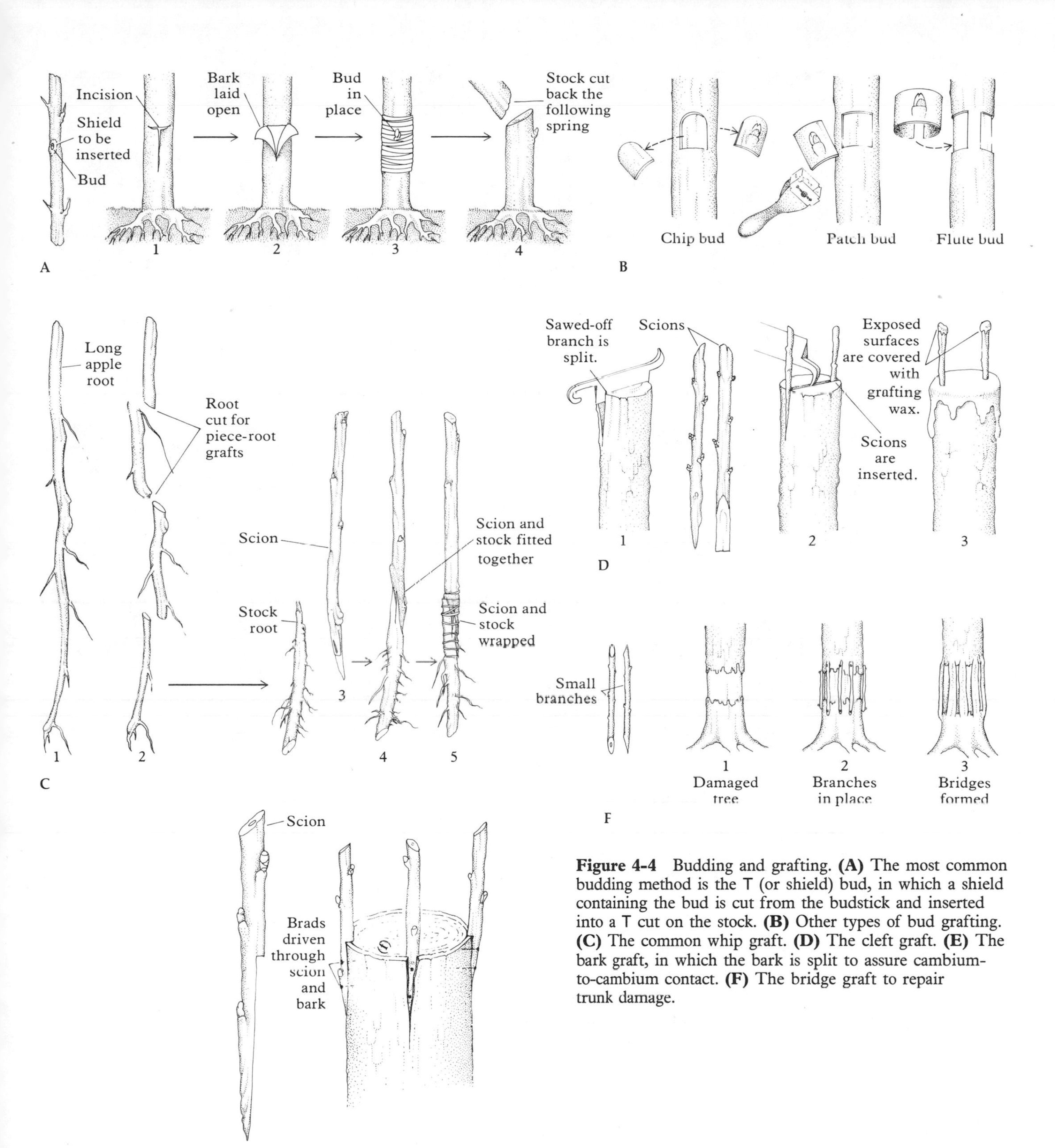

Figure 4-4 Budding and grafting. **(A)** The most common budding method is the T (or shield) bud, in which a shield containing the bud is cut from the budstick and inserted into a T cut on the stock. **(B)** Other types of bud grafting. **(C)** The common whip graft. **(D)** The cleft graft. **(E)** The bark graft, in which the bark is split to assure cambium-to-cambium contact. **(F)** The bridge graft to repair trunk damage.

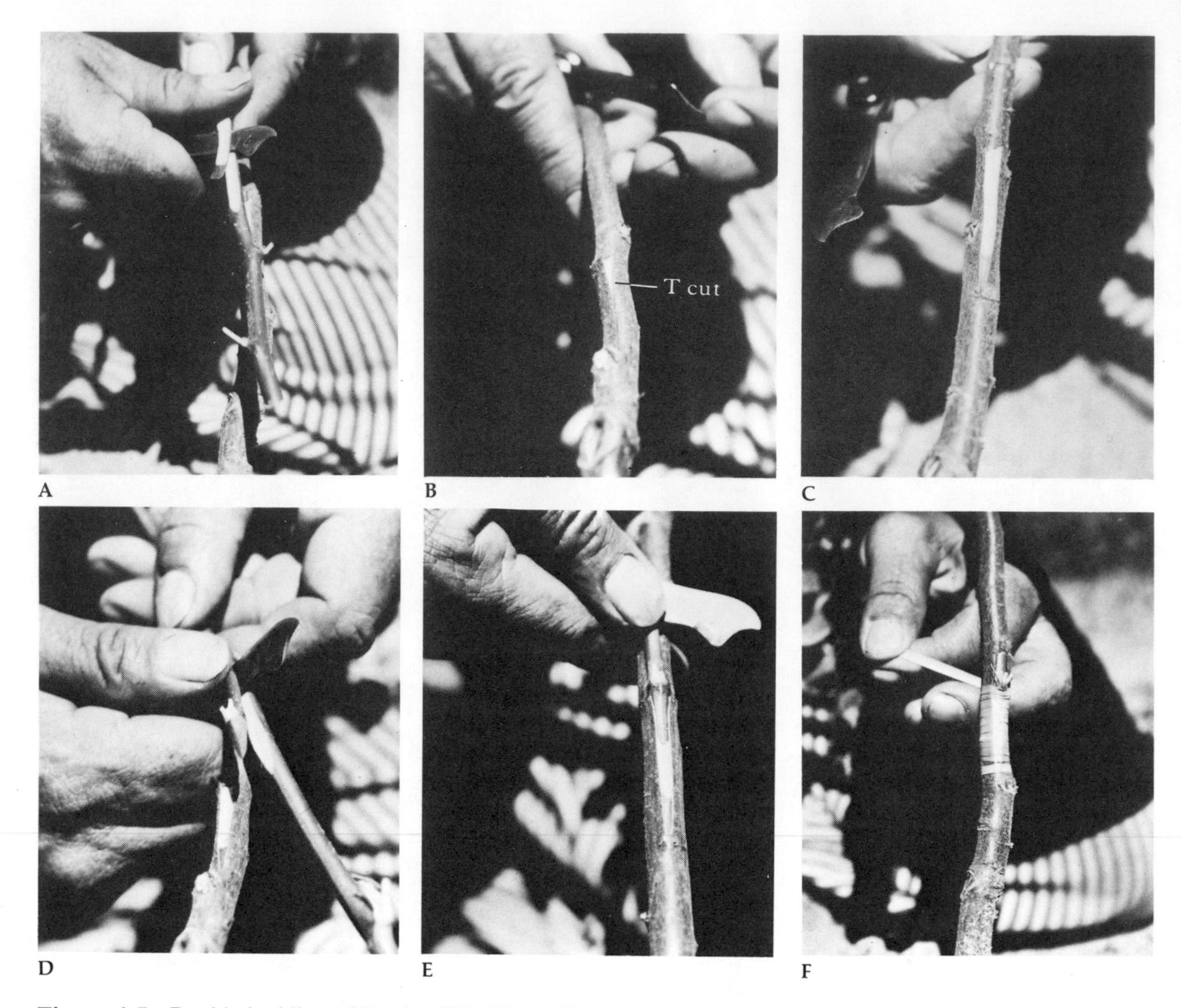

Figure 4-5 Double budding of Bartlett/Old Home/Quince to achieve compatibility of pear on quince. **(A)** A thin plate of Old Home (minus the bud) is inserted in **(B)** a T cut on quince stock. **(C)** The plate is in place. Then, **(D)** the usual shield bud of Bartlett **(E)** is inserted directly over the Old Home plate. **(F)** It is wrapped with a budding band in the usual manner. [Courtesy of Kent Brooks, Carlton Nursery]

It is secured by looping the end under the last wrap. Another convenient wrap for the whip graft is paper masking tape about 1.3 cm wide. It protects the graft during callusing but splits as growth proceeds. Canvas or cloth tape also works but must be cut on one side to prevent girdling as the graft grows. Whip grafts of small plants to be planted with the union underground should be made with cloth tape, which will rot off by microbial action. Rubber grafting bands should not be used underground, as they will not deteriorate and will ultimately girdle the plant. Sunlight on above-ground rubber wraps causes them

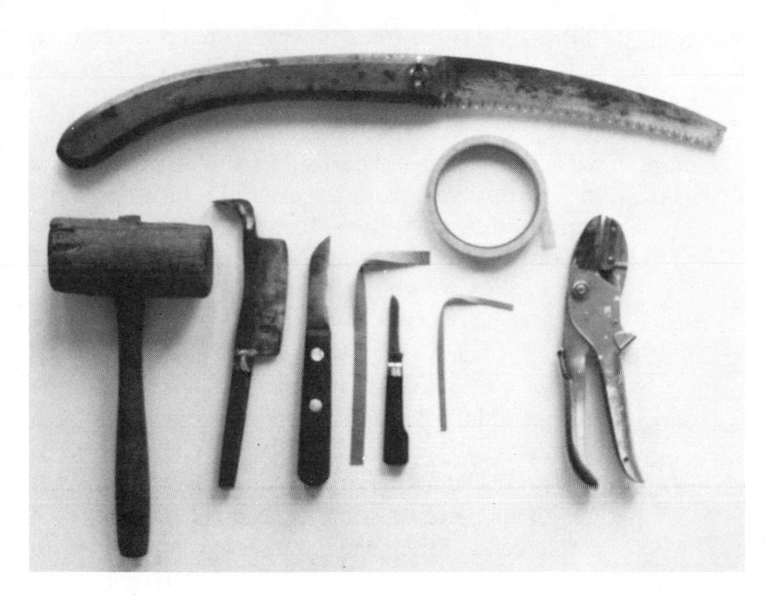

Figure 4-6 Standard equipment used for grafting and budding. The saw (*top*) is used for making large cuts. *Bottom, left to right:* The mallet and the grafting tool are used to make a vertical cleft, as shown in Figure 4-4. The larger grafting knife has a concave blade for making a whip graft. The large grafting band (or the roll of paper masking tape) is used to wrap a whip graft. The smaller budding knife has a convex blade. The smaller band is used to wrap the finished bud. The hand clipper is used for making small cuts in preparation for the grafting operation.

to check and deteriorate with time, preventing girdling.

Paints and waxes of various kinds are used to protect exposed cuts from drying. They should be nontoxic to the plant and durable enough to preclude re-painting. In addition to the regular "grafting wax" preparations, several rubber-base and asphalt-base preparations are also available.

Grafting and budding machines of several types have been used to speed the bench-grafting operation. One type makes three matching slots (Fig. 4-7) for grape grafting. The chip-budding machine (Fig. 4-8) has been used for grapes and walnuts.

COMPATIBILITY The performance of both scion and rootstock depends upon the compatibility of the two components. In general, closely related cultivars and species are compatible, some genera are compatible, but more distantly related plants are almost always graft-incompatible. Some confusion exists about the meaning of the term "compatible." As here used, it means a satisfactory union in the horticultural sense, which takes into account both the physical connection at the graft union and the physiological harmony of the genetic systems. Bartlett pear on most quince rootstocks is weak and tends to break off at the union. Bartlett on Oriental sand pear *(Pyrus pyrifolia)* root, while physically strong at the union, causes a severe disorder of the fruit in which the calyx end turns black (this condition is called "black end").

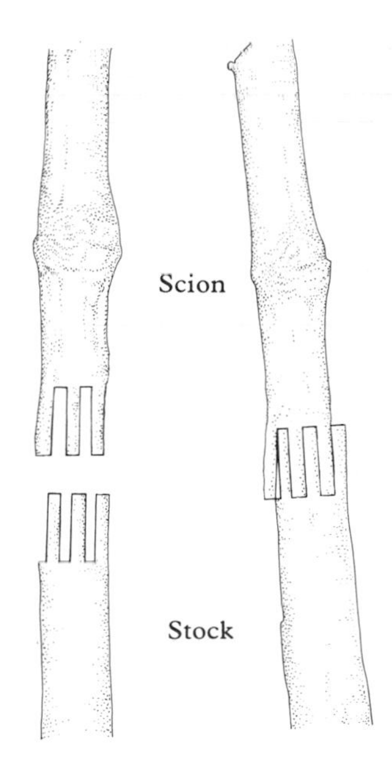

Figure 4-7 The grape-grafting saw cuts two sets of matching slots in scion and stock portions. They may be wrapped but are sometimes callused bare in a moist medium.

In horticultural terms, both rootstocks are incompatible with Bartlett.

Some incompatibilities have delayed effects; the blackline condition of English walnut grafted on black walnut root appears only after twenty or more years of satisfactory performance. Most incompatibilities of orchard trees, however, are apparent at an early age. A number of incompatibilities in trees are induced by either a virus or a mycoplasma (see Chapter 16). Sweet cherry infected with buckskin on *Prunus mahaleb* root and pear infected with pear decline on *Pyrus pyrifolia* root are examples of induced incompatibilities (Westwood, et al., 1971).

Graft unions after a few years' growth may be of three types: (1) both scion and stock are of equal size, (2) the scion overgrows the stock, (3) the stock overgrows the scion (Fig. 4-9). Grafts where the scion and stock are unequal in size are often thought to be incompatible. This notion stems largely from the fact that many incompatible unions show a scion overgrowth. However, many unions with scion overgrowth are perfectly compatible, and the overgrowth simply reflects the genetic tendency of the scion to lay down wood growth by increased cambial activity, for example, Comice pear overgrows all rootstock types, even those with which it is completely compatible.

Figure 4-8 (A) The chip-budding machine **(B)** removes a bud chip from the scion and **(C)** makes a matching slot in the understock. [Courtesy of H. B. Lagerstedt, USDA]

A

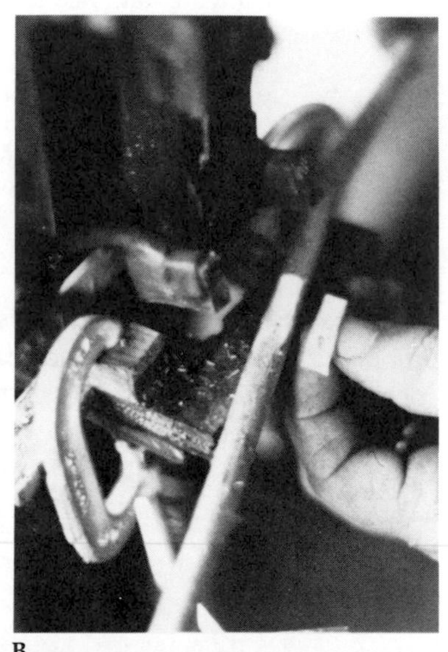
B

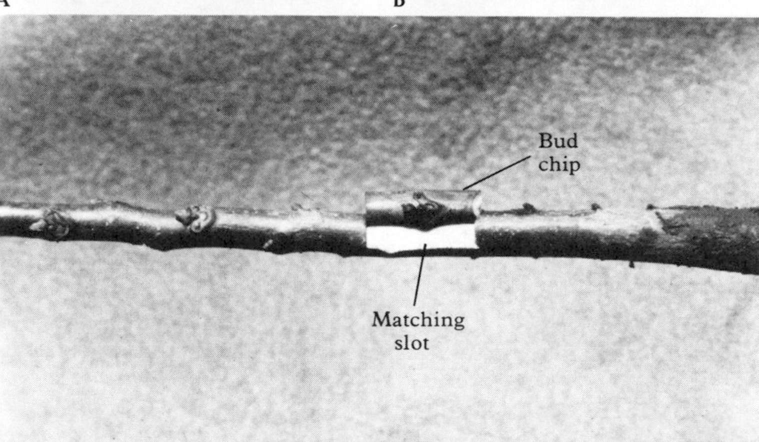

C

ROOTSTOCK PERFORMANCE

Root Function

Roots, unlike tops of plants, have no distinct period of dormancy and are able to grow whenever temperature, moisture, and other conditions are favorable. Also, they are not passive organs that many suppose them to be. They do not merely anchor the tree to earth and absorb water, which then passively carries mineral elements into the plant. The twenty-odd elements essential to life are obtained mainly from the soil by means of active transport mechanisms across the membranes of root cells (Epstein, 1973). This active transport of minerals from the soil solution into the roots is as important to life on earth as is the entry of carbon (in CO_2) from the atmosphere into the leaves. Metabolic energy is required for mineral uptake because the essential elements are accumulated in plants against a concentration gradient—that is, they must be transported from a low concentration in the soil solution to a higher con-

A

B

C

Figure 4-9 The three general types of graft union. **(A)** The stock and scion are of equal size. **(B)** The scion overgrows the stock. **(C)** The stock overgrows the scion. Although the scion overgrows the stock in most incompatible unions, a large number of overgrown unions are perfectly compatible. Overgrowth or undergrowth of the scion is more related to the genetic tendency for growth than to compatibility.

centration in the plant. (Normal passive diffusion in nature occurs from an area of high concentration to an area of lower concentration.) A large number of factors influence root function, some of which are soil O_2 and CO_2 content, soil moisture, temperature, biotoxins, chemical residues, compaction, acidity, extent of mycorrhiza, the specific genetics of the root, and the special physiology of the cultivar-rootstock compound genetic system. The genetics of the root are important because they determine the extent of growth control; tolerance to different soil types and environments; resistance to soil-borne diseases, insects, and pests; graft compatibility with the scion; nutrient uptake and balance; and ultimately such factors as fruit quality and yield efficiency (Chaplin, et al., 1972, Westwood, et al., 1973). These factors will be discussed under the sections dealing with specific rootstocks.

Root distribution is affected by soil conditions as well as genetics. A single root system of a winter-rye plant, excluding root hairs, was found to total 623 kilometers in length, with about 237 square meters of exposed surface. When the root hairs were added in, the total length was 11,263 kilometers and the area was 650 square meters (Epstein, 1973). Studies of tree roots at East Malling, England, indicate that there is a general pattern of distribution that may be altered by soil type, temperature, moisture, etc. (Barlow, 1959). In general, the root system was found to consist of a system of nearly horizontal main roots, some of which were more sloping, at a depth of 25 to 50 centimeters, and "sinkers" descending vertically 1.5 to 3.0 meters, depending on soil conditions. In a coarse sand that was low in nutrients, few roots went deeper than .9 meter, with three-quarters of the roots in the top 30 centimeters. In another soil, poor drainage limited root depth to .9 meter. The ratio of branch weight to root weight is

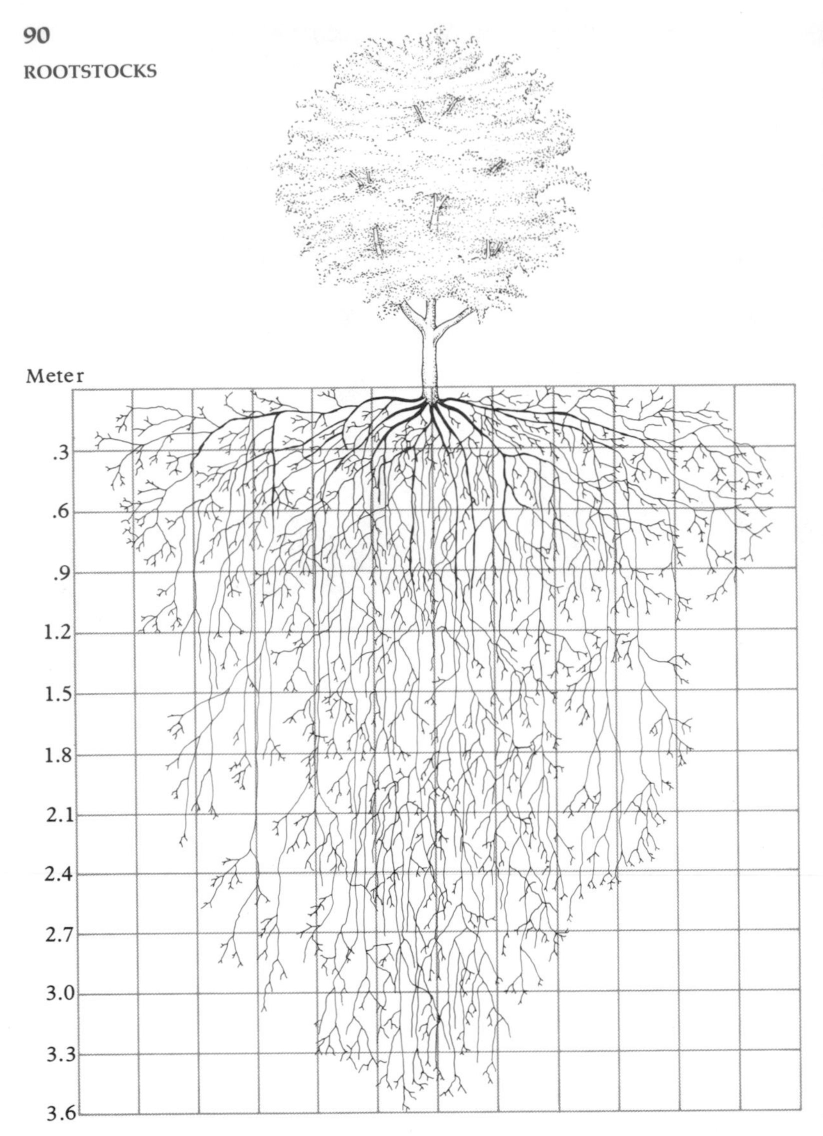

Figure 4-10 Diagrammatic representation of root development of an apple tree in good soil. Roots normally will spread two to three times as far as the top. Downward growth of roots would stop at a permanent water table or at an impervious layer of subsoil. [After Emanuel Epstein, "Roots." Copyright © 1973 by Scientific American, Inc. All rights reserved.]

nearly constant for different fruit plants under the same set of conditions, regardless of cultivar or rootstock. Thus the ratio for dwarf trees was similar to that of large trees under the same conditions. As optimum soil conditions were approached (rich, moist, well-aerated, deep loam), less root weight was needed to support a unit of branch weight. Often the dry weight of the aboveground tree is more than double that of the root. Generally, however, roots spread farther than branches—one and one-half times in clay, twice as far in loam, and three times as far in sand (Fig. 4-10). Even in clay, old trees may have roots spreading two to three times that of the branches. The effect of soil type and internal drainage on root distribution is shown in Figure 4-11. The ash-gray soil (B) limits root growth, because of excess water, and is unsuited for orchards.

Under a permanent cover crop, tree roots grow just below the sod and send up feeder roots into the organic residue at the soil surface, often obtaining nutrients not readily available in the mineral soil below. Also under straw mulch, roots send up rootlets into the rich topsoil and even into the mulch itself. This is due partly to the more favorable moisture at the soil surface, where mulches prevent surface drying.

Root growth can occur at any time of the year, but winter growth is limited by temperature and by waterlogging (Harris, 1926). In early spring, roots grow rapidly as the temperature rises above 6°C. Where winter growth is stopped by low temperature, spring growth under favorable conditions may equal the combined winter and spring growth in areas where winters are mild enough for growth. Root growth is reduced by dry soil and by rapid shoot growth. Roots are of three main types: (1) main roots (white in color at the ends), which extend the root system into new soil, (2) lateral roots arising from main roots and growing to several inches in length, and (3) profuse root hairs .5 to 2.5 millimeters long, which form just back of the growing tips of the other roots. The latter take up water and nutrients from the soil, although water may also enter through the larger roots. Each root hair exudes a tiny droplet of fluid, mostly water, which may coalesce into larger drops. The function of these droplets is not known but may aid in the uptake of some elements from dry soil. The main white extension roots are formed in greatest numbers from April to October and may grow as much as 5 centimeters per week. In midsummer the main-root cylinder back of the white zone turns brownish, and the outer portions may slough off,

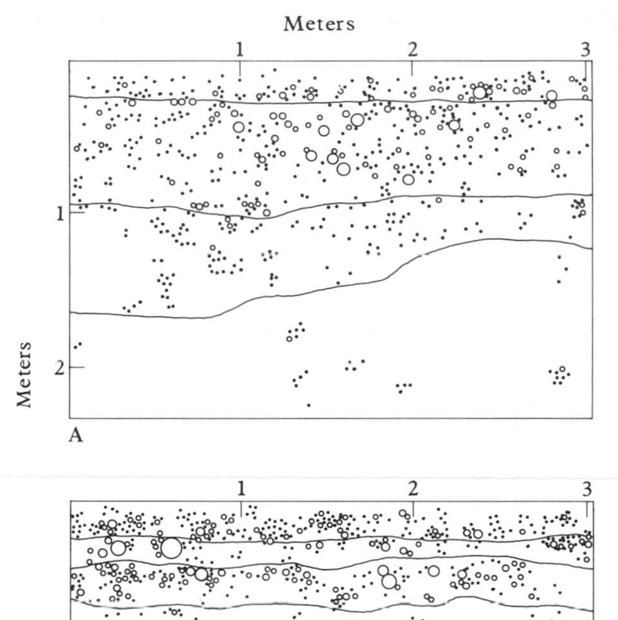

Figure 4-11 Apple root distribution **(A)** in an open, brown, gravelly, sandy loam soil, and **(B)** in a poorly drained, ash-gray loam. The different sized circles represent root sizes as follows: black dot, 0–2 mm; smallest open circle, 2–5 mm; next larger circle 5–10 mm; and so on, in additional gradations of 10 mm. [After Oskamp, 1935]

aided by microorganisms and by diameter growth. It is from these roots that lateral roots arise, 5 to 8 centimeters in length or longer.

Growth Control

The dwarfing rootstock does not necessarily have a shallow root system as earlier supposed. Dwarf apple stock M 9 was found to root as deeply as did vigorous stocks in the same soil. However, some dwarf trees are poorly anchored because their roots are brittle and tend to break under the pressure of wind from one side. Reciprocal grafts of vigorous and dwarfing types indicate that for every grafted system there is a proper balance of top and root that is characteristic of the whole system and not simply of the root.

Tolerance to Different Soils

A key factor in root growth is the extent of soil aeration at different depths. Low O_2 and high CO_2 can reduce or stop root growth and inhibit nutrient uptake. Boynton (1939) found that at soil temperatures of 12.7–21.1°C there was a marked reduction in new rootlets formed on apples when the soil O_2 fell below about 15 percent. At an O_2 level of 10 percent and a CO_2 buildup of 5 to 10 percent, restricted root growth markedly reduced top growth. The small rootlets need a higher level of aeration to function normally than do larger roots. Survival of trees in waterlogged soils is mainly due to a tolerance to low O_2. Plum roots tolerate waterlogging much better than peach or apricot, both of which are injured by hydrolytic release of hydrogen cyanide (HCN) from their own tissues under low O_2 conditions (Rowe and Catlin, 1971). Some species, such as citrus, tend to produce H_2S in their roots under anaerobic conditions, causing injury. Apples generally tolerate wet soils better than do stone fruits, and pear roots are the most tolerant of all deciduous fruits.

Soils may become reservoirs for herbicide and pesticide chemical residues and as such may present special problems for rootstocks. Whether a species will tolerate or be injured by a specific chemical residue depends upon the genetics of the root and the microclimate of the soil. Little is known in this important area of study. Even essential elements, such as copper, boron , manganese, and calcium, can be toxic to roots if they are highly concentrated in the soil. (Some of these problems will be discussed in relation to specific rootstocks.)

Soil Pests

Soil fumigation sometimes is used prior to planting, especially where the land was previously in fruit crops. Soils contain many potential plant pathogens, among which are the Phycomycetes (such as the *Pythium* and *Phytophthora* species, which cause root rot and damping off), bacteria (such as *Agrobacterium*, which causes crown gall and *Verticillium* species, which cause wilt), and pathogenic roundworms, or nematodes, which cause root knots and lesions on many plants. Soil fumigation with methyl bromide, chloropicrin, and other organic biocides often improves the growth of plants during the first year or two. These fumigants are toxic to roots, however, so fumigation should be done enough in advance of planting to permit dissipation of the toxicant.

Healthy plants growing in good soil are found to have abundant mycorrhiza—symbiotic fungi that grow on or in the plant root. The plant supplies the fungus with organic food and the fungus aids the plant in mineral nutrient uptake. Cranberry roots seem never to be free of mycorrhiza. Recently, Benson (1972) found that apples planted in old orchard land with a high arsenic content had relatively few mycorrhiza. As yet the importance of mycorrhiza in fruit growing is unclear. It may be that healthy trees have mycorrhiza because they are healthy rather than the reverse. We know, for example, that earthworms are abundant in a good soil, yet putting earthworms in a bad soil does not make it into a good one.

One of the best ways to cope with soil diseases and pests is to select or breed for genetic resistance. This avoids the problem of control by chemicals, which often creates the added problem of environmental pollution. Resistance to some pests has been found, for example, root aphids of apple and pear (Westwood and Westigard, 1969), nematodes of pear and stone fruits, *Phytophthora* of pears, and so forth. These pests will be discussed under headings that name specific rootstocks.

Yield Efficiency

The rootstock type profoundly affects the performance of a given cultivar. As will be shown, there can be as much as 50 percent or more difference in yield of a given cultivar grown on different rootstocks. Not only yield per tree but yield efficiency, that is, yield per unit of tree size, varies greatly with rootstock. This has been shown for apples, pears, cherries, plums, walnuts, and other fruits and nuts. The reasons for such effects on yield are not always apparent, but they often can be traced to differences in tolerance to adverse soils, in resistance to pests, or in uptake of nutrients. Other effects on physiological balance between root and top, which affects yield, are not as easy to define but are nevertheless inferred by considerable evidence. Some such effects are on flower initiation, some are on fruit set, and others are on fruit growth and ultimate size.

As Chapter 10 explains, cropping efficiency is measured by yield per unit of tree size (expressed as kilograms of fruit per square centimeter of trunk cross-section) or by yield per hectare. Efficiency is related not only to cultivar, climate, and culture, but to the kind of rootstock used. Specific rootstock effects have been noted for apple, pear, sweet cherry, sour cherry, prune, filbert, and walnut. Being one of the fixed factors that cannot be changed without replanting the orchard or vineyard, the efficiency of a stock should be known before the choice is made.

Fruit Quality

Rootstocks can greatly influence fruit quality—as was mentioned above, incompatibility produces black end of pear and makes the fruit useless—but usually the influence is not so dramatic. Effects of rootstock on quality are not always the same with different cultivars. For example, when *Pyrus betulaefolia* root is used for Anjou pear, the resulting fruit have internal necrotic spots in the flesh (a condition called "cork spot"), which lowers the grade of fruit. When the same rootstock is used for Seckel, large, high-quality fruit is produced. The most common rootstock influences on fruit quality are differences in firmness, in levels of organic acids, and in sugar content. The balance of these three factors tends to change the flavor and texture. Apple scions are usually propagated on rootstocks of their own species (*Malus pumila*), so differences in quality are minimal. Plum and prune stocks may induce different amounts of sugar and acids, but some can affect the incidence of internal browning, stem-end shrivel, and gum spot of the fruit. The quality of cherries, peaches, and nuts is not known to be affected greatly by rootstock, even though species other than the scion species are sometimes used.

TYPES OF ROOTSTOCKS

Apple Rootstocks

The specific characteristics of clonal stocks must be known in order to use them to advantage. Important traits to consider are growth control, tolerance to soil and climatic variables, resistance to insects and diseases, prococity and yield efficiency, anchorage, and ease of propagation. Throughout this book the rootstocks developed at the East Malling Research Station, England are designated M 1, M 2, etc., and those developed jointly by East Malling and Merton stations will be designated as MM 101, MM 102, and so forth. The MM series were specifically selected for resistance to woolly aphid (*Eriosoma lanigerum*) (H. B. Tukey, 1964).

GROWTH CONTROL The approximate relative size of trees on various rootstocks is shown in Figure 4-12. A recent introduction (not shown) is M 27, which is only about half the size of trees on M 9. The absolute size of the mature tree on a given root is determined by soil, climate, culture, and the scion cultivar. The inherent vigor of a cultivar will carry through regardless of the rootstock. Thus vigorous types like Gravenstein and Mutsu will be perhaps twice as large on M 9 dwarf stock as will Jonathan or Golden Delicious. Compact or spur mutants are much smaller on dwarfing stocks than are the parent cultivars. Shallow soils and those low in fertility cause trees generally to be smaller than average for a given stock.

ANCHORAGE The most dwarfing of the apple stocks (M 27, M 9, M 26, M 7, MM 106) usually require staking or trellis wire supports, particularly in the early years and where heavy annual cropping is desired. Their poor anchorage is due to the brittle nature of the roots rather than to their being shallow-rooted. Tests show that in deep soils dwarf roots go just as deep as those of vigorous stocks. Some of the more vigorous stocks such as M 2, MM 104, and MM 109 are not as well anchored as seedlings.

SUCKERING Root sprouting or suckering is a problem with some of the clonal apple stocks such as M 9

Figure 4-12 Approximate relative size of apple trees on different rootstocks. The clonal stocks originated in Canada (Robusta-5), Sweden (Alnarp-2) and England (M and MM series). The M 27 stock, not shown, is even more dwarfing than M 9.

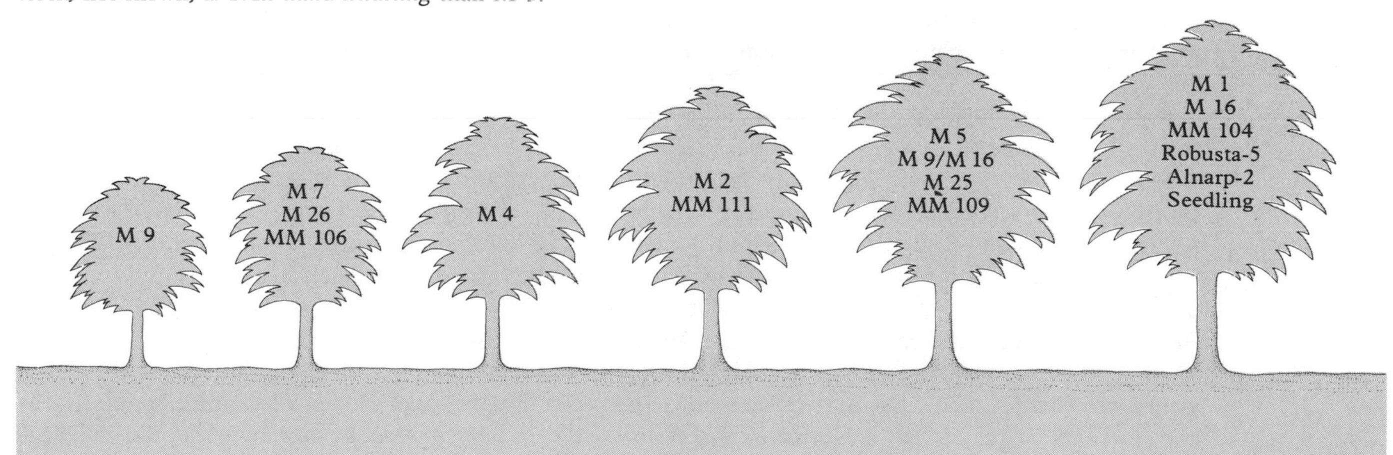

Table 4-3 Tree size (cross-section in square centimeters) at fourteen years, and annual yield in metric tons per hectare (years 10–14) for six apple cultivars grown on eight rootstocks.

Rootstock	Jonathan		Delicious		Golden Delicious		Gravenstein		Rome Beauty		Newtown	
	Trunk Size	Yield	Trunk Size	Yield	Trunk Size	Yield	Trunk Size	Yield	Trunk Size	Yield	Trunk Size	Yield
M 9	92	30.4	95	33.4	124	38.5	287	29.0	117	51.6	129	29.2
M 7	212	28.1	280	25.6	335	40.4	486	30.2	238	45.0	238	30.0
M 4	272	34.2	316	31.8	321	49.0	460	34.8	266	64.3	—	—
M 5	319	27.3	422	37.8	479	41.3	787	34.7	404	53.3	420	28.5
M 2	347	30.2	370	28.3	512	43.8	697	19.0	354	37.6	302	33.4
M 1	366	31.8	435	26.9	432	50.7	954	19.2	374	53.3	405	21.7
M 16	408	28.2	532	27.0	604	51.2	1002	22.0	423	53.3	490	25.2
Seedling	408	31.8	492	22.4	581	47.8	677	24.8	453	44.0	456	19.7

SOURCE: Unpublished data of A. N. Roberts and M. N. Westwood, Corvallis, Oregon.

and M 7. This problem is partly overcome by budding the variety high in the nursery so that the trees can be planted deeper. Planting too deeply in tight, fine-textured soil can lead to poor growth and survival. Care must always be taken to get the bud or graft union 10 to 15 centimeters above ground so that scion rooting above the union cannot occur. Should scion rooting occur, the tree will no longer remain a dwarf and will grow into a full-sized tree in a few seasons. The semidwarfing MM 106 does not tend to sucker and is more desirable than M 7 in that regard.

Table 4-4 Effect of rootstock on bloom, tree size, and yield efficiency of 5-year Starking Delicious apples.

Rootstock	Bloom, 5th Year (%)	Trunk Size (cm²)	Yield Efficiency (g/cm²)
M 9	88	24	1725
Sikkimensis	40	39	409
M 7	78	41	1135
M 4	53	53	636
MM 111	20	58	363
MM 109	33	58	636
M 9/M 16	46	63	590
M 2	41	63	545
M 25	26	63	545
M 1	50	68	772
MM 104	10	72	182
M 16	9	72	272

SOURCE: Roberts and Mellenthin, *Western Fruit Grower* 18: 26–27, 1964.

PERFORMANCE TESTS In a fourteen-year test (Table 4-3), the dwarfing stocks generally were more efficient than seedlings. In a number of cases, the more vigorous clonal stocks (M 5, M 2, M 1, M 16) also were more efficient than seedlings during the last five years.

Roberts and Mellenthin (1964) followed this first test with another test of twelve rootstocks (Table 4-4). Growth was about as expected, but growth control was not a good indicator of precocity. Only four rootstocks induced 50 percent or better bloom in the fifth year. As usual, M 9 was more efficient than larger trees, but M 7 was far more efficient than *Malus sikkimensis,* although the trees were about the same size. Likewise the vigorous trees on M 1 root were more efficient than similar sized trees on M 16 and MM 104. At present in the Northwest it appears that M 9, M 26, M 7, M 4, MM 106, MM 109, M 2, M 25, and M 1 are good stocks for their size. M 26 has performed well under irrigation, but is reported to be susceptible to drought. Each area has its own set of limiting factors. Cold hardiness of rootstocks will be discussed in detail in Chapter 15.

Pear Rootstocks

In contrast to apple rootstocks, which are all *Malus pumila,* pear rootstocks may be of several different species of *Pyrus,* and a few are even in a different genus. For this reason, more problems of graft incompatibility and fruit quality exist with pear. As noted above, pear roots are generally more tolerant of poorly drained, fine-textured soils than any other tree fruit. They grow and produce well in clay soils, which are unsuited to apple or stone-fruit production. Because of the more diverse genetic characteristics of pear stocks, there is wide variation among them in the uptake of mineral nutrients. Relative to the common domestic seedling stock *(Pyrus communis)*, *P. betulaefolia* absorbs more nitrogen, while *P. calleryana* and quince *(Cydonia oblonga)* absorb less nitrogen. Trees on quince are consistently higher in magnesium and lower in boron than those on *P. communis*. Recent studies indicate that the use of *Sorbus aucuparia* root results in as high as 700 ppm manganese compared to the usual 70 ppm with *P. communis*. Trees on the Korean pea pear *P. fauriei* are intermediate, with about 120 ppm manganese. Specific clones of *P. communis* selected for resistance to fire blight also have shown differences in mineral uptake.

PEST AND DISEASE RESISTANCE The new disease called "pear decline" has eliminated susceptible *P. ussuriensis* and *P. pyrifolia* stocks. Pear decline is an induced scion/rootstock incompatibility caused by a mycoplasma transmitted to the trees by pear psylla insects. The pathogen migrates through the phloem downward to the union. If the rootstock is susceptible, the phloem just below the union is killed, effectively girdling the trunk. In time, the roots starve and the top then declines or wilts and dies. Seedling stocks appear variable in their tolerance to decline, except those of *P. betulaefolia* which all seem resistant. The proportion of tolerant stocks in a seedling population varies with the species or seedling type (Table 4-5; see details in Chapter 16).

Fire blight, *Erwinia amylovora*, is the most serious disease of pear and has long confined pear production to the western states, where relatively dry summer weather helps prevent the spread of blight. Reimer (1925, 1950) identified several sources of blight resistance, among which are *P. ussuriensis*, *P. calleryana*, and *P. communis* (Old Home × Farmingdale, etc.). He also selected four *P. betulaefolia* seedlings that are resistant to blight. Cameron, et al. (1969) found that seedlings from these *P. betulaefolia* seedlings also are resistant, even though seedlings of unselected *P. betulaefolia* are usually susceptible. Table 4-6 provides some general information about the most

Table 4-5 General effect of rootstock on the incidence of decline in pear.

Rootstock	Tree Decline		
	Healthy	Moderate decline (%)	Severe decline or death
P. communis			
Bartlett seedling	69	21	10
Nelis seedling	83	11	6
Imported French seedling	70	15	15
Old Home × Farmingdale seedling	91	7	2
Old Home (self rooted)	98	1	1
Bartlett (self rooted)	79	13	8
Anjou (self rooted)	88	12	0
P. calleryana seedling	81	15	5
P. betulaefolia seedling	98	0	2
P. ussuriensis seedling	36	21	43
P. pyrifolia (Serotina) seedling	37	16	48
Quince			
Old Home/Quince A (well drained soil)	90	7	3
Old Home/Quince A (poor drainage)	69	12	19
Old Home/Quince A (very wet soil)	0	2	98
Old Home/Quince A (union deep, OH rooted)	91	9	0

SOURCE: Westwood and Lombard, 1966.

Table 4-6 Relative susceptibility of rootstocks to damage from various causes.

Rootstock	Fire Blight	Cold Damage	Pear Root Aphid	Nematodes
Quince	3	4	0	0
Imported French *Pyrus caucasica*	4	0	4	3
Bartlett seedling	4	0	4	3
Winter Nelis seedling	4	0	4	3
Old Home × Farmingdale seedling	2	0	2	3
Old Home × Farmingdale clones	0	0	1	3
Old Home	0	0	4	3
Bartlett cuttings	4	0	4	3
P. calleryana	0	4	0	0
P. betulaefolia (Reimer's)	0	2	0	3

SOURCE: Oregon Agr. Ext. Fact Sheet 61 by Stebbins et al., 1972.
KEY: 0 = Not susceptible, 4 = Highly susceptible.

common stocks being used for pear. *P. calleryana* appears also to be resistant to *Phytophthora* root rot, oak-root fungus, and crown gall. It is susceptible to lime-induced chlorosis but tolerates wet, poorly drained soils better than any other stock. Quince is susceptible to lime-induced chlorosis and is least tolerant of wet soils (see Table 4-5). Trees on quince root grown in poorly drained and wet soil declined because of the soil condition, while declining trees on *Pyrus* stocks suffered from pear decline induced by a mycoplasma.

GROWTH CONTROL Quince is the standard dwarf rootstock for pear, but recently an Old Home × Farmingdale clone (OH × F51) was selected as a dwarf *P. communis.* Other OH × F clones that are semi-dwarfing are numbers 217, 9, 87, 333, and 69 (Stebbins et al., 1972). All of these have been tested for blight resistance. Other growth-controlling stocks under test in Oregon are hawthorn (*Crataegus* species), mountain ash (*Sorbus* species), and *Pyrus fauriei.* The latter species is tolerant of very wet soils but will not stand prolonged drought.

Standard vigorous rootstocks are found among the domestic seedlings, *P. calleryana,* wild *P. communis* (caucasica) Old Home, and OH × F clones 18, 97, 136, and 340. *P. betulaefolia* is generally more vigorous than any other stock and is best used in poorer soils where adequate growth is hard to obtain. It is a deep-rooted type and as such is tolerant to dry soils, but it is susceptible to lime-induced chlorosis. *P. betulaefolia* should not be used as a rootstock for Anjou (see the discussion of fruit quality at the end of this section). Relative growth control of pear stocks is given in Figure 4-13.

COMPATIBILITY Quince and pear often are incompatible and must be double-worked with a compatible interstem. Old Home is better as an interstem than

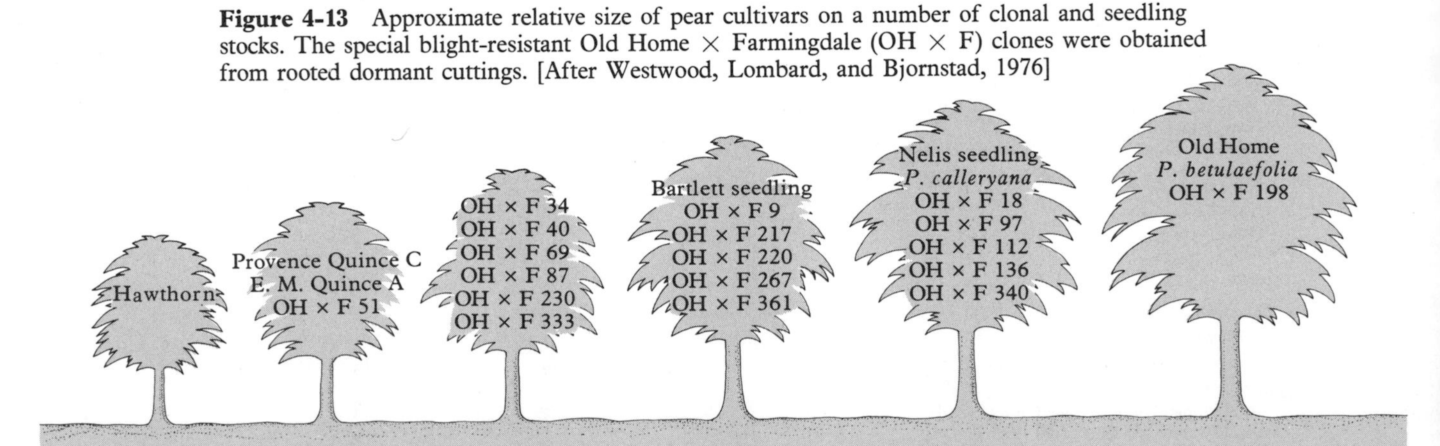

Figure 4-13 Approximate relative size of pear cultivars on a number of clonal and seedling stocks. The special blight-resistant Old Home × Farmingdale (OH × F) clones were obtained from rooted dormant cuttings. [After Westwood, Lombard, and Bjornstad, 1976]

Hardy, which was used for many years. The following list of cultivars indicates compatibility with quince:

Compatible	*Incompatible*
Anjou	Bartlett
Comice	Bosc
Hardy	Seckel
Packham's Triumph	Winter Nelis
Gorham	Eldorado
Flemish Beauty	Clapp Favorite
Maxine	Forelle
Old Home	Farmingdale

The length of the Old Home interstem does not seem to matter, and a very short piece is enough to form a compatibility bridge. Recently, it was found that a Bartlett mutant is compatible with quince. It is called Swiss-compatible Bartlett, P.I. No. 267940. Most cultivars are incompatible with hawthorn, but Old Home has made a suitable interstem for it as well as quince. However, most cultivars are generally more compatible with mountain ash than is Old Home. There is good graft compatibility among the *Pyrus* species, but *Pyrus* grafted to other related genera are often incompatible.

ANCHORAGE Trees on quince roots need support for best performance. The root system is much branched and fibrous, but large roots are brittle and tend to break under the weight of crop and wind. Other dwarfing stocks including hawthorn and mountain ash may be grown without support. However, high-density tree walls do better if supported in the early years by a one- or two-wire trellis, on which lateral branches can be tied at 45° angles both to induce early bloom and to support the crop in the first years of bearing.

SUCKERING Many good pear stocks tend to sucker (develop root sprouts); they include Old Home, some clonal OH × F, quince, imported French, *P. fauriei,* and, to some extent, domestic seedlings. Several OH × F clones that tend not to sucker have been selected.

PERFORMANCE Yield efficiency is good on quince rootstock and on OH × F51 in high-density hedgerows. The use of Old Home as a compatible interstem for quince has consistently resulted in higher yield than when Hardy is used. Of the vigorous stocks, *P. calleryana* induces early yield more than seedling *P. communis.* Clonal OH × F rootstocks, however, were selected for their high-yielding potential, so they are better than unselected seedlings of the species. *P. betulaefolia* appears to be too vigorous for Anjou and usually the yield efficiency is low. Yet *P. betulaefolia* is outstanding as a stock for Seckel and some other weak-growing cultivars.

FRUIT QUALITY Two fruit disorders related to rootstock are hard end or black end, caused by *P. ussuriensis* and *P. pyrifolia* (serotina), and cork spot of Anjou, caused by *P. betulaefolia* stock (see Chapter 13). Fruit quality is generally good on quince, *P. communis, P. calleryana,* and *P. betulaefolia,* except as noted above. Fruit firmness is greater at harvest and during storage when grown on the oriental stocks *P. calleryana, P. pyrifolia, P. ussuriensis,* and *P. betulaefolia* than when grown on quince or *P. communis.*

Flavor and eating quality are related to factors attendant to normal ripening (texture, juiciness, odor, etc.). In addition, the balance between organic acids and sugar determines whether a fruit will taste full-flavored, sweet, flat, or sour. Moderately high acids and sugar plus volatile esters result in full flavor. Low acids with moderate sugars result in flat or bland flavor, while high acids with moderate or low sugars tend to result in sour flavor. Effects of several rootstocks on firmness, acids, and sugar (soluble solids) are shown in Tables 4-7 and 4-8. Firmness and acid content are usually affected more by rootstock than are soluble solids. The principal fruit acids of pear are malic and citric, and these are the ones most affected by rootstock (Table 4-8). At this time the physiological mechanism by which rootstocks influence fruit is not known. Notice that P18 × F results in less of the major acids than either OH × F or French, even though all of these stocks are *P. communis.* Also, note that *P. pyrifolia* root induces lower acids, but that *P. calleryana* results in acid levels similar to *P. communis* (Table 4-7). Acid levels in fruits from *P. communis* varied from orchard to orchard, reflecting effects of soil, fertilizer and culture.

Table 4-7 Effects of several rootstocks and locations on quality factors of Bartlett, Anjou, and Seckel pears.

Location and Rootstock	Quality Rating (0 to 15)[1]	Flesh Firmness (kg)[2]	Acid Content (mg/100 ml juice)	Soluble Solids (%)
		Bartlett		
Medford Station				
Old Home × Farmingdale	11.5	9.72	388	15.0
P18 × Farmingdale	11.2	9.99	264	14.5
Imported French	11.4	10.3	316	14.8
Mira Vista				
P. calleryana	10.2	6.81	328	14.4
Imported French	9.0	6.40	249	13.7
Hill Orchard				
Quince	9.0	5.54	276	14.0
P. pyrifolia (serotina)	8.0	10.8	211	14.4
Imported French	7.9	6.81	281	14.5
		Anjou		
Bear Creek				
P. calleryana	8.5	6.17	184	
P. betulaefolia	6.7	6.63	154	
P. pyrifolia (serotina)	7.3	6.81	162	
P. ussuriensis	7.1	6.67	180	
P. communis (French)	8.1	5.90	194	
		Seckel		
Bear Creek				
P. calleryana	7.4	6.40	150	
P. betulaefolia	8.0	7.35	147	
P. pyrifolia (serotina)	7.2	6.76	140	
P. ussuriensis	7.1	5.22	174	
P. communis (French)	7.2	6.22	143	

[1]0 to 4 = Very poor, 11 to 15 = Very good.
[2]Firmness determined on a peeled fruit cheek with a U.S. (Magness-Taylor) tester fitted with a 5/16-inch head.

Table 4-8 Effect of rootstock on individual organic acids of Bartlett pear.

Location and Rootstock	Shikimic	Quinic	Malic	Citric	Tartaric	Total
	(mg acid per 100 g tissue)					
Medford Station						
P18 × Farmingdale	0	23	48	96	6	173
Old Home × Farmingdale	16	30	116	164	8	334
French seedling	14	27	78	139	6	264
Mira Vista						
P. calleryana	12	18	76	182	12	300
P. communis (French)	10	26	74	186	10	306
Hill Orchard						
P. pyrifolia (serotina)	18	46	83	98	16	261
P. communis (French)	18	41	115	127	11	312

Cherry Rootstocks

The stocks employed most commonly for propagating cherries are mazzard *(Prunus avium),* mahaleb *P. mahaleb),* and Stockton Morello *(P. cerasus).* In spite of the many tests conducted with cherries on these and other rootstocks, a considerable difference of opinion still exists among growers, nurserymen, and investigators as to which one is most satisfactory. Probably no one root is superior for all situations. Some of the controversy over cherry-rootstock compatibility and performance may be due to the presence of one or more viruses.

Much of the literature pertaining to the controversy over the relative merits of mazzard and mahaleb stocks was reviewed by Coe (1945). In his own trials in Utah, he found that mahaleb was superior to mazzard and morello for five different cultivars of sweet cherries. Coe's test plot was on a coarse, gravelly loam with good drainage and aeration, where the mahaleb-rooted trees outgrew and outproduced those on the two other stocks.

But other workers have reported that although sweets on mahaleb roots may be larger at first, mazzard-rooted trees usually overtake and surpass them in good cherry soils. Many growers prefer mazzard and believe that mahaleb has a dwarfing effect. In general, better results with mazzard are reported in the United States.

Mazzard is used more generally for sweets in England, South Australia, and Victoria. On the other hand, sour cherries *(P. cerasus),* particularly Montmorency, appear to do better on mahaleb than on mazzard. And in good cherry soils, mahaleb also is preferred to morello because of the dwarfing effect of the latter. In Vermont, however, Early Richmond on mazzard grew better and produced more fruit than similar trees on mahaleb.

The best rootstock for sweet cherries in California depends both upon local soil and climatic conditions and upon the cultural practices employed (Norton et al., 1963d). Mahaleb is the preferred stock for dry climates and sandy or gravelly soil, while Stockton Morello helps adapt sweet cherries to wet clay soils. If morello is to be used, it should be kept in mind it is a symptomless carrier for necrotic rusty mottle virus. In California, mazzard stocks appear to be intermediate between mahaleb and morello in their sensitivity to high water tables.

In recent years it has been recognized that certain clonal selections or strains of these stocks may have superior qualities that justify their propagation by vegetative means. Selection of the proper strain may thus be more important than a choice between mazzard, mahaleb, or morello.

For ease of propagation and economy of production, nurserymen prefer mahaleb rootstocks for cherries. It is a cheaper stock in the nursery than the mazzard and a better-looking tree can be delivered to the fruit grower. Moreover, mazzard seems to be more susceptible to aphid and leaf spot in the nursery and the seedlings are more tender to cold. Also, it is agreed generally that cherries on mahaleb are hardier in the orchard and less subject to winter killing than when rooted on mazzard. But this made little difference in the injury sustained by cherry orchards in the severe freeze of 1933–34 in New York State. In controlled experiments in the laboratory, mazzard roots were killed at $-10°$ to $-11°C$, whereas mahaleb roots were killed only at $-15°$. The opinion is expressed rather generally that mahaleb is more resistant to drought than mazzard, though no actual experiments have been performed to show this. It is known that mahaleb is more resistant to zinc deficiency than is mazzard.

The relation of rootstocks to nutrition has been reviewed recently by Westwood and Bjornstad (1970a). They report that Badanin found that a high tannin content of root bark correlated well with chlorosis resistance, and that organic fertilizers, $(NH_4)_2SO_4$, or $FeSO_4$ (as a spray) helped to correct chlorosis. Malycenko using sour cherry on mahaleb rootstock, found that injured roots regenerated more readily when phosphorus and potassium fertilizers were applied early in the growing season. Kirkpatrick reported that the kind of rootstock for Montmorency influenced the uptake of mineral nutrients. Potassium uptake was less with mahaleb than with mazzard root. Hintze found that F12/1 mazzard rootstock (introduced by East Malling Research

Station) was best and that both potassium and nitrogen fertilizers were needed for highest performance. The matter of compatibility of stock and scion is one of prime importance in grafting. This subject was reviewed thoroughly by Argles (1937).

Roberts (1962), found that differences exist between different clones of mahaleb with regard to growth, flowering, and yield of Montmorency tart cherry. Tree size ranged from very vigorous (P.I. 193703) to dwarf (P.I. 193695). Trees on F12/1 were slightly more vigorous than the most vigorous mahaleb clone. Yield was greater with several of the semidwarfing stocks than with either the dwarf or vigorous stocks. Although Montmorency on the vigorous F12/1 mazzard was late coming into bearing, these trees high-worked on F12/1 branches are very strong and withstood 100 mile-per-hour winds in 1962. Trees low-worked (at ground line) in the same plot were torn apart by the wind.

Table 4-9 gives some general characteristics of the three main stocks. Sweet cherry cultivars were compared on F12/1 and mazzard seedling by Roberts (1962), who found that the two are similar but that F12/1 is slightly more vigorous. The F12/1 root and frame stock does very well for sweet cherry in the Willamette Valley. Except for the Corum cultivar, low-worked trees do poorly because of infection by bacterial gumosis, *Pseudomonas syringae,* in trunks and crotches. F12/1 frame stock is fairly resistant to gumosis and protects Napoleon (Royal Ann) and other susceptible cultivars from this disease. Mahaleb stock is not well adapted to clay loam soils but is satisfactory in more coarse-textured, well-drained soils. Stockton Morello rootstock is a clone of *P. cerasus* and is the best dwarfing stock now known for sweet cherry. Trees on this stock will begin to bear earlier than those on mazzard stock. They are also more tolerant of heavy, poorly drained soil than other

Table 4-9 Characteristics of cherry stocks in California.

Trait	Rootstock Type		
	Mazzard	Mahaleb	Stockton Morello
Soil preference	Sandy loam	Sandy; does not tolerate wet soils	Loam or clay loam; most tolerant of wet soils
Compatibility	Good	Some overgrowth if grafted high above ground	Scion overgrowth but strong; some cultivars weak
Anchorage	Very good; shallow roots injured by tillage	Very good; roots deeper than others	Good; roots smaller in diameter than others
Peach tree borer	Less susceptible than Mahaleb	Moderately susceptible	Moderately resistant
Pocket gophers and mice	Moderately resistant	Highly susceptible	Moderately resistant
Oak root fungus	Moderately resistant	Susceptible	Susceptible
Crown rot	Susceptible	Susceptible	Moderately resistant
Crown gall	Moderately susceptible	Moderately resistant	Moderately resistant
Verticillium wilt	Susceptible	Susceptible	Susceptible
Buckskin virus	Susceptible to both strains	Susceptible to Napa Valley strain	Susceptible to both strains
Bacterial canker	Variable resistance	Root resistant	Root resistant
Root-knot nematodes			
M. incognita	Immune	Resistant	Immune
M. javanica	Resistant	Susceptible	–
Root-lesion nematodes			
P. vulnus	Susceptible	Moderately resistant	Susceptible
Nutritional		Less susceptible to Zinc deficiency than Mazzard	
General	Preferred stock in soils too heavy for Mahaleb	Trees slightly smaller, bear earlier and heavier than Mazzard	Tree semi-dwarf, earliest bearing; trees hard to establish

SOURCE: Norton et al., 1963d.

stocks. The resistance of Stockton Morello to bacterial gumosis makes possible its use in moist marine climates. However, high limb-worked trees with *P. cerasus* framework are weak and may split out under crop loads or wind if not cross-braced or supported. Low-budded Corum on Stockton Morello should result in a semidwarf that is healthy and early bearing. This stock results in a tree about one-third to one-half the size of mazzard. Yield during the first eight years has been more efficient than with standard stocks. However, Napoleon seems to be less than completely compatible with morello, and anchorage on this stock is sometimes poor and trees may lean in the wind. A recent study (Ryugo, 1975) indicates that the *P. cerasus* stock Vladimir is more dwarfing than Stockton Morello and does not sucker as badly (Fig. 4-14). Figures 4-15 and 4-16 indicate relative tree sizes for sweet and sour cherry rootstocks.

Figure 4-14 Sour cherry, *Prunus cerasus,* as a dwarfing stock for sweet cherry. This young sweet cherry on Vladimir sour cherry is both dwarfed and precocious. [From Ryugo and Micke, 1975]

Prune Rootstocks

Historically, prunes have been propagated mainly on peach seedling rootstocks. For years seedlings of the freestone cultivar Lovell have been used. Recently nurseries have also used seedlings of Halford peach. In California, several special peach stocks have been

Figure 4-15 Approximate relative size of sweet-cherry cultivars on a number of clonal and seedling rootstocks, including *Pr. avium, Pr. mahaleb, Pr. cerasus* and mazzard × mahaleb (M × M) hybrids.

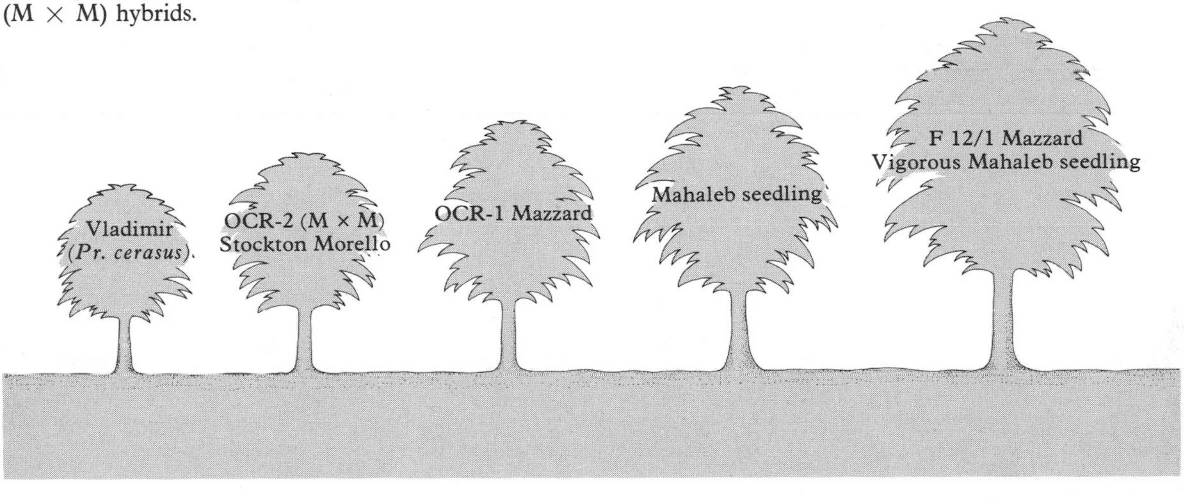

Vladimir (*Pr. cerasus*)
OCR-2 (M × M)
Stockton Morello
OCR-1 Mazzard
Mahaleb seedling
F 12/1 Mazzard
Vigorous Mahaleb seedling

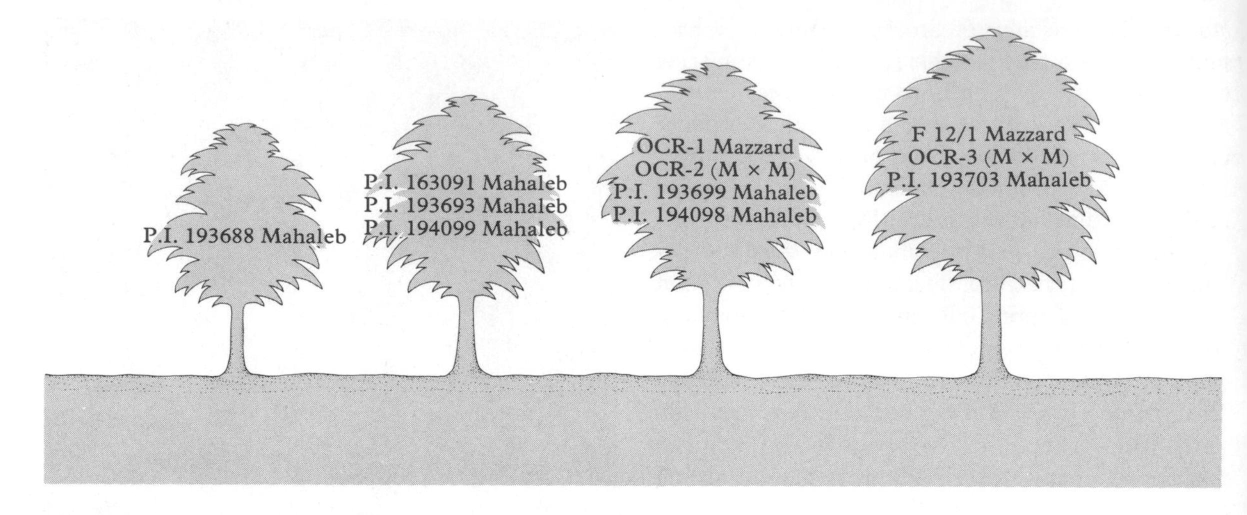

Figure 4-16 Approximate relative size of Montmorency sour cherry on a number of clonal stocks, including *Pr. avium* (mazzard), *Pr. mahaleb,* and hybrids of mazzard and mahaleb (M × M). [After Westwood, Roberts, and Bjornstad, 1976]

Table 4-10 Characteristics of several plum and prune rootstocks.

	Myrobalan *P. cerasifera*	Myrobalan 29-C	Marianna 2624 *cerasif × munsoniana?*	Peach seedling *P. persica* (Lovell)	Nemaguard seedling (peach)	Rancho resistant seedlings (peach)	S-37 peach seedlings	Almond *P. amygdalus*	Apricot *P. armeniaca* Royal seedlings
Soil preference	Wide adaptation	Wide adaptation	Wide adaptation	Sandy loam	Sandy loam	Sandy loam	Sandy loam	Sandy	Sandy loam
Compatibility	Most cultivars	Most cultivars	Most cultivars	Most cultivars[1]	Most cultivars	Most cultivars	Most cultivars	Incomp. some cultivars	Incomp. some cultivars
Anchorage	Good	Shal.[2] rooting	Shal.[2] roots	Good	Good	Good	Good (in U.S.)	Very good	Good
Peach tree borer	Sus.	Sus.	Sus.	Very sus.	Very sus.	Very sus.	Very sus.	Sus.	Mod. res.
Pocket gophers and mice	Sus.	Sus.	Sus.	Mod. res.	Mod. res.	Mod. res.	Mod. res.	Very res.	Very sus.
Oak root fungus	Sus.	Sl. res.	Mod. res.	Sus.	Sus.	Sus.	Sus.	Sus.	Sus.
Crown gall	V. sl. res.	Sl. res.	Mod. res.	Sus.	—	—	—	Very sus.	Mod. res.
Crown rot	Mod. res.	Mod. res.	Mod. res.	Sus.	Sus.	Sus.	Sus.	Very sus.	Sus.
Verticillium wilt	Mod. sus.	Mod. sus.	Mod. sus.	Very sus.	Very sus.	Very sus.	Very sus.	Very sus.	Very sus.
Bacterial canker	Gen. sus.	Mod. sus.	Very sus.	Mod. res.	—	—	Sus.	Mod. res.	Sus.
Nematodes									
Meloidogyne incognita	Gen. sus.	Immune	Immune	Sus.	Very res.	Immune	Immune	Sus.	Mostly immune
M. javanica	Gen. sus.	Immune	Immune	Sus.	Res.	Sus.	Sus.	Sus.	Mostly immune
Pratylenchus vulnus	Sus.	Sus.	Sus.	Sus.	—	—	Sus.	Sus.	Res.
Nutritional problems	Res. +B							Res. +B	Sl. res. chlorosis

SOURCE: Norton et al., 1963c.
KEY: Sus. = Susceptible, Res. = Resistant, Sl. = Slightly, +B = Boron toxicity.
[1] Not compatible with Robe de Sergeant and Sugar.
[2] Shallow roots first three or four years.

used because of their resistance to nematodes. Seedlings of apricot, almond, and plum also are used there in some orchards.

Table 4-10 indicates some of the known characteristics of plum and prune rootstocks. Soil preference of these stocks is related to texture as well as drainage. Myrobalan and marianna plum stocks tolerate heavy soils and poor drainage better than do peach and apricot stocks. But even plum stocks do not do well in very fine-textured (heavy) and wet soils. Peach, almond, and apricot roots grow better in coarse-textured, well-drained soils. Almond root is very sensitive to wet soils. Almond stocks are the most resistant to excess boron in the soil, followed by myrobalan plum. Conversely, peach and apricot stocks are most efficient in taking up boron from boron-deficient soils. Almond is most resistant to lime-induced chlorosis, and peach is least resistant. Generally trees on peach root develop less bacterial canker of the trunk than those on plum root (Norton et al., 1963c). The nature of this influence is poorly understood. It might result from nutritonal differences related to rootstocks.

When new trees are to be replanted in old orchard land where trees with peach rootstocks were planted, new peach stocks should not be used. A toxin is released in the soil by old peach roots that inhibits the growth of new peach roots on replants. A suitable plum rootstock will grow better than peach in such situations.

Recently, eleven trials with Italian, Early Italian, and Brooks prunes, started between 1963 and 1966, indicate several differences in nutrient uptake as related to rootstock (Table 4-11). Trees with plum rootstocks had greater leaf nitrogen, potassium, manganese, and zinc and slightly less boron and magnesium than those on peach. Plum clones, Myrobalan 29-C, Myrobalan B, and St. Julien A, were more efficient in the uptake of calcium. There were positive correlations between nitrogen and calcium; nitrogen and magnesium; nitrogen and boron; nitrogen and zinc; calcium and magnesium; calcium and boron; and magnesium and boron for most of the stocks. There was a negative correlation between potassium and magnesium for Myrobalan 2–7 and the three marianna clones. Myrobalan B and Marianna 2623 and 2624 had a negative correlation for potassium and calcium, whereas St. Julien A had a positive correlation. Fruit samples taken from some of the plots indicated that those from myrobalan were slightly firmer and had slightly less soluble solids than did those from other stocks. This probably indicates a slight difference in maturity.

Table 4-12, which uses peach rootstocks as a standard, shows that trees on all myrobalan clones are

Table 4-11 Effects of rootstock on nutrient levels in leaves of Italian prune.

Rootstock	(percent dry wt.) N	K	Ca	Mg	(ppm) Mn	B	Zn
Peach	2.13	1.98	1.79	.46	55	36	17
Myrobalan 29-C	2.55	2.26	1.76	.37	90	33	24
Myrobalan B	2.47	2.00	2.23	.53	80	33	21
Myrobalan 2-7	2.46	2.58	2.20	.47	88	34	19
Marianna 4001	2.49	2.03	1.84	.44	90	34	22
Marianna 2623	2.30	2.22	1.92	.44	78	34	20
Marianna 2624	2.36	2.17	1.79	.43	118	32	22
St. Julien A	2.49	2.82	2.02	.37	89	33	20

Source: Chaplin et al., 1972.

Table 4-12 Effect of rootstock on tree size and yield (1969 and 1971) of Italian prune (5 to 8 years old).

Rootstock	Tree Size as % of Peach Percent	Yield (mT/ha)	Yield as % of Peach Percent	Yield Efficiency Index[1]	Tree Survival (%)[2]
Peach	(100)	17.0	(100)	333	91
Myrobalan 29-C	120	15.7	92	256	88
Myrobalan B	162	25.1	148	300	93
Myrobalan 5Q	111	14.1	83	240	98
Myrobalan 2-7	138	22.2	130	286	84
Marianna 4001	142	26.2	154	373	92
Marianna 2623	105	17.7	104	336	86
Marianna 2624	90	14.8	87	320	77
Marianna 2625[3]	88	16.6	98	366	79
St. Julien A	102	17.3	101	313	79
Brompton[3]	94	9.4	55	193	73
Michaelmas[3]	110	17.5	103	330	76
Damas C[3]	132	28.0	165	460	100
Common Mussel[3]	124	28.7	169	513	58
Ackermann's Marunke[3]	82	7.2	42	193	81

Source: Westwood et al., 1973.
[1] Yield per unit of trunk cross-sectional area (g/cm^2).
[2] Peach is used as the standard for survival.
[3] Represented by only one or two plots. Others from 4 to 7 plots.

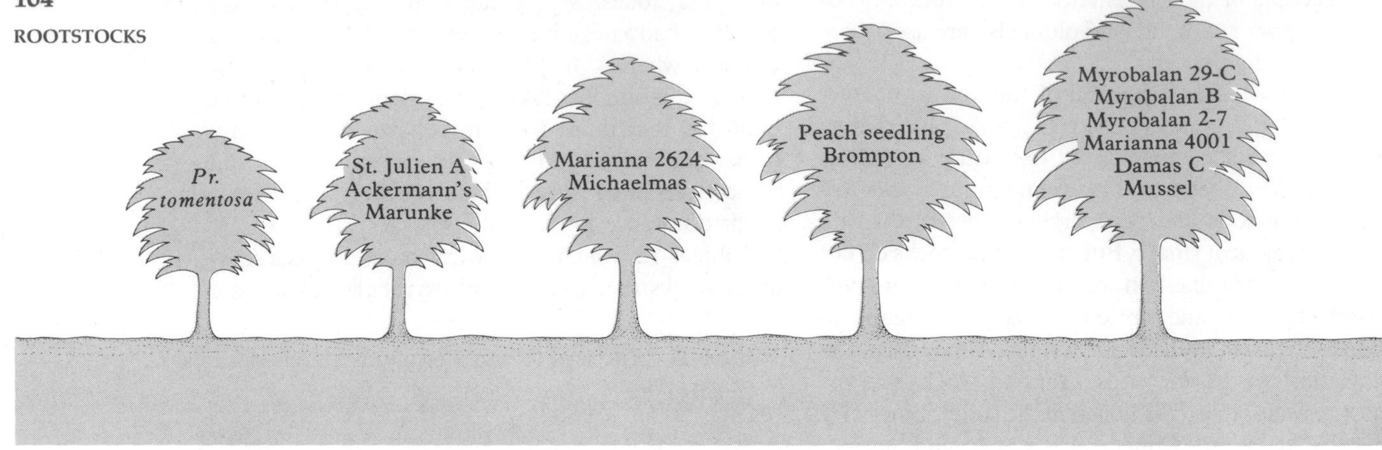

Figure 4-17 Approximate relative size of plum and prune cultivars on a number of rootstocks. The most common stocks used are peach, myrobalan, and marianna. The dwarf *Pr. tomentosa* stock induces a high incidence of *Pseudomonas* canker on the scion trunk. [Westwood et al., 1973]

larger than peach and that three of four marianna and six other plum clones produce trees smaller than peach. (Relative tree size with different plum stocks is shown in Figure 4-17.) Note that some clones resulted in more yield and some less than peach; also note that yields were not primarily related to tree size. This is best seen by looking at yield efficiency, that is, yield per unit of tree size. Of the fully replicated plots with myrobalan and marianna clones, only Marianna 4001 was more efficient than peach. Of the partially replicated rootstocks, Marianna 2625, Damas C, and Common Mussel were appreciably better than peach and deserve further study.

In recent years, losses of young trees from bacterial canker have been observed. The survival rates shown in Table 4-12 reflect all losses, including those from canker. Cameron (1971) reported that several plum stocks resulted in more bacterial canker of Italian trunks than peach root. Yet the general rate of tree survival is about equal with peach, myrobalan, and marianna stocks. But myrobalan seedling, Common Mussel, and tomentosa stocks did not survive as well as others. These three were shown by Cameron (1971) to have a high incidence of trunk canker. Conversely, the stocks showing the least canker generally survived better. Yet Myrobalan 29-C and Marianna 4001, which are susceptible to canker, actually survived as well as peach, which is least susceptible. Apparently trees on some stocks are capable of outgrowing trunk canker better than others. Marianna 4001 was the only stock more vigorous than peach that also was more efficient in producing fruit than peach. Its high survival rate (92 percent) warrants its further use in commercial trials.

Peach and Apricot Rootstocks

As indicated in Table 4-13, peach cultivars do best on peach rootstock but at times are worked to apricot or almond seedlings. All listed peach stocks are sensitive to wet, poorly drained soils. Even though apricot and almond roots are more resistant to lime-induced chlorosis than peach, neither are recommended because of questionable compatibility with peach (Norton, et al., 1963b).

Other stocks used to dwarf peach trees are *P. tomentosa* and regular peach seedling root with an inter-

Table 4-13 Characteristics of peach and apricot rootstocks in California.

Rootstock	Soil Preference	Compatibility	Anchorage	Peach Tree Borers	Gophers and Mice	Oak Root Fungus	Crown Rot	Crown Gall	Verticillium Wilt	Bacterial Canker	Nematodes: *M. incognita*	Nematodes: *M. javanica*	Nematodes: *Pratylenchus vulnus*
For Peach Scions													
Peach													
Lovell sdlg.	W.D. S. loam	V. good	Good	Sus.	Mod. res.	Sus.	Sus.	Sus.	Sus	Sus.	Sus.	Sus.	Sus.
Nemaguard sdlg.	W.D. S. loam	Good	Good	Sus.	Mod. res.	Sus.	Sus.		Sus.		Res.	Res.	
Rancho Res. sdlg.	W.D. S. loam	Good	Good	Sus.	Mod. res.	Sus.	Sus.		Sus.		Res.	Sus.	
S-37 sdlg.	W.D. S. loam	Good	Good	Sus.	Mod. res.	Sus.	Sus.		Sus.	Sus.	Res.	Sus.	Sus.
Okinawa sdlg.	W.D. S. loam	Good	—	—	—	—	—		—		Res.	Res.	
Apricot													
Royal sdlg.	W.D. S. loam	Fair–poor	Good	Mod. res.	Sus.	Sus.	Sus.	Mod. res.	Sus.	Sus.	Most res.	Most res.	Res.
Almond sdlg.	W.D. sandy	Fair	V. good	Sus.	Sus.	Sus.	Sus.	Sus.	Sus.	Mod. res.	Sus.	Sus.	Sus.
For Apricot Scions													
Apricot													
Royal sdlg.	W.D. S. loam	V. good	Good	Mod. res.	Sus.	Sus.	Sus.	Mod. res.	Sus.	Sus.	Most res.	Most res.	Res.
Peach													
Lovell sdlg.	W.D. S. loam	Good	Good	Sus.	Mod. res.	Sus.	Sus.	Var. sus.	Sus.	Var. sl. res.	Sus.	Sus.	Sus.
Nemaguard sdlg.	W.D. S. loam	Good[1]	Good	Sus.	Mod. res.	Sus.	Sus.	—	Sus.	—	Res.	Res.	—
Rancho Res. sdlg.	W.D. S. loam	Good[1]	Good	Sus.	Mod. res.	Sus.	Sus.	—	Sus.	—	Res.	Sus.	—
S-37 sdlg.	W.D. S. loam	Good[1]	Good	Sus.	Mod. res.	Sus.	Sus.	—	Sus.	Sus.	Res.	Sus.	Sus.
Plum													
Myro. sdlg.	Wide range	Fair	Good	Sus.	Sus.	Sus.	Mod. res.	Var. sl. res.	Mod. sus.	Var. sus.	Mostly sus.	Mostly sus.	Sus.
Myro. 29-C clone	Wide range	Fair	Good-shal.	Sus.	Sus.	Sl. res.	Mod. res.	Sl. res	Mod. sus.	Mod. sus.	Res.	Res.	Sus.
Marianna 2624 clone	Wide range	Fair	Good-shal.	Sus.	Sus.	Mod. res.	Mod. res.	Sl. res.	Mod. sus.	Sus.	Res.	Res.	Sus.

SOURCE: Norton et al., 1963a, 1963b. Data are based on limited testing.
KEY: W.D. = Well Drained, S. = Sandy, Var. = Variable, Sus. = Susceptible, Res. = Resistant, Sl. = Slightly, Mod. = Moderately, V. = Very
[1]Graft compatibility probably the same as Lovell sdlg.

Figure 4-18 Approximate relative size of peach cultivars on several rootstocks. Peach seedling is the usual rootstock used. Some plum species stocks are not compatible with peach. [After A. N. Roberts and M. N. Westwood, unpublished data]

stock of *P. subcordata.* These stocks should be used only on a trial basis. Relative size of some of these are shown in Figure 4-18.

Apricot is the best rootstock generally for apricot cultivars on good soils, but the peach and plum stock listed in Table 4-13 are also satisfactory. However, peach rootstock should not be replanted after peach in old orchard land because of poor establishment and growth. Plum stocks are used for apricot only when the soil is too wet for peach or apricot root (Norton, et al., 1963a).

Nut Rootstocks

Juglans regia is native to the Carpathian Mountains eastward to Manchuria and Korea. In Oregon, selections of *J. regia* such as Manregian seedlings are used as rootstocks. The use of various species of black walnut stocks has been discontinued because the unions tend to develop a delayed incompatibility known as black line. In California, the northern California black *J. hindsii* is still used because it is tolerant to salts in the soil or irrigation water and because of its resistance to oak-root fungus, *Armillaria mellia.* Also, hybrids of *J. regia* × *J. hindsii,* called "Paradox stocks," are used because they seem to be better than *J. hindsii* on poor, hilly soil and because they seem more resistant to crown rot and root lesion nematodes.

Filberts are propagated commercially on their own roots by tip layerage. Such root systems sucker badly and tend to uproot in high winds. Deeper rooted, nonsprouting Turkish seedlings are being tried in Oregon, but none has yet been recommended.

Almond usually does well on almond root, but peach root is also satisfactory on soils slightly wet for almond or in sandy soils where nematodes may be a problem. Almond root is more resistant to lime-induced chlorosis and sodium or boron toxicity than is peach root. Also, trees on peach are not as long-lived as those on almond.

Pecans do not root easily from cuttings, so seedling pecans are used as rootstocks. Seedlings of a hickory species *Carya aquatica* have been tested for wet soils. But the nuts of pecan are usually smaller when grown on hickory than on pecan stocks.

GENERAL REFERENCES

Argles, G. K. 1937. A review of literature on stock-scion incompatibility in fruit trees with particular reference to pome and stone fruits. *Imperial Bureau Hort. and Plantation Crops Tech. Communication* 9:1–115.

Day, L. H. 1953. *Rootstocks for stone fruits.* Calif. Agr. Bull. 736.

Epstein, E. 1973. Roots. *Scientific American,* May, pp. 48–58.

Hartmann, H. T., and D. E. Kester. 1975. *Plant propagation.* 3rd ed. Prentice-Hall, Inc., Englewood Cliffs, N.J.

Martin, J. T., and D. J. Fisher. 1965. Surface structure of plant roots. *Ann. Rep. Long Ashton Res. Sta.,* pp. 251–254.

Snyder, J. C., and R. D. Bartram. 1970. *Grafting fruit trees.* Pacific Northwest Bull. 62.

Tukey, H. B. 1964. *Dwarfed fruit trees.* The Macmillan Company, New York.

5

Establishing the Planting

SPACING

In planting an orchard, one primary objective is to develop maximum bearing surface per hectare in a minimum of time. The kind of rootstock used and the arrangement of the trees are both very important in achieving this objective as well as in sustaining the yield of the orchard. Either vigorous or dwarf rootstock may be used, but the selection of tree spacing and row width must take into account the inherent vigor of the stock and the ultimate size of the trees. The trees must be arranged in such a way that pest control and other orchard practices can be done easily and efficiently when the trees are mature. The conventional square pattern wastes space and also makes efficient filler removal difficult. Various rectangular patterns prove to be more satisfactory than the square pattern because more trees (and hence more early bearing surface) can be put in each row. The distance between rows should be selected for most efficient land use, either permanently (for dwarfs or standard trees) or temporarily (for standard trees when filler removal is planned).

STANDARD VIGOROUS TREES Table 5-1 gives the number of trees per hectare for various planting distances. Vigorous trees should be planted at medium density in a filler plan. Depending upon the known growth characteristics of the cultivar and the depth and fertility of the soil, 400 to 1000 trees per hectare can be planted. Cultivars that are both vigorous and do not normally bear for several years must be given

Table 5-1 Trees per hectare at various in-row and between-row spacings.

In-row Spacing (m)	Spacing Between Rows (m)											
	2.0	2.5	3.0	3.5	4.0	4.5	5.0	5.5	6.0	6.5	7.0	7.5
1.0	5000	4000	3333	2857	2500	2222	2000	1818	1667	1538	1429	1333
1.5	3333	2670	2222	1905	1667	1481	1333	1212	1111	1026	952	889
2.0	2500	2000	1667	1428	1250	1111	1000	909	833	769	714	667
2.5	2000	1600	1333	1143	1000	889	800	727	667	615	571	533
3.0	1667	1333	1111	952	833	741	667	606	556	513	476	444
3.5	1428	1143	952	816	714	635	571	519	476	440	408	381
4.0	1250	1000	833	714	625	556	500	455	417	385	357	333
4.5	1111	889	741	635	556	494	444	404	370	342	317	296
5.0	1000	800	667	571	500	444	400	364	333	308	286	267
5.5	909	727	606	519	455	404	364	331	303	280	260	242
6.0	833	667	556	476	417	370	333	303	278	256	238	222
6.5	769	615	513	440	385	342	308	280	256	237	220	205

NOTE: One meter = 3.28 feet. Trees per acre × 2.47 = Trees per hectare.

wider spacing than precocious ones. However, some combination of trunk girdling and spraying with flower-inducing chemicals often can be used to achieve higher yield at closer plantings. The spacing in the row should be closer than that between rows. As the young trees grow, they fill the open spaces within the row, forming a "tree wall" in one direction, with enough room between rows to get through with equipment.

A sample planting plan is shown in Figure 5-1A. The initial spacing of 3 × 2 meters is used as long as it is practical; then, after perhaps two or three years of pruning back the filler trees (the black circles) and allowing the permanent trees (the white circles) to fill the space, the fillers are removed. The new spacing (Fig. 5-1B) is 4 × 3 meters, and the direction of equipment travel is now changed by ninety degrees. An advantage of a rectangular pattern over a square pattern is that, by removing alternate trees in every row and changing the direction of travel, additional width is obtained in the drive row when fillers are removed. (This can only be accomplished, however, when the in-row spacing is greater than half the between-row spacing.) Such a system allows for more space for the trees as they mature and for adequate space for equipment. In a 3.7 × 2.7 meter Bartlett pear planting, trees on the best rootstocks produced 42 metric tons of fruit per hectare (19 tons/acre) in the sixth year. Some "hedge pruning" in the direction of travel was necessary to permit the equipment to get through, but at seven years all trees had adequate room to grow. In the hypothetical plot shown in Figure 5-1B, a second removal (of the trees now indicated by black circles) might be necessary after 15 to 20 years, or sooner, depending upon the kind of tree planted. The nice thing about the rectangular plan is that as many removals as are needed can be done in an orderly way. Selection of the proper distances at the planning stage not only accomplishes the objective of maximum bearing surface at an early age, but permits one to sustain maximum bearing surface per hectare during several filler removals. However, it is imperative that filler trees be removed before serious crowding occurs; overcrowding will shade out the bearing area and lower yield and fruit quality.

With the availability of a wide variety of growth-controlling rootstocks, it is now possible to plant permanent high-density plantings of many different fruits. High densities should not be attempted without dwarf rootstocks because excess vigor invites overpruning, which reduces yield and quality. The use of chemical retardants alone to control growth is expensive and at times unsuccessful.

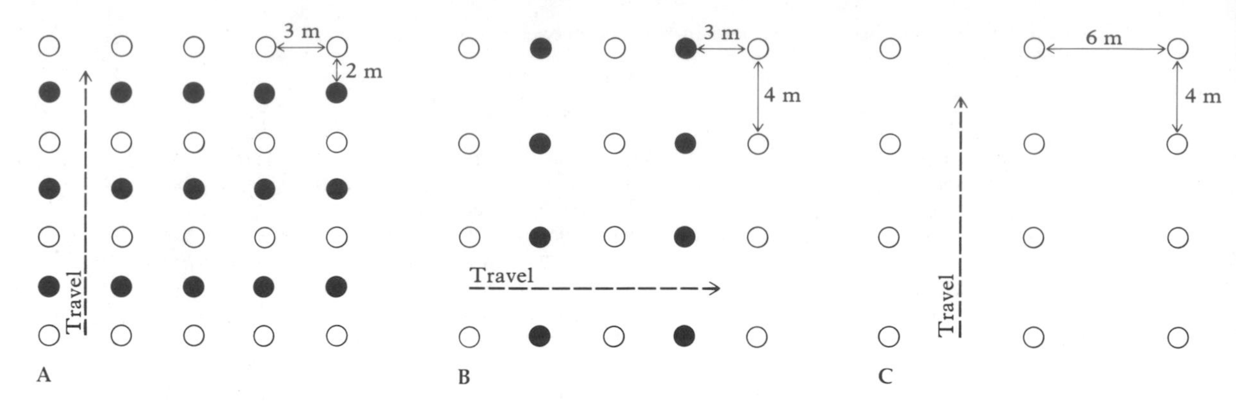

Figure 5-1 An example of a multiple-filler system with orderly tree removal at intervals in the life of the planting. **(A)** The initial planting is a rectangular pattern spaced 2 × 3 m. After crowding occurs, filler trees (dark circles) are removed, resulting in **(B)** a new rectangle of 3 × 4 m. When crowding occurs the second time, the semipermanent trees (the dark circles) are removed, leaving **(C)** permanent trees at 4 × 6 m.

One common fallacy is that tree spacing must conform to the requirements of the existing farm equipment in one's possession. On the contrary, one should determine the most desirable distances at which to plant trees and then choose equipment of suitable proportions to operate efficiently in the orchard. If well planned, the new orchard will outlast several sets of equipment, so equipment should be tailored to the orchard—not the other way around. Recently, orchard tractors, sprayers, tillers, and so forth, have been designed for narrow hedgerow culture. One full-powered tractor on the market has a 1.3 meter overall width. Another 25 horsepower tractor has a width of only .9 meter. But, of course, with any projected planting, one must be certain the right equipment is available before planting.

Another poor practice in current use is the renewal of old orchards by interplanting young trees alternately with the old trees in the row. The intention is to get the new trees established and then prune back the old trees as the young ones grow, ultimately removing the old ones altogether. In practice, however, this renewal method has failed. The young trees are shaded out, choked out by weeds, run over by harvesting equipment, or given improper irrigation. Part of this comes about because the required care and culture of very young trees is quite different from old trees. Also, old orchards often have a compacted soil layer just below the surface of the tillage zone. Such a layer can best be broken up by subsoiling with deep chisels after old trees are removed and before new ones are planted. Thus, the best way to replant is to rotate out several rows of old trees each year and to replant with better spacings. Interplanting old orchards would still be inadvisable even if none of the above-listed faults existed, because in most old orchards spacing is wrong for the best land utilization. Interplanting commits the new planting to this same unsuitable spacing. As an example, let us consider a block of Bartlett and Bosc pears planted at the Medford Experiment Station in 1934 at the standard square spacing of 7.6 × 7.6 meters (70 trees/acre). The soil is a tight adobe clay in which trees do not grow rapidly. After 37 years these trees had not yet filled the space allotted them. To renew this plot by interplanting would be to heap one mistake upon another, because it would perpetuate the same incorrect spacing. The actual way this orchard is being renewed is by bulldozing out a section at a time, subsoiling, and then replanting to better spacings. In all, 1.2 hectares have been pulled (210 trees) and planted back in a rectangular pattern of 3.8 × 1.5 meters

(2000 trees). It is expected that by the time the new plot is 6 years old it will exceed the top yield of the old plot at 30 years.

DWARF TREE SPACINGS Trees on fully dwarf rootstocks can be planted at 1700 to 3000 trees per hectare. Trees in the row are usually best at 1 to 1.5 meters apart. Rows can be between 3 and 4.5 meters apart, depending upon cultivar and soil. Most fully dwarf trees are weakly anchored and should be placed on a wire-post trellis by the second or third year. Experience has indicated that upright central-leader-type training is best for small trees, because it increases the effective depth (height) of the tree wall, increasing yield per hectare. The principal support of the trees is provided by tying the central leaders to the wire. Also the wires may be used to support willowy fruiting branches during the first few seasons of bearing. Ultimately the mature dwarf tree, when short-pruned to support its crop on sturdy wood, can be supported by a single wire tie at a height of 2 or 2.5 meters.

The high initial cost of dense plantings can quickly be recovered by early high yields and low per-tree production cost. For example, 2500 trees planted on one hectare would cost one-tenth as much to irrigate, cultivate, and spray as the same 2500 trees planted on 10 hectares. Such savings greatly increase the profits during the early years. Also, dense plantings make heating for frost protection economical in the early years because of the higher yield potential. The initial cost of trees can be reduced over present prices if growers are willing to order nursery trees in advance under contract. An excellent grower can save even more by planting unworked rootstocks and budding them to the desired cultivar.

SMALL FRUITS Plant densities vary greatly among the vines and shrubs comprising the cultivated small fruits but, in general, require many more plants per hectare than orchard species. The highest densities are with the smallest species such as strawberry (Fig. 5-2) and cranberry. Table 5-2 gives some commonly used spacings for the different small fruits. Table 5-3 gives the number of plants per hectare for a number of spacings.

Figure 5-2 The spacing of strawberries varies a great deal from place to place. **(A)** The single-row system is typical of Oregon, where plantings are cropped several years before being replaced. **(B)** Multiple-row systems are used in Florida, where the plants are cropped for one season only. A four-row bed is shown. [A: Courtesy of F. J. Lawrence; B: Courtesy of J. R. Stang]

A

B

Table 5-2 Spacings and plants per ha for a number of small fruits.

Kind	Spacing Range (m)	Plants per Hectare
Cranberry	.45 × .45–.15 × .15	47,840–430,550
Blueberry	1.5 × 2.7–1.2 × 2.4	2,390– 3,360
Blackberry	3 × 3 –.75 × 2.0	1,077– 6,150
Black raspberry	1.8 × 3.0–.75 × 2.0	1,656– 6,150
Red raspberry	1.2 × 3.0–.75 × 1.8	2,691– 7,176
Grape	2.7 × 3.0–2.0 × 2.7	1,196– 1,707
Currant and gooseberry	1.5 × 3.0–.60 × 2.4	2,152– 6,726
Strawberry single row	.3 × .9–.20 × 1.1	35,879– 46,205
Strawberry double row	.20 × 1.1	92,410

NOTE: One meter = 3.28 feet. Plants per acre × 2.47 = Plants per hectare.

Table 5-3 Plants per hectare for small fruits.

In-row Spacing (m)	Spacing Between Rows (m)										
	.61	.91	1.22	1.52	1.83	2.13	2.44	2.74	3.05	3.35	3.66
.15	107637	71758	53818	43055	35879	30754	26909	23919	21527	19570	17939
.30	53818	35879	26909	21527	17939	15377	13455	11960	10764	9785	8970
.46	35879	23919	17939	14352	11960	10252	8970	7974	7176	6523	5980
.61	26909	17939	13455	10764	8970	7687	6726	5980	5382	4892	4485
.76	21527	14352	10764	8611	7176	6150	5382	4784	4304	3914	3588
.91	17939	11960	8970	7176	5980	5125	4485	3986	3588	3262	2990
1.07	15377	10252	7687	6150	5125	4393	3845	3417	3076	2795	2562
1.22	13455	8970	6726	5382	4485	3845	3363	2990	2691	2446	2244
1.37	11960	7974	5980	4784	3986	3417	2990	2659	2392	2174	1994
1.52	10764	7176	5382	4304	3588	3076	2691	2392	2152	1957	1794
1.68	9785	6523	4893	3914	3262	2795	2446	2174	1957	1779	1631
1.83	8970	5980	4485	3588	2990	2562	2244	1994	1794	1631	1495
1.98	8280	5520	4139	3311	2760	2365	2071	1841	1656	1505	1379
2.13	7687	5125	3845	3076	2562	2197	1922	1707	1537	1399	1282
2.29	7176	4784	3588	2871	2392	2051	1794	1594	1436	1305	1196
2.44	6726	4485	3363	2691	2244	1922	1683	1495	1344	1223	1122
2.59	6331	4220	3165	2533	2110	1809	1584	1406	1265	1151	1055
2.74	5980	3986	2990	2392	1994	1707	1495	1329	1196	1087	996

NOTE: One meter = 3.28 feet. Plants per acre × 2.47 = Plants per hectare.

ORCHARD LAYOUT

A planting to be laid out in a regular pattern (that is, a pattern based on square, rectangle, or triangle) is started by establishing a straight base line; usually, this is next to a fence or roadway. Then, lines at right angles to the base are established at both ends and one or two places in the middle of the plot. Right angles are easily established by using three chains (or ropes or wires) whose lengths are in 3:4:5 proportions. For example, one could use ropes of lengths 30, 40, and 50 meters. One would put the 40 meter length along the base line, then place the 30 meter one at approximately a right angle, and finally lay the 50-meter piece to close the triangle. The 30 meter piece is adjusted in either direction so as to just touch the end of the 50 meter piece. In this arrangement, the 30 meter piece is at a right angle to the base line. Next, one would place upright markers along the base line and the right-angle lines for sighting to extend these lines (see Fig. 5-3).

From this point on, any desired row and tree spacing can be established by using a steel tape or marked rope to measure off the proper intervals. Small stakes can be used to mark the locations for tree holes, but if a power auger is to be used to dig the holes, then a small amount of lime placed on the spot is better because a single tractor operator can then dig the holes without the bother of a stake, which, if not removed, might hinder the positioning of the auger.

Contour layout for sloping land requires the use of a surveyor's level and rod. The first row is at the highest elevation, and is staked out on the level (that is, all points on the line are at the same elevation). Next, the steepest slope along this first row is found (see arrows, Fig. 5-4) and the distance that has been selected as the minimum distance between rows is measured down the slope. From that point is laid out the next row on a level line as before. Moving away from this steepest slope to less-steep slopes, the rows will be wider apart. Wherever the distance between two adjacent rows becomes twice the minimum distance, a short contour row is laid out between them from that point to the end of the plot.

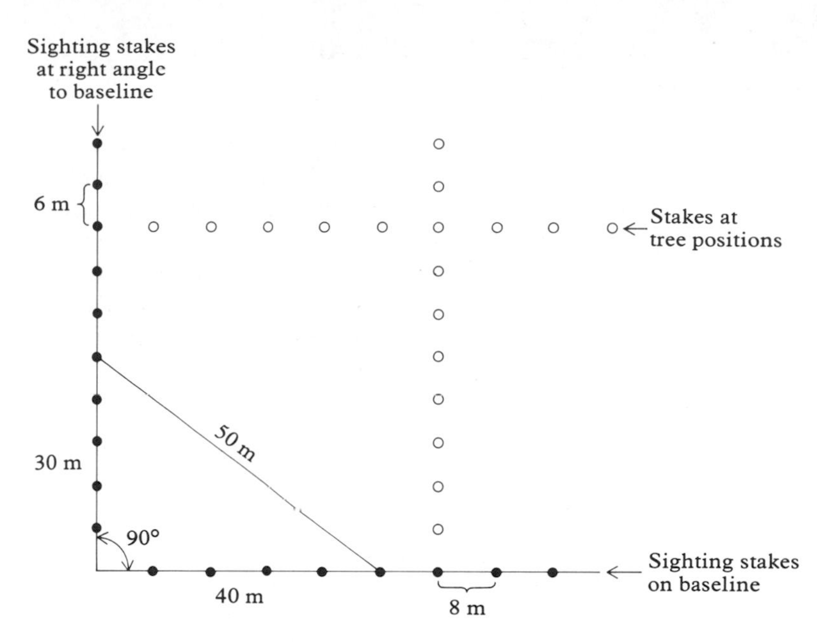

Figure 5-3 Diagram showing the use of 30:40:50 m ropes to establish a line at right angles to the base line. From these two lines a few measurements can be made to place sighting stakes from which the entire planting can be laid out.

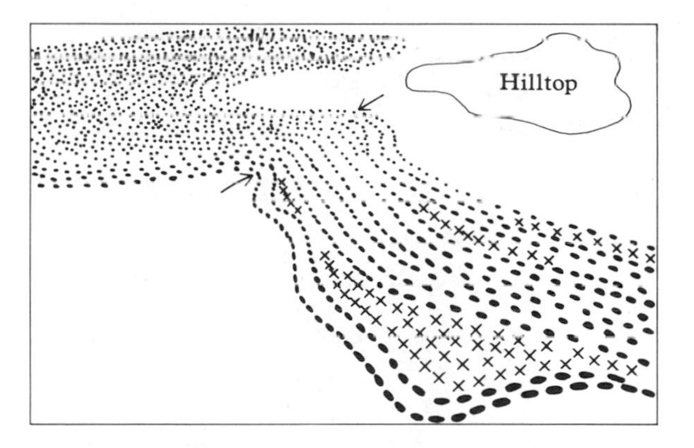

Figure 5-4 Contour planting is often used on upland slopes or on rolling land above frosty valley-floor sites. Arrows indicate the steepest slope, which is the starting point for laying out the contour rows. Wherever the distance between two contour lines exceeds twice the chosen distance between rows, short rows (indicated by X's) are used to fill the gaps.

PLANTING AND EARLY CARE

In areas with mild, moist winters it is best to plant trees in late fall. (Some root growth occurs at soil temperatures of 2° to 7°C.) In areas with cold or excessively dry winters, early spring planting is usually best. The main object in planting is to protect the roots from drying or freezing and to get them into firm contact with moist soil. Tree holes should be large enough to accommodate the root system with a minimum of root pruning. However, long roots can be cut back enough to balance the root system and to allow planting without crowding or twisting roots to get them into the hole. If an auger (Fig. 5-5) is used in wet, heavy clay soil, the inside surface of the hole glazes and seals off, sometimes preventing root penetration. Sometimes the roots grow in circles and the tree becomes "pot bound." Sometimes irrigation water saturates the soil in the hole and does not drain out readily, creating a low-oxygen condition similar

Figure 5-5 (A–D) Digging tree-planting holes with an auger. In wet soil with a moderate clay content, the rotary motion of the auger causes the walls and bottom of the holes to be glazed; this prevents free drainage of water from the area after an irrigation. Note the glazed wall in D.

A B C D

to that of a high water table. This undesirable glazing of holes can be prevented by digging the holes when the soil is dry or only slightly moist. If this is not practical, part of the glaze can be removed by slicing the edges with a shovel at planting time.

Tree roots should be kept in water or be moist prior to planting and, unless the soil is wet, should be watered after planting. Clean soil (preferably topsoil) should be put around the roots and firmed in with the feet after the hole is about half full of soil. If they are to be watered, the water should be put in just after the soil is firmed around the roots. The hole should be filled the rest of the way up with loose soil after the water has soaked in. Trash, weeds, manure, and other organic debris should be kept out of the hole because they might damage the tree roots. Above all, chemical fertilizers should *not* be put in the hole at planting time because the roots can be killed by such treatment. If the soil is poor, a light surface application of nitrogenous fertilizer can be used. If general growth and leaf color are poor, analysis of leaf samples in August will indicate fertilizer needs. To prevent sunscald of trunks, plant the trees slanted slightly to the southwest. Tree planting should not be complicated by such unnecessary devices as planting boards or other tree positioners. Trees planted 6 or 8 inches out of line present no problem in later care. When the soil is loose enough, trees can be planted by kicking the soil into the hole. This eliminates the need for a shovel, and frees the hands to position the tree and thus increases the ease of planting. Semimechanized planting also is practiced (Fig. 5-6).

Figure 5-6 Semimechanical tree planting for high-density hedgerows. The tractor plows a furrow into which the trees are placed by the men on the rear-mounted platform, after which the two plows at the rear cover the roots of the trees. Such a team can plant several thousand trees per day. [Courtesy of Cal Bosch, *Goodfruit Grower*]

If the orchard is being planted in an old orchard site, there may be a harmful amount of lead arsenate residue in the soil. If leaf analysis for peach indicates 2 ppm arsenic, and if total arsenic in the top 0.30 meter of soil is over about 30 ppm, then corrective measures are indicated. Peach is more sensitive to arsenic than some other trees, but all are somewhat sensitive. Peach symptoms for arsenic toxicity are poor growth and pale green leaves with numerous "shot holes" in them. If it is known before planting that the soil contains arsenic, then virgin soil should be hauled in to place around the roots. As the roots grow into the old soil, surface applications of $ZnSO_4$ (4.5 kilograms per tree) and $(NH_4)_2SO_4$ (2.3 to 3.2 kilograms per tree) will help by reducing the amount of arsenic uptake and by making the tree tissues more tolerant to arsenic. As the trees get older, they naturally become more tolerant to a given level of arsenic.

The most common faults observed in establishing the newly planted tree are improper irrigation and lack of weed control. These two faults cause more stunting and loss of trees than any others during the first two years. Irrigation during the first year should be applied often (at least every 2 to 3 weeks), but each irrigation should be of short duration. While the root system is still very small, it may deplete the moisture immediately surrounding each root, and even though good moisture is found at the surface close to the tree, the soil around the roots may be dry. Thus, frequent irrigations of *short duration* will replenish the dry areas around the roots without waterlogging the soil. Inexperienced growers often apply too much water at one time and drown the roots (that is, the excess water creates an oxygen deficiency), or they let them become too dry—or they commit both errors.

Weed growth around a young tree robs it of soil moisture, nutrients, and sunlight, all of which are needed by the tree for the rapid growth so essential to early flowering and cropping. Weeds can be kept out by tillage or by surface mulches of sawdust, bark chips, or black plastic. In the second and succeeding years, chemical herbicides can be applied to prevent weed growth. It is usually hazardous to use chemical weed control the first season because of possible injury to unestablished trees.

TREE TRAINING

Pruning for Training

Because of the distinctly different objectives of pruning to train a young tree and pruning the bearing tree, these two types of pruning will be considered separately. Training will be discussed here, and pruning mature trees will be discussed in Chapter 6. The term "training" is defined as the cutting away of portions of a tree to obtain the desired shape and framework (Fig. 5-7).

At planting time the tree is cut back to .75 to 1.0 meter if it is an unbranched whip. If it is a branched tree, it is reduced to three or four wide-angled branches, each of which is cut back to one-half or one-third its length. Or, if a central leader tree is desired, all side branches are cut off and the leader headed

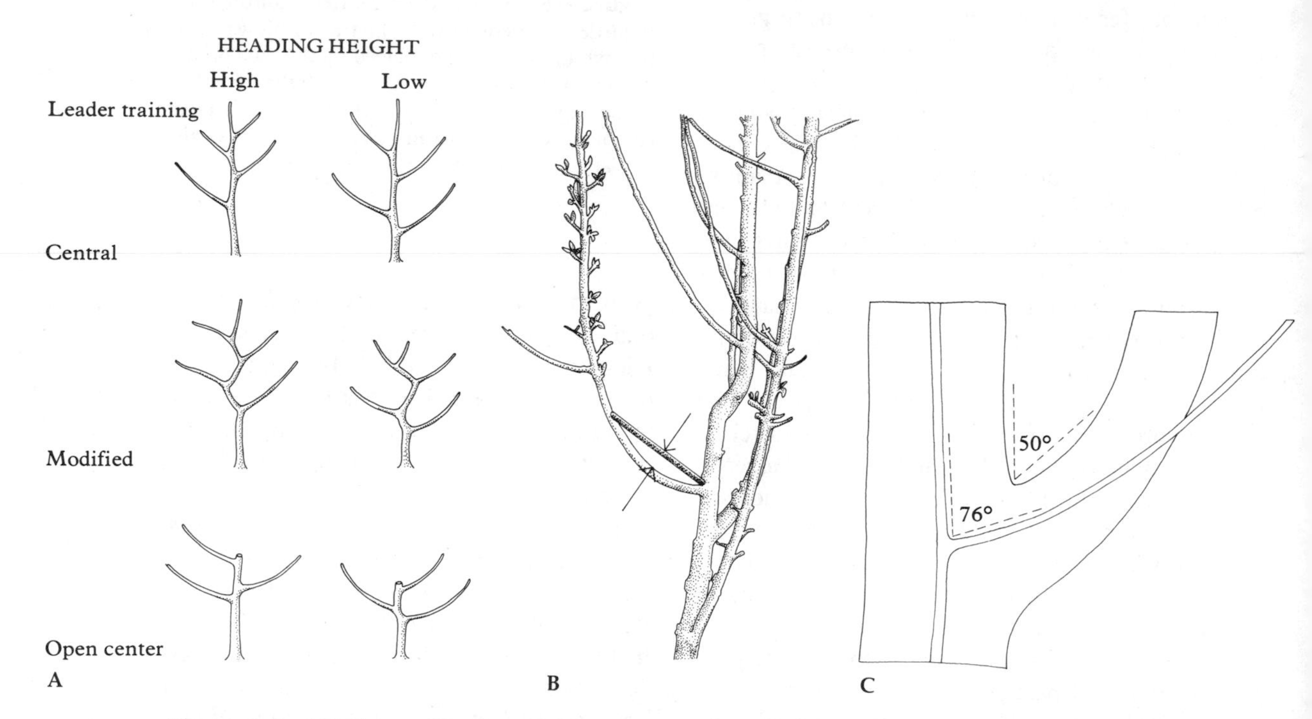

Figure 5-7 **(A)** Types of leader training and height of head. **(B)** Desirable wide-angled branch *(arrows)* on a tree trained to a low-headed modified leader. **(C)** The narrowing of branch angles with age. [A: From Jules Janick, *Horticultural Science,* 2nd ed. W. H. Freeman and Company. Copyright © 1972; B,C: After Verner, 1955]

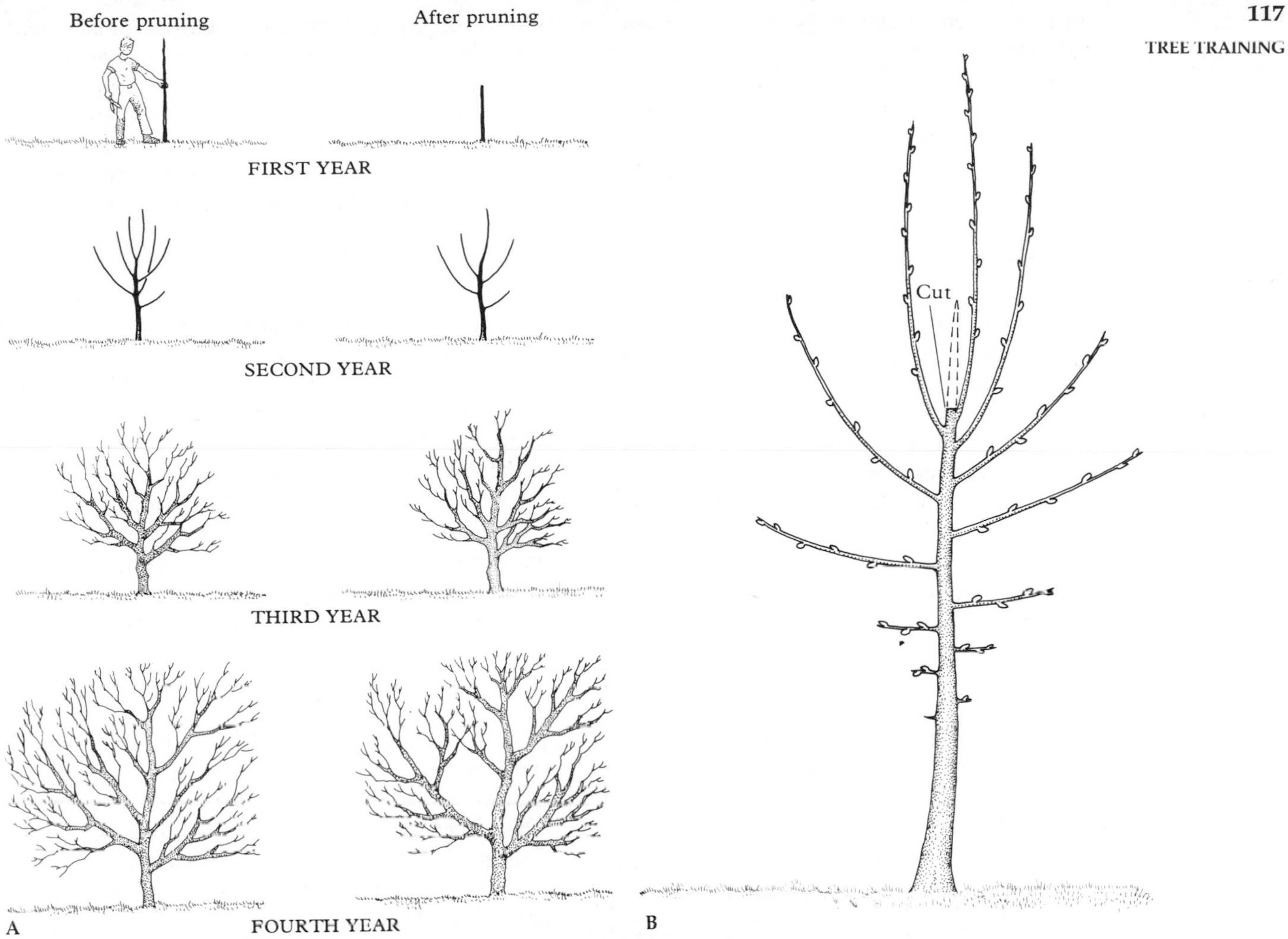

Figure 5-8 **(A)** Generally accepted practice of training a modified leader tree during the first four seasons. **(B)** The pattern of growth expected when the unbranched tree is headed back at planting and allowed to grow without further training the first season. The upper branches are the most vigorous and have the narrowest crotch angles; the dominance of the upper branches creates the hormonal balance that results in the wider branch angles and shorter growth of the lower branches. [A: After Snyder and Luce, 1950]

back to the desired height (Fig. 5-8). It is important that the newly planted tree be pruned back enough so that the top is in good balance with the root system, which was pruned during the digging operation in the nursery. In general, the permanent conventionally spaced orchard tree is developed so that three or four wide-angled side branches form the framework of the tree, with the uppermost branch becoming the modified leader. However, trees to be spaced closely in a hedgerow or those that are to be temporary fillers need a different sort of initial training. Hedgerow trees, especially those on dwarf roots, should be trained to a central-leader system to encourage upright growth to a height of 3 to 4 meters. Except for initial pruning to balance top and root, filler trees should not be trained, but should be left unpruned until bearing starts. Any training cuts made on a filler tree will delay bearing and reduce total yield.

Hormones in Training

Verner (1955) has shown that growth in apple is controlled by endogenous hormones and that pruning can alter the normal growth pattern by altering the hormone balance. His work indicates that auxin (indoleacetic acid, or IAA), produced in the vigorous upper shoot tips and young leaves, moves downward in the phloem and modifies the growth of shoots, buds, and branches below the point of origin of the auxin. The relatively large amount of auxin produced and transported downward, in concert with other hormones, affects the growth below the dominant upper shoots in the following ways:

1. Lateral bud growth is usually inhibited.
2. Crotch angles between the trunk and side branches are increased. Such angles are quantitatively related to the amount of auxin supplied from shoots directly above.
3. Elongation of side branches is depressed.
4. The upward curvature normally occurring at the tip of a horizontal branch (negative geotropism) is delayed, resulting in a more spreading tree.

These growth characteristics, related to natural auxin supply, can be altered by treatments such as pruning, girdling, or the application of hormones.

After girdling, latent buds begin to grow *below* the girdle and form narrow-angled crotches. Growth of side branches below the girdle is increased and the direction of growth is changed to the vertical. When the natural auxin supply is experimentally reduced by girdling, all of its effects can be restored by external application in lanolin paste of a synthetic auxin such as indolebutyric acid (IBA).

The several effects (lateral bud inhibition, cell elongation, cambial stimulus, and root initial stimulus) appear in some way to be related to the movement, asymmetric distribution, and concentration of auxin. The same concentration of auxin appears to stimulate cell division in some tissues (cambium) while inhibiting cell divison in others (lateral buds). One possibility is that each kind of tissue has a different IAA-oxidase activity that creates a different effective concentration of IAA in that tissue. Another partial explanation is that when IAA is placed at the base of a stem, IAA production and transport within the stem are greatly reduced. Thus, when a supply of IAA produced by a strong branch moves down the trunk to the base of a branch below, it results in a wider crotch angle and seems also to limit growth of that branch by reducing synthesis and movement of the branch's own IAA. The phenomenon of phototropism (the bending of a shoot tip toward light) appears to result from the rapid lateral migration of IAA in the subapical region to the "shady" side of the shoot, where it stimulates greater cell elongation and hence the bending toward the light source.

The tree shown in Figure 5-8 indicates the type of growth obtained when a newly-planted whip is planted and the main stem is cut back. The topmost three or four shoots are upright and vigorous in growth (and similar to a shoot from an apical bud). Thus, when an apical bud is removed, the lateral buds nearest the end respond as if they were apical buds. If a lateral branch has four or more vigorous branches above it on the trunk, then it will develop desirably wide crotch angles. Branches with such angles are much stronger and less susceptible to winter injury than are narrow-angled ones. If the auxin supply from above is in sufficient amount, then it tends to build up on the upper side of the crotch, stimulating greater cell growth on the upper (dorsal) side and making the crotch angle wider.

Delayed Heading

In order to take advantage of the known response in growth to pruning, two stages of pruning were developed. The first is dormant heading back of the unbranched whip, which evokes the growth shown in Figure 5-8 if allowed to go unchecked for the entire growing season. But if, after growth in the spring has advanced to the point at which the topmost upright shoots are about 15 to 20 centimeters long, the trunk is again headed back to remove the shoots with narrow crotch angles, the laterals below will be stimulated to grow more rapidly while retaining their wide crotch angles. All of these remaining branches are left on the tree until the next dormant season, when the main scaffolds are selected and the other branches are pruned back severely, but not removed entirely. The growth resulting from this type of pruning furnishes a source of auxin in the crotch area that serves to maintain the wide crotch angles of the main scaffolds (Figs. 5-9 and 5-10).

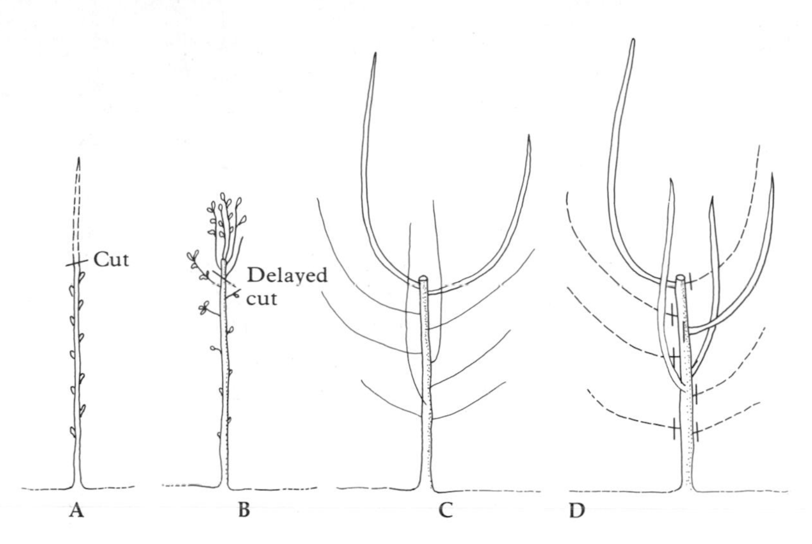

Figure 5-9 Delayed heading sequence. **(A)** Heading back at planting. **(B)** Delayed heading (when new growth is about 15–20 cm long) to remove narrow-angled shoots and to stimulate growth of the wide-angled branches below. **(C)** Growth at the end of first season. **(D)** First dormant pruning to retain only four strong branches. This procedure permits the primary training to be complete after the first season.

Trunk Renewal

Trunk renewal is the cutting back of the trunk of a newly planted tree a few inches above the graft union. When growth starts, one strong shoot is selected and the others are rubbed off. The vigor of this single shoot is such that it develops lateral shoots the first season. Such branches have wide crotch angles from which to choose a strong framework the first dormant season. This is a useful practice on trees that do not produce satisfactory growth the first season.

Heinicke (1975) proposes a system of training young apple trees to get the high-density orchard off to a good start (Fig. 5-11). This method results in a central-leader tree with well-spaced wide-angled branches after 4 years. While the method works, it is expensive and the number and severity of training cuts tends to delay early production. By contrast, full dwarf trees supported by a trellis wire do not need as much detailed training to perform well.

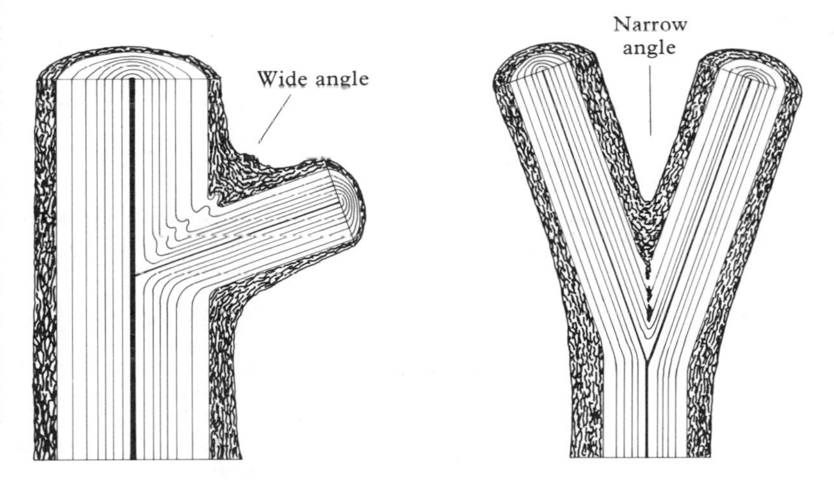

Figure 5-10 Longitudinal section of a strong wide-angled branch (left), contrasted to a narrow-angled crotch (right), which is weak because of bark inclusions and discontinuous xylem. [Adapted from Eames and MacDaniels, *An Introduction to Plant Anatomy,* Copyright © 1947 by McGraw-Hill, New York. Used with permission of McGraw-Hill Book Company]

New Evidence on Growth

New evidence indicates that not all buds or branches of a tree respond in the same way to imposed or natural inhibitory or stimulatory mechanisms. Some buds remain inactive, some grow into short spurs and others grow into vigorous shoots. The growth potential of a bud (seen in its stored nutrients, water supply, mineral nutrients, growth hormones) interacts with various inhibitory conditions to produce a net growth that reflects the balance of opposing forces. There are several growth patterns that are incompletely understood, but they are believed to be directed by the interplay of hormones, gravity, light, and vigor (the capacity to grow).

Recent evidence supports the hypothesis that branch angles are influenced by the mechanical pressure exerted on the branch base by tissues growing in the crotch area (Jankiewicz, 1966). When apple trees were grown in a horizontal or an upside-down position, the same types of crotches developed, indicating that the influence on branch angle is independent of orientation with respect to gravity. As the bud begins to grow, the crotch angle between the young branch and the main axis enlarges rapidly while the branch tissue is not yet lignified. Later, this process of angle enlargement ceases, indicating the relative rigidity of the system after secondary cell walls are formed.

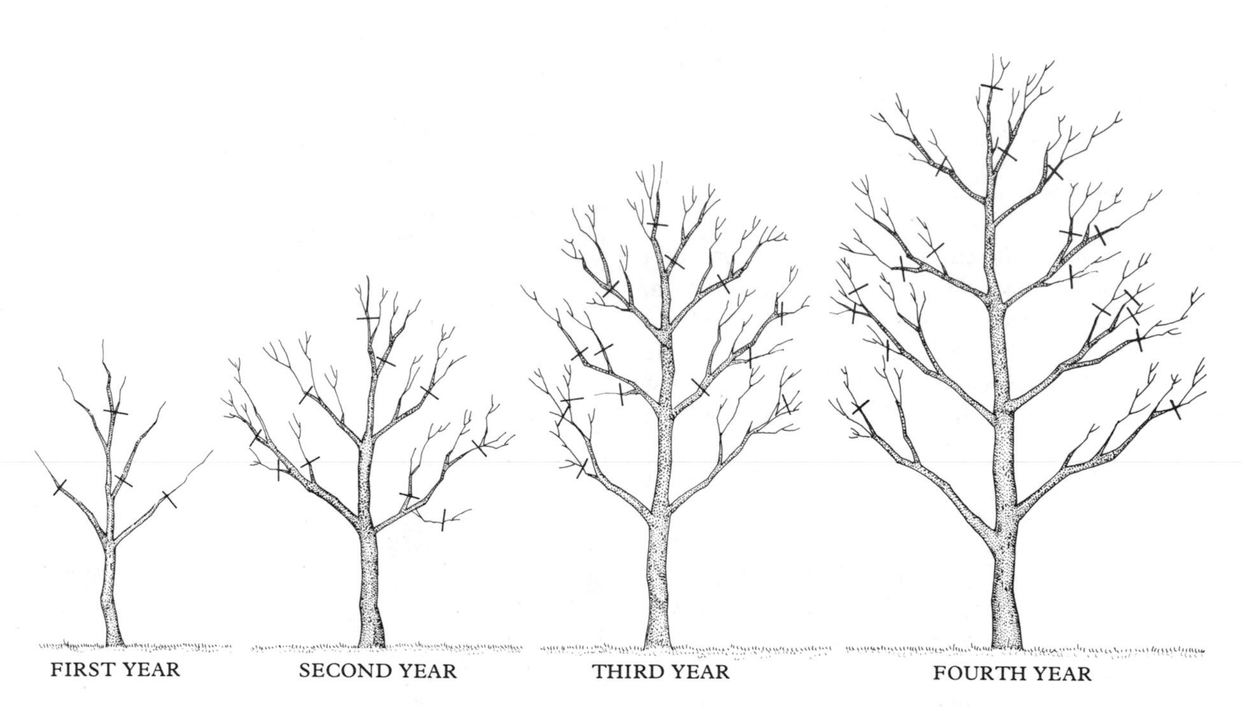

Figure 5-11 The method of Heinicke (1975) for training high-density trees during the first five years requires much more pruning than is normally done on young trees but may be satisfactory if trees are on a dwarf, precocious rootstock. The following description is taken from the original USDA figure— *First year:* Remove all competing shoots. Head back the terminal shoot. *Second year:* Select and head the lateral branches. Remove unnecessary laterals. *Third year:* Spread the branches, remove forked terminals to a single shoot, and head that shoot. Head side shoots. *Fourth year:* Spread the branches, remove forked terminals to a single shoot, and head that shoot. Head side shoots. *Fifth and following years:* If the tree has filled its allotted space, head back where necessary into two-year-old wood to an unheaded side shoot. Avoid heading cuts into one-year-old shoots until the tree is fruiting well. [After D. R. Heinicke, *High Density Apple Orchards—Planting, Training, and Pruning.* USDA Agricultural Handbook No. 458]

Figure 5-12 Use of chemicals to train Italian prune trees. **(A)** No treatment, showing some natural branching and few flower buds. **(B)** 1000 ppm SADH, showing little terminal branching but many short lateral spurs with many flower buds. **(C)** 500 ppm ethephon, showing killing of some terminal buds *(arrows)* followed by some branching with many flower buds on one-year wood. **(D)** 100 ppm cytokinin, showing much lateral branching, vigorous growth, and very few flower buds.

A newly discovered class of chemicals called morphactins tend to alter a plant's normal responses to gravity and light. Normally, plant roots grow downward toward the earth's center of gravity (positive geotropism), and shoots grow upward away from it (negative geotropism) as well as toward a source of light (positive phototropism). When germinating seeds were treated with solutions (6×10^{-5} M) of morphactins, roots and shoots lost their capacity to respond to either gravity or unilateral illumination (Khan, 1966). The roots and shoots grew in several directions. These findings indicate that the response of plants to gravity and light is probably a chemical reaction.

Chemical Training

Decapitated shoots treated with 250–1000 ppm of indolebutyric acid (IBA) in lanolin resulted in wide branch angles of all shoots below, whereas removal of the downward flow of auxin by girdling resulted in narrow-angled branches. Also, disbudding to allow only a few branches to grow reduces the total auxin supply and results in narrow angles on the few branches left.

There is some evidence that chemicals, such as SADH, maleic hydrazide (MH), triiodobenzoic acid (TIBA), and ethephon will cause desirable branching if applied to young trees at the proper time in the spring. SADH is applied at 500 ppm, TIBA at 50 ppm and ethephon at 300–500 ppm when shoots are 15 to 20 centimeters long, and MH at 1200 ppm in early July (Fig. 5-12, unpublished data of author). Cytokinins also are known to cause lateral buds to break in both apple (Williams and Stahly, 1968) and plum (Fig. 5-13, unpublished data of author).

CARE OF THE YOUNG ORCHARD

After the orchard is established and trees are trained to the desired shape and conformation (this should be complete at the end of the second year), only minimal corrective pruning should be done. During this

Figure 5-13 Cytokinin treatment on apple buds. The lower five buds on the plant at right were treated with cytokinin; the plant at left was untreated. Note the lateral spurs produced by cytokinin [From Williams and Stahly, 1968]

Bringing the Orchard into Bearing

Many cultivars of peach, plum, apricot, and filbert naturally begin to bear the third or fourth year after planting. Some cultivars of apple (e.g., Golden Delicious, Jonathan), pear (e.g., Bartlett), cherry, and walnut also bear at an early age. A number of cultivars and types, however, begin to bear only after 6 to 10 years unless they are specifically induced to flower and fruit earlier.

The adverse effect of pruning on early yields is found on all fruit and nut crops. Potential bearing surface is reduced and the time of first flowering is delayed. In a study on peaches, severe pruning greatly reduced yields during the first 5 years as compared to slight or average pruning. Table 5-4 shows that the highest yields were obtained with the least pruning (corrective) and the lowest yields resulted from severe pruning. In Idaho a study showed that unpruned Italian prune trees out-yielded pruned ones during the first 14 years (Verner and Franklin, 1960).

Under many climatic conditions, young Delicious apple trees seldom yield fruit as soon as tree size is sufficient to support a crop. Such trees are slow to initiate flowers, and when blossoming does occur, yield is low because of poor set. Tests indicated that pruning reduced yield and girdling increased it on young Delicious trees.

Trunk or branch girdling can be used to induce flowering and heavier cropping in some tree crops. Table 5-5 shows that ringing (making a single cut

Table 5-4 Effects of pruning on yield of Elberta peach trees 4 years old.

Pruning Method	Total Yield (boxes/ha)	Average Size (cm diam.)
Only corrective cuts	351	5.72
Moderate thinning out	277	5.72
Thinning out plus moderate heading back	200	6.04
Severe thinning and heading	121	6.04

SOURCE: Ashton et al., 1950.

period of developing the bearing surface, the least pruning is the best. Studies with cherry, apple, pear, and peach indicate that the highest yields during the first 5 years come from trees that received no pruning during that time.

Young trees seldom need fertilizer before they start to bear if planted on good soil, but if growth and poor leaf color indicate the need, small amounts of fertilizer containing only nitrogen should be applied on the surface around the tree. Keep the band out about 15 centimeters from the trunk to prevent burning of bark by direct contact with the fertilizer salt. Trees do not need phosphorus fertilizers and seldom need potassium until bearing begins. A good rule of thumb is to use nothing but nitrogen unless a mineral analysis shows a need for other elements.

Table 5-5 Effect of ringing 5-year Delicious apple trees (at 7 and 25 days past full bloom) on yield and tree size.

Treatment	Year				Tree Size 1964		
	1961	1962 (boxes per ha)	1963	Total	Trunk (cm cir.)	Spread (m)	Height (m)
Control	267	941	1922	3131	40	4.48	4.91
Ringed 1961 (F.B. + 7)	452	1376	1868	3697	40	4.24	4.85
Ringed 1962 (F.B. + 25)	267	1040	2189	3496	40	4.76	4.60
Ringed 1961 and 1962	452	1547	2081	4080	38	4.60	4.27

SOURCE: Burkhart and Westwood, 1964.

through the bark completely encircling the trunk) young apple trees increased yield, altered tree shape, and reduced tree size. Ringing 7 days past full bloom increased fruit set on the small amount of bloom present in 1961 and also increased floral initiation for the following year. Ringing in 1962 also increased set for that year and increased the bloom the following year. Ringing two years in succession increased yields both years, even though the trees were smaller. The smaller, spreading, more productive trees resulting from repeated ringing were desirable because the trees were planted 6.1 × 6.1 meters in a filler system, and early heavy bearing was the goal. The less vigorous, ringed trees responded well to the heavier pruning required to keep the trees confined to the 6.1 × 6.1 meters spacing. In another test, trees in three Delicious orchards were ringed and compared to controls (Table 5-6). Current-year fruit set was increased by ringing, as was next year's flowering.

The influence of pruning, nitrogen fertilizer, and ringing of young Delicious trees was examined during 3 years. A factorial analysis was made of the data to see how these variables interacted. Table 5-7 shows the significant main effects and interactions.

All of the above-cited tests were done with standard scions on seedling rootstocks. Tests also were done on 5-year Delicious/M 9 dwarf root, in which the trees were planted in hedgerows (4.58 × 1.22 meters) and were flowering heavily but were not setting a crop. The poor light intensity during the post-bloom period (Corvallis, Oregon) was thought to be limiting set. In this test, fruit set and bloom were increased by both ringing at full bloom and by more vigorous pruning to admit more light early in the season (unpublished data of author).

Table 5-6 Effect of ringing on current year set and following year bloom of young Delicious apple.

Treatment	Current Year (% set)	Following Year (% bloom)
Ringed (petal fall)	47	82
Control	28	38

SOURCE: Batjer, 1961.

Table 5-7 Some effects of pruning, ringing, and nitrogen on yield (during 3 years) of Delicious apple beginning at age 3.

Treatment	Yield (boxes/ha/3 yrs.)
Main effects	
Moderate pruning	946
No pruning	1616
Moderate N	1433
No N	1124
Ringing	1033
No ringing	615
Interactive effects	
No N, no ringing + pruning	554
N + ringing, no pruning	1779

SOURCE: Batjer and Westwood, 1963.

The following points should be remembered regarding the care of nondwarf young apples to favor early bearing:

1. Don't prune after initial training.

2. Ring fillers or other vigorous trees to induce early fruiting.

3. Add nitrogen if you plan to ring or if trees lack vigor.

4. Don't ring a weak tree.

5. Ring early (full bloom to petal fall) and use only a single cut (no bark removal).

Stone fruits generally should not be ringed because of severe side effects and poor healing. The response of pears ringed 4 to 5 weeks past bloom is similar to that of apples, but care should be taken not to infect the trees with fire blight during ringing. In theory, any woody plant can be induced to flower more heavily by ringing, but care should be exercised with species and cultivars for which no data are available. Severe winter injury may occur to trees the winter following ringing, particularly late ringing in which a section of bark is removed.

Temporary filler trees that will remain in the orchard only 10 to 12 years should not be trained and should remain unpruned until they are in heavy production. Early-bearing cultivars should be used as fillers, but even so, every practice should be directed toward maximum yield during the years prior to removal. A study by Degman (1961) indicated that Golden Delicious and Jonathan apple orchards bore more if fillers were cut back heavily rather than removed outright when they began to crowd the permanent trees. At 9 years (when the fillers were beginning to crowd), he removed all the fillers from some sections of the orchard. In other sections he merely cut back the side branches of the fillers to permit the permanent trees to grow and spread normally. During the next 4 years the following annual yields were obtained:

Fillers removed	1764 boxes/hectare
Fillers hedged back	2758 boxes/hectare

Not all cultivars would continue to bear after severe pruning, as did these Golden Delicious and Jonathan trees, but by using ringing to induce flowering, stepwise restrictive pruning rather than early tree removal is clearly the better way to handle fillers. Often, when growth-controlling rootstocks are available, permanent medium or high-density plantings are preferable to filler systems. Sometimes, however, when dwarf stocks are not available or where large trees are not a problem, filler systems are warranted. This is true of walnuts, pecans, filberts, prunes, and perhaps pears.

Four chemicals—SADH, ethephon, CCC, and TIBA—have increased bloom the next year when applied to young trees 15 to 50 days past full bloom (see Chapter 14).

HIGH-DENSITY SYSTEMS

The term "density" has been used in various ways in referring to orchard systems, but it usually means number of trees per hectare or acre. ("Density," in this sense, should not be equated to bearing potential, because bearing potential is affected by tree size and age.) Cain (1970) developed a model for optimum tree density based upon tree spread and tree numbers. Maximum density for a given spacing is calculated from tree spread at the time when increases in tree spread are slow and trees can easily be held to a given diameter with minimal pruning. Such a model is most applicable to medium- or low-density plantings, in which closer spacing would limit light penetration or would result in moisture deficiency in nonirrigated soils.

Bearing Surface

Recently, a method was reported for estimating bearing density in trees of any age or size by simply measuring trunk diameters and converting those values to the cross-sectional area of a trunk (Westwood and Roberts, 1970). The average value, with the number of trees per acre, can be used to determine the bearing potential or density of the orchard. The following steps are used to determine the potential bearing surface for a given crop:

1. Measure in centimeters the trunk diameter, or circumference, of forty to fifty trees in the block

(record zeros for missing trees). Calculate the average trunk size.

2. Find from Table 10-2 (Chapter 10) the cross-sectional area (in square centimeters) for the average trunk size obtained in Step 1.

3. Multiply the average cross-section (in square centimeters) by the number of trees per hectare for the block in question. Compare this value with the maximum bearing surface for that fruit as shown in Table 10-1 (Chaper 10). For example, a value of 44,450 for dwarf Bartlett pear is compared with the table value 69,450 maximum, so the orchard in question has about 64 percent of its potential bearing surface.

Light

The direction of the row and the planting pattern were found by Ferguson (1960) to affect the amount of light interception by tree models in temperate latitudes. With triangular tree placement in rows, there was a negligible effect of row direction, but tree walls in a regular rectangular pattern with rows running east and west received 9 percent more radiation in late summer than did rows running north and south. This could conceivably increase quality and color of late maturing fruit during this period. However, Lombard and Westwood (1975) found by actual test that hedgerow pears at 43° N latitude produced 30 to 48 percent more fruit from north-south than from east-west rows. In this test, the difference was due to poor fruit set on the north sides of the east-west rows.

Heinicke (1964) showed that tree size also affects the amount of light received by the leaves. By measuring light in different areas of dwarf, semidwarf, and standard trees, he found that the larger trees had more shading of leaves and that effective leaf area per acre was less than with dwarf trees. When considering only foliage that received at least 30 percent of full-sun exposure, dwarf trees at spacings with some overlap of canopy had about one-third more effective foliage per hectare than large trees. Light distribution in tree systems not only affects yield efficiency but also affects fruit color, quality, and size. A critical point to remember is that there are essentially only two populations of leaves on a tree: those that are directly shaded by one or more other leaves, and those that are exposed to direct sunlight or diffuse sky light. Shading by a single leaf will lower the intensity of light reaching a shaded leaf to the extent that it is not functioning to produce excess photosynthate for the system. In practice, most leaves are not totally shaded but are a mosaic of shaded and illuminated areas. For a given leaf, this pattern changes with air movement and with changes in the sun's angle. Thus, the photosynthetic efficiency of a given tree is, in part, a function of the proportion of its total leaf surface exposed to direct light multiplied by the duration of that exposure during the day.

Pears and Apples

In 1957 some tests were started at Corvallis exploring dense plantings of apples and pears. One plot of Bartlett and Comice pear on quince stock was planted at 6.1 × 2.44 meters (667 trees/hectare). That was the spacing to give 270 trees/acre recommended by the East Malling Station. A second planting of pear was put in at 4.58 × 1.22 meters (1794 trees/hectare) to demonstrate that yields can be reduced by planting too closely. After 8 years the total yield per tree was the *same* in the high-density plot as in the lower one. Yield per hectare in the higher-density plot was nearly three times that of the standard spacing. When crowding occurred in the close planting, we learned that more severe pruning increased set without stimulating excessive growth because the dwarf quince root, together with root competition and heavy cropping, kept growth down.

Further pruning tests with hedgerow Comice/quince showed that heading-back pruning reduced the total bearing surface but resulted in 31 percent fruit set, compared with only 13 percent set with long thinning-out pruning. This sharp difference in fruit set resulted in higher yields per tree with short pruning even though the trees were smaller. Yield of

Delicious apple on M 9 in dense hedgerows also was increased by more severe pruning, while Golden Delicious tended to overset regardless of the severity of pruning. The foregoing illustrates an important difference between dwarf and vigorous rootstocks. Heavier pruning improves the yield of dwarf trees, but the same degree of pruning on standard trees stimulates too much growth and reduces yields. This difference is of utmost importance to the degree of success obtained with a dense planting. While it is possible to maintain a medium density planting (500 to 750 trees/hectare) on vigorous roots by heavy pruning and nitrogen starvation, high densities of 1500 to 2500 trees per hectare require a dwarf root for best results.

Several stages of orchard work can be mechanized when dwarf tree walls are used. Completely mechanized pruning can be done both in topping and in vertical hedging with mowers. Such pruning works well on dwarf trees but produces an undesirable growth response in vigorous standard trees. Low, uniform tree walls lend themselves either to single-level platform harvest or possibly to completely mechanized harvest by over-row machines with vibrators and mobile catching belts. Such machines, however, have not been developed for apple and pear, although they have worked well for caneberries and grapes.

To use dwarf trees, one must properly compensate for their natural weaknesses. None of the fully dwarf rootstocks are well anchored, and they thus must be supported by stakes or a wire trellis. A one- or two-wire trellis with the trees tied only at the central leader is minimal but satisfactory. When the trees are very young, side branches may be tied up to the wire to prevent breakage. This permits leaving more fruit on the trees without serious breakage. As the trees mature, however, stub pruning eliminates the need for supporting branches, and the support of the central leader keeps the tree from leaning or falling over.

What have been our results with several dense spacings? In a plot of .4 hectare (1 acre) Red and Golden Delicious on M 9 were planted in 1956 at 4.58 × 1.22 meters, 4.58 × 1.83 meters, and 4.58 × 2.44 meters. Surprisingly, the closest spacing (1792 trees/hectare) produced the highest yield for both cultivars during the mature years as well as in the early years. Clearly, when dwarf rootstocks are used, the choice of spacing is crucial to the ultimate yield potential (Table 5-8). Experimental plots using more dwarfing systems and higher tree densities are being tried. These include 3 × 1 meter spacing for dwarf pears and 1.8 × .6 meter for compact mutants of Golden Delicious on M 26 rootstock. In all such tests, yield per hectare during the first few years is much higher than at wider spacings even though yield *per tree* is less. At the end of 3 or 4 years' growth, these very-high-density systems can reach full bearing potential, compared to 6 to 8 years for medium to high densities, and 10 to 15 years for low-density plantings.

The question of whether a given high-density

Table 5-8 Effect of in-row spacing on yield of apples on M 9 root planted in rows 4.58 meters apart.

Variety	In-row Spacing		Annual Yield in Metric Tons per hectare[1]			
	Meters	Feet	First 6 yrs.	Second 6 yrs.	Third 6 yrs.	Total 18-yr. Yield (m ton/ha)
Starking	1.22	4	7.0	55.4	65.9	767
	1.83	6	6.3	51.6	61.4	715
	2.44	8	5.6	38.6	52.5	581
Golden Delicious	1.22	4	16.1	67.9	98.6	1098
	1.83	6	13.0	60.5	91.5	994
	2.44	8	9.0	59.4	92.1	965

Source: Westwood et al., *J. Am. Soc. Hort. Sci.* 101:309–311, 1976.
[1]One ton per acre equals 2.24 metric tons per hectare.

planting is feasible depends upon such things as the availability of suitable orchard and harvesting equipment, availability and cost of trees, trellises, and irrigation systems. Also in climates such as in South Africa and central Washington, where the sun is very bright, the more open, exposed dwarf tree results in too much sunburning of fruit of some cultivars.

From the economic point of view, a grower might question whether he could afford 3500 trees per hectare (1400 trees/acre). He should consider these facts in arriving at an answer: (*a*) Early profits per tree are much greater in high-density plantings because of much lower fixed costs in land, taxes, irrigation, tillage, etc., for each tree; A hectare with 3500 trees costs only one-tenth as much to care for as 10 hectares with 350 trees per hectare. (*b*) Frost protection is feasible the second and third years on high-density plots. Standard spacings take much longer to develop enough bearing surface to make heating profitable. (*c*) Tree costs can be reduced by buying layered rootstocks at a low price per unit and planting directly in the orchard, and cultivars can be budded at the end of the first summer.

The compact dwarf tree is a complex compound genetic system and requires more exacting care than does the standard tree, but such dwarfs are essential to success in high-density plantings. The genetic dwarfing of the stock cannot be replaced by the use of seedling-rooted trees sprayed with growth retardants or otherwise treated to suppress natural vigor. In soils and climates that result in low vigor, care should be taken to use the correct rootstock to obtain the vigor needed for a given spacing. In such cases medium density trees with slightly more vigor than needed may be controlled with chemical regulants and heavy cropping. Only the best growers should consider high densities.

Stone Fruits

Recent studies of rootstocks for Italian prune indicate a number of possible dwarfing stocks; the most dwarfing of these stocks is *Prunus tomentosa,* the Chinese bush cherry. The trees, however, suffer from a high incidence of bacterial canker of the trunks. Another potential dwarf stock is *P. subcordata,* the Pacific wild plum. Neither of these has been tested enough to suggest them for use. Two growth-controlling stocks of *P. domestica* show promise as semidwarfs, but they attain about three-fourths the size of trees on peach, or about half the size of trees on Marianna 4001, a very vigorous stock. They are Black Damas C and Common Mussel. Both have yield efficiencies (fruit yield per square centimeter of trunk size) better than Italian grown on peach root, which is the standard stock in the Northwest. A new dwarf rootstock from the East Malling Station is Pixy. It has not been tested widely. Thus far, the prospect for true high densities with prunes has not been realized. Where the fruit are harvested mechanically, it is doubtful that dwarf trees are any better than vigorous ones on efficient rootstocks. For example, Italian prunes on vigorous Myrobalan 29-C and peach seedling stocks were planted at the moderately high density of 1120 trees per hectare (3.66 × 2.44 meters). At 5 years, as the trees began to crowd, target yields of 34 metric tons per hectare were obtained. At 6 years, alternate filler trees were removed (after previous hedging to provide permanent trees with more room to grow). Yet seventh-year yield was 29 metric tons per hectare, with only 560 trees.

High density of peaches is being studied in Oregon in a minor way. The use of *P. subcordata* as an interstem, using Lovell peach seedling root, results in a half-sized peach tree that is adapted to tree-wall culture. Some plum rootstocks and *P. tomentosa* are dwarfing roots for peach and seem to offer the best hope of getting a full dwarf. With some of these dwarfs, there seems to be a problem of mineral deficiencies without a special fertilizer program. No suggestions for commercial plantings can be made at this time.

Very little work has been done on dwarfing and high densities for cherry, except for the older work showing that Stockton Morello (*P. cerasus*) rootstock for sweet cherry produces a tree about one-third standard size. However, some cultivars are not completely compatible—for example, Napoleon. Some new work is being done with dwarf selections of mazzard (*P. avium*) × mahaleb (*P. mahaleb*), but no suggestions can be made at this time (see Chapter 4).

Nuts

No dwarf rootstocks are used at the present time, but somewhat higher densities are being tried for filberts on their own roots (Lagerstedt, 1972b). These higher densities have resulted in higher yield per hectare in the early years, but such trees not on dwarfing roots will ultimately crowd at these densities, necessitating some kind of filler removal. Filler removal and special pruning were discussed earlier.

Martin et al. (1969) reported that walnuts at 15.24 × 9.15 meters produced as much per tree in the early years and more per hectare than did those at 15.24 × 12.2 meters or 15.24 × 15.24 meters (standard spacing). Long-term net monetary returns were greatest from the most dense planting. As a result of these tests, California growers are now planting at 9.15 × 9.15 meters; with nondwarf rootstocks, some tree removal will be necessary as crowding occurs.

Present Outlook

Present research programs seek to develop practical, intensive orchards with high stable yields of quality fruit. European dwarf plantings achieve intensive culture with high inputs of human labor. The goal in the U.S. is to achieve intensification with a high degree of mechanization to take the place of expensive and scarce hand labor.

One intriguing possibility is the achievement of partial climate control in dwarf orchards. Plastic covers, supported by a wire grid at the top of the trellis-wire posts, could be spread over an orchard. Such covering in some climates would protect completely against spring frosts. It would also increase orchard temperatures during the day, which can aid pollen-tube growth and ultimate fruit set. Agriculture is now moving into an era in which the need for production and the constraints of costs and technical inputs create an increasing need for climate modification and control to prevent crop losses due to perhaps a single vagary of weather. We can look forward to complete climate control over crop areas at some time in the future. At that time, only the most efficient cropping systems will find a place in agriculture.

GENERAL REFERENCES

Cain, J. C. 1970. Optimum tree density for apple orchards. *HortScience* 5:232–234.

Ferguson, J. H. A. 1960. A comparison of two planting systems in orchards as regards the amount of radiation intercepted by the trees. *Neth. J. Agr. Sci.* 8:271–280.

Heinicke, D. R. 1975. *High density apple orchards—planting, training and pruning.* USDA, Agr. Handbook No. 458.

Verner, L. 1955. *Hormone relations in the growth and training of apple trees.* Idaho Agr. Exp. Sta. Res. Bull. 28.

6

Cultural Practices

This chapter deals with soils, fertilizer practice, and pruning. Soil-related factors are discussed as an integrated biotic-edaphic system. The effects of soil moisture, weed control, nutrient availability, and fertilizer practice on growth and performance are considered. Pruning practices are discussed in relation to the growth, flowering, and fruiting habits of different species and in relation to the inherent vigor of the rootstock. Pruning here refers to the cutting of mature bearing trees or vines and does not include pruning for the purpose of training a young nonbearing plant (which was discussed in Chapter 5). The objectives of the two practices are entirely different.

SOILS

Soil Type and Texture

Soil serves plants by supplying minerals and water for growth and as the medium in which to anchor the root system that keeps the plant upright. Soil is a complex biophysical mixture of inorganic minerals, organic matter, and living organisms; soil contains air and water in various proportions. Soil particles vary in size from >256 millimeters (boulders) to <.002 millimeters (clay) (Table 6-1). Textural classifications

Table 6-1 Classification of soil particles.

Particle	Diameter (mm)
Boulders	>256
Cobbles	256–64
Pebbles	64–4
Gravel	4–2
Fine gravel	2–1
Coarse sand	1–0.5
Medium sand	0.50–0.25
Fine sand	0.25–0.10
Very Fine sand	0.10–0.05
Silt	0.005–0.002
Clay	<0.002

SOURCE: USDA and Wentworth.
NOTE: The International Classification (Atterberg) System refers only to soil particles under 2 mm: Coarse sand, 2.0–0.2 mm; Fine sand, 0.2–0.02 mm; Silt, 0.02–0.002 mm; Clay, <0.002 mm.

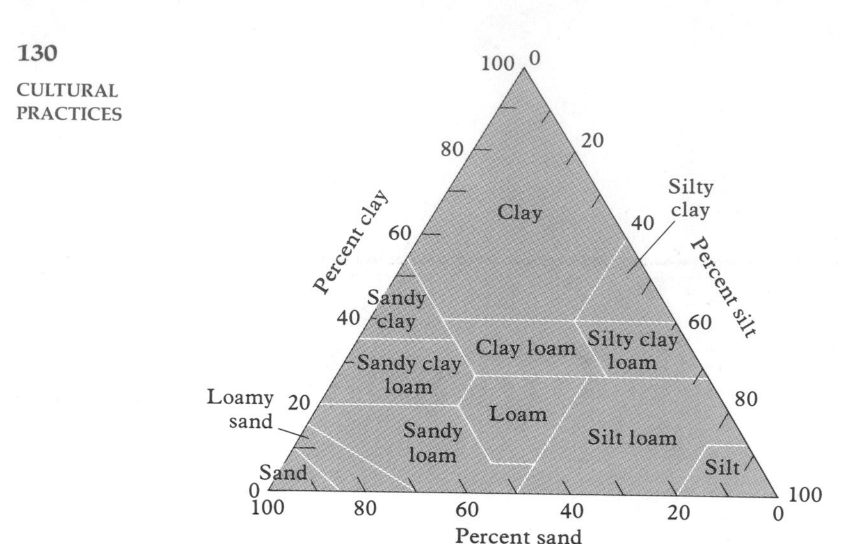

Figure 6-1 This "texture triangle" shows the relative percentages of sand, silt, and clay in each textural class. As used here, "loam" refers to a soil with more or less equal proportions of sand, silt, and clay. [After USDA]

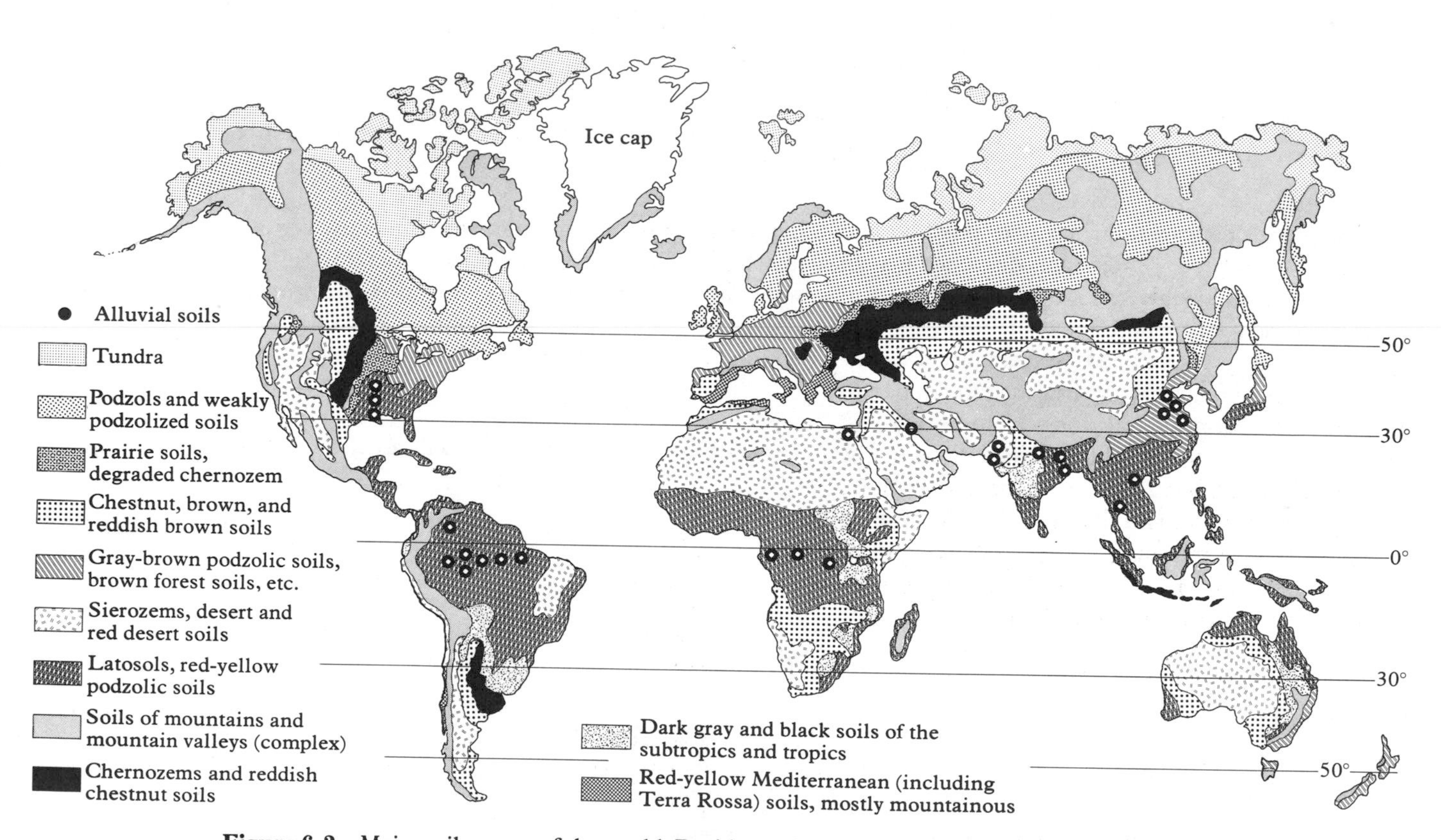

Figure 6-2 Major soil groups of the world. Deciduous fruits may be grown on a wide variety of soils if they are of suitable pH, depth, and have good internal drainage. Many desert soils are suitable with irrigation if their salt content is not excessive. [After USDA]

are listed in Figure 6-1. Soil type, depth, and degree of internal drainage are important in determining which kinds of fruit can be grown. The major soil groups of the world are given in Figure 6-2.

Soil Moisture

Much work has been done to determine the best irrigation practice for fruit species growing in various soils. Studies have shown, for example, that the growth rate of pear fruits is reduced if the available soil moisture in clay soil falls below 50 percent, based on a permanent wilting percentage (PWP) = 17 and field capacity (FC) = 32 percent. PWP is the percentage of moisture at which the plant wilts and will not recover when placed in a saturated atmosphere. FC is the percentage of moisture retained against gravity in a saturated soil. In one test 70 percent available moisture was needed in Meyer clay adobe soil for maximum growth of Anjou pear fruit (Work, 1939). Why then is "available" moisture in the soil not always available? Root distribution in the soil is uneven, so that soil near a root may be drier than a portion of the soil containing no roots. The percentage soil moisture as determined by a soil sample is merely an average and does not specifically indicate the moisture content near roots.

Formerly, water uptake and movement in plants was explained by diffusion pressure deficit, that is, $DPD = OP - TP$, where OP is osmotic pressure (which lowers diffusive energy) and TP is turgor pressure (which raises diffusive energy). The term "DPD" has been abandoned because it is no longer used in thermodynamics and because diffusion no longer is considered the only factor in water movement along a gradient. Water potential (Ψ) expressed in bars or atmospheres has replaced the old terminology. The adsorption of water on cell surfaces and colloids, and dissolved solutes alter the liquid structure of water so that thermodynamically free energy is lowered. If pure free water has a potential (Ψ) of zero, then plant water will have Ψ less than zero, except during guttation, when Ψ will be positive. The expression $\Psi = \Psi_p + \Psi_s$ now represents the component potentials in plant-water relations, where Ψ_p = potential arising from turgor, and Ψ_s = solute or osmotic potential (less than free water and thus negative). Two other potentials to be considered are Ψ_m (matrix) and Ψ_z (gravitational). Both are negative, that is, they lower the capacity of water to do work, but Ψ_z is small and usually disregarded. However, Ψ_m must be considered because it accounts for the water retained by surface forces of cell structures and of cytoplasm. Nonvacuolated meristematic tissue, for example, retains much of its water by matrix forces. As Ψ is lowered, water loss from a system is reduced as dehydration becomes critical. In the woody plants we are dealing with, leaf water potential Ψ may drop to −100 bars. In such cases Ψ_m becomes a significant component (Leopold and Kriedemann, 1975).

The aqueous phase in plants functions in both biochemical and biophysical systems and is dynamic rather than static. Water provides a suitable medium for biochemical reactions and is a structural component of macromolecules. Thus, both the chemical and physical functions of water depend on the energy status of its molecules, expressed by the general term "plant water potential (Ψ)."

Evapotranspiration from leaves lowers Ψ in leaf tissue, which results immediately in transmission of this lowered Ψ to the root by way of tension in the vascular system. Thus the demand for water in distressed tissue sets up a water-potential gradient that favors water flow to the area of lowest potential (Ψ). However, as shown in Figure 6-3, water-uptake rates lag behind transpiration during the diurnal cycle.

As shown in Figure 6-4, soil type has a significant influence on Ψ at a given soil water content. In a clay soil, the wilting point may vary between 16 and 23 percent soil moisture, depending upon the transpiration rate. Thus, both water content and soil water Ψ at wilting are influenced by dynamic factors (Gardner, 1960). In Work's test, apparently the root density or the capillary conductivity of the soil were not high enough to meet the transpiration demands and still permit optimum fruit growth. In a more recent study of factors affecting foliage efficiency of apple trees (Westwood et al., 1960), Golden Delicious fruit were

Figure 6-3 **(A)** The daily rise in transpirational water loss precedes by several hours the generated water uptake by the tree roots. **(B)** This dendrograph shows that greater stress occurs during daylight in the upper trunk near the leaves, resulting in greater daily shrinkage than in the lower trunk, which is nearer the source of water. [A: After Kramer, 1937; B: After Martin H. Zimmermann, "How Sap Moves in Trees." Copyright © 1963 by Scientific American, Inc. All rights reserved.]

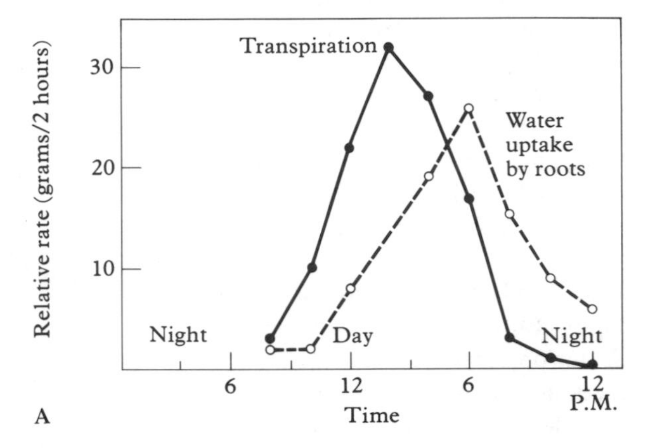

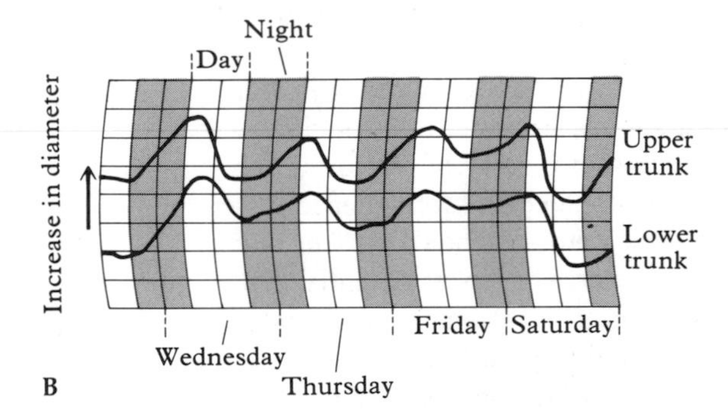

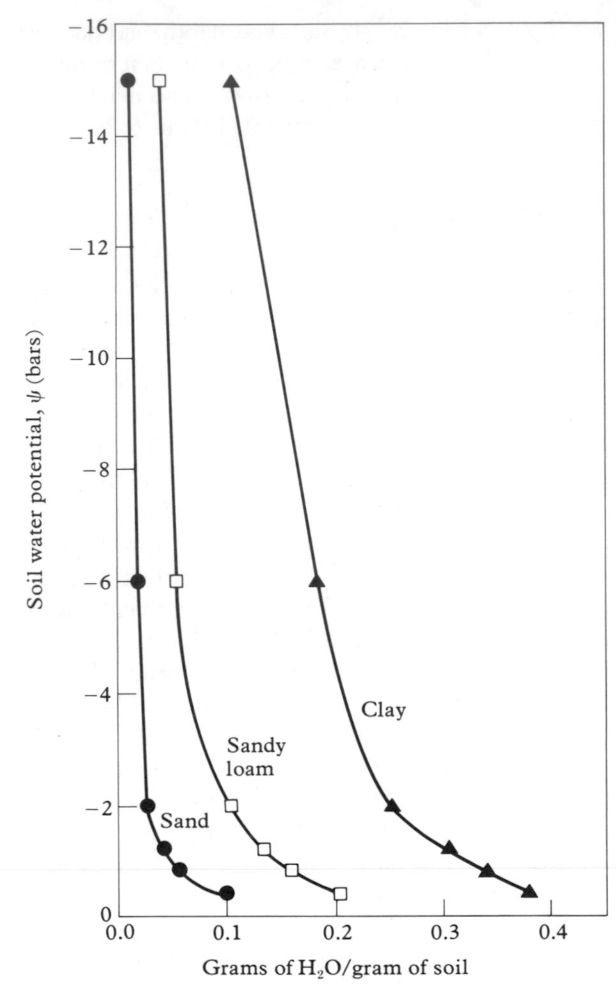

Figure 6-4 Clay soils have more water-holding capacity than sandy soils, but in both the water potential (ψ) declines exponentially as the soil water content is reduced. [After Slayter and McIlroy, 1961].

24 percent smaller when grown with sub-optimal soil moisture, as compared to ample soil moisture. At no time during the growing season, however, was there any visible wilting on trees in the sub-optimal moisture plot. Tukey (1959) has shown that moisture is pulled from the fruit during high transpiration in the diurnal cycle, causing the fruit to shrink. But, under normal conditions, turgor is restored and growth occurs at night.

With relatively coarse-textured soils, several studies indicate no reduction in tree or fruit growth until soil moisture is reduced to just above the permanent wilting percentage. Also, apples grown in a mixed soil consisting of equal parts of field soil, sand, and peat are not affected until four-fifths of the available moisture is used up. In contrast, however, other studies with coarse and medium-textured soils, under field conditions, indicate tree and fruit growth are reduced before the permanent wilting percentage is approached. In some situations, some areas of the root zone are drier than others, and the roots in moist soil are not able to supply the needs of the whole tree. It should be stressed that in a soil wet to field capacity, water does not move readily into the drier soil nearby. When the water content is below field capacity, the capillary conductivity is sufficiently small that redistribution of moisture in the soil is usually negligible.

Root distribution varies considerably with soil type and depth. Aldrich et al. (1935) found that pear roots were unevenly distributed by depth in clay adobe soil and that moisture extraction was positively correlated with root concentration in the different zones. In citing his own and other research tests, Work (1939) stated that tree-root distribution is much more extensive in coarse-to-medium textured soils than in fine-textured ones, and also that capillary movement is better in coarse than in fine-textured soils. Thus, in clayey soils the complete use of available moisture is never realized, due to both limited root distribution and limited capillary movement. Under such conditions, the soil in contact with the roots is probably much drier than soil only a short distance away. Roots grow more slowly in clay soil, and water is absorbed mainly by the growing root tip. This would explain why irrigations have resulted in better fruit growth during periods of stress, even though the *average* soil moisture was well above 50 percent available. This does not imply that capillary movement of water through the soil directly adjacent to roots is not important, but that, under conditions of stress, capillary conductivity may be too slow to permit efficient removal of water by the plant roots. Aldrich et al. (1935) found that roughly twice as much available water was extracted from the .3 to .6 meter zone as from the .9 to 1.2 meter zone. As pointed out by Kramer (1937) and Gardner (1960), there are two aspects of water availability: the ability of the plant to absorb water with which it is in contact, and the ease with which water moves in to replace that used by the plant. Water movement by capillary conductivity in soils becomes very small when the water content decreases below field capacity and becomes extremely small as soil water Ψ decreases, but even at low Ψ it does not reach zero. At -1 bar Ψ, Chino clay has a capillary conductivity of about 10^{-3} cm/day. At -10 bars it is reduced to slightly over 10^{-5} cm/day. Roots in dry soil do not grow to seek water but grow only in moist (above PWP) soil. Thus moisture, even one centimeter away, may be unavailable to a root.

The closure of stomata during the day is related indirectly to soil moisture. As soil moisture is depleted, stomata of leaves close earlier in the day, and fruit growth is reduced. Early stomatal closure is related both to low soil moisture and to a high rate of evapotranspiration. Both fruit growth and total carbohydrate production are reduced when stomata close early in the day. Temporary water stress in leaves results in moisture withdrawal from fruit during the day.

Soils of different textures and containing different amounts of organic matter have different water-holding capacities and different amounts of available moisture for plant growth. Table 6-2 indicates some of the properties of three general soil types.

Table 6-2 The relationship of soil type to water holding capacity and moisture availability to plants.

Soil Type	Water Content: At Field Capacity (%)	Water Content: At Permanent Wilting (%)	Percent Available Water
Sandy loam	12	4	8
Loam	24	12	12
Adobe clay	38	19	19

With mature pear trees under irrigation in a clay soil, soil moisture was lowest in compacted soil, intermediate in cultivated soil, and highest under alfalfa or straw mulch. Yields were directly related to soil moisture (Westwood et al., 1964). In another plot of Anjou pears, soil moisture and yield were highest under clean cultivation. Sod plots had lowest moisture. There was a clear indication that additional irrigation water should have been used with the sod in order to supply adequate moisture to both grass and trees. In another plot at the same site (clay soil), younger Bartlett and Bosc trees given the same irrigation as the larger Anjou trees appeared not to suffer with sod culture even though the soil contained slightly less moisture than cultivated soil (Westwood et al., 1964).

In summary, several factors must be considered in evaluating tree performance as related to moisture supply. The major factors are (*a*) soil type, (*b*) irrigation frequency and depth of water penetration, (*c*) extent of root distribution, and (*d*) the intensity and duration of external moisture stress, such as wind, temperature, and humidity. All of these factors are closely interrelated, and a single one cannot be evaluated apart from the others.

Figure 6-5 Sprinkler irrigation in caneberries. This method of irrigation is well adapted to land that is too uneven for furrow irrigation. [Courtesy F. J. Lawrence]

Irrigation

The main irrigation methods used are sprinkler, furrow, border (flooding), and—recently—trickle or drip irrigation. Furrow irrigation is unsuited to steep slopes except where contouring is done rather than running the furrows straight down the slope. Sprinklers are adapted to a wide range of soil types and to various topographies and slopes. They are especially useful on rolling land that cannot be leveled or on steep slopes with erosible and shallow soils (Fig. 6-5).

Sprinklers may be used for frost protection and heat control as well as for irrigation. A single overtree system can be designed to do all three. Three common types of sprinklers are fixed-head, perforated pipe, and rotating head. Sprinklers should be designed to apply water uniformly at a rate that will not cause runoff and with a pump and water-supply capacity to handle maximum irrigation needs for the climate and crop. Pressure to operate sprinklers can be supplied by gravity or by pump. The rising costs of electricity and fuel for pumping water in recent years is of concern to many fruit growers.

Trickle, drip, or daily-flow irrigation has come into use in recent years (Fig. 6-6). (The term "trickle" was coined in England, "drip" in Israel, and "daily flow" in Australia.) This type of irrigation is based on the concept of providing a continuous supply of moisture to only part of the root system. This reduces moisture stress continuously, as contrasted to conventional irrigation, which corrects an existing moisture stress. Trickle irrigation applies water to about 25 percent of the root system under low pressure at the low rate of 3 to 7.5 liters per hour per plant to maintain near field capacity in the soil zone near the plant. Water is delivered by an emitter at each plant. The system should provide equal delivery from all emitters; to accomplish this, pressure, friction loss, and changes in elevation must be taken into consideration. The system includes a pump, a filter, a time-clock-operated solenoid valve, a pressure regulator, plastic hose for major distribution, microtubes as individual emitters, and tensiometers to measure soil moisture. Microtubes sometimes become plugged with algae, so some orchardists are using larger emitters. Some advantages of trickle irrigation are (*a*) reduced water consumption, (*b*) even distribution on steep, rocky hillsides, (*c*) easier use of equipment, and (*d*) easier weed control. Because of the many differences in crops, soil types, water quality, and method of application, details of the best systems for local fruit areas should be obtained from local extension agents or experiment stations.

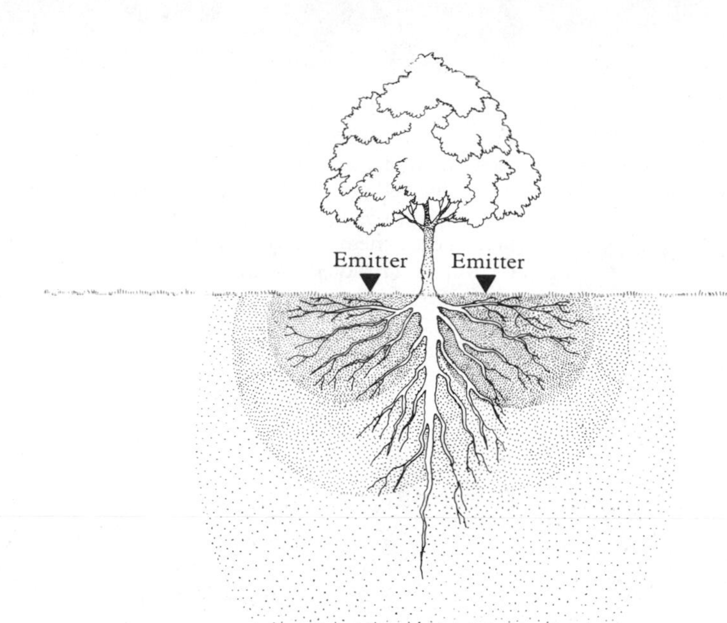

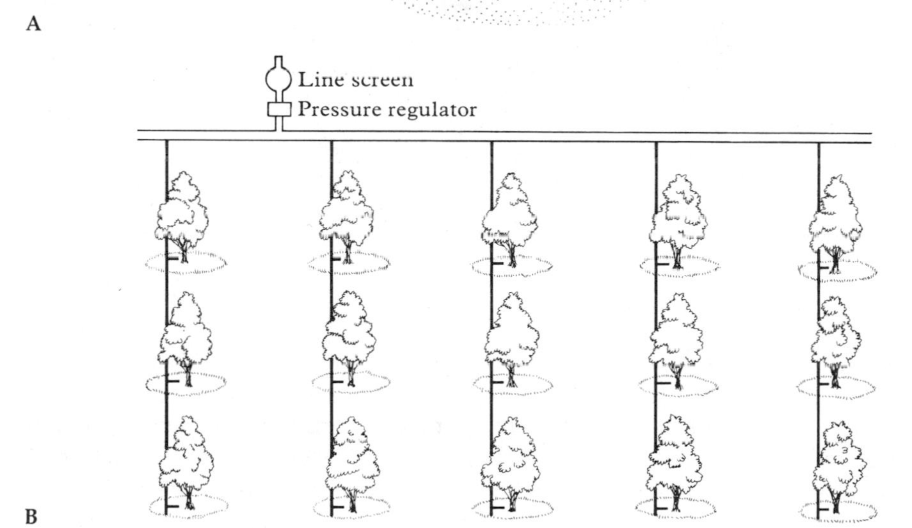

Figure 6-6 **(A)** Pattern of wetting from trickle irrigation **(B)** A trickle system outlined for an orchard. The main line, with screen and pressure regulator, ties into the header line, which feeds half-inch laterals along the rows, with an emitter at each tree. [A: After Shearer et al., 1974; B: After Kenworthy, 1972]

Soil Management and Weed Control

Soil management in fruit and nut plantings is aimed primarily at weed control, but it also is concerned with two other goals that are closely related to weed control: (*a*) retaining good soil structure by preventing compaction and (*b*) the possible use of cover crops to improve fertility and to prevent erosion.

Weed control may be achieved by mechanical means (tillage, flailing, mowing, etc.), use of chemical herbicides, planted cover crops, or a combination of two or more of these methods (Fig. 6-7). The choice

A

B

C

Figure 6-7 Weed-control practices. **(A)** Combination chemical control in the rows, with a mowed grass sod between rows. **(B)** Weed control over the entire orchard floor with nonresidual chemicals. **(C)** Complete orchard-floor control with residual-action chemicals. The old method of discing or other tillage (not shown) is found to cause greater soil compaction than nontillage methods.

of method is determined by the nature of any critical problems in the specific planting.

Cover crops are not often used in nonirrigated plantings because they compete with the fruit plant for moisture. However, where irrigation is done, a permanent cover crop may be used to prevent erosion and to provide a firm base for orchard equipment when the soil is wet. Clover, alfalfa, or other legumes can be used to enrich the soil with nitrogen. This may be of value in view of increasing fertilizer costs, but one disadvantage is that the amount of nitrogen fixed per hectare is not easily controlled. Too much nitrogen too late in the season might delay fruit maturity and prolong plant growth, predisposing the plants to early winter freeze damage. Another disadvantage is that rodents proliferate in legume cover crops. Of the nonlegumes, a mowed grass sod is often used between rows, with chemicals used to control weeds in the row where the trees or vines are growing. Grasses such as orchard grass, blue grass, or fescues are often used.

Herbicidal chemicals may be used alone for weed control, but often they are used in combination with tillage, mowing, or other mechanical means. In using herbicides, these general rules should be followed:

1. Weeds are killed best under conditions for fast germination and growth, except for pre-emergence chemicals.

2. Young weeds are killed more easily than older established ones.

3. Poor results may result from unusual temperatures or rainfall during or soon after treatment.

4. Fine textured (clay) soils require higher rates of herbicide than coarser (sandy) ones.

5. Equipment should be accurately calibrated and herbicides should be accurately measured to insure that the right amount of toxicant is applied uniformly over the land.

6. Learn the characteristics of the herbicide being used—for example, the evaporation loss, the rate of movement into the soil, the kind of plants killed and the stage at which they are killed, and the decomposition rate of the residue.

7. Use only herbicides registered for a given use and apply them separately unless the registration indicates they can be mixed.

8. All herbicides are dangerous if not handled properly. Store them in a locked place and destroy empty containers. Use safety procedures, and follow the recommendations *on the label.*

General characteristics of several common herbicides are given in Table 6-3. The use of an herbicide on a particular crop in a given area should be done under the detailed direction of the county agricultural agent.

Nutrient Availability

The nutrients available to plant roots depend upon many factors, among which are climate, plant species, rootstock type, soil type, total nutrients in the soil, soil moisture, soil oxygen content, humus content, soil pH, and base saturation. Figure 6-8 presents an overall view of the processes going on in the soil and the relationships with plants growing in the soil.

Positively charged ions (cations) are held in the soil on the surface of negatively charged clay particles and humus. These cations consist largely of calcium, magnesium, potassium, sodium, and hydrogen. The first four raise the pH value of the soil and make the soil more basic (alkaline); hydrogen ions lower the pH value and make the soil more acid. The ability of a soil to take up and release cations is called its cation-exchange capacity (CEC). This capacity is expressed as milliequivalents (m.e.) of cations adsorbed per 100 grams of soil. When all sites on the soil clay particles and humus are filled by hydrogen, it is hydrogen saturated. When the sites are filled by bases (calcium, magnesium, etc.), it is base saturated. If a soil with a CEC of 20 contains 15 m.e. of bases, then it is said to be 75 percent base saturated. One m.e. of a cation is ionically equal to one m.e. of another, but the weights are different because of their different molecular

Table 6-3 Characteristics of herbicides used in orchards and berry fields.

Chemical	Amount per Hectare (kg or liters, actual active ingredient)	Time and Remarks
Simazine	.9–1.8	Late fall to early spring. Lower rates in sandy soils.
Diphenamid	6.7	After weeds removed—at planting. Need moist surface.
Diuron	3.6	Lower rates in sandy soil.
CIPC	13.5	Winter.
Dichlobenil	6.7	Irrigate in if above 21°C.
2,4,D amine	1.1–2.2	Any time. Repeat as needed. Early bud stage of weeds.
2,4,D	1.1–2.2	Any time. Repeat as needed. Early bud stage of weeds.
Dalapon	5.6	In spring. Repeat as needed. For grass control.
Mineral spirits	Undiluted	When vines dormant. Use as spot treatment.
Dinitro general	2.1	Any time on small weeds. Cover foliage—don't wet trunks.
Aromatic weed oil	151 liters	Any time on small weeds. Cover foliage—wet trunks.
Paraquat	.6–1.2	Any time except if nuts are on the ground. Weeds 2–15 cm high.
Trifluralin	.6–2.2	Prior to planting. Mix with soil.
Summer spray oil	1.9 liters	Spray to 45 cm on raspberries to remove leaves and spurs.
Chloroxuron	4.5	Any time except later than 60 days from harvest.
Terbacil	1.8–3.6	Early spring. Lower rate on sandy soil.

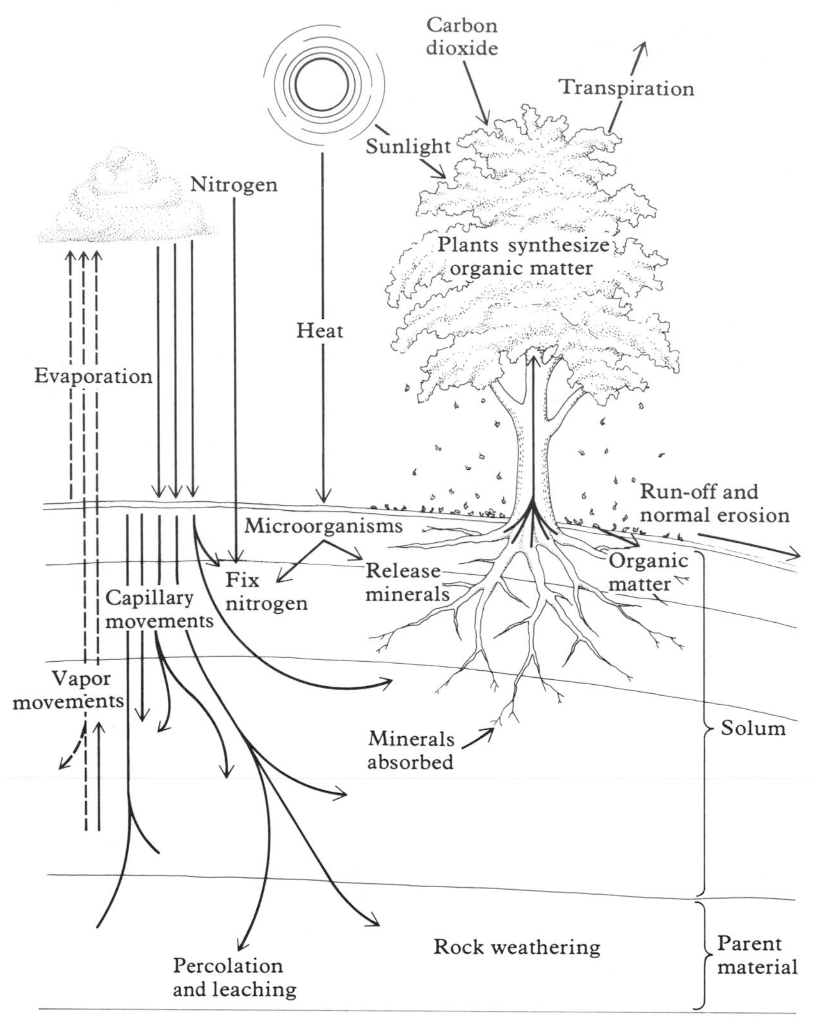

Figure 6-8 Both physical and biological processes are essential to the life of the soil. Plants remove water and mineral nutrients from the soil and combine them with carbon dioxide from the air to form complex organic substances. Much of this organic matter eventually returns to the soil, where it decomposes to release nutrients for a new cycle. If intensive agriculture is practiced, additional minerals are often needed to get best use of the land and incident solar energy. [After Charles E. Kellogg, "Soil." Copyright © 1950 by Scientific American, Inc. All rights reserved.]

weights. For each of the following elements, 1 m.e. per 100 grams of soil has the weight shown below:

Element	*Weight per Hectare (in kg)*
Hydrogen	221
Calcium	449
Magnesium	269
Potassium	875
Sodium	516

The diagrammatic clay particle in Figure 6-9 shows one possible type of base saturation.

Soil pH is a measure of the active H ion in the soil, which determines acidity and alkalinity. A pH of 7.0 is neutral, and each full unit below 7.0 is ten times more acid than the next higher value. Thus, pH 6.0 is ten times and pH 5.0 is one hundred times more acid than pH 7.0. Likewise pH 8.0 is ten times more alkaline than is pH 7.0 (Fig. 6-10).

The availability of mineral nutrients in soil as influenced by pH is as follows:

Nutrient	*pH Range for Best Availability*
Nitrogen (N)	5.8–8.0
Phosphorus (P)	6.5–7.5
Potassium (K)	6.0–7.5
Calcium (Ca) and Magnesium (Mg)	7.0–8.5
Sulfur (S)	6.0–10.0
Iron (Fe)	4.0–6.0
Manganese (Mn)	5.0–6.5
Boron (B)	5.0–7.0
Copper (Cu) and Zinc (Zn)	5.0–7.0
Molybdenum (Mo)	7.0–10.0

NITROGEN **(N)** Because N is not a breakdown product of soil minerals, most of it is in the organic matter of the soil. This organic matter is decomposed by microorganisms, releasing nitrogen in ammonia (NH_3), which in turn is oxidized by bacteria to NO_3^-. In this form, N is readily taken up by plant roots or is lost by downward leaching. NH_4^+ and urea may be absorbed by some plants, e.g., cranberry and blueberry.

PHOSPHORUS **(P)** Total P in soils is variable but low. Available P comes from breakdown of soil minerals or organic matter, but at any one time it constitutes only about .5 to 1 percent of the total soil P. Besides organic forms, P occurs mostly as the slightly soluble calcium phosphate or apatites.

POTASSIUM (K) Except for sands, most mineral soils are high in total K, containing as much as 2 to 3 percent in the surface 30 centimeters. Most of it is in the form of micas and feldspars and is unavailable. The readily available forms are the K salts dissolved in the soil water and the K ions adsorbed on the soil colloids. Another form of K, termed "slowly available" is held tightly in the lattice structure of the clay complex. Part of this K becomes available as the more soluble forms are used up.

CALCIUM (Ca) AND MAGNESIUM (Mg) Acid soils are often limed to increase pH. Limestone varies in both $CaCO_3$ and $MgCO_3$ content but is mostly $CaCO_3$. One type is 80 percent $CaCO_3$ and 14 percent $MgCO_3$. Dolomitic lime contains more Mg than other forms. Trees usually do best at a soil pH above 5.5 and below 8.0. When the pH (and the base saturation) is low, improved soil structure, aeration, and root growth are obtained by liming.

SULFUR (S) The supply of S in a soil may come from SO_4^{--} in the irrigation water, residues of pesticide sprays (e.g., lime sulfur and wettable sulfur), organic matter and S-containing fertilizers such as $(NH_4)_2SO_4$ and K_2SO_4. Applied inorganic forms of S are quickly converted to organic forms by microorganisms.

BORON (B) Available soil B may be reduced by crop removal, leaching, and reversion to less available forms. Low soil moisture decreases B availability. Overliming also reduces availability.

SODIUM (Na) AND SALINITY A high soil content of soluble salts adversely affects crop yields. Poor physical condition (crumb structure) of the soil occurs when the proportion of Na^+ to other exchangeable bases is high. Tree fruits generally have a low tolerance to high salt content. Excess salt can be leached

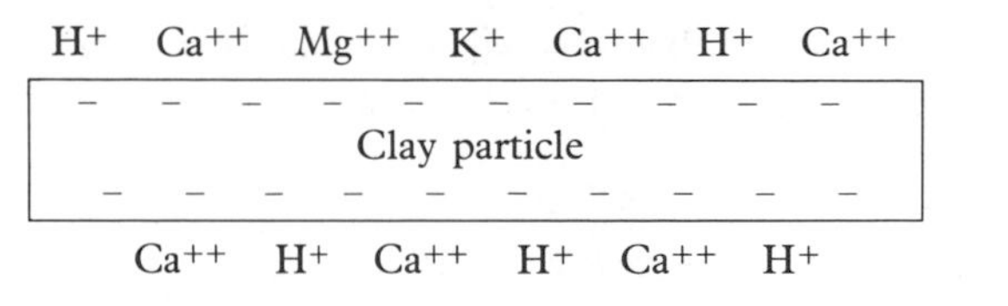

Figure 6-9 This drawing represents a clay particle with twenty negative (−) charges. Five charges are saturated with hydrogen. Fifteen charges are saturated with bases, which make it 75 percent base saturated. Of these fifteen charges: twelve are saturated with Ca (= 60 percent Ca saturation); two are saturated with Mg (= 10 percent Mg saturation); one is saturated with K (= 5 percent K saturation). [After Alban, 1958]

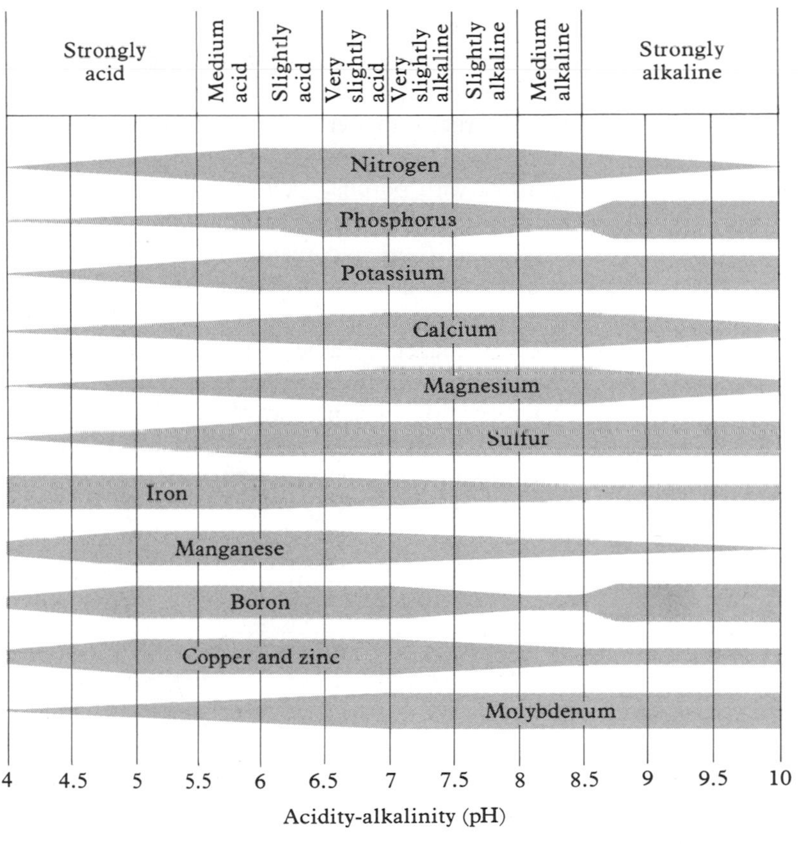

Figure 6-10 Availability of nutrients to plants is affected by the condition of the soil. The more soluble a nutrient is under a particular condition of soil acidity or alkalinity, the thicker is the horizontal band representing the nutrient. Solubility is directly related to the availability of the nutrient in an ionic form that is assimilable by the plant. [After Christopher J. Pratt, "Chemical Fertilizers." Copyright © 1965 by Scientific American, Inc. All rights reserved.]

from a soil with good quality water, provided enough water is applied, and good drainage is provided. When excess exchangeable Na^+ is removed, an amendment such as gypsum ($CaSO_4$) is added to increase base saturation and improve soil structure.

ORGANIC MATTER Organic matter is constantly changing in the soil as it passes through the various steps of decomposition, but the total amount usually remains nearly constant. This is because of additions in the form of leaves, stems, roots, and dead soil organisms. Organic matter comes only from the decomposition of plant and animal residues. These residues are converted to CO_2, H_2O, mineral elements, and other fairly stable organic compounds. The latter comprise most of the organic matter in soils. Oxidation of organic matter is slow in cool soils with relatively low O_2 supply, but the process is rapid in hot climates with porous open soils having a good O_2 supply.

Humus is organic matter in the soil that has undergone considerable decomposition. It is not a homogeneous, clearly defined compound of known chemical makeup. Rather it is a dark heterogeneous mass that includes residues of plant and animal origin, together with substances synthesized by soil microorganisms. Humus formation depends in part on incorporation of mature plant material containing lignin. Humus has a C:N ratio of about 10:1 in contrast with straw (80:1) and green legume (12–20:1). Green-manure crops tilled into the soil furnish a more immediate source of nutrients. The N of humus appears to be bound in proteins and possibly other complexes, which decompose only slowly. This is advantageous in crop production because with humus decomposition there is a slow release of available N for use by plant roots. Humus contains 3 to 6 percent N and 55 to 58 percent C.

All of the facts presented above are of value in evaluating the nutritional needs of trees, but by themselves they are not adequate. When used along with leaf-tissue analysis, these facts can be of value, especially where problems of high salt, gross nutrient imbalance, or soil pH imbalance exist.

Nutrient Uptake

In order for minerals to be absorbed from the soil by plants, they must first pass from the soil solution into root cells. To do this, they must pass through the plasma membrane, traverse the root cortex via cytoplasmic strands connecting the cells, then unload into the xylem vessels (Fig. 6-11). But to do this, the ions must once again pass through the plasma membrane, this time of the stelar cells. Once outside the cell and

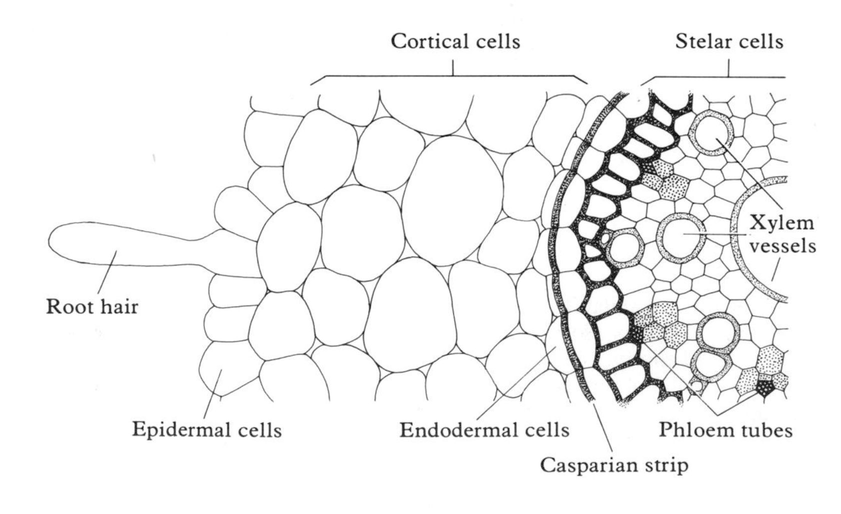

Figure 6-11 The casparian strip prevents mineral ions from moving by diffusion through the porous wall spaces of cortical cells (*left*) into the wall spaces of stelar cells (*right*) and thence into the xylem vessels of the stele. The strip, connecting the walls of adjacent endodermal cells, is impregnated with waxlike substances and therefore is impermeable both to water and to ions in solution. [After Emanuel Epstein, "Roots." Copyright © 1973 by Scientific American, Inc. All rights reserved.]

into the porous cell wall space, ions are free to diffuse into the xylem, and hence to leaves and other parts of the plant. It is not known with certainty how this passage is accomplished, but recent evidence presented by Epstein (1973) indicates that carrier mechanisms in the plasma membrane form complexes with the mineral ions. With each membrane passage, loading and unloading takes place via the carrier complex (Fig. 6-12). A general schematic for the entire system is shown in Figure 6-13. Because ion accumulation takes place against a concentration gradient, metabolic energy must be expended by the plant in the form of respiration. Thus it is important that the soil contain enough air to provide the oxygen necessary for such respiration. About 5 percent of the dry weight of plants is made up of minerals absorbed from the earth. This concentration is many times greater than

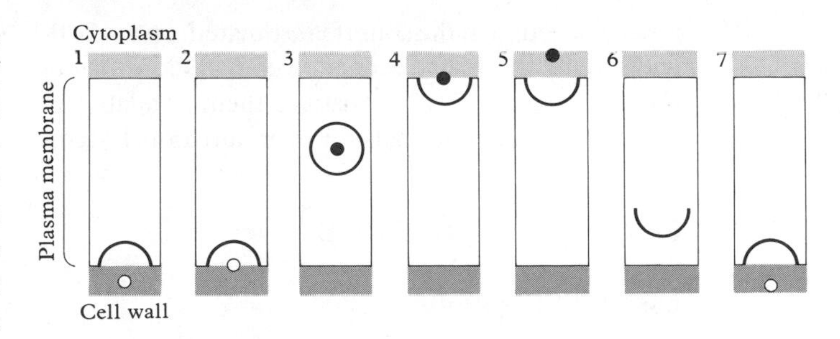

Figure 6-12 Carrier mechanisms that transport mineral ions across the plasma membrane and into the cytoplasm of plant cells are thought to operate as shown in this schematic diagram. **(1)** When the carrier is located at the outer surface of the membrane its conformation is such that **(2)** it can form a complex with a nearby ion. **(3)** Carrier and ion then migrate across the membrane. **(4)** At the inner surface the carrier changes conformation and **(5)** releases the ion into the cytoplasm. **(6, 7)** The carrier then repeats the cycle. [After Emanuel Epstein, "Roots." Copyright © 1973 by Scientific American, Inc. All rights reserved.]

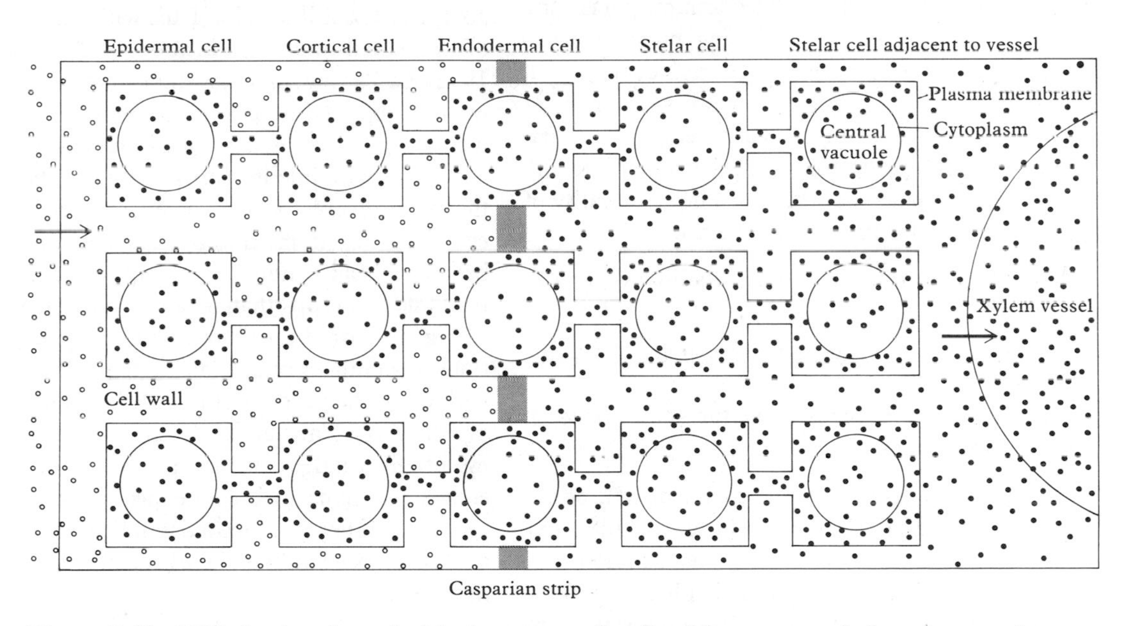

Figure 6-13 Diffusing ions (*open dots*) in the porous cell walls of the cortex reach the xylem vessels of the stele in spite of the barrier presented by the Casparian strip. First (*left*), carrier mechanisms transport the ions through the plasma membrane into the cytoplasm of epidermal and cortical cells. The transported ions (*solid dots*) then move from cortex to stele along the many strands of cytoplasm that run from cell to cell, bypassing the Casparian strip. When the ions reach a cell near a xylem vessel (*right*), "reverse" mechanisms are thought to transport ions across the membrane once more. Free to diffuse at random, many ions enter the xylem vessel. [After Emanuel Epstein, "Roots." Copyright © 1973 by Scientific American, Inc. All rights reserved.]

is usually found in the soil. It is estimated, on a worldwide basis, that plant roots absorb 5 billion tons of minerals annually, thus making them available to humans and other animals for their nutritional needs.

FERTILIZER PRACTICE

Essential Elements

The status of essential mineral nutrients in plant tissue, when properly understood, provides a useful tool for the orchardist, because the amount of each element in the plant determines plant performance. To be in proper balance and intensity for maximum yield, each element should be within specified concentration limits. Below these limits, there is a deficiency of the element; above them, there is an excess, possibly even a toxic excess. Soil tests, including determination of base saturation (calcium, magnesium, potassium, sodium), are of value in estimating nutrient availability and adverse pH (acidity) or salt content. But since leaves are the important synthesis centers of plants, tests of leaves more nearly reflect the nutritional status of a plant than do tests made on the soil. One feeds the plants, not the soil. The essentiality of an element is based upon these criteria:

1. The plant cannot complete its life cycle without it.
2. No other element can completely substitute for it.
3. It must be found essential for a wide range of higher plant species.

On this basis, nitrogen (N), phosphorus (P), potassium (K), calcium (Ca), magnesium (Mg), sulfur (S), iron (Fe), boron (B), copper (Cu), manganese (Mn), zinc (Zn), molybdenum (Mo), and chlorine (Cl) are essential elements, in addition to the important nonmineral elements carbon (C), hydrogen (H), and oxygen (O).

N and S are constituents of both cytoplasmic and nuclear proteins. N also is found in coenzymes, hexoseamines, base units of nucleic acid, and in cytokinins and auxins. N is much more mobile in plants than is S.

A key role of P is in energy exchange by breaking and forming high-energy bonds, such as those found in adenosine triphosphate (ATP); it also is found in phytic acid, coenzymes, fatty materials (phospholipids), and nucleic acids. P is not needed in large amounts in trees because it is used over and over and not very much is removed with the crop. A full crop of pome or stone fruits removes only about 10 kg of P per hectare from the soil. P is quite mobile in plants.

Ca functions as a cross-linkage in the pectates of the middle lamella, which binds cells together. It also is needed in enzymatic reactions (such as in the reduction of NO_3^-), provides a balance of anions and cations in the plant, and serves an important role in the stabilization of cell membranes. It is usually found in combination with pectates, as oxalate crystals and as phosphate. Calcium crystals in many tissues of the tree bind Ca and make it less available for use in tissue; this might explain why apple and pear trees with abundant Ca still develop fruit with bitter pit, a disorder related to low Ca in the fruit (Stebbins et al., 1972). Ca has low mobility in plants, but some movement is obtained by lowering pH and PO_4^{---} or by increasing Na in tissues containing calcium oxalate.

K is very mobile in plants and is needed to satisfy the requirement for a monovalent cation in several enzymatic reactions. It also helps maintain the cation balance. Usually more K is removed from the soil by fruit and nut crops than is any other mineral.

Mg is a part of the chlorophyll molecule and functions in a number of enzyme systems. It also is found with Ca in the pectates of middle lamellae.

The trace elements Fe, B, Cu, Mn, Zn, and Mo are essentially catalytic in nature and serve in various enzyme systems to mediate necessary biochemical reactions. They are needed only in small amounts and can (especially B and Cu) become toxic if present in excess.

Tissue Analysis

In order to use leaf analysis as a meaningful tool, one must know the essential elements and how they function in tissue, be able to recognize in the field common toxicities and deficiencies, and understand the complex relationships of nutrition to climate, soil type, soil moisture and special needs of crop species. Three of the factors affecting nutrient levels of leaves are:

1. Nutrient availability as related to soil condition, pH, aeration, and available moisture.

2. Nutrient ion interactions (i.e., the enhancing or depressing effects of one ion—in the soil or in the plant—on accumulation and utilization of other ions).

3. Seasonal fluctuations in leaf nutrient content as affected by crop load, climate, and type of rootstock.

LEAF SAMPLING Leaf analysis has been widely used for many years as a means of (*a*) determining the nutrient element associated with a nutrient disorder, and subsequent prescription of a corrective measure, and (*b*) as a means of estimating fertilizer needs prior to the occurrence of nutrient disorders.

Leaf nutrient levels (shortage, below normal, normal, above normal, and excess) have been established from fertilizer experiments. Leaf nutrient levels, however, may vary throughout the season and from leaf to leaf on the tree, as well as from tree to tree in the orchard. Standards developed are valid only for leaves collected during the proper period, from the proper part of the plant, and handled in a standard fashion. It is important to follow the procedures listed below for a leaf sample to be considered valid:

1. Sample only plantings old enough to bear a commercial crop or where a nutritional problem is suspected. It is good practice to run analysis on new plantings during the second season to avoid overuse of fertilizers and to detect imminent deficiencies.

2. Collect leaf samples during mid-July through August (in the Northern Hemisphere).

3. A single sample should not represent an area of more than 2 hectares. Two samples from each uniform 2-hectare area would give less chance of misrepresentation. For diagnosing a trouble spot, take a composite sample from five affected trees or vines and five nonaffected ones. Two samples from each trouble spot would be still better. In the absence of a trouble spot, take the sample from five plants of average vigor spaced evenly through an area of less than 2 hectares. Cross the area in another direction for the second sample.

4. Include only one cultivar or strain in a sample and preferably only one rootstock type.

5. Mark or map each plant or area sampled for future resampling.

Unless leaves are unusually small, fifty leaves are enough for a sample. Select all leaves from the periphery of trees at shoulder height or higher, or from a standard position on vines and bush fruits. Collect ten leaves per tree from *shoots* randomly selected from all sides of the tree. Select only one leaf from a shoot. Except for diagnosing a trouble spot, collect leaves that are free of disease or other damage. Remove the leaves with a downward pull so that the petiole remains on the leaf. From all crops except walnuts, pick leaves from the middle of the current season's *terminal shoots* of about average vigor. With walnuts, take one mid-leaflet from a leaf in the middle of a fruiting terminal. If five leaflets are present, select one leaflet immediately below the terminal leaflet. If seven leaflets are present, take one leaflet from the second pair of leaflets below the terminal leaflet (Fig. 6-14). With some plants such as grape, petioles rather than leaf blades are used in the analysis. Take spur leaves only if no other leaves are available. If the samples are contaminated with soil, spray, or other residues that would interfere with the analyses (particularly the micronutrient analyses), they should be cleaned by washing in a detergent solution and rinsed with soft water. The leaves should be washed while still fresh, and this should be done quickly (for one minute or less) to prevent loss of nutrient elements. Nearly all nonionic detergents are satisfactory for this procedure. The leaves should then be oven dried at 80°C. They may be air dried if a drying oven is not available. Do not place the leaves in a damp location where they may mold or otherwise spoil. Submit the dried sample to a laboratory for analysis along with pertinent information as to soil type, cultivar, fertilizer practice, and special problems. Local agricultural extension agents or experiment stations should be contacted to find out the services available in a given area.

SEASONAL TRENDS IN LEAF NUTRIENTS Leaf elements are usually expressed in percentages or as parts per million (ppm) on a dry weight basis. Dry weight of leaves, however, is not constant. Leaves of stone fruits (peach) and pome fruits (apple) increase in dry weight per square centimeter of surface area during the season up to about harvest, after which there is a rapid decline in dry weight in early fall. Percentages of N and P in leaves for both fruits is highest early in the season, relatively stable during midseason, and declines rapidly in late season. The trend for K is similar but does not change as much seasonally as N and P. Mg increases seasonally in peach leaves but shows no change in apple leaves. Ca increases seasonally in both peach and apple leaves. Considering all elements and their seasonal changes, the most stable period seems to be during August. For this reason, most leaf sampling for diagnostic purposes is done between late July and the end of August. Seasonal levels of several nutrients for peach leaves are shown in Figure 6-15 and for apple leaves in Figure 6-16.

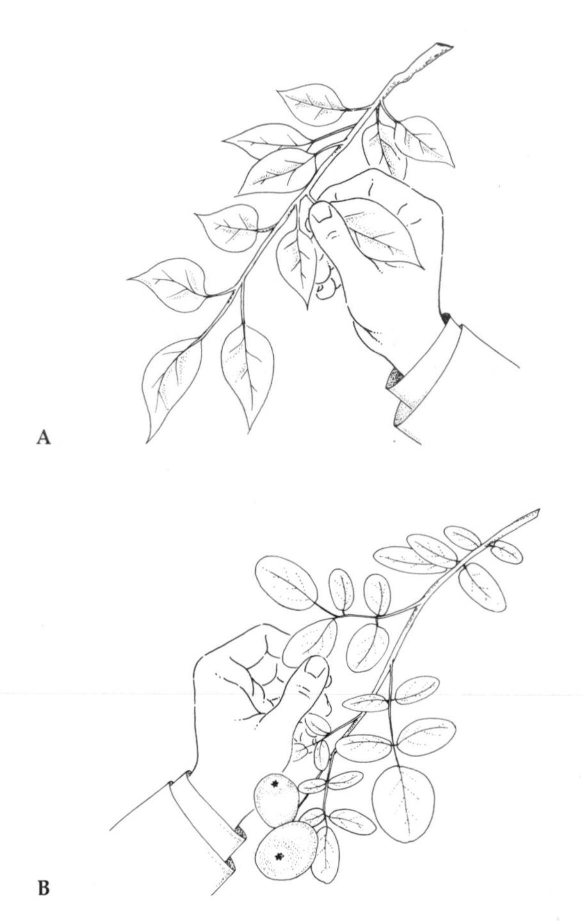

Figure 6-14 **(A)** Leaf sampling for mineral analysis should utilize the leaves in the mid-region of current shoots for most species. **(B)** With walnuts, a mid-leaflet from the middle of a fruiting terminal shoot should be used.

PEACH FRUIT: SEASONAL CHANGES A separate analysis of flesh, endocarp (stone), and kernel tissue of the peach show striking differences during the season. There is an increase in the amount of all nutrients in the flesh tissue (measured in milligrams per fruit). The nutrients increase differentially in the following order (from most to least): $K > N > P > Ca > Mg > B$. Kernel (embryo) tissue gains seasonally in all nutrients except for Ca, which declines prior to fruit maturity (Figs. 6-17, 6-18). At harvest, amounts of elements (in milligrams per kernel) were found to rank in the following order: $N > K > P > Mg > Ca > B$. Endocarp tissue gained nutrients (measured in milligrams per stone) for a time, but, after pit hardening, lost nutrients rapidly (all except Ca, which did not decline). At their highest levels, the order was as follows: $(N = K) > P > (Ca = Mg) > B$. The outward migration of nutrients from the endocarp after

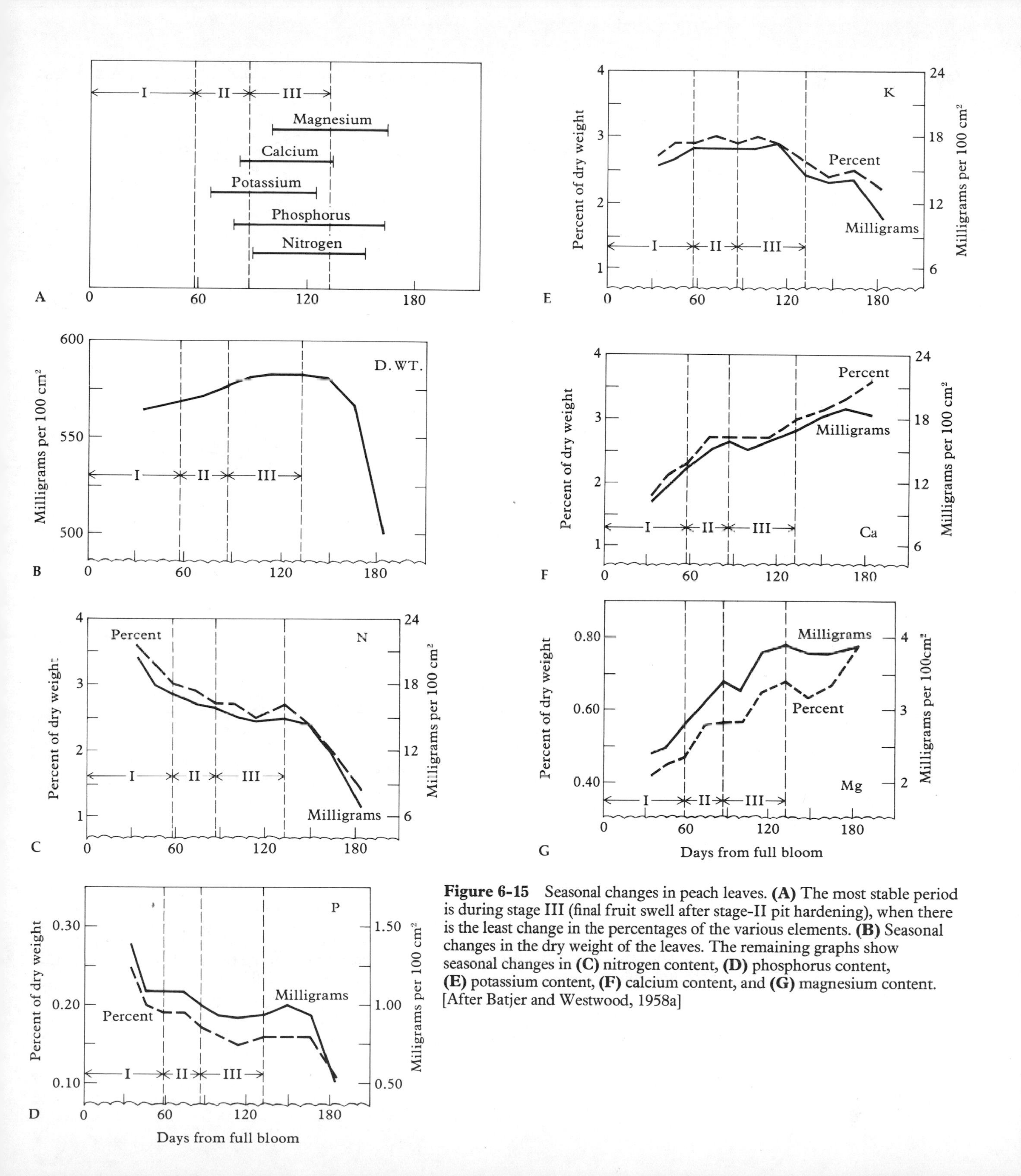

Figure 6-15 Seasonal changes in peach leaves. **(A)** The most stable period is during stage III (final fruit swell after stage-II pit hardening), when there is the least change in the percentages of the various elements. **(B)** Seasonal changes in the dry weight of the leaves. The remaining graphs show seasonal changes in **(C)** nitrogen content, **(D)** phosphorus content, **(E)** potassium content, **(F)** calcium content, and **(G)** magnesium content. [After Batjer and Westwood, 1958a]

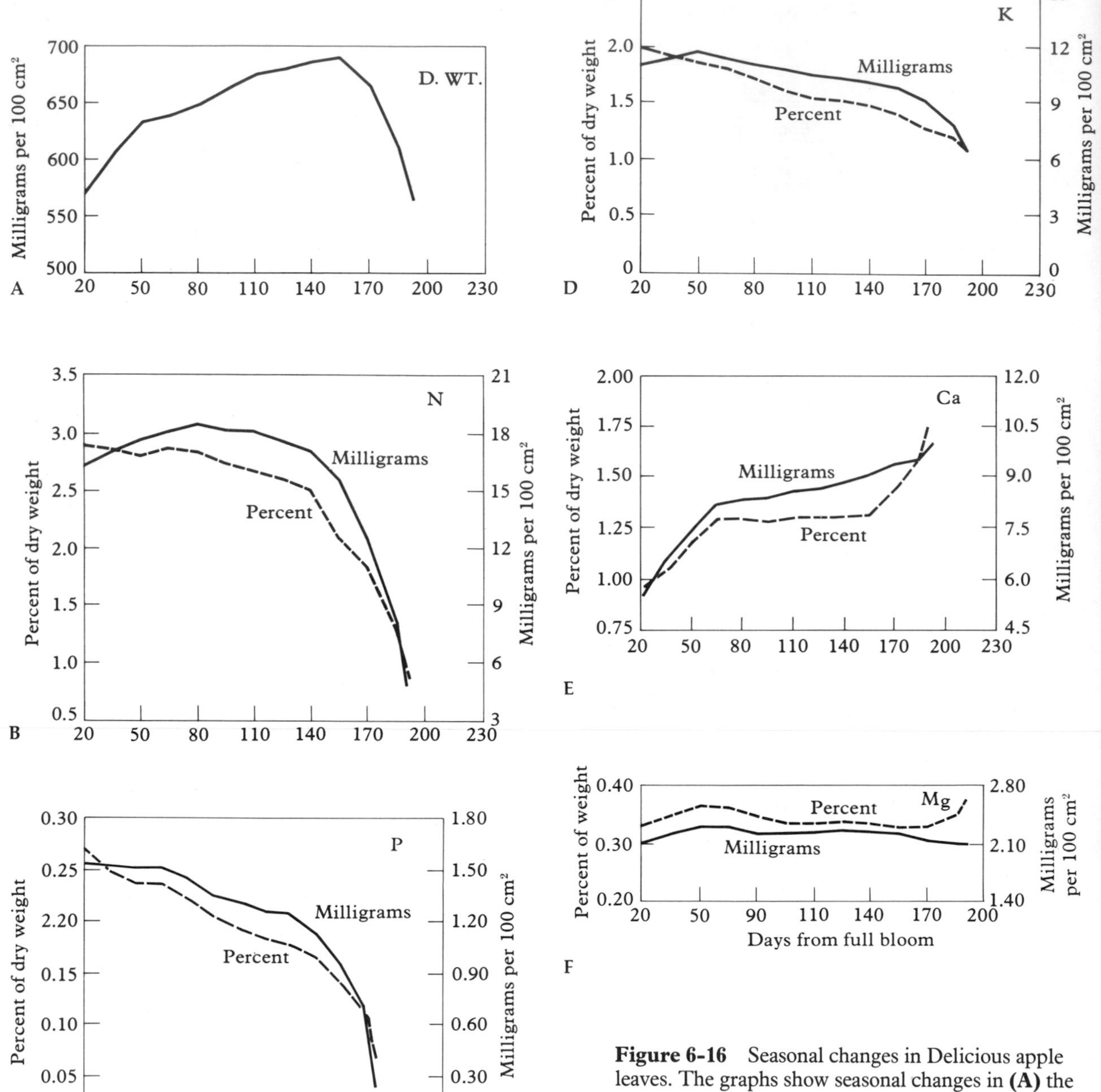

Figure 6-16 Seasonal changes in Delicious apple leaves. The graphs show seasonal changes in **(A)** the dry weight of the leaves, **(B)** nitrogen content, **(C)** phosphorus content, **(D)** potassium content, **(E)** calcium content, and **(F)** magnesium content. [After Rogers et al., 1953]

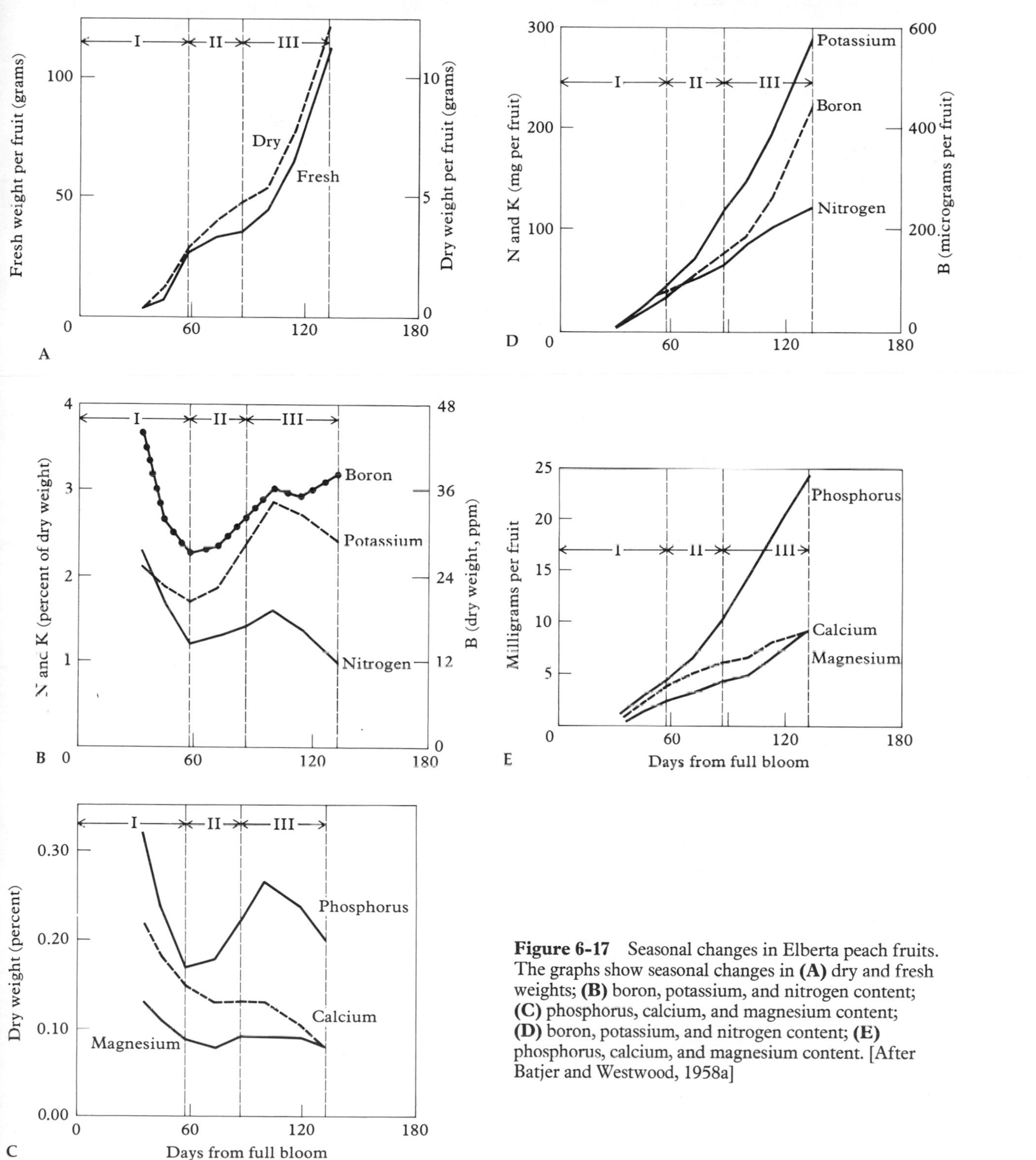

Figure 6-17 Seasonal changes in Elberta peach fruits. The graphs show seasonal changes in **(A)** dry and fresh weights; **(B)** boron, potassium, and nitrogen content; **(C)** phosphorus, calcium, and magnesium content; **(D)** boron, potassium, and nitrogen content; **(E)** phosphorus, calcium, and magnesium content. [After Batjer and Westwood, 1958a]

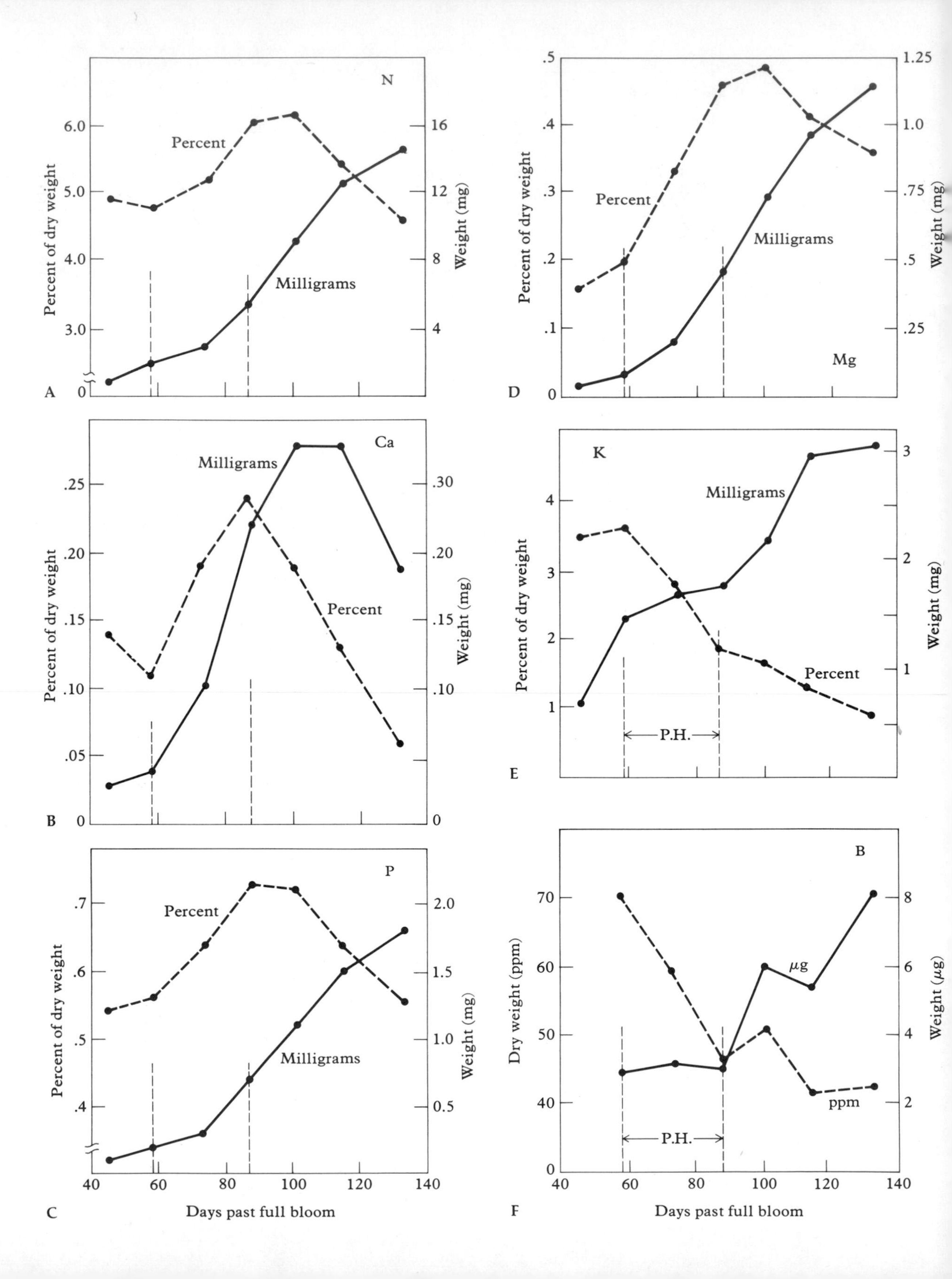
A
N
Percent
Milligrams
Percent of dry weight
Weight (mg)
B
Ca
Milligrams
Percent
Percent of dry weight
Weight (mg)
C
P
Percent
Milligrams
Percent of dry weight
Weight (mg)
Days past full bloom
D
Mg
Percent
Milligrams
Percent of dry weight
Weight (mg)
E
K
Milligrams
Percent
P.H.
Percent of dry weight
Weight (mg)
F
B
μg
ppm
P.H.
Dry weight (ppm)
Weight (μg)
Days past full bloom

Figure 6-18 Seasonal changes in peach kernels (seeds). The graphs show seasonal changes in **(A)** nitrogen content, **(B)** calcium content, **(C)** phosphorus content, **(D)** magnesium content, **(E)** potassium content, and **(F)** boron content. Except for calcium, all elements increased in total amount per seed during the season. The declines in percentages were the result of the increasing dry weight of kernels during the season. Vertical dotted lines denote the period of pit hardening (P.H.). [After Batjer and Westwood, 1958a]

pit hardening was sufficient to supply the needs of the rest of the fruit to the extent of 20 percent for N, P, and Mg; 9 percent for K; and 6 percent for B. Figure 6-19 shows these trends graphically.

APPLE FRUIT: SEASONAL CHANGES Apple fruit flesh, like peach flesh, shows increases in amounts of all nutrients during the season. For apples, the order of differential increase is slightly different: K > N > Ca > P > Mg > B (Fig. 6-20). Even though the concentration (percentage) of all elements declined seasonally (Fig. 6-21), there was a continuous inward migration (in milligrams per fruit). This occurs because fruit growth (dry-weight increase) proceeds faster than nutrient accumulation.

LEAF NUTRIENT LEVELS OF FRUIT TREES Tables 6-4 and 6-5 show levels of nutrient elements that represent the range from below normal to excess or toxic levels for several orchard species. Normal levels of nutrients are considered to be adequate levels, while above-normal levels are considered to be levels at which there is more than the amount needed for high yield and quality. The best levels within the range, where no symptoms of deficiency or excess show, usually must be determined locally for each cultivar and for each climate; however, leaf levels at which a response to fertilizer may be expected are quite similar for a given cultivar throughout the fruit districts of the world. One important variable is light intensity: areas with high light intensity permit greater utilization of nutrients than ones in low light. For example, the use of N by plants is greater under high intensity light than under low intensity; also, S deficiency and Zn deficiency may develop under high-intensity but not low-intensity light conditions. However, deficiencies occur in some areas because of the soil rather than the climate. Some areas of minor nutrient deficiencies in the U.S. are given in Figure 6-22.

Mineral Interactions

The supply of one nutrient can greatly affect the uptake or utilization of another. Thus, an increase in soil K can depress the uptake of Mg and Ca. Raising the P level tends to depress N intake and vice versa. If soil B is low enough, merely adding more N can induce severe B deficiency by depressing B uptake. The proper formation of Ca pectate is inhibited if B is too low. Thus some mineral interactions involve the absorption process, others involve translocation, and still others involve utilization in plant tissues.

The two general ion interactions are:

1. Antagonism, when one ion counteracts or depresses the effect of another.
2. Synergism, when an increase in one ion increases the concentration of another.

An external interaction involves the precipitation of Fe^{++} when high concentrations of PO_4^{---} are present. Plant roots may thus be "coated" with insoluble Fe, an element that is needed in the tissues. Antagonisms may or may not be reciprocal. Some can be beneficial, as when excess SO_4^{--} is given to young trees to depress the uptake of arsenic, which is toxic (Thompson and Batjer, 1950), or when Ca or Mg is added to irrigation water to reduce the uptake of soluble Cu. Mn antagonizes Fe in plants by converting it to the improper ionic state:

$$Fe^{++} \xrightarrow[\text{excess}]{\text{Mn}} Fe^{+++}.$$

Ferrous iron (Fe^{++}) is the form needed in leaves, so excess Mn creates a deficiency of functional Fe and "Iron deficiency chlorosis" results. Or, if Mn is deficient, Fe^{++} may be present in toxic amounts and "Manganese chlorosis" results. Also if Mn is too high, N uptake is reduced. Uptake of B is reduced by high N, but increased B levels do not depress N uptake. All

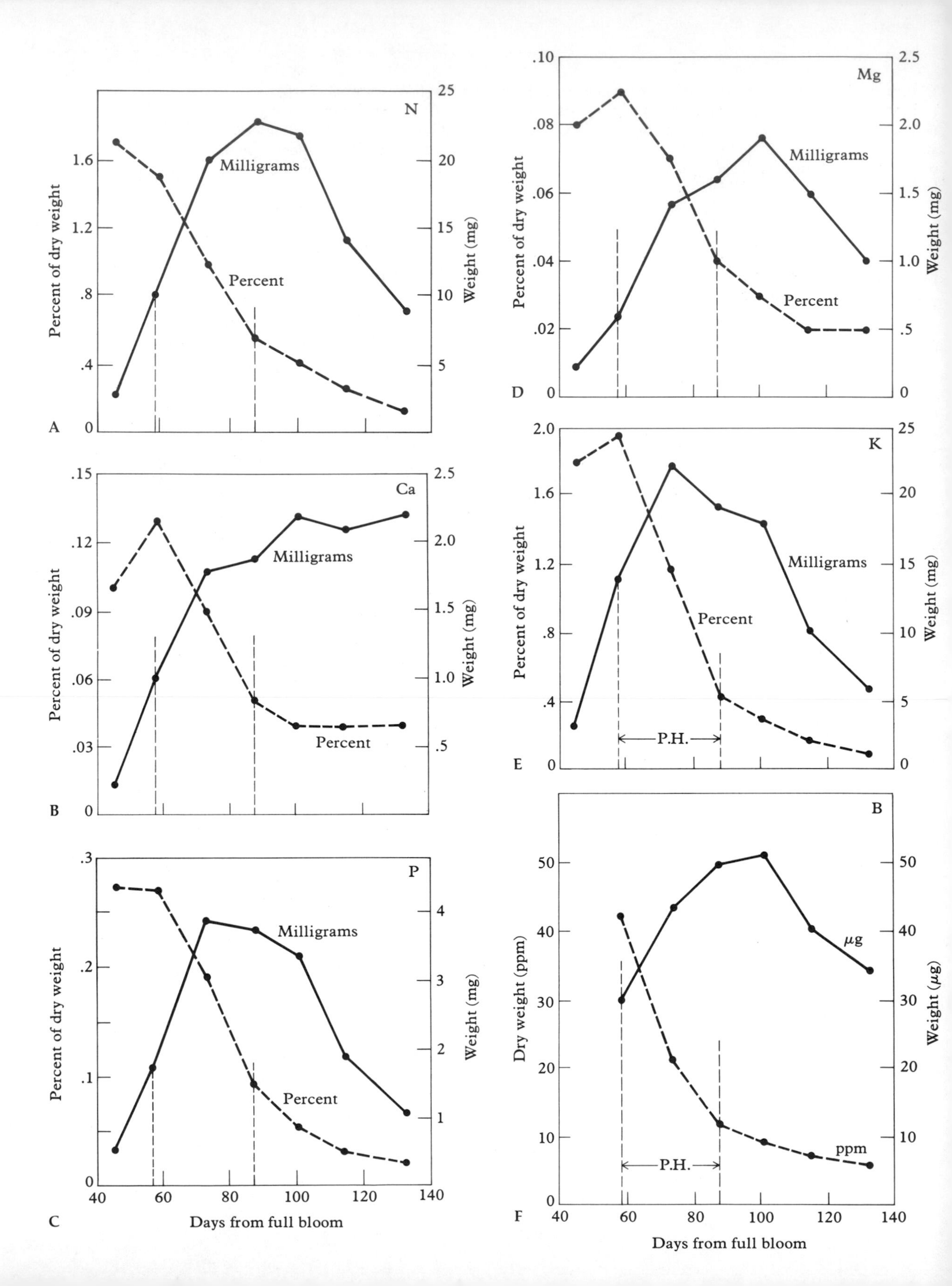

N
Milligrams
Percent
Percent of dry weight
Weight (mg)
A
Ca
Milligrams
Percent
B
P
Milligrams
Percent
C
Days from full bloom
Mg
Milligrams
Percent
D
K
Milligrams
Percent
P.H.
E
B
μg
ppm
P.H.
Dry weight (ppm)
Weight (μg)
F
Days from full bloom

Figure 6-19 (opposite page) Seasonal changes in peach endocarps (stones). The graphs show seasonal changes in **(A)** nitrogen content, **(B)** calcium content, **(C)** phosphorus content, **(D)** magnesium content, **(E)** potassium content, and **(F)** boron content. All elements increased until near the end of pit hardening, after which there was a dramatic outward migration of all elements except calcium. [After Batjer and Westwood, 1958a]

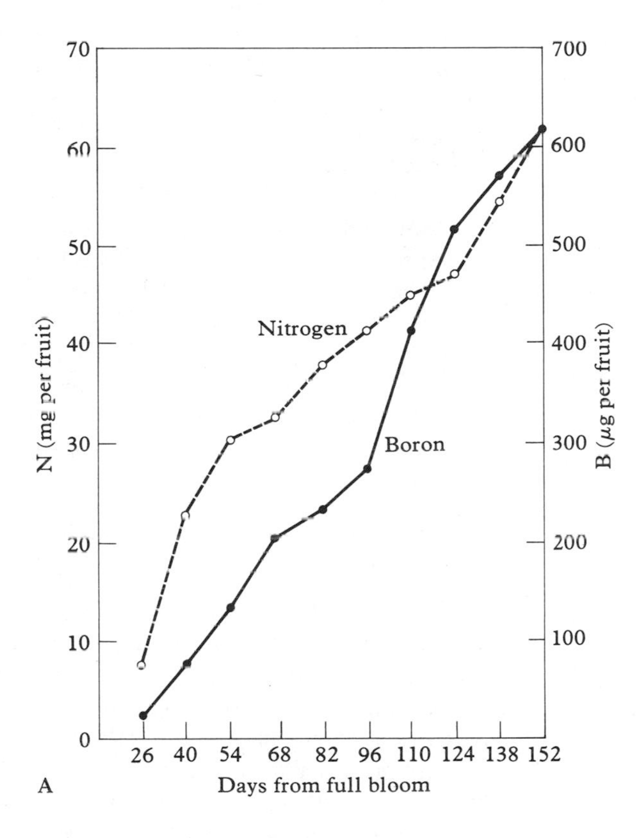

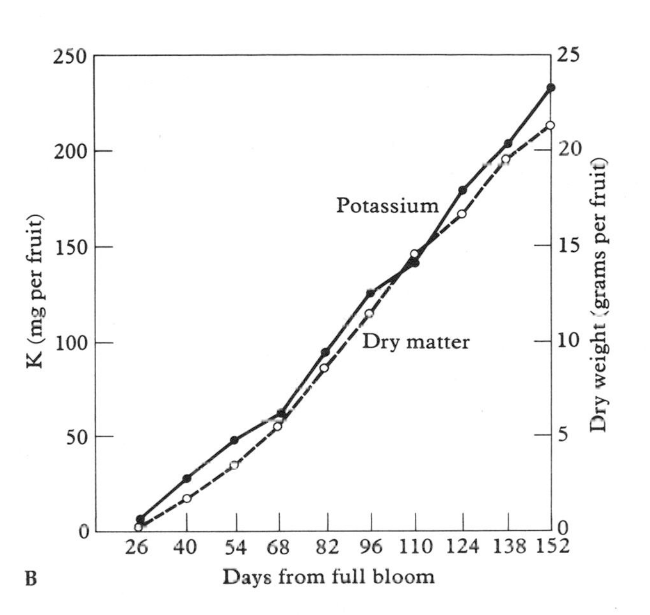

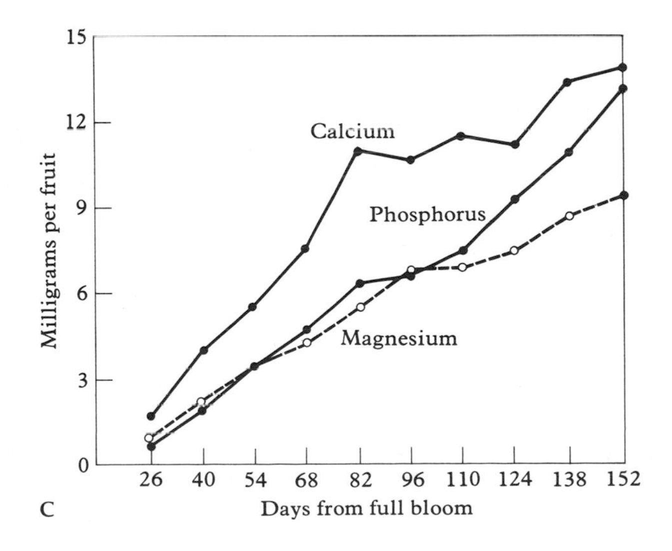

Figure 6-20 Seasonal increase in nutrients in Delicious apple fruit. The graphs show **(A)** nitrogen and boron content, **(B)** potassium and dry-matter content, **(C)** calcium, phosphorus, and magnesium content. [After Rogers and Batjer, 1954]

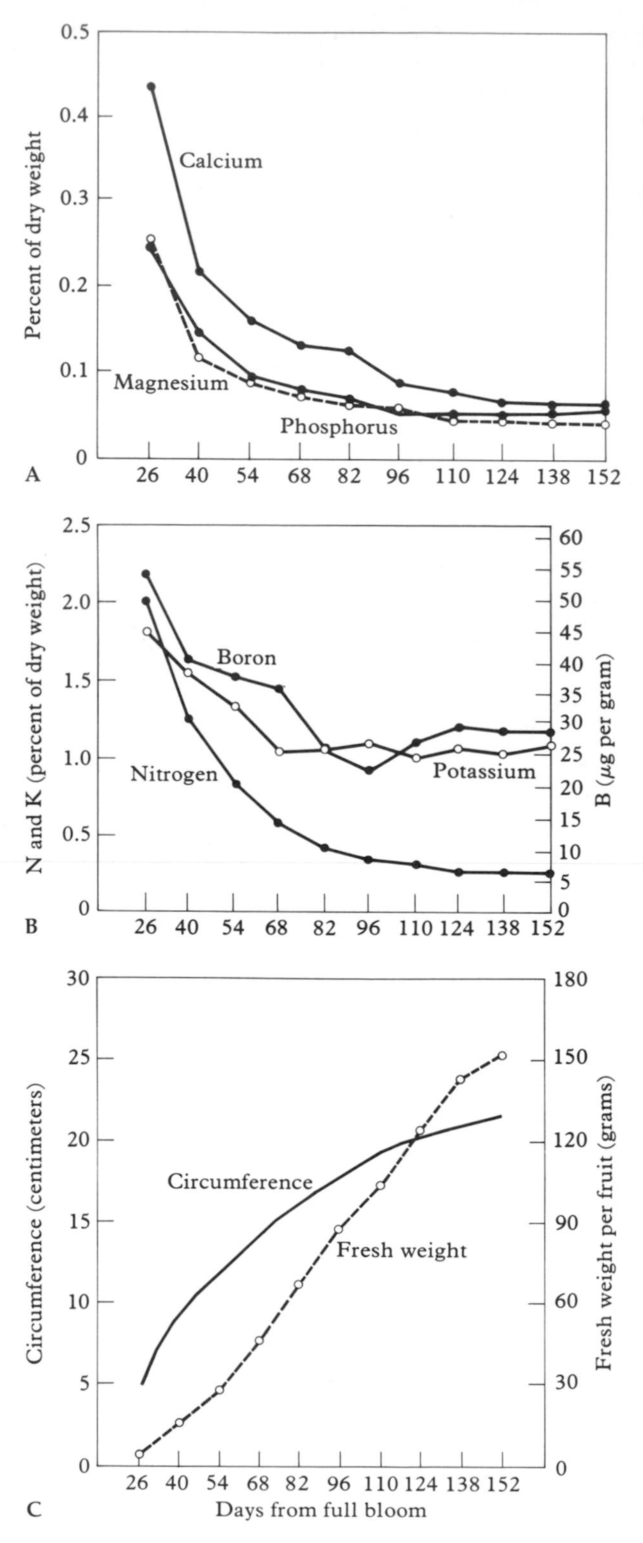

Figure 6-21 Seasonal decline in the concentration (percentage) of nutrients in Delicious apple fruits as the size and fresh weight of the fruits increase. The graphs show: **(A)** calcium, magnesium, and phosphorus content; **(B)** boron, potassium, and nitrogen content; **(C)** fruit growth. [After Rogers and Batjer, 1954]

types of chlorosis (deficiencies of Fe, Mn, Mg, Cu, or K) or leaf yellowing and necrosis appear to result from the blockage of protein synthesis, presumably caused by improperly functioning enzymes. The resulting high buildup of amino acids in leaves may be toxic by itself.

Interactions may involve more than two elements, or more than one plant organ. For example, K was shown to depress Mg at high N levels but not at low N levels. Fruit analysis may be a better indicator of Ca deficiency than leaf analysis. Often, a fairly complete leaf analysis is needed in order to see the balance of all nutrients as well as to see the levels of each in different tissues. Comparing levels of the various nutrients permits one to better evaluate fertilizer needs.

Fertilizers

Nutrient availability in the soil is the net effect of several interacting factors, some of which are in the soil and some of which are related to the plant itself. These interactions between tree and soil often require both soil and tissue analysis to develop sound fertilizer practice in particular situations. The use of fertilizers by country is shown in Figure 6-23. Unfortunately, fertilizer use worldwide is more related to economic status than to the needs of crop soils.

The principal element added and which is most often deficient in all soils is N. Other elements should be added only if there is a clear need for them. P usually is not deficient and should be applied only when a demonstrated beneficial response exists. K and Mg fertilizers should be applied in orchards where analysis and symptoms indicate a need. A study done in central Washington with peach and apple indicates the amounts of N, P, and K needed each year to re-

Table 6-4 Nutrient levels in leaves of several species during late July and August, indicating the range from below normal to excess.

Nutrient Level[1]	(percent d. wt.)					(ppm d. wt.)				
	N	K	P	Ca	Mg	Mn	Fe	Cu	B	Zn
Apple										
BN	1.5	.9	.08	.20	.18	20	40	1	30	10
N	2.0	1.2	.12	1.0	.24	25	50	4	35	18
AN	2.3	3.0	.30	2.5	1.0	200	400	50	80	100
Ex	3.5	4.0	.70	3.0	2.0	450	500	100	100	200
Pear										
BN	1.9	.4	.08	.20	.18	20	40	1	30	10
N	2.2	.7	.12	1.0	.24	25	50	4	35	18
AN	2.4	3.0	.30	2.5	1.0	200	400	50	80	100
Ex	3.5	4.0	.70	3.0	2.0	450	500	100	100	200
Cherry										
BN	1.7	1.0	.08	.20	.18	20	40	1	30	10
N	2.3	1.2	.12	1.0	.24	25	50	4	35	18
AN	2.6	3.0	.30	2.5	1.0	200	400	50	80	100
Ex	4.0	4.0	.70	3.0	2.0	450	500	100	100	200
Peach										
BN	2.0	1.0	.08	.20	.18	20	40	1	30	10
N	2.8	1.5	.12	1.0	.24	25	50	4	35	18
AN	3.8	3.0	.30	2.5	1.0	200	400	50	80	100
Ex	4.5	4.0	.70	3.0	2.0	450	500	100	100	200
Plum and Prune										
BN	1.7	1.0	.08	.20	.18	20	40	1	30	10
N	2.2	1.4	.12	1.0	.24	25	50	4	35	18
AN	2.5	3.0	.30	2.5	1.0	200	400	50	80	100
Ex	3.5	4.0	.70	3.0	2.0	450	500	100	100	200
Filbert										
BN	1.8	0.4	.08	.20	.18	20	40	1	30	10
N	2.2	0.7	.12	1.0	.24	25	50	4	35	18
AN	2.5	2.0	.30	2.5	1.0	200	400	50	80	100
Ex	3.5	3.0	.70	3.0	2.0	450	500	100	100	200
Walnut										
BN	2.0	.9	.08	.20	.18	20	40	1	75	10
N	2.3	1.2	.12	1.0	.24	25	50	4	90	18
AN	2.8	2.0	.30	2.5	1.0	200	400	50	100	100
Ex	4.5	3.0	.70	3.0	2.0	450	500	100	150	200
Pecan										
BN	1.6	.4	.08	.20	.18	140	40	1	40	10
N	2.3	1.0	.12	.7	.30	200	75	4	60	18
AN	2.8	1.5	.30	1.5	1.0	500	150	50	100	100
Ex	3.0	2.0	.70	3.0	2.0	1000	300	100	600	200
Almond										
BN	1.5	1.0	.08	.20	.25	20		2	30	10
N	2.4	1.5	.12	1.00	.50	75		10	35	25
AN	3.0	3.0	.30	2.5	1.00	200		50	80	100
Ex	4.0	4.0	.70	4.0	2.0	450		100	100	200

SOURCE: In part from Childers, 1966.
[1]BN = Below Normal, N = Normal, AN = Above Normal, Ex = Excess.

Table 6-5 Approximate range of leaf content of nutrients in healthy strawberry plants that fruited satisfactorily.

N (%)	P (%)	K (%)	Mg (%)	Ca (%)	Mn (ppm)	B (ppm)	Zn (ppm)	Cu (ppm)	Fe (ppm)
2.35	0.18	1.10	0.28	1.25	129	111	58	6.2	70
to	to	to	to	to	to	to	to	to	to
2.93	0.24	1.70	0.34	1.48	170	170	73	7.0	80

SOURCE: Data compiled by Carter R. Smith, Rutgers University.

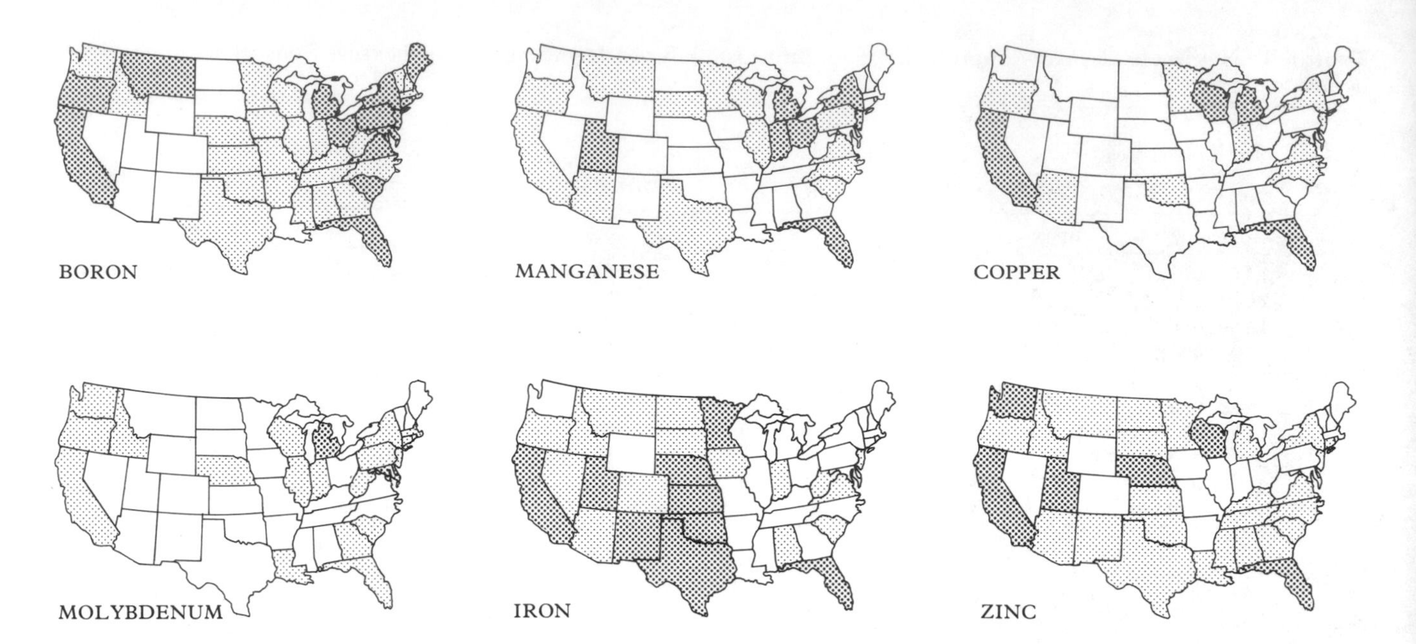

Figure 6-22 Nutrient deficiencies of various minor elements in the U.S. Dark shading indicates states reporting a severe deficiency; light shading indicates states reporting a moderate deficiency; white indicates no reported deficiency. Not all areas of a shaded state are necessarily deficient in the designated element. [From Christopher J. Pratt, "Chemical Fertilizers." Copyright © 1965 by Scientific American, Inc. All rights reserved.]

Table 6-6 Estimated annual kilograms per hectare removed of nitrogen, phosphorus, and potassium by mature trees.

Tree	Nitrogen	Phosphorus	Potassium
Apple	39	10	71
Peach	76	11	96

Table 6-7 Nitrogen, phosphorus, and potassium content (% dry wt.) of leaves of apple and peach after 10 years of complete (NPK) versus nitrogen-only (N) fertilizers.

Leaf	Nitrogen Content (%)		Phosphorus Content (%)		Potassium Content (%)	
	N Fertilizer	NPK Fertilizer	N Fertilizer	NPK Fertilizer	N Fertilizer	NPK Fertilizer
Apple	2.74	2.74	.19	.19	1.73	1.66
Peach	2.97	2.93	.22	.20	2.00	2.02

Source: Batjer, 1954.

place that removed by the crop (Table 6-6). Batjer (1954) also showed that complete "NPK" fertilizers were no better than N fertilizers in central Washington soils (Table 6-7). These data indicate that only when the plant needs more than it is getting from the soil is application of the element necessary. Complete NPK fertilizers cost much more than N alone and should not be used unless all three are needed. Table 6-6 shows that a very small amount of P per ha is removed with the crop. The normal weathering processes in the soil make those amounts available even in soils quite low in P. Also, trees and shrubs store

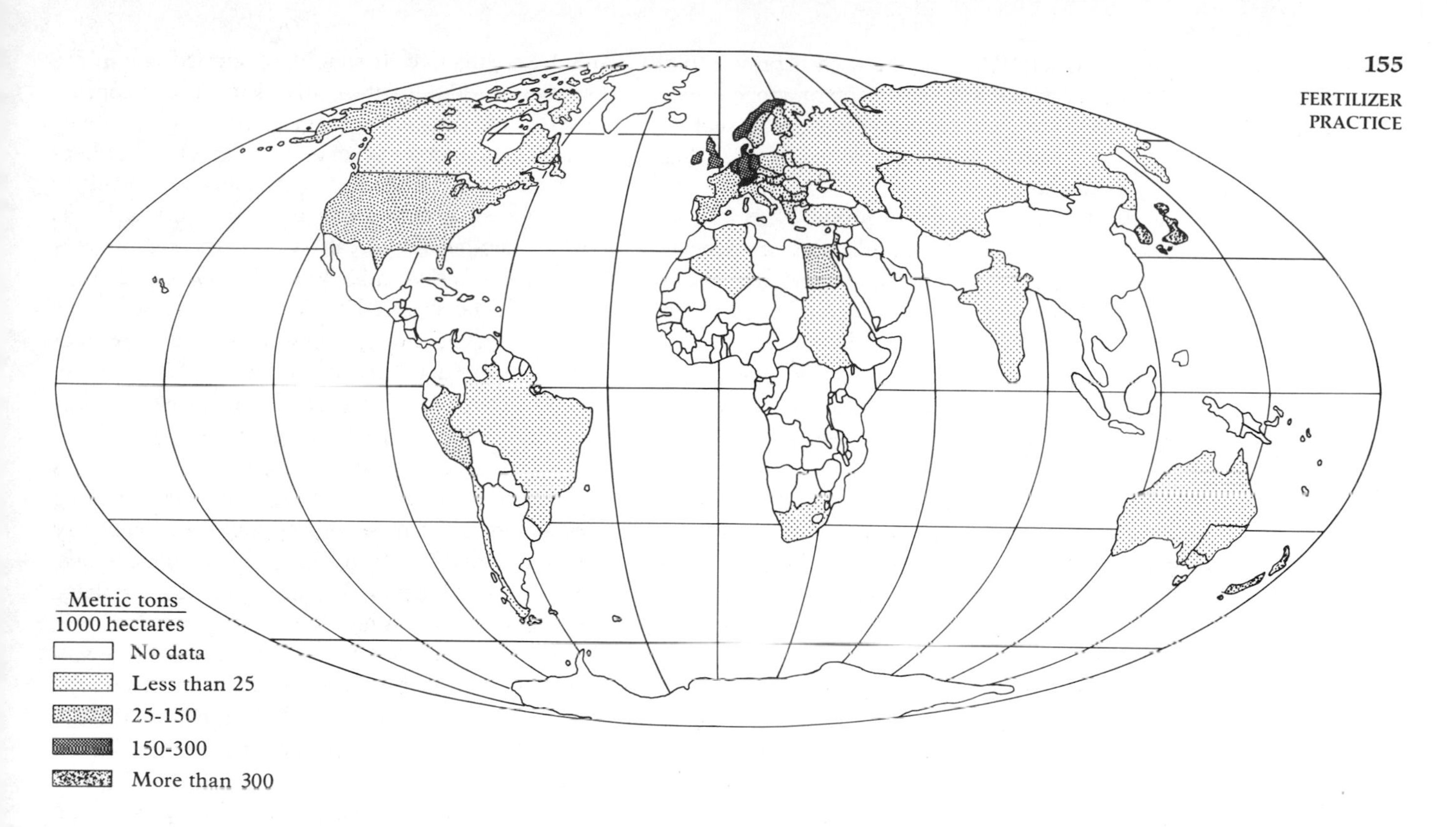

Figure 6-23 Annual world use of chemical fertilizers on arable land. Note that the countries with the heaviest use of these fertilizers are those which are principal exporters of food and fiber. [From Christopher J. Pratt, "Chemical Fertilizers." Copyright © 1965 by Scientific American, Inc. All rights reserved.]

minerals in their tissues from one season to the next, so that part of the requirement is met from this reservoir. Recently, however, walnut trees in California responded to applications of P. More intensive cropping worldwide probably will bring about more cases of P deficiency.

Some fruit growers believe that complete fertilizers result in better colored, firmer, and better-quality fruit than straight N fertilizer. There is no experimental evidence to support this view. Excess N can reduce fruit color and firmness, and quite often a grower who uses NPK fertilizer uses less N per hectare than a grower who uses only N. This fact could account for a difference in fruit color. Because fruit color is reduced by overuse of N, growers should be careful not to apply more than is needed for adequate growth and fruit production.

The cost per kg of N is an important factor in choosing among the different N fertilizers. Also, the effect of applied fertilizers on soil pH and element balance must be considered (Table 6-8). Soils which have pH 5.5 or lower should not be fertilized with materials which lower pH appreciably, or else they should be limed as needed to neutralize the soil and raise the pH. $Ca(NO_3)_2$ can be used if only a small rise in pH is desired. Soils of pH 7.5 or above usually respond best to acid-forming fertilizers such as $(NH_4)_2SO_4$, which neutralize some of the alkali (OH^-) in the soil.

Of the elements essential to plants, N, B, Zn, Fe, K, Mg, Mn, and in a few cases P and S, are deficient in some soils. It is best not to use pre-mixed fertilizers because each orchard requires different elements and different proportions of them. Thus one should apply the nutrients separately *as needed.* Do not apply a nutrient unless there is good evidence that it is needed. Leaf analysis is the most reliable indicator.

NUTRIENT SPRAYS Not all elements need be applied to the soil. Some, such as Zn, are better sprayed directly on the tree and absorbed by the leaves. Leaf content of Zn often varies greatly and may not indicate whether or not a deficiency exists. Visible symptoms, however, are quite well defined: leaves are small, narrow, with a striped irregular chlorotic pattern between the veins, which remain green. Often there is a rosette of leaves at the tips of shoots, with bare wood below the tips. Applications of Zn to calcareous soils are not effective because the soil quickly converts it to an unavailable form. A spray (applied to dormant trees) of 16.8 kg of Zn per hectare (45 kg of 36 percent $ZnSO_4$ crystals, or 47.3 liters of liquid $ZnSO_4$) is effective. It should be applied late in the dormant period but before any visible green appears from the buds. It can be applied either as a dilute or a concentrated spray as long as the amount per hectare is constant. On trees very deficient or difficult to correct (such as with sweet cherry), apply twice as much Zn per hectare. Zn also can be sprayed on trees in leaf, but this should be done *after harvest,* and the rate should be lowered to 11 kg per hectare. For severe Zn deficiency, the fall spray plus the dormant spray is most effective. Do not use the fall spray on apricots—they are injured by it. After obvious deficiency has been corrected, a maintenance spray of 2.2 kg Zn (6.7 kg of 36 percent crystals) per hectare is applied annually after harvest. Young nonbearing trees in summer can be sprayed any time deficiency shows up with 227 g Zn (681 g of 36 percent crystals, $ZnSO_4$) per 378 liters (100 gallons) of water. Foliage should be wet thoroughly.

Boron as sodium pentaborate (which is more soluble than boric acid) is applied as an annual spray at 6 to 11 kg per hectare (1.1 to 2.2 kg of actual B) to correct B deficiency. A B deficiency may occur at leaf levels below 30 to 40 ppm (see Table 6-4). Dilute sprays are applied at the rate of 227 to 454 g of sodium pentaborate (Solubor) per 378 liters of water and usually are sprayed on after bloom in the spring. However, pear trees affected by blossom-blast, a temporary B deficiency in early spring, should receive a spray of 900 g/378 liters in the fall after harvest but while leaves are still green or at the pre-pink stage in early spring. Recent evidence (Callan and Thompson, 1976), however, indicates that the fall spray is most effective for all species except filbert.

Iron deficiency chlorosis is a disorder indicative of poor soil aeration, high Ca and P content of the soil, or other mineral imbalances. The ultimate correction lies in favorably altering the soil, but sprays can give temporary recovery. Fe chelates and other similar materials are sprayed on the leaves. Two sprays are usually needed; one about 4 weeks after bloom and the second 3 weeks later. The rate depends upon the commercial material used, so the directions on the label should be followed. In Idaho, injection of

Table 6-8 Amounts of nitrogen in different fertilizers and effects of these chemicals on soil pH.

Fertilizer (NPK)[1]	% N	Fertilizer (kg N per m ton)	Effect on Soil pH
Urea (45–0–0)	45	450	Lowers slightly
NH_4NO_3 (33–0–0)	33	335	Lowers slightly
$(NH_4)_2SO_4$ (21–0–0)	21	210	Lowers
$Ca(NO_3)_2$ (15.5–0–0)	15.5	155	Raises slightly
Mixed (16–20–0)	16	160	
Mixed (12–20–8)	12	120	
Mixed (10–10–5)	10	100	
Mixed (5–10–10)	5	50	

[1]NPK in % N, % P_2O_5, and % K_2O in the given fertilizer.

Fe chelate in water into the soil around chlorotic peach trees corrected the disorder for 2 or 3 years (Kochan, 1962). The best chelate for this was sodium ferric ethylenediamine di-(o-hydroxphenylacetate), (EDDHA-Fe).

N is usually applied to the soil; urea can be sprayed on at the rate of 2.3 kg/378 liters of water where a deficiency exists. Apples respond well to such a spray if it is applied early in the spring after growth has started. Urea should not be sprayed on bearing apples later than first cover (May) because later sprays inhibit color development of the fruit. Pear, peach, and a number of other fruits do not utilize foliar sprays of urea.

Correction of Mg deficiency may be obtained by a spray of $Mg(NO_3)_2$, in which 2.3 kg of $Ca(NO_3)_2$ and 2.3 kg $MgSO_4$ are mixed in 378 liters of water. The $CaSO_4$ precipitates, leaving $Mg(NO_3)_2$ in solution. The deficiency appears as an interveinal chlorosis and/or necrosis and appears on the older leaves first. If the deficiency persists, many of these older leaves drop by midseason, leaving much bare wood below tufts of leaves at the ends of shoots. At times, this spray may cause injury, and if repeated can cause injury to both trees and fruit.

Sulfur deficiency is easily corrected by soil application of a SO_4 salt, or it can be corrected by several kinds of sprays containing S. The symptom is pale, yellow-green leaves, similar to that of N deficiency. Usually the first leaves formed in the spring are more normal than those developing later.

ROOTSTOCKS Rootstocks have been shown to influence the nutritional status of a number of deciduous fruit and nut species. Recent work with pear indicates that pear on quince takes up less N and B, and more Mg than does pear on pear root. Pear on mountain ash *(Sorbus aucuparia)* stock absorbs less B but six times more Mn than does pear on pear stock. Evaluations of plum rootstocks in California have shown myrobalan and marianna to be widely adapted to different soil types and moisture conditions, and peach to be best suited to well-drained sandy-loam soils. French and President plums on almond *(P. amygdalus)* and myrobalan roots are less effective than on peach in the absorption of B, and hence would be less susceptible to B toxicity. Plantings of Italian prune *(P. domestica)* established on seedling peach *(P. persica)* and clonal Myrobalan 29-C, B, 2-7 *(P. cerasifera);* Marianna 4001, 2623, 2624 *(P. cerasifera* × Munsoniana?); and St. Julien A *(P. insititia)* rootstocks in seven orchard sites have provided new data on nutrient uptake (Chaplin et al., 1972). Leaf samples during three seasons showed that plum stocks had greater leaf N, K, Mn, and Zn, and slightly less B and Mg than those on peach. Plum clones Myrobalan 29-C, Myrobalan B, and St. Julien A, were more efficient in the uptake of Ca. There were positive correlations between N and Ca, N and Mg, N and B, N and Zn, Ca and Mg, Ca and B, and Mg and B for most of the stocks. There was a negative correlation between K and Mg for Myrobalan 2-7 and the three marianna clones. Myrobalan B and Marianna 2623 and 2624 had a negative correlation for K and Ca, whereas St. Julien A had a positive correlation.

One study with apple showed that M 1 and M 7 generally resulted in high nutrient content of several cultivars, while M 13 and M 16 tended to be low. Other stocks were intermediate (Awad and Kenworthy, 1963).

There are some striking differences in nutrient uptake attributed to rootstocks. Some of the high cost of fertilizers can be saved by using rootstocks that are efficient in nutrient uptake. But there are many other traits to consider in choosing a stock. Dwarfing, pest and disease resistance, tolerance to adverse soil conditions, and hardiness are all important. The effect on nutrition should be considered as only one factor in the choice.

PRUNING

Pruning is a dwarfing process and can be used to maintain any desired tree size. Removal of a branch removes not only stored carbohydrates, but reduces the potential leaf surface as well. From these reductions comes a loss in root growth as well. The dwarf-

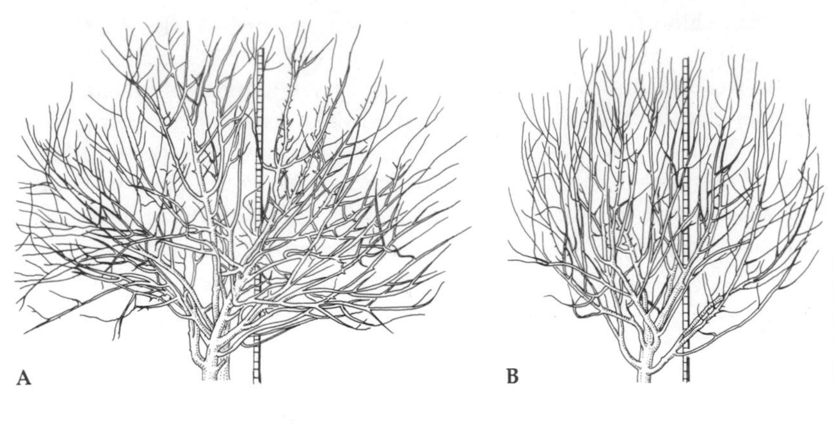

Figure 6-24 Effect of pruning on the size of six-year-old apple trees. **(A)** This tree was not pruned. **(B)** This tree received moderate pruning. [After Batjer and Westwood, 1963]

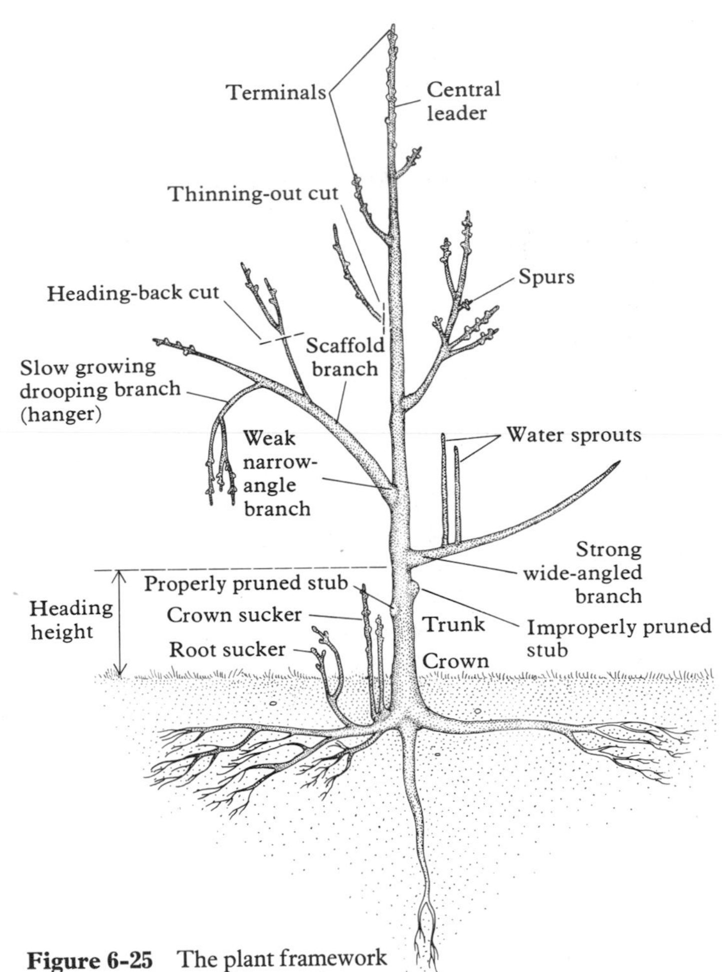

Figure 6-25 The plant framework indicating terms used in pruning.

ing effect of four years of moderate pruning versus slight corrective pruning of apple is shown dramatically in Figure 6-24. Pruning increases fruit size, nitrogen per growing point and stimulates growth near the cut. Excessive pruning reduces fruitfulness—this is especially true with young vigorous trees that may already be developing too much vegetative growth. Large cuts result in excessive stimulation of sprouts near the cut, while well-distributed small cuts spread the stimulus better over the entire tree. Heading-back cuts (cutting part of branches off, see Fig. 6-25) strongly stimulate the local area within a few centimeters of the cut. Thinning-out cuts (removal of branches back to their point of origin) tend to have less local stimulus (see Fig. 6-26).

The severity of pruning varies with tree age and vigor and with kind of fruit. Old trees in low vigor can be pruned more heavily than younger and more vigorous ones of the same kind. The rejuvenation of old trees is better accomplished by many small cuts than by a few large ones; the latter result in uneven stimulation. The kind of fruit is important in determining the best type of pruning because of the different flowering habits and the different ultimate fruit size at harvest. Fruits that initiate their flowers on one-year wood and that have naturally large fruits (e.g., peach) respond better to heavy pruning than do types which set flowers on older spurs and that have naturally small fruits (e.g., cherry). Thus we see the importance of a thorough knowledge of flowering and fruiting habits in determining pruning practices. It is poor management to attempt to use heavy pruning to avoid fruit thinning, because yields will be reduced drastically. One should have clearly in mind the reason for pruning or else one should not prune.

In areas where winter killing of trees may be experienced, dormant pruning should be done as late as possible in the winter. Brown et al. (1964) reported that Anjou pears pruned prior to a severe freeze were severely injured, but trees not pruned before the freeze showed only minor injury. This generally represents the relationship between early pruning and winter injury in fruit plants. The stimulation of cellular activity near the cut from wound-healing reactions and the initiation of growth of latent buds causes the tissue to lose its hardiness. The greatest degree of hardiness is found in tissues where no cell divisions are occurring and which have a low rate of metabolic activity. De-hardening occurs rapidly when cells are stimulated by high temperature, pruning, or other external factors. The importance of time of pruning for a given orchard would have to be determined by the grower himself after reviewing the climatic history of the location.

Pruning Apple and Pear

The large-fruited pomes bear their crops from mixed buds mainly on terminals of short spurs or twigs. The apple has a determinate inflorescence of five flowers, while the pear has an indeterminate inflorescence of seven to ten flowers. Because short spurs on wood two years old or older are needed for cropping, pome fruits are pruned rather lightly to develop such a spur system. Heavy pruning results in the growth of too many long, unfruitful shoots that ordinarily would have remained short and fruitful (Fig. 6-27).

The large size of the pome fruits (with the exception of commercial crab apples) permits heavier prun-

Figure 6-26 Pruning mature apple and pear trees to maintain the desired framework and to sustain good fruiting wood may be done by **(A)** thinning out more shoots and branches toward the end of major branches and **(B)** cutting upper limbs back to more upright branches, making the lower limbs more productive by keeping the top narrower than the bottom. **(C)** To reduce the height of a tall tree, cut whole limbs from the top, making the cuts flush with the bark of the lower limb. **(D)** Undesirable narrow-angled branches should be removed while the tree is young, because later breakage from crop load may split the entire tree. [After Stebbins, 1976]

Figure 6-27 Moderate thinning-out pruning of Bartlett pear trees **(A)** at 5 years and **(B)** when mature, is conducive to good cropping. **(C, D)** The severely pruned younger tree will be delayed coming into bearing and will have much less bearing surface than **(E, F)** a similar tree pruned very lightly. [From Batjer et al., 1967]

ing of fruiting points than is desirable with species that produce small fruits. The large size of a mature apple, for example, permits a full crop with only 5 to 8 percent of the flowers setting fruits. With smaller fruits, a much higher percentage of fruit set is necessary for a full crop, and a great many more fruiting points are needed per tree. Much of the pruning on standard apple trees consists of thinning out dense areas and removing weak, thin, unproductive wood in areas of the tree where light distribution is poor. The degree of pruning required depends upon the vigor in a particular orchard. More than about 20 to 40 centimeters of annual shoot growth is undesirable, and excess pruning is often the reason for too much shoot growth.

Pruning pear trees is, in general, similar to pruning apple trees. Westwood et al. (1964) cited a number of pruning studies that produced the following findings.

Moderate to heavy pruning of Anjou pear induced a much higher percentage set than did light or no pruning. Even though heavy pruning reduced the bearing surface, the heavier set resulted in more fruit per tree with pruning than with none. Trees grown in clay soil performed well with light pruning if ample irrigation water was applied; but when soil moisture was deficient, heavy pruning was better than light (in that it produced more fruit). Also, Anjou trees on shallow soil yielded much more with heavy pruning than did more vigorous trees in deep soil, possibly due to a greater supply of N per growing point. A given surface area of pear shoot leaves causes fruits to grow better than a comparable area of spur leaves does. Thus pruned mature trees have an advantage because they have proportionately more shoot-leaf area than do unpruned trees.

Further studies showed (*a*) that pruning one portion of an Anjou tree does not improve set on unpruned leaders of the same tree (Table 6-9) but (*b*) that not pruning a portion of the tree while pruning another portion does seem to reduce set on the pruned portion. In contrast to Anjou pears, Bartlett and Bosc pears yield more when unpruned than when pruned, and yield reduction is directly related to severity of pruning. However, the maximum yield of fruit of 6 centimeters in diameter and larger is attained by light pruning of Bartlett and heavy pruning of Bosc (Table 6-10). These data indicate that the highest yield of well-sized fruit probably would come from a combination of mild pruning and some fruit thinning. Both apples and pears usually respond best to moderate thinning-out pruning to facilitate good distribution of sunlight and to permit adequate spray coverage.

Table 6-9 Effect of pruning whole trees or leaders on set and yield of Anjou pear.

Pruning treatment		Fruit Set (no./100 clusters)	Yield (boxes)
Moderate pruning		34	8.98
No pruning		11	7.31
3 leaders pruned	Same tree	21	9.29
1 leader unpruned		7	
1 leader pruned	Same tree	17	7.88
3 leaders unpruned		6	

SOURCE: Westwood et al., 1964.

Table 6-10 Effect of pruning on fruit size and yield of Bartlett and Bosc pear trees.

Pruning	Bartlett		Bosc	
	Total Boxes	Yield Above 6-cm Dia.	Total Boxes	Yield Above 6-cm Dia.
None	10.8	3.7	9.3	4.2
Light	9.5	5.0	8.1	4.8
Heavy	7.2	5.2	7.7	6.6

SOURCE: Westwood et al., 1964.

Pruning Cherries

The sweet cherry requires less pruning than any other tree fruit because it bears on short spurs (rather than on shoots stimulated by pruning), and because of the small size of the fruit, necessitating a great many fruiting points per tree for maximum cropping. Very little pruning is done on sweet cherries the first several years after training. On vigorous nonbearing trees, it may be desirable to do mild heading-back of shoots to induce greater branching. Such an operation may be done in training by tipping shoots after they have grown 20 to 25 centimeters in summer, but this often results in narrow-angled crotches. Dormant heading results in wider crotch angles.

Mature sour cherries may need some pruning because of a tendency to produce too many flowers and not enough leaves. A minimum of about 15 centimeters of annual growth is needed to maintain the proper flowering-vegetative balance so that enough leaves are produced to bring about satisfactory fruit size and quality.

Pruning to adapt trees to harvesting by mechanical shakers is necessarily heavier than normal pruning. Pendulant lower limbs are removed to give the operator a clear view for attaching the grasping clamp of the harvester. Willowy branches are headed back some, so that the fruit is borne on relatively stiff wood. The more rigid system created by this kind of pruning better transmits the energy from the shaker to the fruit and facilitates fruit removal.

Pruning Plums

Most cultivars of prune plums *(P. domestica)* bear well with little or no pruning during the first 12 years. Like the sweet cherry, the fruit are borne laterally on short spurs. As the bearing area becomes dense, older trees will need some thinning-out pruning to facilitate spraying and to allow better light penetration.

Japanese hybrids and Japanese plums *(P. salicina)* are pruned much more heavily than European cultivars. Their fruits are large and their branches brittle. Pruning back the fruiting branches helps prevent breakage from the heavy-cropping tendency of these cultivars.

Pruning plums and prunes to facilitate mechanical harvesting is similar to that for cherries, except that plums are much easier to remove by shaking. Such pruning is a compromise between high yields and efficient removal by mechanical shakers.

Pruning Peaches

During the first 5 years, nonpruned peach trees bear the heaviest crops (Ashton et al., 1950). But when annual growth is less than 30 to 40 centimeters, some pruning is needed. Because of the large size of the peach fruit (requiring fewer fruiting points for a full crop), and because the best fruiting wood is on one-year shoots, this tree is pruned more heavily than others. Pruning may be by thinning-out, heading-back, or both. A study done in Utah indicated that thinning-out pruning gives the best yield of fruits of a commercial size once the trees reach early mature age; severe pruning increases size slightly but reduces yield. A further test showed that poor light distribution inside the tree canopy is associated with heading-back pruning and that fruit blush color is poor with this method of pruning (Table 6-11 and Fig. 6-28).

The four basic types of pruning used on peach trees are defined as follows:

1. Corrective. Only broken, dead, or strongly interfering branches are removed.

2. Thinning-out. Initially, the secondary branches are thinned-out to form an open framework. All of the weak fruiting shoots are removed, plus about half the remaining ones. No heading-back is done.

3. Conventional. Thinning-out is done and fruiting shoots are headed back to about half their length.

4. Severe. Weak shoots plus 50 to 75 percent of others are removed. Remaining shoots are headed back to 10 to 15 centimeters.

Table 6-11 Effect of pruning method on fruit color, size, and yield of 7-year Elberta peach trees.

Pruning Method	Percent Blush Color	Yellow Color Rating[1]	Yield (kg)	Fruit Dia. (cm)
Corrective	54	3.6	119	5.5
Thinning-out	58	3.7	108	5.8
Conventional	42	3.3	85	6.0
Severe	27	3.2	50	6.2

SOURCE: Westwood and Gerber, 1958.
[1] 4 = Amber-yellow; 3 = Green-yellow.

With corrective pruning, fruit color and yield are good but size is poor. Also, the fruit is borne in a thin shell of bearing surface at the periphery of the tree. Thinning-out pruning results in good light distribution in the tree, and good fruit color and yield; fruit size is smaller than with more severe pruning, but yield is considerably higher. The data in Table 6-11 are not important in absolute terms but indicate the relative responses to be expected.

Bulk pruning by mechanical hedgers and mowers has been practiced to a considerable extent recently. The peach lends itself to such methods, although, as noted above, proper hand pruning is better.

Pruning Almonds

Almond trees are pruned much as peach trees are but less severely because many more fruits per tree are needed for a full crop. Bulk pruning or large cuts are sometimes done every 2 or 3 years rather than more detailed pruning each year.

Pruning Apricots

Apricot pruning is similar to peach pruning but is generally lighter. When pruned more heavily, the apricot develops larger fruit, which, in part, compensates for the reduction in bearing surface.

Pruning Filberts

Filbert female flowers are borne on buds of the past season's growth, as are peach and almond flowers. Some pruning is needed to obtain 15 to 20 centimeters of new growth each year, but a large bearing surface is required for a full crop due to the small size of the

A

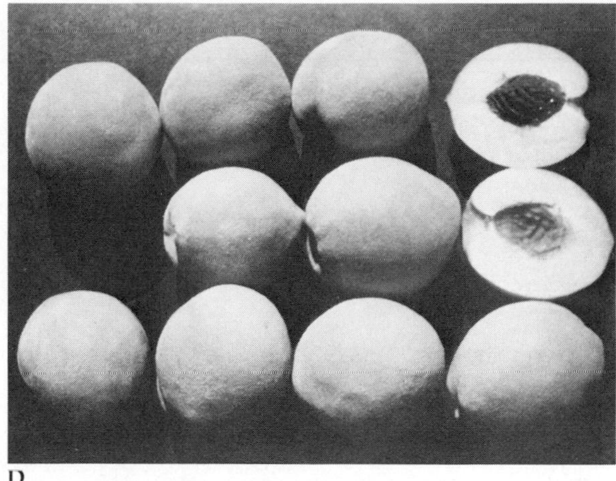
D

B

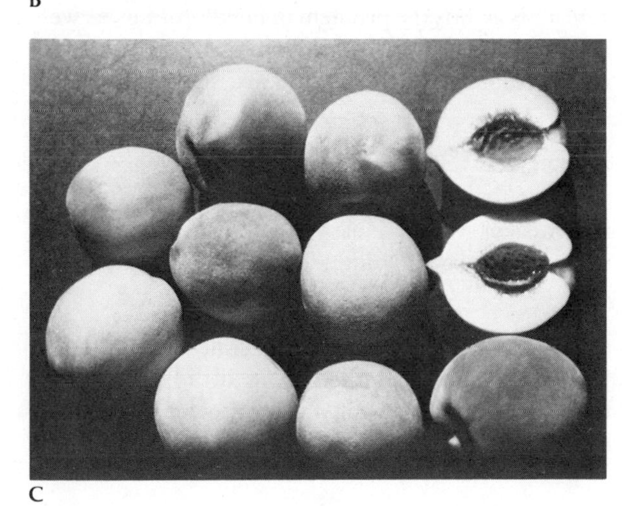
C

Figure 6-28 Peach trees are pruned by **(A)** thinning-out cuts throughout the tree rather than by **(B)** severe thinning plus heading back. **(C)** Properly pruned trees produce more fruit of better color than do **(D)** severely pruned ones.

nut. As with almond, bulk pruning of filbert can be done every 3 or 4 years rather than annual pruning. This practice, however, results in too vigorous growth and poor yield the first year after pruning.

Pruning Walnuts and Pecans

Walnut is usually trained to a modified or central leader with five or six main laterals distributed up and down and around the main stem. After this minimal training, very little pruning is done. Female flowers are borne at the ends of short current season's terminal shoots (some cultivars bear also on lateral shoots). Coastal orchards in California typically have about 10 to 15 centimeters of shoot growth per year, while those of interior valleys have 20 to 30 centimeters. Overly long shoots are usually unfruitful. Some thinning-out pruning is usually desirable, especially on heavy-bearing cultivars. Nut crops (and fruits also) that are not hand harvested need not be pruned to limit the height of the tree as is needed with other fruits grown on vigorous rootstocks. The pecan is a large upright tree with well-formed branches. Very little pruning is done except to top them to limit tree size.

Pruning Dwarf Trees

Trees on dwarf roots can be pruned more severely than standard trees, without danger of making them overvegetative. This tendency makes the dwarf tree ideally suited to bulk mechanical pruning in dense hedgerows or treewalls. For convenience in picking from a trailer or platform, the trees should be pruned so that fruit in the center of the tree can be picked at arm's reach while standing on a platform at the edge of the treewall. In order to develop enough bearing surface for high per-hectare yields, dwarf trees should be trained to a central-leader system and topped only after reaching a height of 3.0 to 3.7 meters. Such a system in Golden and Red Delicious apples on M 9 root planted in a 4.6 × 1.2 meter pattern yielded nearly 5000 boxes per hectare in the tenth year. A similar planting of Bartlett pear on Angers Quince root yielded 58 metric tons per hectare. Figure 6-29 shows the equipment used for mechanical pruning of dwarf treewalls and the resulting "modified Christmas-tree" shape of the system. The one shown is on M 9 stock, is 20 years old, and has had bulk pruning for the last 8 years. The lower portion of the treewall is 1.8 meters wide and the top is 0.6 meter wide. This shape permits the best distribution of sunlight to the system. With vertical sides, the upper branches and leaves would shade the lower ones.

Dwarf or semidwarf trees of compact (or spurred) mutants likewise can be pruned more heavily by hedging or other nonselective pruning. Such trees flower more heavily and set more fruit than standard trees with similar pruning. The proper combinations of compact mutants and clonal rootstocks should make mechanical pruning routine in a few years.

Pruning Small Fruits

GRAPE Fruiting is from lateral shoots arising from one-year canes. Old canes are cut back to the main trunk after they have fruited. With vinifera types, the one-year fruiting canes are cut back to three to five buds, while labrusca types are left much longer, at about forty-eight buds per vine. Or, if the "balanced" system of pruning is used, labruscas are adjusted to forty-eight buds per 1.36 kilogram of pruning removed, and two additional buds are left per each 114 grams of prunings over 1.36 kilograms. Several pruning methods are shown in Figure 6-30.

CANEBERRIES *Rubus* species bear on biennial canes that grow up the first year, then fruit and die the second. Pruning consists of three steps: (1) removal of canes after fruiting, (2) thinning-out and heading-back of new canes during dormancy, and (3) summer tipping or topping of current season's shoots. The extent of summer tipping and dormant heading depends upon the kind and cultivar of caneberry and the training system used (Fig. 6-31).

Recent studies indicate that blackberries may be economically grown by cutting half the acreage back to the ground with mechanical pruners and only fruiting the other half of the acreage (Sheets et al., 1972). Thus only canes are produced in a given field the first year to produce a large crop the second season. This avoids the problem of hand pruning, as well as that of extensive damage to new canes by mechanical harvesters, which would reduce next year's crop.

BLUEBERRY The crop is borne on buds of last year's wood, so the object of pruning is to remove the older and weaker wood. Heavy pruning, however, will severely reduce cropping.

CRANBERRY Usually, excess runners are cut out, leaving the uprights to bear fruit. If the uprights are too thick and tall, the bog may be mowed off short to permit regrowth.

A

B

C

D

Figure 6-29 Mechanical pruning. **(A)** The sickle-bar mower may be used **(B)** to top hedgerows at the desired height and **(C)** to taper the sides. Mechanical pruning can be supplemented by hand pruning if desired. **(D)** An end-view of the hedgerow shows the modified "Christmas tree" shape.

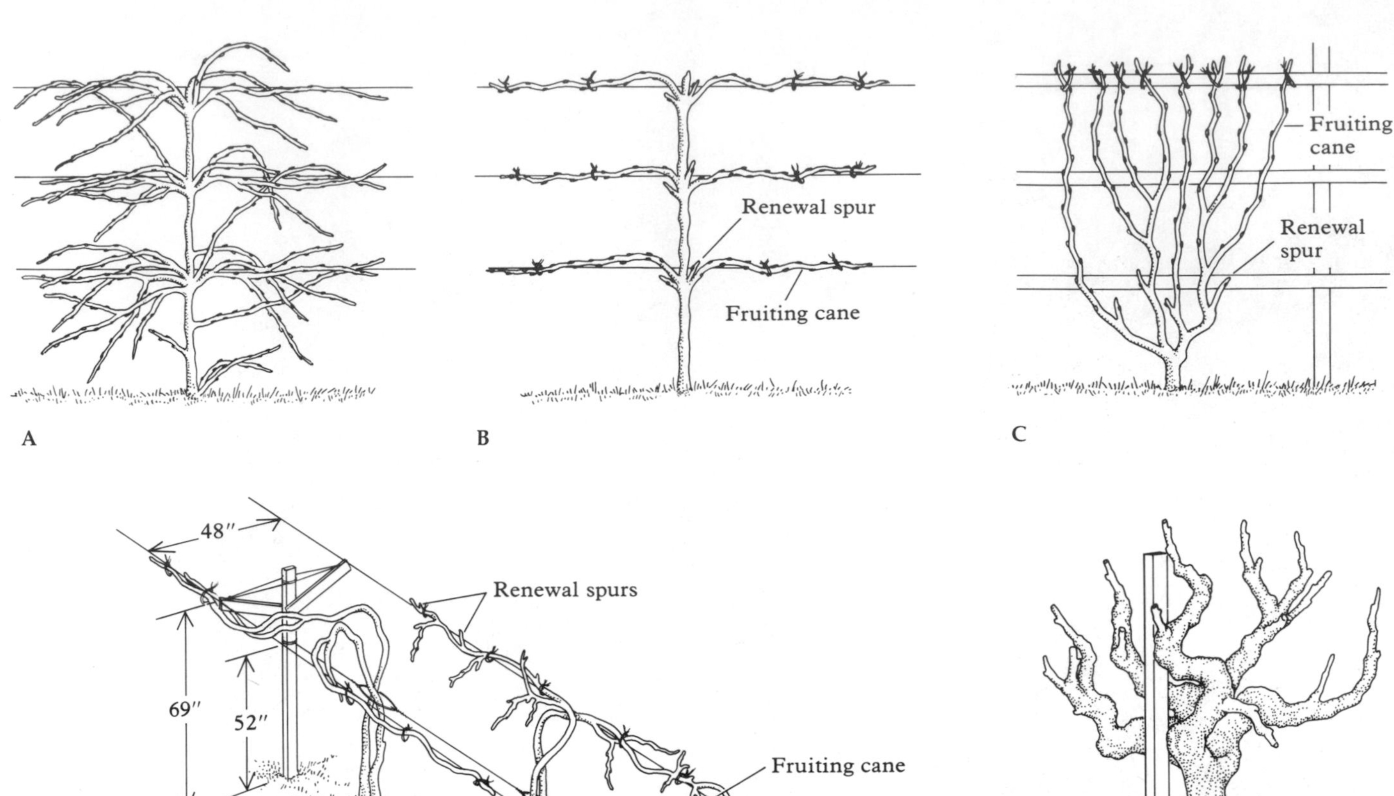

Figure 6-30 Grape pruning systems. **(A)** Unpruned. **(B)** Pruned to six-cane Kniffen system. **(C)** Fan system. **(D)** Horizontal-arm spur system. **(E)** Geneva double-curtain system, with a lower center wire to support the trunks and two upper side wires to support the fruiting canes. **(F)** The head system with spur pruning, used primarily for vinifera cultivars in California. [A, B, C, D: After Jules Janick, *Horticultural Science*, 2nd ed. W. H. Freeman and Company. Copyright © 1972; E: After New York State Agricultural Experiment Station Bulletin 811, 1967]

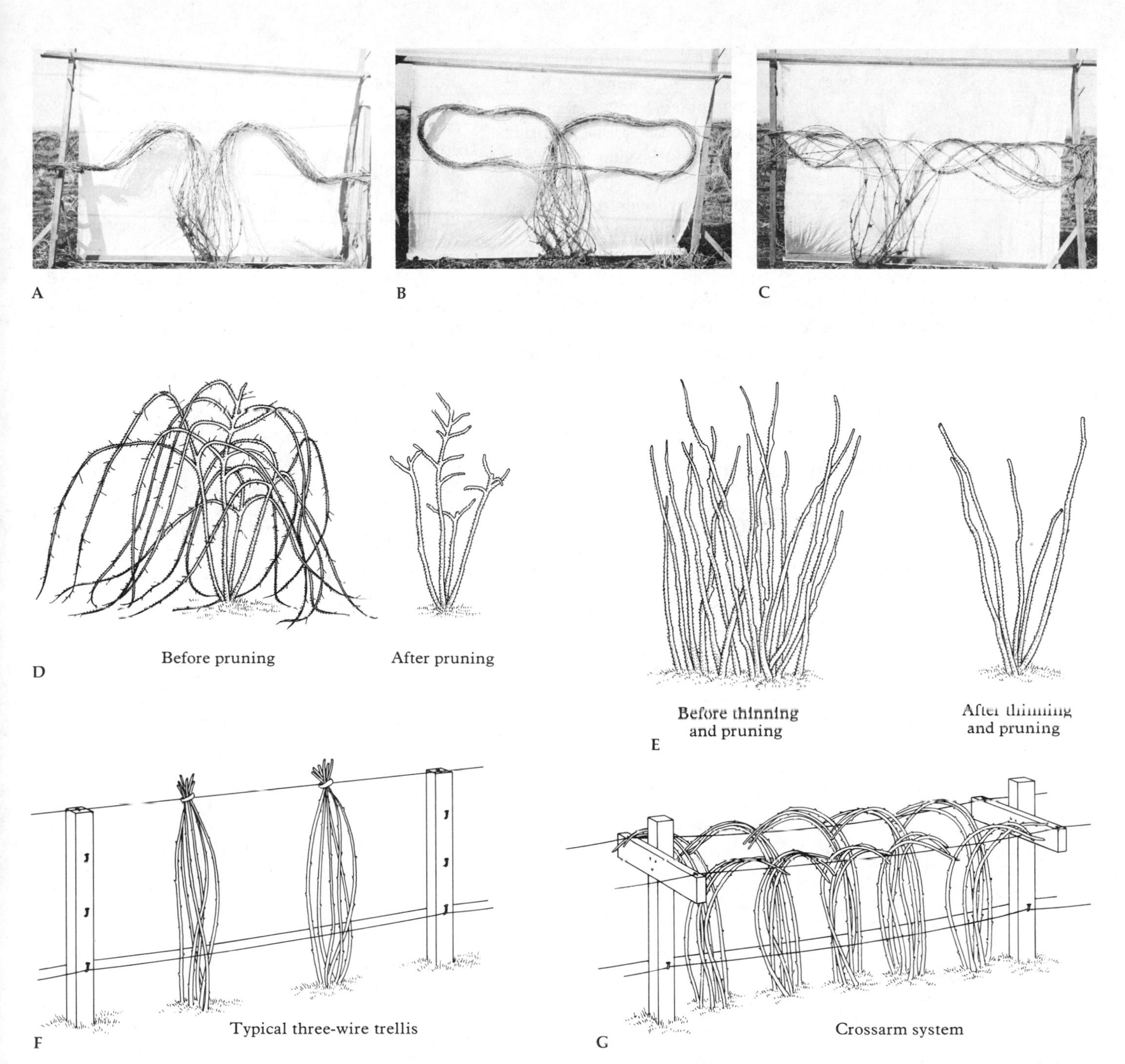

Figure 6-31 Pruning and training systems for caneberries. **(A, B, C)** Blackberry and other trailing *Rubus* species. **(D)** Black raspberry and upright blackberry. **(E, F, G)** red raspberry. [A, B, C: Courtesy Oregon Agricultural Experiment Station; D, E: After USDA; F, G: After Garren and Sprowls, 1975]

CURRANT AND GOOSEBERRY Wood older than 3 years produces inferior fruit, so pruning is done to remove such wood and to maintain a balance between 1-, 2- and 3-year wood on the bushes. Black currant fruits mostly on 1-year wood, so pruning is done to leave mostly new wood.

Details of pruning specific cultivars of fruits or nuts should be obtained from the numerous local experiment station and extension bulletins or from county extension agents and farm advisors from the local area. Such details obtained locally, together with the general concepts presented here on the flowering and fruiting habits of species, should provide a firm basis for good pruning practice.

GENERAL REFERENCES

Brison, F. R. 1974. *Pecan culture.* Capital Printing, Austin, Texas.

Childers, N. F., Ed. 1966. *Fruit nutrition.* Horticultural Publications, New Brunswick, New Jersey.

Dana, M. N., and G. C. Clingbeil. 1966. *Cranberry growing in Wisconsin.* Wisconsin Agr. Circ. 654.

Gauch, H. G. 1972. *Inorganic plant nutrition.* Dowden, Hutchinson, & Ross, Stroudburg, Pennsylvania.

Kramer, P. J. 1969. *Plant and soil water relationships: A modern synthesis.* McGraw-Hill, New York.

LaRue, J. H., and M. Gerdts. 1973. *Growing plums in California.* Calif. Agr. Exp. Sta. Circ. 563.

Ourecky, D. K., and J. P. Tomkins. 1974. *Raspberry growing in New York State.* N. Y. Agr. Exp. Sta. Ext. Bull. 1170.

Scott, D. H.; A. D. Draper; and G. M. Darrow. 1973. *Commercial blueberry growing.* USDA, Farmers' Bull. 2254.

Schuster, C. E., and R. E. Stephenson. 1940. *Soil moisture, root distribution, and aeration as factors in nut production in western Oregon.* Oregon Agr. Exp. Sta. Bull. 372.

Stebbins, R. L. 1971. *Growing prunes.* Oregon Agr. Ext. Circ. 773.

Stebbins, R. L. 1976. *Training and pruning apple and pear trees.* Pacific Northwest Ext. Pub. 156.

USDA. 1970. *Growing raspberries.* Farmers' Bull. 2165.

Verner, L.; W. J. Kochan; D. O. Ketchie; A. Kamal; R. W. Braun; J. W. Berry, Jr.; and M. E. Johnson. 1962. *Trunk growth as a guide in orchard irrigation.* Idaho Agr. Exp. Sta. Res. Bull. 52.

Woodroof, J. G. 1967. *Tree nuts.* Two vols. Avi Pub. Co., Westport, Connecticut.

7

Flowering

JUVENILITY

Flowering is an essential stage of fruit and nut production, and an understanding of factors affecting flowering is important in determining optimal production practices. Starting at the juvenile stage (when the plant cannot flower), this chapter will discuss various factors and practices that hasten the transition from juvenile to adult. For adult plants, flowering habits for different species will be outlined, followed by a consideration of practices affecting initiation and development of flowers. An understanding of the flowering process permits wiser use of rootstocks, fertilizers, pruning, girdling, and growth-regulator chemicals.

Juvenility can be defined as that physiological state of a seedling plant during which it cannot be induced to flower. This state is followed by a transition phase in which flowering can occur but not as readily as later, when the plant grows into the adult phase. In woody plants, the juvenile phase may be very short or very long depending upon both environmental and genetic factors. The juvenile period can often be shortened by increasing the growth rate of the young seedling, because a minimum size must be attained to reach the adult state (Zimmerman, 1972). An extreme example of this is provided by alpine trees and shrubs that have remained juvenile for more than 100 years because very poor growth has prevented their reaching the required minimum size.

Seat of Juvenility

The young seedling is juvenile in all of its tissues until it grows into the transition phase. However, attainment of the transition and ultimately the adult phase does not alter the juvenile tissues. They remain *in situ*

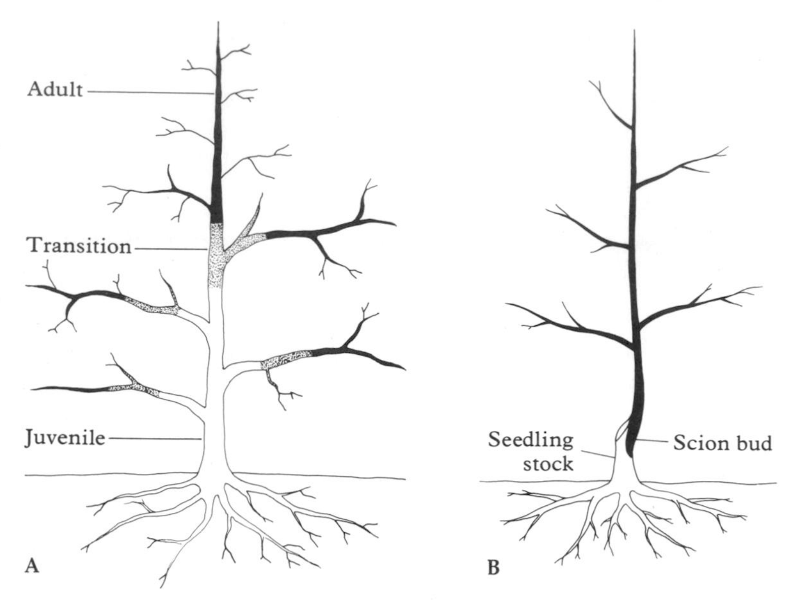

Figure 7-1 **(A)** The seedling tree has a juvenile zone at the crown and lower region of the tree, a transition zone in the mid-region, and an adult zone at the top and periphery. **(B)** The grafted or budded nursery tree, however, is entirely adult above the bud union.

at the base of the tree for its entire life (Fig. 7-1A). Thus if one cut off an 80-year-old seedling tree at the lower trunk, the new growth from latent buds at the base would be juvenile until it once more grew into the adult phase. In contrast to the seedling plant, commercial cultivars budded to seedling stocks are entirely adult above the bud union (Fig. 7-1B).

Marks of Juvenility

Besides the inability to flower, juvenile plants are distinguished by several physiological and morphological traits, such as glabrous, lobed leaves, creeping stems, thorny stems, semi-evergreen habit (in deciduous species), ease of rooting of stems and less ribonucleic acid (RNA) in their tissues (Ali and Westwood, 1968). This last is possibly the reason for the many other differences noted, because RNA is the carrier of genetic information from the cell nucleus to the cytoplasm. Apparently, certain genetic information is "turned off" in juvenile plants; such information is needed to produce hormones, enzymes, etc., which must be present for flowering and adult morphology to occur. Some plants can flower while juvenile leaves are still present; some plants have no distinct juvenile morphology, yet still have a clearly defined juvenile period.

Transition Phase

Transitional tissues are found between the purely juvenile and adult sections of stems of seedling trees (Fig. 7-1A). Characteristics here are intermediate between juvenile and adult. Flowering can occur on transitional stems but not as readily as on adult stems. The juvenile phase is characterized by all of the purely juvenile traits; the transition phase by some juvenile and some adult traits; the adult phase by only adult traits. Again it should be emphasized that the time interval between the juvenile and the adult phase depends largely on the growth rate of the plant and the minimum plant size required to achieve adulthood (Zimmerman, 1972).

Rejuvenation

Once the adult phase has been reached, the adult portions of most woody plants do not revert back to the juvenile, although the adult plant produces seed, which are themselves embryonic juvenile plants. However, adventitious buds are initially juvenile. Such buds can be induced by removing all normal buds, including latent buds, from stems or by forcing adventitious shoots from a severed exposed root. With some plants, shading or treatment with growth regulators, such as gibberellins, induce reversion, but this is not generally true of deciduous fruit species.

Heavy pruning of old, weak orchard trees is thought by some to induce rejuvenation. Such pruning brings about *invigoration* but does not return the tree to the juvenile state. Another false notion held by some is that budded or grafted nursery trees or young orchard trees are in the juvenile phase. Since they are propagated with adult buds of the cultivar, for example,

Redhaven peach, all of the growth from such buds is adult tissue, in contrast to the ungrafted seedling tree of Figure 7-1A. The nonflowering phase of budded trees is termed the *"vegetative adult"* phase. This phase is very short in some cultivars, such as Golden Delicious apple, and quite long in other cultivars, such as Northern Spy. Plant breeders are interested in shortening the period from seeding to first flower and often refer to this entire period as the "juvenile phase." In fact, this period includes both the juvenile phase and the adult vegetative phase. What the breeder refers to as a "long juvenile period" might instead be a long vegetative adult phase, which we know to be quite variable within and between species. In breeding work, "precocity" refers to a short total period from seed to first flower. In fruit growing, "precocity" refers to a short vegetative adult phase.

The retention of juvenility is important in plant propagation because plants root better from juvenile cuttings or explants than from adult tissues of the same clones. Also, rooting of stocks by layerage is better with juvenile plants. There is evidence that juvenile tissues contain a number of rooting cofactors that are absent or present in low concentration in adult tissues (Hess, 1961; also a review by Westwood, 1972).

FLOWERING HABIT

Apple

Apples have two kinds of buds: vegetative and mixed. The inflorescence is determinate and domestic forms contain five (or sometimes six) flowers. At the base of each inflorescence are several lateral vegetative buds. Flower buds are usually borne terminally on shoots or short spurs, but some are borne on lateral buds of one-year shoots (Fig. 7-2). Floral initiation occurs in early summer for next year's crop. Young developing fruit on nearby twigs tend to inhibit floral initiation of some cultivars, resulting in biennial bearing.

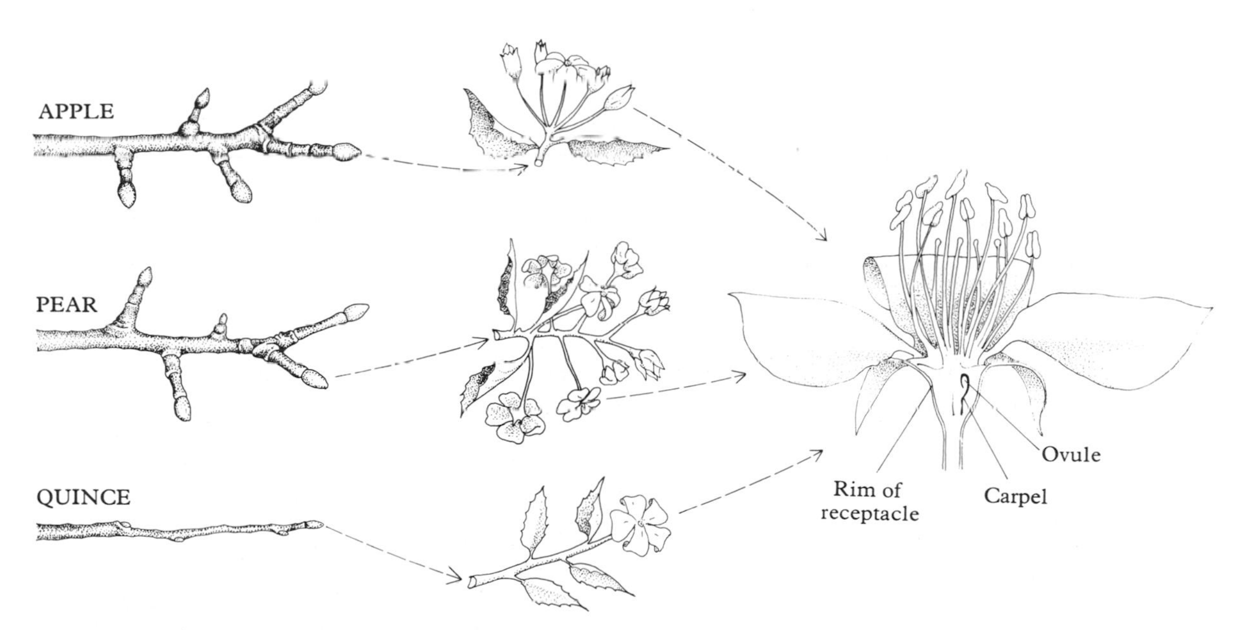

Figure 7-2 Pome fruit flowering habits. **Apple** flower buds on two-year and older spurs; inflorescence is determinate with five flowers. **Pear** flower buds on two-year and older spurs; inflorescence is indeterminate with six to eight flowers. **Quince** flower buds at the terminals of short or long one-year shoots; flowers are solitary. All pome species have epigynous flowers, that is, flowers with inferior ovaries.

Pear

Pear buds and inflorescence are similar to those of apple with two distinct differences. The pear inflorescence contains seven or eight flowers and is indeterminate; that is, the side or lateral blossoms open first and the terminal bloom opens last (Fig. 7-2). Floral initiation of pears occurs about 60 days past full bloom. Flower buds are formed on terminals of shoots and short spurs 2-years-old and older. Most pears are not alternate bearing and tend to flower every year. However, Hardy and Comice on quince rootstock tend to flower lightly following a heavy crop year.

Quince

The quince flower bud contains only one flower, borne at the end of a short current-season's shoot (Fig. 7-2). Thus, floral initiation takes place immediately prior to anthesis rather than during the previous season as it does in apple and pear.

Plum and Prune

Plums initiate flowers in lateral buds of both current-season shoots and new growth on older spurs (Fig. 7-3). Initiation occurs mostly in late summer but has been noted as early as July 5 and as late as September. Each bud produces one to three flowers but no leaves. All terminal buds are vegetative.

Many *P. domestica* cultivars are completely or partly self-fertile, while others, such as German Prune and President, are self-sterile and require cross-pollination (Table 3-7). Most American and *P. salicina* plums are self-sterile. Exceptions are the salicina cultivars Santa Rosa, Beauty, Climax, and Methley, which are self-fruitful. Most plums flower annually, but Sugar Prune is distinctly alternate bearing—unless heavy early blossom thinning is done in the "on" year.

Apricot

Floral initiation in apricot occurs during late summer on both current season's shoots and on short older spurs. The solitary flowers have no basal leaves and are borne only on lateral buds. Most cultivars are self-fertile, but Riland and Perfection are self-sterile (Fig. 7-3B).

Peach

Peaches produce solitary flowers from unmixed axillary buds of last season's growth, so 40 to 50 centimeters of new growth each year are needed to maintain good cropping (Fig. 7-4). Floral initiation begins in midsummer and continues for several weeks. Enough flowers are formed each year for good annual cropping, but there is a tendency of trees to bloom lighter following a very heavy crop year. The peach is precocious and often blooms the second or third year after planting.

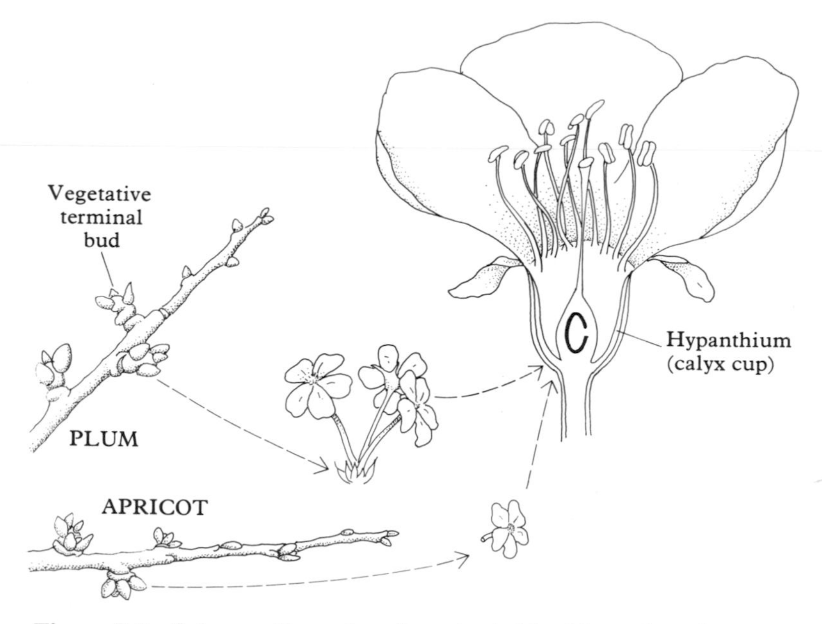

Figure 7-3 Subgenus *Prunophora* flowering habits. **Plum:** lateral, unmixed flower buds on two-year and older spurs; the inflorescence is a few-flowered fascicle. **Apricot:** fruit buds on one-year and older wood; flowers are solitary. Both species have perigynous flowers with a single pistil.

Almond

The flower bud of almond is similar to that of peach, being solitary, without leaves, and always in lateral rather than terminal position (Fig. 7-4). Many more flowers are normally formed than are needed to produce a full crop, but a much heavier set is required than with peach because the peach is very much larger at maturity. Almonds are not thinned, and heavy fruit set does not result in alternate bearing on healthy trees. Production can be as high as 4.5 metric tons per hectare but is usually much less. The inshell yield per bearing hectare in California is about 1700 kilograms.

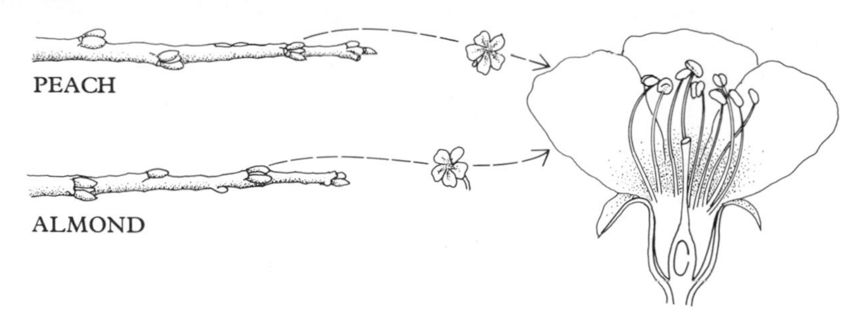

Figure 7-4 Subgenus *Amygdalus* flowering habit. **Peach** and **almond** flower buds, borne laterally on one-year shoots. Flowers are solitary, perigynous, with a single pistil.

Sweet Cherry

Flowers of sweet cherry are borne in unmixed clusters of two to four laterally on short spurs or near the base of longer shoots (Fig. 7-5). Floral initiation takes place in July, after the crop is harvested. In contrast to plums, apricots, and peaches, the cherry initiates flowers only on buds in which the attendant leaves opened relatively early in summer. Flowers normally have but one pistil each, but following very hot summers two pistils per flower may form, resulting in double fruits, an undesirable condition.

All sweet cherry cultivars except Stella are self-sterile, and there are several intrasterile groups as well, in which none of the group will cross-pollinate any other member of the group. Bing, Lambert, and Napoleon comprise such a group. Van, Black Republican, Black Tartarian, and Corum will cross-pollinate cultivars of that group. It is important to establish compatibilities for any new cultivar introduced.

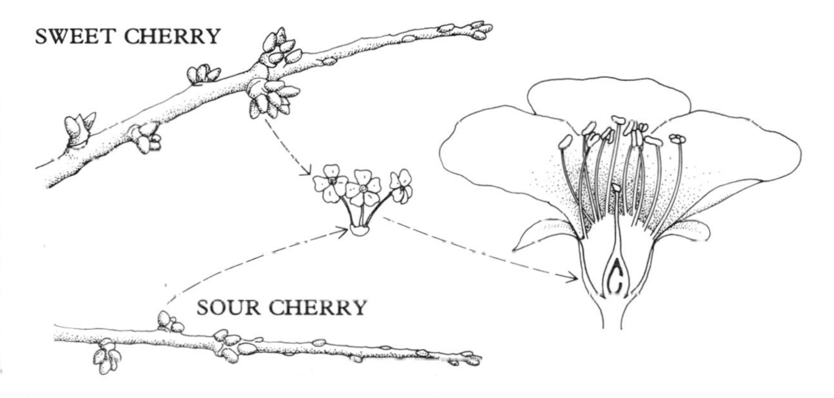

Figure 7-5 Subgenus *Cerasus* flowering habit. **Sweet cherry** and **sour cherry** flower buds. Both are borne on two-year wood or at the base of one-year shoots. They are found only on lateral, unmixed buds. Inflorescence is a few-flowered fascicle, and the flower is perigynous.

Sour Cherry

Sour cherry has buds with two to four flowers, borne laterally on both shoots and spurs (Fig. 7-5). Trees low in vigor tend to overflower, leaving too few leaf buds to properly size the crop. Flowers open late enough in spring not to be often killed by frost.

Fig

Like the mulberry, fig fruit develops from the entire inflorescence—including flower parts, sepals, peduncle, and ovaries. Mulberry flowers, however, are on the outside surrounding the peduncle (a syncarp), while the fig inflorescence is "inside out," with the succulent peduncle on the outside and the flowers on the inside (Fig. 3-4). The fig fruit is called a "syconium."

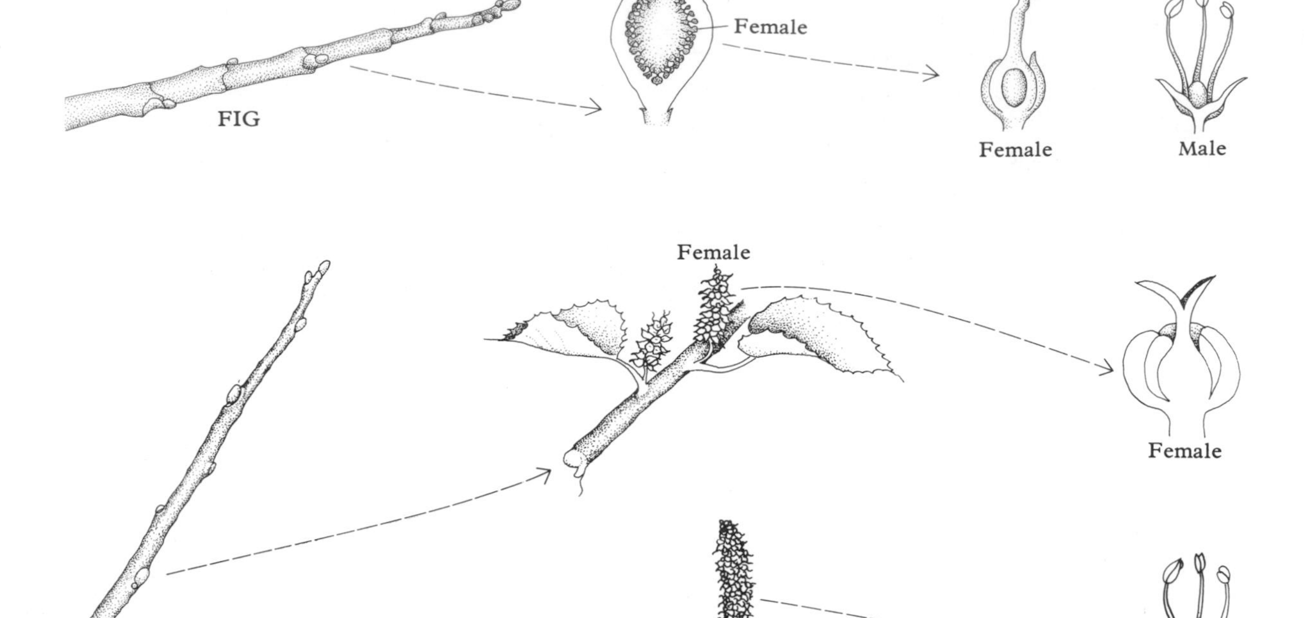

Figure 7-6 Flowering habits of **fig** and **mulberry.** The fig inflorescence is borne at the nodes of one-year shoots beside a vegetative bud that may or may not develop into a new shoot. The mulberry is monoecious or dioecious; its inflorescence is a many-flowered catkin borne at the axils of leaves on new spring growth. The female flower of the mulberry contains two pistils.

Flower initials typically form at the axil of each leaf, where a central vegetative bud is accompanied by two flower buds (Fig. 7-6). Initiation occurs in late summer for next year's first crop. Some cultivars develop only one fig at a node, while others develop fruit from both buds. Second-crop flowers are initiated in the same season in which the figs develop.

Persimmon

Floral initiation of persimmon appears to begin in July in buds that in the next season will be lateral buds at the axils of new leaves (Fig. 7-7). *D. kaki* has a very low chilling requirement, as buds grow well after warm winters in southern California.

Filbert or Hazel

The filbert or hazel nut is monoecious; that is, the staminate and pistillate flowers are borne on the same tree. Male flowers are borne on catkins from unmixed lateral buds. Female flowers are borne in clusters on lateral buds of last year's growth and are terminal on a short shoot that develops from the mixed bud. They

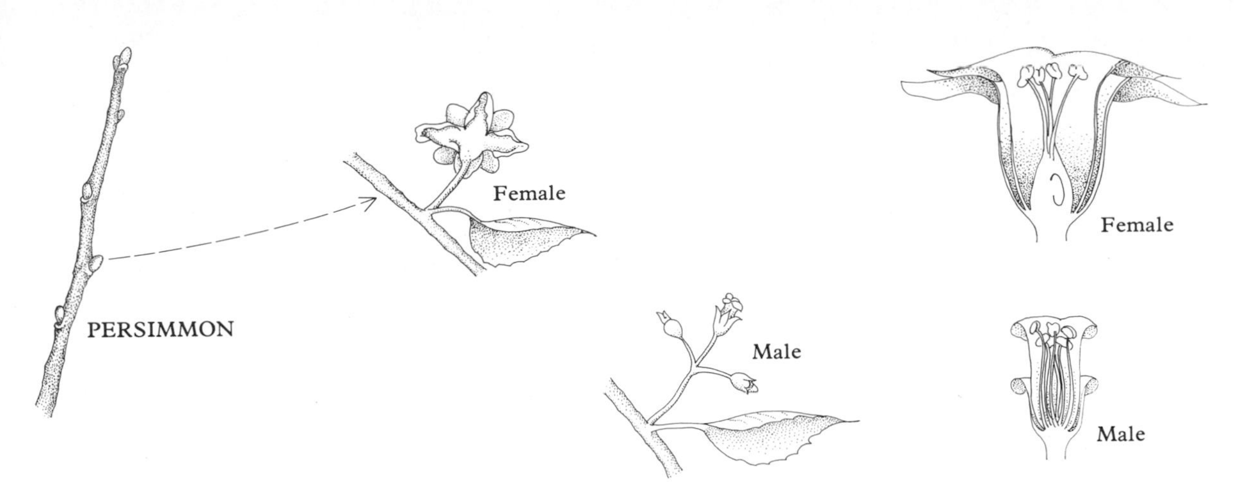

Figure 7-7 Flowering habit of the **persimmon,** which is dioecious. Female flowers are usually solitary; male ones are in cymes; both are borne at leaf axils on new growth. The female flower is hypogynous.

have no petals and appear only as small red tufts (stigmas) protruding from the bud in winter. Both male and female flowers are initiated the previous summer (Fig. 7-8).

Walnut, Pecan, and Hickory

Walnuts, pecans, and hickory nuts have similar flowering habits (Fig. 7-8). They are monoecious, with the male catkins borne in lateral buds of last year's growth and female flowers borne terminally (and sometimes laterally) on current season's growth. As with other wind-pollinated nuts, the flowers have no petals and consist of large reflexed stigmas attached to an ovary.

Pistachio

The pistachio is dioecious and thus about one-sixth of the trees in an orchard should be male-flowering trees. The apetalous female flowers are wind-pollinated and are borne laterally in panicles (Fig. 7-8). Heavy cropping causes initiated flower buds on the same branch to drop during the summer, resulting in biennial bearing (Crane, 1973; see Fig. 7-9).

Chestnut

Chestnuts also are monoecious. Both kinds of flowers are borne on current-season shoots. The male catkins are found on lateral buds along the lower portion of the shoot. The female flowers are located at the base of one or more male catkins, near the tip of the new shoot. Three nuts are produced in each "bur," or spiny involucre (Fig. 7-8).

Grape

The grape produces flower clusters from a compound bud, or "eye," which is more complex than the mixed bud of the pome fruits. Each eye, borne on last year's cane growth, contains a number of buds, but only the central bud usually develops. Floral initiation begins in midsummer for the flowers that open the next

FILBERT

Female

Male

WALNUT

Female

Male

Female

Male

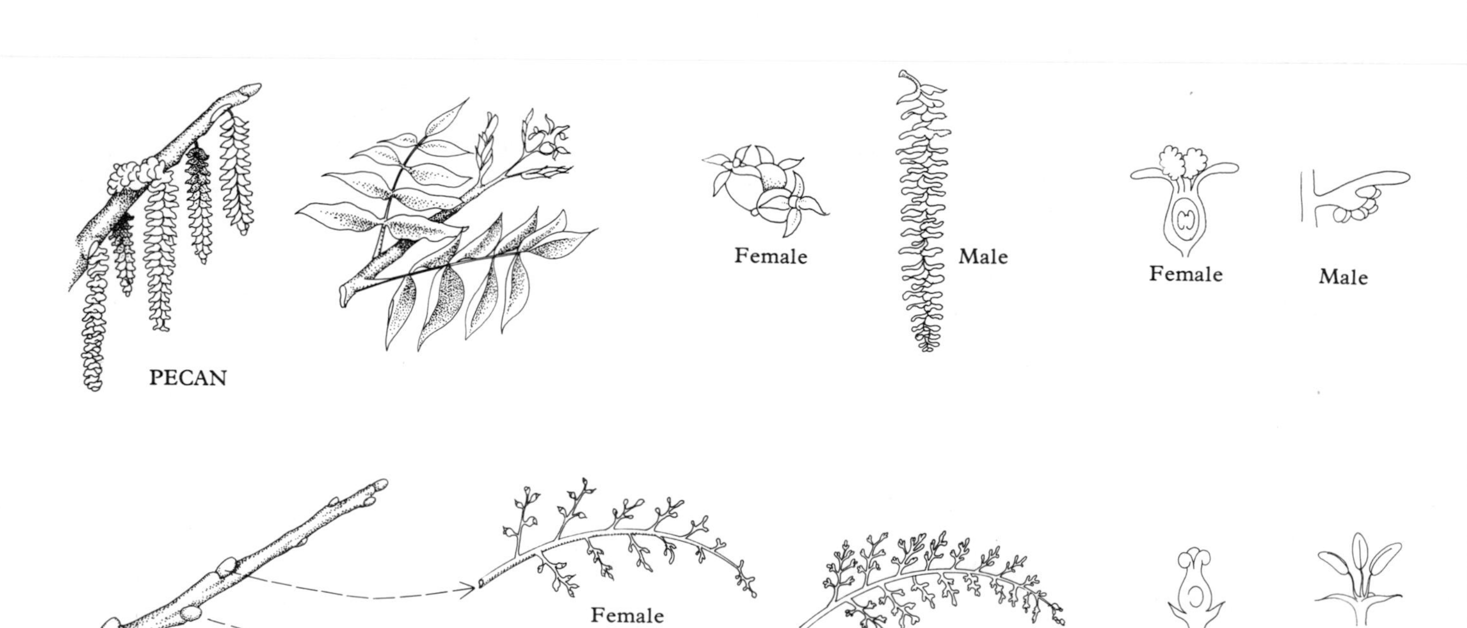

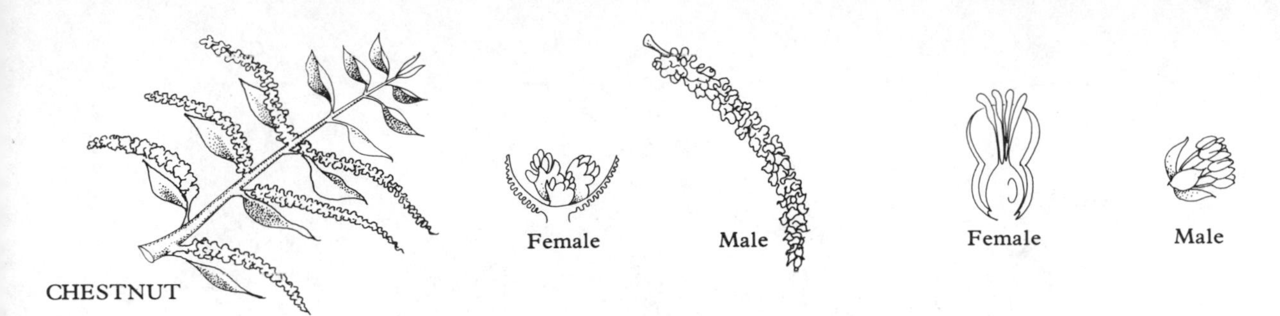

Figure 7-8 Flowering habits of tree nuts. The **filbert** is monoecious; male and female flower buds are borne in the lateral positions on one-year shoots. The female inflorescence is borne in a mixed bud that grows into a short shoot with the flower cluster at its terminus. The **walnut** is monoecious; male catkins are borne laterally on one-year shoots; the female inflorescence occurs, after a period of vegetative growth, as a terminal raceme. The **pecan** is monoecious and is similar to walnut. Male catkins are borne laterally near the end of last season's growth; the female inflorescence is a few-flowered terminal spike. The **pistachio** is usually dioecious; both male and female flower buds occur laterally on one-year shoots; inflorescence is a panicle; the female flower contains a single ovule. The **chestnut** is monoecious; both male and female catkins are borne in axils of leaves of new spring shoots; female flowers, usually three, are borne at the base of the upper male catkins; the ovary is six-celled.

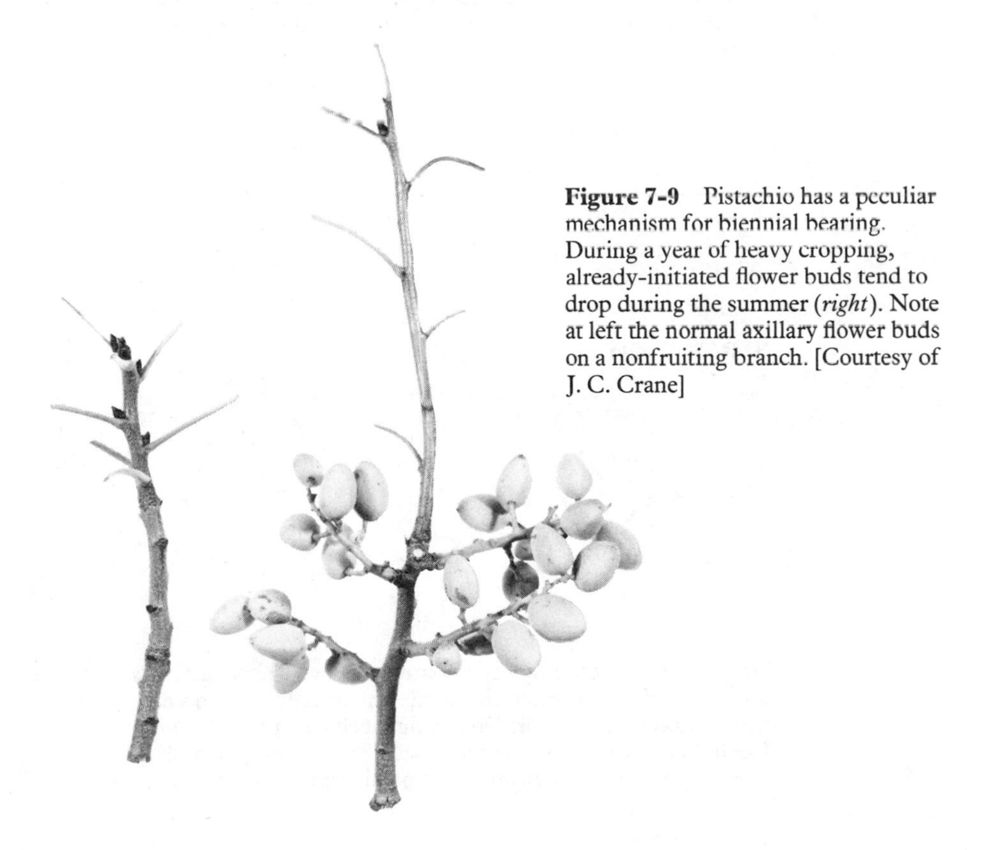

Figure 7-9 Pistachio has a peculiar mechanism for biennial bearing. During a year of heavy cropping, already-initiated flower buds tend to drop during the summer (*right*). Note at left the normal axillary flower buds on a nonfruiting branch. [Courtesy of J. C. Crane]

Figure 7-10 Flowering habit of **grape.** Flower buds are at the nodes of one-year canes, with the racemose inflorescences appearing opposite the leaves as the new shoot develops. Hypogynous flowers are usually perfect, sometimes dioecious; the two-celled ovary is superior.

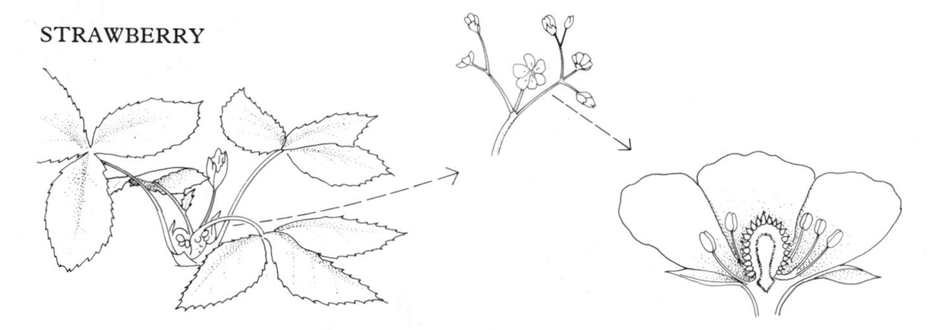

Figure 7-11 Flowering habit of **strawberry.** The flower buds are borne at the axils of leaves on the crown; the inflorescence is cymose, with the early central flowers opening first and being much larger than later ones. The flowers are polygamo-dioecious, the male flowers being larger and showier. There are many pistils per flower; the ovaries are superior.

spring. Following bud break, a shoot and leaves appear first. The two or three flower clusters per shoot appear opposite the leaves. If a shoot does not produce flowers, tendrils—which are simply sterile inflorescences—appear opposite each leaf. The grape inflorescence is a racemose panicle with primary and secondary branching to accommodate the many flowers (Fig. 7-10).

Strawberry

The strawberry inflorescence is produced terminally but later is laterally displaced by the uppermost axillary bud, so that the terminal bud appears in an axillary position. Depending upon the cultivar, the cluster contains few-to-many flowers. The earliest flowers and largest fruits are in the pseudolateral position, the flowers in terminal position being late and smaller (Fig. 7-11). Spring-bearing cultivars initiate flowers during short days of the previous fall.

Blackberry and Raspberry

Blackberries, raspberries, and other *Rubus* species have biennial stems that produce flower buds along the upper ends the first year and then fruit and die the second year. The buds may be on the main cane or on laterals developed by earlier topping. The inflorescence is a several-flowered raceme borne at the terminal of a short shoot that also bears leaves (Fig. 7-12). Floral initiation occurs in late summer when cane growth has stopped.

Currant and Gooseberry

Currant and gooseberry flower buds are borne on lateral buds of 1-year shoots or on short 1-year branches from 2- or 3-year wood. The inflorescence is a raceme often reduced to one flower in gooseberry. The flower cluster bears no leaves (Fig. 7-13). Floral initiation occurs in late summer.

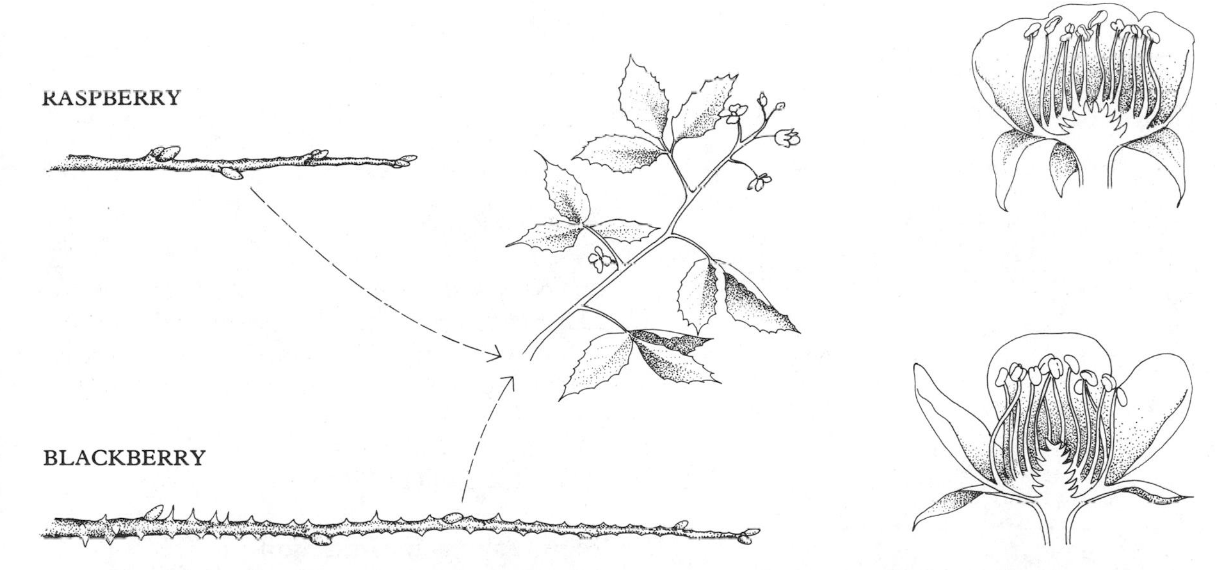

Figure 7-12 Flowering habit of *Rubus.* The mixed flower buds of both **raspberry** and **blackberry** are borne on one-year canes. The indeterminate racemose inflorescence is terminal on a short shoot arising from the flower bud. The many-pistilled flowers are perfect, and the ovaries are superior.

Petal

Sepal

CURRANT

or

GOOSEBERRY

Figure 7-13 Flowering habit of **currant** and **gooseberry.** The unmixed flower buds are lateral on one-year shoots or spurs. Inflorescence is a raceme in the currant and is usually solitary in the gooseberry. The perfect flowers have inferior ovaries.

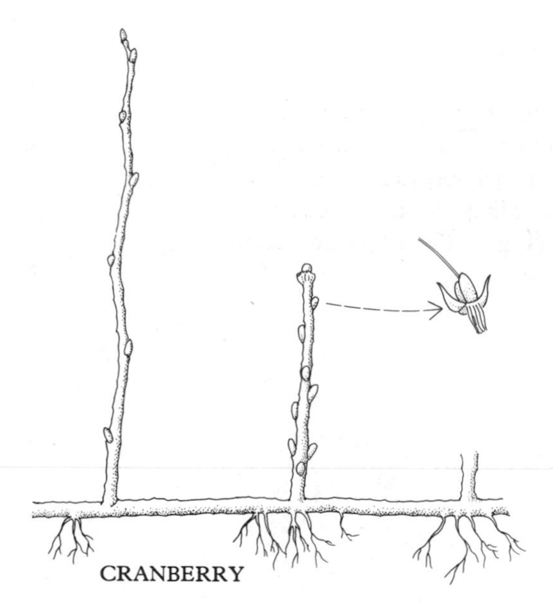

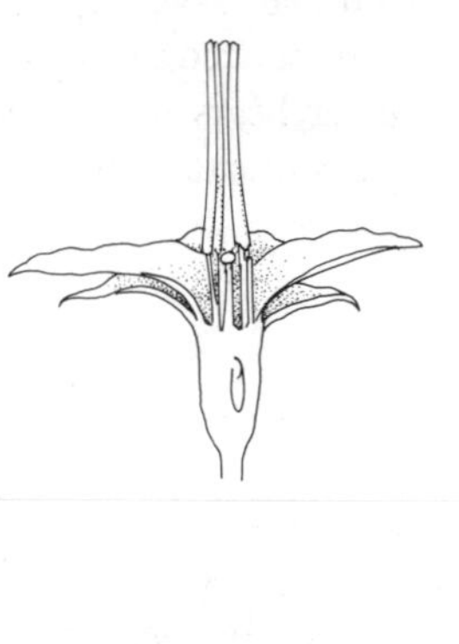

Figure 7-14 Flowering habit of **cranberry** and **blueberry.** Flower buds are either lateral or terminal on one-year shoots. Blueberry inflorescence is a raceme; the cranberry flower is solitary. The epigynous flower is perfect and has a four- to ten-celled inferior ovary.

Cranberry and Blueberry

Cranberry and blueberry flower buds are unmixed buds, mostly borne laterally on shoots of the previous season. As Figure 7-14 shows, the inflorescence is a dense raceme (blueberry) or solitary (cranberry).

FLOWER INITIATION

Requirements for Initiation

Both external and internal requirements for floral initiation vary with species. Generally, deciduous perennials initiate flowers just after shoot growth ceases and when leaves are mature. Seedling plants, which are chiefly of interest to plant breeders, must outgrow the juvenile stage and reach the adult phase before flowering can occur. Budded, grafted, or self-rooted cultivars are adult at the start and thus are capable of flowering under favorable conditions. Presumably the interaction of specific physiologic and environmental factors provides the proper balance of endogenous hormones to cause initiation. Responses of annuals, biennials, and perennials to a given environment are quite different because their life cycles are different. Initiation in temperate-zone species is much more dependent on seasonal variation in climate than it is in tropical plants, which have fewer problems in surviving adverse climates. All plants, however, depend on environmental cues—for example, light, temperature, and day length—to put the plant in synchrony with the season so that it is reasonably programmed to complete its life cycle in the proper season. Various kinds of growth regulants can be used to trigger initiation. Some chemicals (GA, for example) that trigger initiation in one species may inhibit it in another. Chemical treatments, as well as such practices as grafting, girdling, and pruning, usually affect initiation indirectly through their effects on one or more processes in a plant's complex physiological system.

Early work on factors influencing flowering indicated that a carbohydrate build-up, as well as the presence of healthy leaves exposed to light, were required for floral initiation. When spurs are defoliated prior to normal initiation, no flowers are formed—even though the carbohydrate level of the spurs is equal to that of leafy spurs which initiate flowers. This and other evidence leads to the conclusion that some hormonal substance, produced in the leaves and transported to the buds, is required for floral initiation (Cjlachjan, 1937). Apparently, other important internal conditions occur concomitantly with the observed favorable carbohydrate/nitrogen balance, and such changes control flowering. Recently, flowering has been induced in several kinds of tree fruits by application of growth regulating chemicals. One of them (SADH) is a growth retardant while another (TIBA) is not. Both have induced flowering of young apple, pear, and cherry trees and have helped prevent biennial bearing in apples. The fact that these chemicals induce floral initiation on spurs currently bearing fruit (in the "on year") suggests that hormonal balance rather than the carbohydrate/nitrogen ratio actually controls flowering. The fruit on untreated alternate-bearing cultivars seem to produce an unfavorable hormone balance or perhaps even to produce a specific flowering inhibitor. Fruits and seeds produce GA, and SADH is a GA antagonist. Also, fruits produce auxin, and TIBA inhibits the normal downward polar transport of auxin in the plant; TIBA possibly also inhibits polar transport of flowering hormones.

Biennial bearing of most tree fruits results from poor flower initiation during a heavy crop year. Chan and Cain (1967), working with the apple cultivars Spencer Seedless and Ohio 3, removed seeded fruit at intervals after bloom and observed that 65 percent of the flower inhibition by seeded fruit occurred within the first 3 weeks after pollination, while seedless fruit did not inhibit flower initiation. Therefore, developing embryos appear to produce the inhibitory factor to flower initiation, rather than carbohydrate/nitrogen balance and nutritional competition in the fruits, as has been assumed for many years. GA, produced by young seeds, is well known as an inhibitor of flower initiation in many plants (Dennis, 1967 and Dennis and Edgerton, 1966).

Florigen

Florigen is the name given a hypothetical hormone, as yet not isolated and identified, that is said to be responsible for changing a vegetative bud into a flower initial. Much of the evidence for this flowering hormone comes from photoperiod-sensitive herbaceous plants, but some work has been done with woody perennials. Florigen appears to be synthesized in relatively mature leaves and translocated in the phloem to the nearby bud, where initiation occurs. The movement of the flowering stimulus from leaves to buds is quite local and is not detected some distance from the leaves. Studies with apple showed that buds on defoliated spurs did not initiate flowers even though there were leaves on nearby spurs.

Jackson and Sweet (1972), however, cite evidence which renders the florigen hypothesis inadequate because of the many known alternative pathways by which flowering can be initiated in a given species. They also point out that no specific flowering hormone has ever been found in plants and suggest that flower initiation, "like that of any other organ, is dependent on a necessary interaction in space and time of endogenous hormones and assimilates and that these are not substances specifically connected with flowering." There thus seems to be a delicate hormonal balance regulating seed development, fruit set, and floral initiation. The increase in flowering by sprays of TIBA, SADH, CCC, and so forth (Chapter 14), may result from a shift in the hormonal balance to overcome the inhibitory effect of GA. The greater return bloom the year following chemical blossom and fruit thinning probably results from the lowered GA level when fruit (and seed) are removed from the tree prior to the period of floral initiation.

Time of Initiation

The responses of plants to day length are thought to result from the balance between two forms of a pigment called phytochrome. Phytochrome is light sensitive and changes from one form to the other when alternately exposed to red (660 nanometers) and far-red (730 nanometers) light. Thus, the length of the day and the night determines the phytochrome balance; in some species (such as strawberry), this balance determines whether or not flowers will be initiated. Most fruit and nut species, however, do not initiate flowers in response to photoperiod, although other functions, such as the onset of dormancy, do appear to be controlled by phytochrome. These species appear to initiate flowers in response to seasonal physiologic age (that is, days from full bloom) and to proper light intensity and quality, leaf surface, nutrition, pruning, and so forth. Most deciduous fruits start to initiate flowers at the end of the grand growth period for shoots, when the leaves near the buds are mature. Exceptions are the nonwoody strawberry, the fig, and to some degree the peach—all of which continue vegetative growth late in the summer. Most deciduous fruit species initiate flowers during one season, but bloom occurs the following year (Table 7-1). Some fruits, such as the cherry and apricot, initiate flowers after the current crop is harvested, so good care of trees after harvest helps insure a good bloom the following year.

Filbert, walnut, and pecan are monoecious and initiate their staminate flowers earlier than the pistillate ones. Female flowers of walnut and pecan appear to be initiated in the spring of the season they open, but physiologic conditions the previous year seem to determine the extent of such initiation.

Practices Affecting Initiation

Vigorous young trees tend to be overvegetative, and flowering may be less than desired. Any pruning of such trees, no matter how light, will tend to reduce the number of flowers initiated. Young trees on selected clonal rootstocks often flower earlier than those on seedling stocks. The rootstock influence on flowering is not always related to the dwarfing tendency. For example, trees on the vigorous M 1 or on the dwarf M 7 flower earlier and heavier than trees on vigorous seedling or on the dwarfing stock *Malus sikkimensis.*

Both the time and intensity of initiation can be altered by fertilizers, rootstocks, pruning, and other practices. For example, summer pruning of the current season's apple shoots can cause short spur growth on which flowers are initiated late in the summer of the same year. Girdling of the trunk or main branches prior to the usual period of initiation can increase initiation under some but not all conditions. Other physical treatments, such as the bending of branches or root pruning may enhance the flowering of young trees. (These practices are discussed in some detail in Chapter 5.)

As mentioned earlier, several growth regulators can increase flowering, both on young nonbearing

Table 7-1 Time of flower initiation and anthesis of some deciduous fruits and nuts.

Kind	Beginning of Initiation Period	Flowers Borne On	Season of Anthesis Relative to Season of Initiation
		Stone Fruits	
Almond	Mid-Aug.–mid-Sept.	Lateral buds, 1-yr. shoots	Next spring
Peach	Late June–late July	Lateral buds, 1-yr. shoots	Next spring
Apricot	Early Aug.	Lateral buds, 1-yr. shoots + 2-yr. spurs	Next spring
Prune	Late June–mid-Aug.	Lateral buds, 1-yr. shoots + 2-yr. spurs	Next spring
Plum, Japanese	Mid-July–early Aug.	Lateral buds, 1-yr. shoots + 2-yr. spurs	Next spring
Plum, wild goose	Early Sept.	Lateral buds, 1-yr. shoots + 2-yr. spurs	Next spring
Cherry, sweet	Early July	Lateral buds, 2-yr. spurs	Next spring
Cherry, sour	Mid-July	Lateral buds, 2-yr. spurs	Next spring
		Pome Fruits	
Apple	Mid-June–mid-July	Terminal buds, 2-yr. spurs	Next spring
Pear	Early July–early Aug.	Terminal buds, 2-yr. spurs	Next spring
Quince	Early spring before anthesis	Terminal shoots, current growth	Same spring
		Small Fruits	
Blackberry	Late Aug.	Lateral buds, 1-yr. canes	Next spring
Raspberry	Sept.–Nov.	Lateral and terminal buds, 1-yr. canes	Next spring
Strawberry	Sept. (short days)	Crown buds	Next spring
Strawberry, fall bearing	Early summer (long days)	Crown buds	Same summer
Currant	July	Lateral buds, 1-yr. canes	Next spring
Gooseberry	Aug.	Lateral buds, 1-yr. canes	Next spring
Cranberry	Mid-Aug.–mid-Sept.	Lateral buds, 1-yr. canes	Next spring
Blueberry	Late fall	Lateral buds, 1-yr. canes	Next spring
Grape	Mid-June	Lateral buds, 1-yr. canes	Next spring
		Other Fruits	
Fig—first crop	Late summer	Lateral buds, 1-yr. shoots	Next spring
Fig—second crop	Early summer	Lateral buds, current growth	Same season
Persimmon	July	Lateral buds, 1-yr. shoots	Next spring
		Tree Nuts	
Filbert, female	July–Sept.	Lateral buds, 1-yr. shoots	Next winter
Filbert, male	May	Lateral buds, 1-yr. shoots	Next winter
Walnut, female	Feb.–April	Terminals of current shoots	Same season
Walnut, male	Early summer	Lateral buds, 1-yr. shoots	Next spring
Pecan, female	Early spring	Terminals of current shoots	Same season
Pecan, male	Early summer	Lateral buds, 1-yr. shoots	Next spring
Pistachio	Late April	Lateral buds, 1-yr. shoots	Next spring

trees and on heavy alternate-bearing trees as well. The most important of these chemicals are SADH, TIBA, ethephon, and CCC. Effects of TIBA on flower initiation and branch angle are shown in Figure 7-15. Also, the use of chemical thinning agents such as DNOC, NAA, NAAm, Sevin, and Morestan indirectly increase flowering by thinning fruit early in the season by removing much of the flowering inhibitor that is produced by young seeds. The use of TIBA to increase flower initiation is incompletely under-

Figure 7-15 TIBA applied as a spray prior to floral initiation not only increases greatly the number of apple flowers initiated but induces wider and more desirable branch angles. The nearest tree in the photo was sprayed. The second tree is an unsprayed control. [Courtesy of M. J. Bukovac]

stood but is effective with some species (Fig. 7-15). Details of use of growth regulators are covered in Chapter 14.

FLOWER DEVELOPMENT

Following the initiation period—that is, the period of change from vegetative to floral primordia—flowers develop within the bud rather rapidly. In most species, all of the flower parts are distinctly formed by the time the tree goes into winter dormancy (Fig. 7-16). A very slow development takes place during winter, followed by rapid development in the spring, leading up to the opening of flowers and the shedding of pollen (anthesis). During this post-initiation period, many factors determine the extent and quality of flower development. Known factors affecting flower development are: age of wood; position on the tree; temperature; water; carbohydrates, nitrogen, and other nutrient elements; growth regulators; and winter chilling.

Contrary to earlier reports, nitrogen fertilizers often increase flower initiation and development. Williams (1965) recently found that late summer applications of nitrogen resulted in much stronger flowers the following spring. The better fruit set obtained was attributed to longer-lived embryo sacs.

Growth regulators have striking effects on flower development. Auxins applied early tend to inhibit

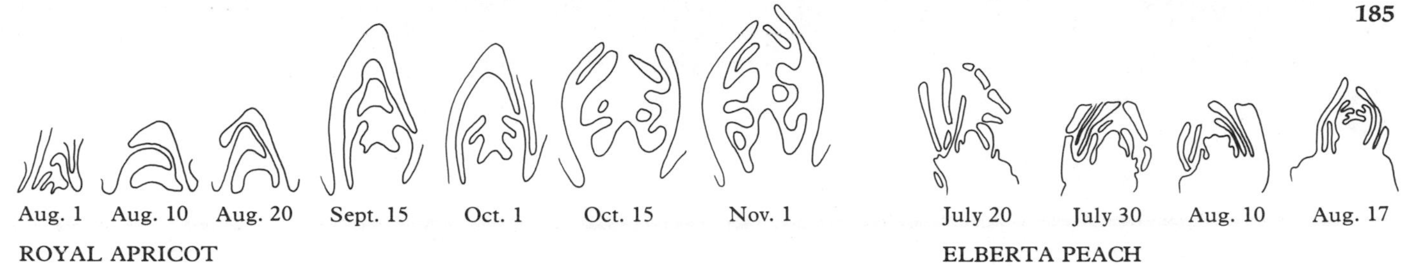

Figure 7-16 Floral development of **apricot** and **peach** from the early period of initiation through late summer and fall. Before entering the dormant period, the floral primordia have developed to the stage where the petals, sepals, anthers, and pistils are clearly evident. [After Tufts and Morrow, 1925]

initiation, yet, if they are applied later, they can enhance flower development. GA not only inhibits initiation but may also cause partial reversion of flower primordia to vegetative structures during development.

Normal flower development requires adequate mineral elements in the proper balance. Any severe imbalance or deficiency of an element can hinder proper development. For example, boron deficiency in pear causes flowers to wither and die just before and during anthesis.

Environmental factors affect flower development. Any severe heat or moisture stress will hinder normal flower growth. Inadequate winter chilling may limit cell divisions and spring development or even cause the flower buds simply to drop from the tree.

GENERAL REFERENCES

Jackson, D. I., and G. B. Sweet. 1972. Flower initiation in temperate woody plants (a review). *Hort. Abstracts* 42:9–25.

Tufts, W. P., and E. B. Morrow. 1925. Fruit bud differentiation in deciduous fruits. *Hilgardia* 1:3–14.

Zimmerman, R. H. 1972. Juvenility and flowering in woody plants (a review). *HortScience* 7:447–455.

8

Pollination and Fruit Set

POLLINATION

Sexual reproduction and seed development in fruits and nuts hinges on pollination, the transfer of pollen from the anther to the stigma. After reaching the stigma, the pollen grain germinates and grows a tube that extends down the style; fertilization occurs when the male nucleus from the pollen tube unites with the egg cell in the embryo sac. After this, a seed may develop along with the fruit or nut. Some have thought that the type of pollen used affects the quality of the fruit on the pollinated plant. Careful studies, however, have revealed that this effect, called "metaxenia," does not occur in deciduous fruits. But if the seed is the portion eaten, then the pollen source (or pollinizer) may affect quality. For example, sweet almond trees pollinated with bitter-almond pollen produce bitter nuts. The terms "pollinator" and "pollinizer" are often confused. *A pollinator is an agent of pollen transfer,* for example, bees, insects, or man. *A pollinizer is the plant cultivar that produces the pollen.*

In most fruits, pollination is a prerequisite to fruit set. Peaches, prunes, cherries, plums, almonds, and apricots must all have seeds, or the fruit will not develop. Normally, apples and pears must have seeds, although in a few cultivars seedless fruits will develop. Some oranges, *Malus* species, and other plants will set seeds by apomixis—that is, without fertilization of an egg cell. Some fruits, such as the fig, will set seedless fruit either naturally (this is termed "parthenocarpy") or with growth-regulator sprays; normally, they set seeded fruits from cross pollination.

In some kinds of plants, the stimulus of pollination alone, that is, pollination without fertilization, is required for fruit growth (this is termed "stimulative parthenocarpy"). Fruits have been caused to develop by treating the pistils, the female portion of the flower, with an extract of pollen (not necessarily pollen of the same species as the pistils). Auxin extracted from apple seed has caused tomato fruit to set parthenocarpically. In recent work by Crane (1969), sweet cherry and apricot fruits were set parthenocarpically using gibberellins and auxins. Completely normal development was not obtained, however. It is conceivable that if the proper chemicals can be found, all fruits can be made to set without pollination (Figs. 8-1, 8-2). Fruits such as strawberry, raspberry, blackberry, currant, gooseberry, blueberry, and cranberry —all of which are multiseeded—require good pollination, fertilization, and seed development for large, regularly shaped fruits. In these fruits, berry size for a given cultivar is correlated with the number of seeds per fruit.

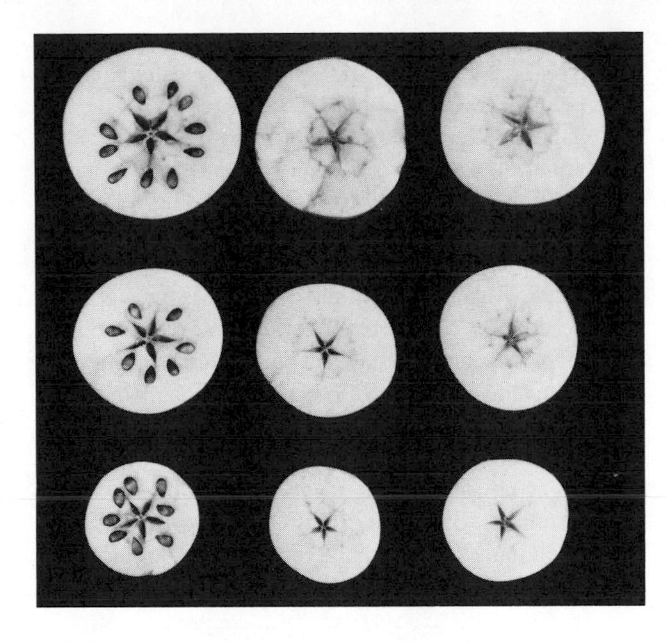

Figure 8-1 Wealthy apple fruit produced by pollination (*left*), by treatment with gibberellin (*center*), and by treatment with extracts from apple seed (*right*). [From Dennis, 1967.]

Wind Pollination

There are two principal ways pollen is transferred from anther to stigma. It is carried by insects, or it is blown by the wind. Some species, such as strawberry and Italian prune, are pollinated both by wind and

Figure 8-2 *Top row:* Parthenocarpic development of Royal apricot after two treatments of flowers with GA. *Bottom row:* Normal open-pollinated, seeded fruit. [After Crane, Primer, Campbell, 1960]

insects. Mulberry, chestnut, pecan, filbert, and walnut are pollinated by the wind. These nut species have two types of flowers: the male flowers (catkins) and the female flowers that produce the nuts. In a wind-pollination system, the pollen grains are very small and light. Some windblown pollen is only .03 millimeters in diameter. Such small grains fall slowly through the air—at the rate of 2.54 centimeters per second. Thus, it takes about 36 seconds for a grain to fall vertically .91 meter. Even in a mild breeze such pollen can travel hundreds of meters. Stigmas of female flowers in a wind-pollinated plant are relatively large compared to stigmas in other flowers, so that the chance a pollen grain will land on the stigma and germinate is greater than it would be otherwise. The number of pollen grains formed in the catkins is tremendous. It is estimated that 250,000 pollen grains are produced for each female egg cell of the filbert.

FILBERT The wind-pollination systems developed by natural selection in the wild do not always work well in the artificial situation of an orchard. Filberts imported from the Mediterranean area to western Oregon and Washington bloom during wet, cold weather, when rain clears the air of pollen grains. The excess moisture destroys pollen viability, and even if the weather turns sunny the pollen is not long lived—germinability may last for only a few hours. Some filbert cultivars, such as Barcelona, are self-sterile, so pollinizers are required. Daviana, the best pollinizer for Barcelona, sheds its pollen during the peak bloom period of the Barcelona female flowers. Other cultivars, such as DuChilly, often shed pollen too late to do much good. Still others shed pollen too early. A better pollinizer for Barcelona is being sought. It should be noted that the female flowers of filberts do not all become receptive at the same time, and that all of the pollen is not shed at once but is released in short bursts during relatively dry periods over a period of several weeks (Fig. 8-3). Release of pollen from the catkin is the result of a natural drying process. If it is released while the wind is not blowing, it remains on the outside of the catkin until wind or air currents carry it away.

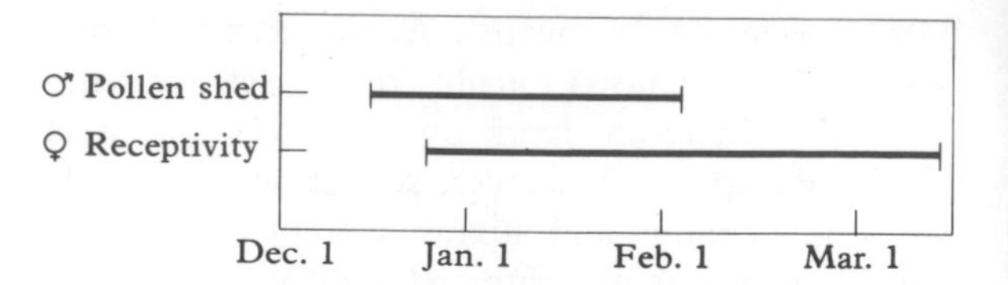

Figure 8-3 Diagram of the time course of filbert flowering, showing the period of pollen shedding and the period of receptivity of the female flowers. There is no problem of poor overlap, even though this species blooms in winter.

Each female floret has two egg cells and each inflorescence about ten florets, so there is a potential of twenty or more nuts per cluster. The usual cluster size of two nuts is apparently due to some other limitation than pollination, because several pollen tubes are found in every style examined. Not all pollen reaching a stigma will germinate, as only about 40 percent of the pollen of some cultivars is viable. Over 80 percent of the pollen of Daviana is viable. Whether or not this factor is important in commercial nut production is at present unresolved. Even if the pollen germinates, the pollen tube may not grow enough to effect fertilization, because pollen from some cultivars will not effect fertilization of another cultivar because of incompatibility. Finally, if fertilization does take place, there may be lethal genes present that cause the embryo to die before it is fully developed. Recent evidence, however, indicates that set may be limited by sub-optimal levels of boron in the young nuts. Nut set occurs very late in the season for filbert. The pollen tube lies dormant from January to mid-June, when the ovary matures. After fertilization, very rapid growth occurs; during this period, considerable embryo abortion also occurs. This abortion may be connected to an incipient deficiency of boron because of a higher need by the embryos at that time. Foliar sprays of boron in late May seem to improve nut set and yield.

WALNUT The following terminology is useful in a discussion of walnut pollination:

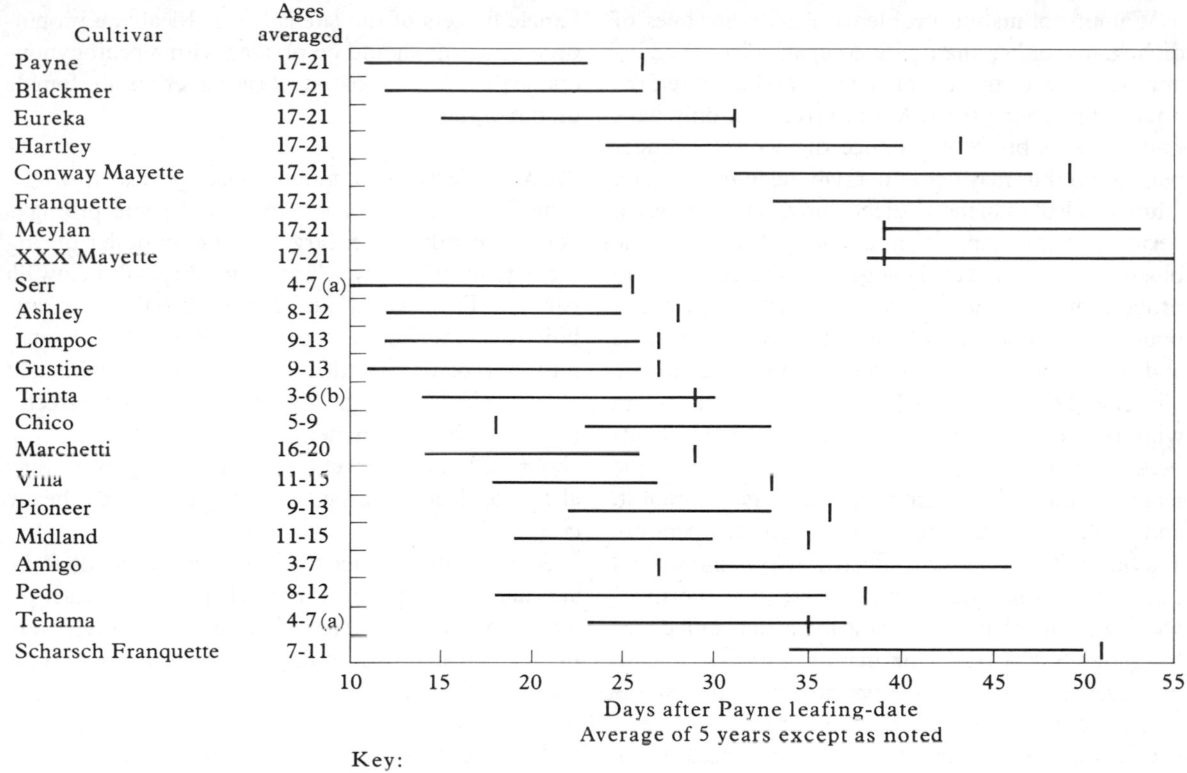

Figure 8-4 Comparison of pollen-shedding period with peak receptivity of pistillate bloom for a number of walnut cultivars at Davis, California. Note that some are protandrous and others are protogynous. [After Forde and Griggs, 1975]

Monoecious. Having pistillate (female) and staminate (male) flowers on the same plant (Fig. 8-4).

Dichogamy. Production of male and female flowers at different times on a monoecious plant. (Dichogamy insures cross-pollination in nature.)

Catkin. Spike which bears the staminate flowers and pollen-producing anthers.

Protandrous. Dichogamy in which the male flowers develop before the female flowers.

Protogynous. Dichogamy in which the female flowers develop before the male flowers.

Regia (English) walnut cultivars are self-fertile and most are protandrous; however, protogynous cultivars exist, of which Meylan, XXX Mayette, and Chico are examples. Experiments, starting over 100 years ago, have shown all cultivars of regia walnuts to be self-fertile and interfertile. (Most species within the genus *Juglans* will also cross.) Controlled crosses between cultivars have set from 10 to 90 percent with an average of 49 percent set. Open-pollinated control trees set an average of 24 percent.

Walnut pollination problems usually are ones of dichogamy rather than of incompatibility. Dichogamy varies with the age of the tree and is more pronounced in young trees. Mature trees not only have more catkins but also produce them over a longer period and thus may better overlap the female bloom. Climate also influences dichogamy. During warm periods, catkins respond more rapidly than the female blossoms. This makes dichogamy more complete in protandrous cultivars—this is an asset for protogynous cultivars because of the better overlap of male and female flowers. Continuous cold weather in late winter and early spring delays blooming, while warm winters tend to hasten blossom development. Temperature and humidity have a direct effect on the length of the bloom period: hot, dry days shorten it, and cool, humid days lengthen it. The temperature and humidity are, in turn, affected by the amount and intensity of sunlight, orchard cover, soil moisture, wind, and air drainage. Dichogamy also is influenced by geography. Coastal or marine climates tend to increase protogyny; more protandry occurs in the interior valleys. This phenomenon can be observed within a single cultivar, such as the Franquette, when traveling from the coast inland. A normally protandrous cultivar would be more apt to self-pollinate when located on the coast. In California valleys, Hartley begins producing catkins later than average and is a shy producer. Grown in cooler districts or at higher elevations, the Hartley bears catkins earlier and more abundantly. A mature walnut tree will produce an average of 5000 catkins, each of which will have between 1,000,000 and 4,000,000 pollen grains. Pollen viability has averaged somewhat less than 25 percent. The potential pollen production from a mature tree is estimated at from 5 to 20 billion grains of which 1.25 to 5 billion would be viable. A Russian study states that normal trees bear seven to eight times more catkins than pistillate flowers.

A partial solution to dichogamy and the walnut-pollination problem is obviously to plant two protandrous cultivars that are selected so that the male bloom of one overlaps the female bloom of the other. However, this makes no provision for pollinating the female flowers of the late cultivar. Ideally, a protandrous cultivar should be planted with a protogynous one so that the male bloom of each overlaps the female bloom of the other.

PECAN Pecan is a monoecious species in which some cultivars are protandrous and some protogynous. The catkins shed large amounts of pollen during the daytime when the relative humidity falls below 85 percent. There is little evidence for self-incompatibility in pecan, but in some cases more self-pollinated nuts fall before maturity than do cross-pollinated ones. Apical pistillate flowers may be imperfectly formed. They seem not to be receptive and drop about pollination time. Other pistillate flowers may also drop heavily, especially the year after a heavy crop.

As with many other tree crops, pecan tends to be alternate bearing, and this tendency varies with region. For example, alternate bearing is more severe in New Mexico than in the southeastern states. This may be due to the thinning action in the southeast of the pecan casebearer. Nut thinning by hand, however, has not prevented alternate bearing. Earlier chemical thinning has been suggested to try to prevent low "off year" bearing following a heavy crop year. However, chemical thinning is not likely to act in the same way on flower initiation as it does with apple because the time of floral initiation for the two species is very different. Apple initiates its flowers for next year's crop about 6 weeks after full bloom; chemical thinning causes heavier floral initiation for the "off year" crop. Pecan, on the other hand, initiates its female flowers in the current season just before anthesis, so there are no nuts developing when floral initiation occurs. The heavy drop of female flowers following a heavy crop year suggests that heavy cropping has a deleterious effect on the whole tree system. Rather vigorous pruning prior to a heavy "on year" crop has resulted in better "off year" yield. Such pruning not only reduces the "on year" yield but also causes more vegetative growth, which may improve the hormonal balance of the tree the next year.

The pecan is somewhat like the filbert in that there

is a delay of several weeks between pollination and fertilization. At about 12 weeks after pollination, when the nut is nearly full sized, the embryo begins a rapid growth period that lasts 6 to 8 weeks.

CHESTNUT Chestnut species are monoecious, and all types are self-sterile. The chestnut is reported to be wind pollinated, but some believe it to be insect pollinated.

PISTACHIO This species is dioecious, and the pollen is windborne. Staminate trees tend to shed pollen before pistillate flowers are receptive. About one staminate tree per six female trees is said to provide adequate pollination.

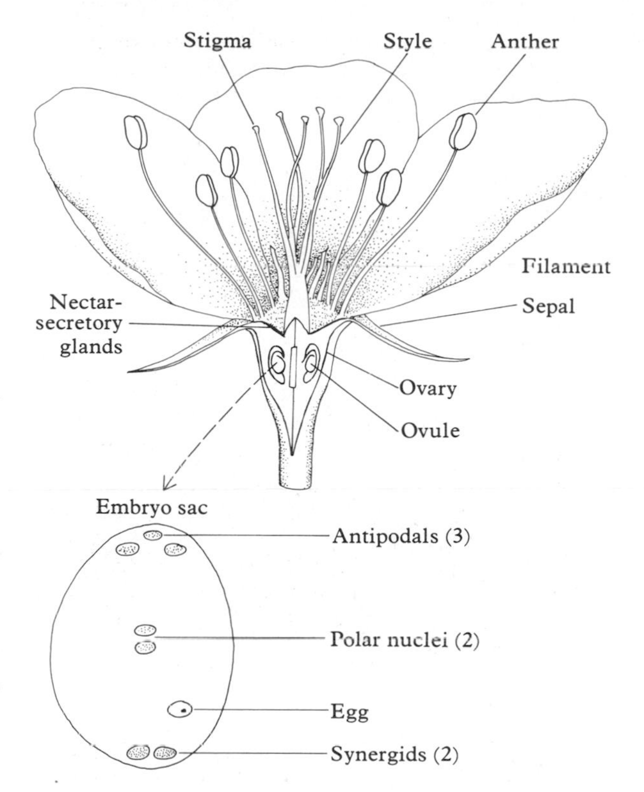

Figure 8-5 Diagram of an apple flower, showing: the stigma on which the pollen germinates, the style through which the pollen tube grows, and the embryo sac prior to fertilization. [After MacDaniels and Heinicke, 1929]

Insect Pollination

All of the plants with showy flowers are insect pollinated. Cherries, plums, prunes, apples, pears, peaches, and nectarines are all pollinated by insects, mostly bees. Commercial honey bees do most of the pollination in orchards, although some pollination is done by wild bees and wasps, and occasionally other insects. Bumble bees are more effective pollinators for blueberry than are honey bees. Most of the insect-pollinated deciduous fruits have both male and female parts in the same flower (Fig. 8-5). Apples and pears have five carpels, each having two egg cells, making a total of ten possible seeds. Five pistils, styles, and stigmas protrude from the carpels. The petals and odor of the flowers attract insects. The anthers shed their pollen as the insects brush against them (Fig. 8-6).

Factors Affecting Pollination

COMPATIBILITY Cultivars may be self-fertile, partially self-fertile, or self-sterile. Information about specific cultivars is given in Chapter 3. Proper provision for pollinizers should be made for main cultivars that are partially or completely self-sterile. Genetic incompatibility results in little or slow pollen-tube growth, so that fertilization fails to take place. In nature this is a genetic mechanism to assure outcrossing of plants, thus bringing together germ cells of potentially greater genetic diversity and giving the offspring some survival advantage. In fruit production, however, where only one or two cultivars are grown, incompatibility is a problem rather than an advantage.

POLLEN VIABILITY Some cultivars produce infertile pollen that germinates poorly, if at all. Triploid apples are examples of cultivars whose pollen has poor via-

Figure 8-6 Bees are important for the pollination of many species. **(A)** A strong hive is needed to pollinate an acre of a self-sterile cultivar. **(B)** A bee pollinates a flower as it gathers nectar and pollen. [Courtesy of R. W. Henderson]

A

B

bility. Under field conditions at moderate temperatures, high humidity, and high light intensity, pollen has a short life and normally is viable for only a few hours. However, at low humidity, low light intensity, and subfreezing temperatures, pollen may remain viable for several years.

POLLINIZER PLACEMENT In areas where poor weather during the bloom period often reduces bee activity, close spacing of pollinizers will result in greater fruit set (Westwood and Grim, 1962). In extremely poor weather, only the trees immediately adjacent to a pollinizer will set fruit. Usually there are enough pollinizers if every third tree in every third row is a pollinizer. But small-fruited, self-sterile types—such as cherry, which require a high percentage set for a full crop—require more pollinizers and more pollinator insects than do large-fruited, self-fertile types. The best situation is to have equal numbers of the pollinizer and the main cultivar. This is possible if the pollinizer is a good market cultivar.

The bloom period of the pollinizer should slightly precede that of the main cultivar for optimum pollination. The sequence of bloom, however, can vary from year to year. For example, Black Republican, which is a sweet cherry pollinizer, blooms too early to pollinate Lambert during some seasons (Fig. 8-7). In hedgerows bees tend to work along the row rather than across rows; for this reason, it is desirable to mix the pollinizer with the main cultivar in each row. But this presents a problem in keeping the two cultivars separate at harvest. One solution is to use primitive, small-fruited species. These are not harvested, so the problem of segregating fruit is eliminated. For apple, *Malus floribunda* and *M. aldenhamensis* have been used. For pear, the tiny-fruited *Pyrus betulaefolia* is being tried.

The pollinizer should be pollen-compatible with the main cultivar. In selecting a pollinizer, be sure to check for time of anther dehiscence (the opening or bursting of the anther). Dehiscence should start in the pollinizer before it starts in the main cultivar.

WEATHER EFFECTS ON BEES Bees do not fly well in rain, strong wind, or at temperatures below 10°C. Cold or windy sides of trees do not set much fruit because bees tend to work the warmest portions of trees and those most sheltered from the wind. Bees also tend to work the unshaded portions of trees more

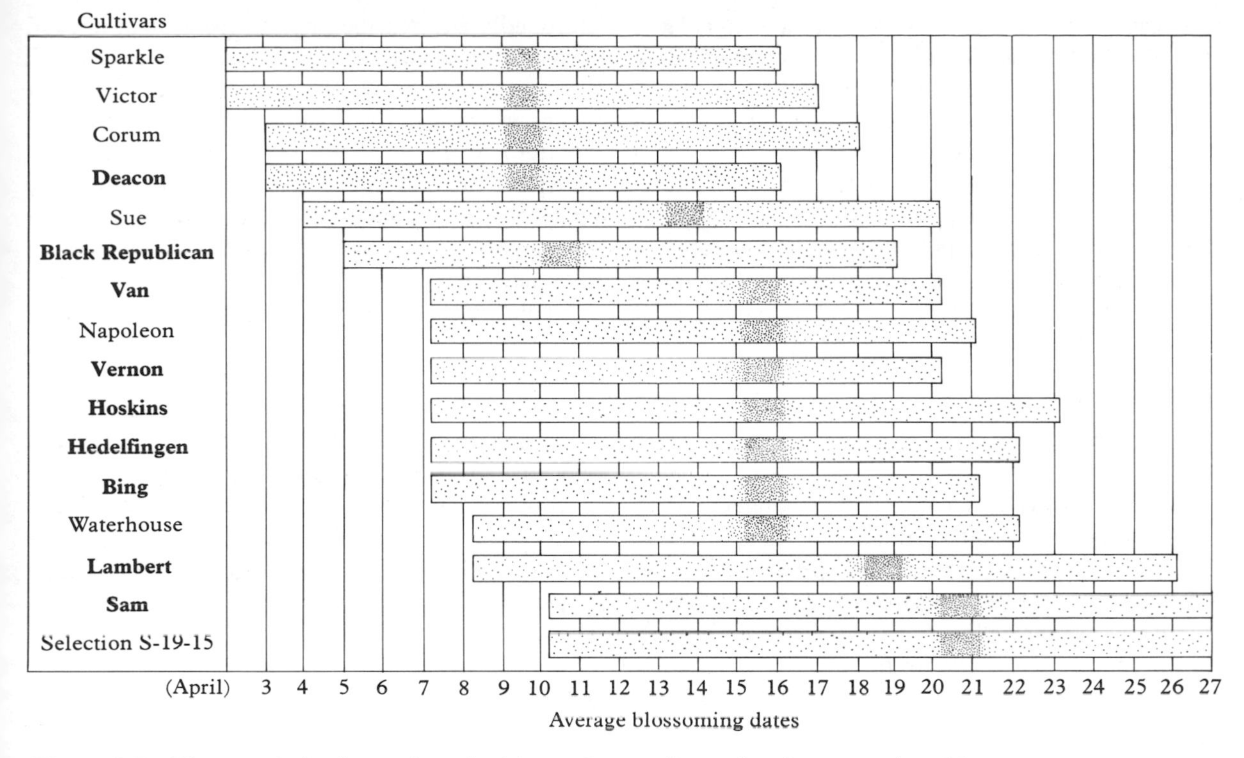

Figure 8-7 Bloom periods of a number of cultivars of sweet cherry, showing some whose bloom periods do not overlap well and thus would not be good cultivars to plant for cross-pollination. Dark cultivars are indicated by **boldface** type. This study was made in Corvallis, Oregon. [After Zielinski, Sistrunk, and Mellenthin, 1959]

than the shaded portions, which helps to account for poor fruit set on the inside limbs.

Bees use the sun for orientation during flight. It is a curious fact, for which no explanation has been found so far, that the position of the sun is correctly used by the bees even when it is hidden behind an unbroken layer of clouds and when, in addition, the hive is placed in surroundings totally unfamiliar to the bees. In territory that the bees know well, they are experts in using landmarks. It appears that infrared rays from the sun penetrating the clouds may guide the bees.

Temperature is an important influence on bee flight. When hives are protected from wind and placed in the sun, bees become active sooner because they are warmer. Distributing hives (two per hectare) throughout the orchard, rather than grouping them at the edge, also helps, particularly in rainy or windy weather. Any competing blossoms on the weeds or ground cover should be mowed or destroyed before bees are brought into the area, and the bees should be brought in only after about 20 percent of the fruit blossoms have opened.

FRUIT SET

With pollination complete, the pollen tube traverses the style, penetrates the micropyle, and fertilization of the egg takes place. Hormonal stimulus from the young developing embryo (or from elsewhere in parthenocarpy) prevents the fruit from abscissing and causes the ovary and adjacent tissues to enlarge into the developing fruit. Fruit set is accompanied by wilting of the petals and, in many plants, the shedding of the anthers and calyx. In most plants, not all flowers set fruit, even though every flower or floret is pollinated and the plant is in good health. The extent of this natural shedding varies with species. Large-fruited types such as apple may shed 95 percent or more of their flowers and young fruits, while a small-fruited species like blueberry may shed only 20 to 30 percent of them.

Ovule Longevity

Williams (1965) has shown that ovule longevity is an important determinant of fruit set and that, if fertilization does not take place within a specific period, the embryo sac becomes nonviable and fertilization cannot occur even if pollination and pollen-tube growth have occurred. He describes the effective pollination period (EPP) as the longevity of the ovules minus the time lag between pollination and fertilization (Fig. 8-8). Lombard et al. (1972) found that temperature profoundly affected the rate of tube growth in pear. At 5°C, tube growth required 12 days while at 15°C, required only 2 days. They determined that pear ovules were viable for 11 days. Thus at 5°C the EPP was zero, while at 15°C it was 9 days. The slow growth of tubes at low temperature is thought to be the principal reason for low set of fruits and nuts during cool spring seasons, even though no freezing of ovules occurs.

Ovule longevity, however, is not constant; it varies with species, temperature, and nutritional status. Williams (1965) found that soil application of nitrogen in late summer to apple trees resulted, the following spring, in strong embryo sacs that continued cell divisions and remained viable twice as long as those of controls (Fig. 8-8). This suggests that cultural practice may be adjusted to compensate for low temperature and slow pollen-tube growth, permitting a longer EPP.

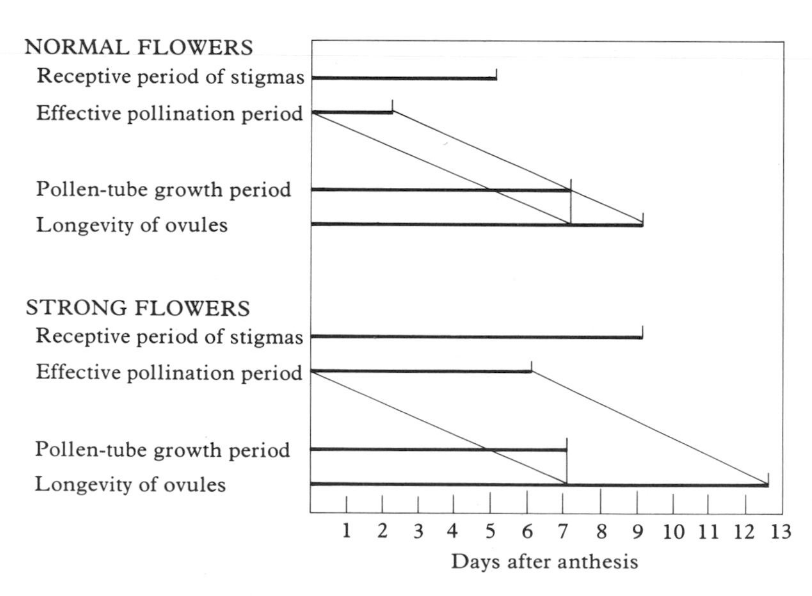

Figure 8-8 Diagram showing the effective pollination period (ovule longevity minus the period of pollen-tube growth) for normal and strong flowers. [After Williams, 1965]

Fertilization

Dennis (1967) found that extracts from immature apple seeds contained substances with GA activity. When such extracts were applied to unpollinated blossoms of the same variety, seedless fruits set and developed to maturity. Evidence indicates that gibberellin, produced in young embryos just after fertilization, is responsible for fruit set in the apple (Fig. 8-1) and apricot (Fig. 8-9).

Studies with the tomato showed that diffusible auxin was not present in the flowers at open bloom, but significant amounts could be obtained after the plants were treated with gibberellin (GA). Early growth of the ovary in plants treated with GA corresponds closely with growth after fertilization. Similar amounts of diffusible auxin were present in both GA-treated and pollinated plants after 22 days (Sastry and Muir, 1963). Effects of auxin on fruit set and development of apple are shown in Figure 8-10.

An interesting study by Melnick et al. (1964) showed that when young fertilized ovules of a number of different species, genera, and families were transplanted onto placenta of pepper fruit, each developed into mature viable seeds. One of the species used was

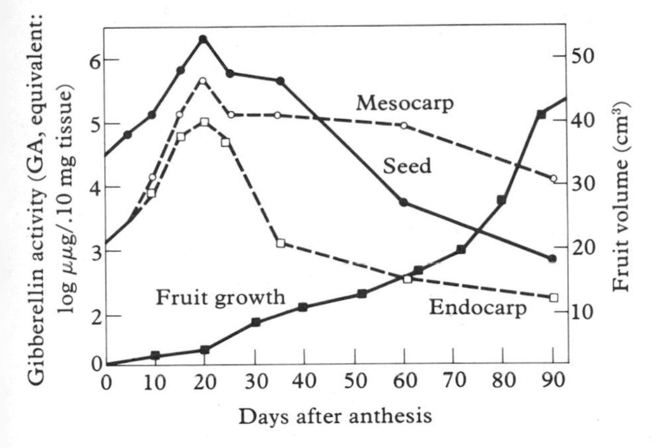

Figure 8-9 Relation of apricot fruit growth and GA activity in different parts of the fruit. [After Jackson and Coombe, *Science*, Vol. 154, pp. 277–278. Fig. 1, 14. Copyright © 1966 by the American Association for the Advancement of Science]

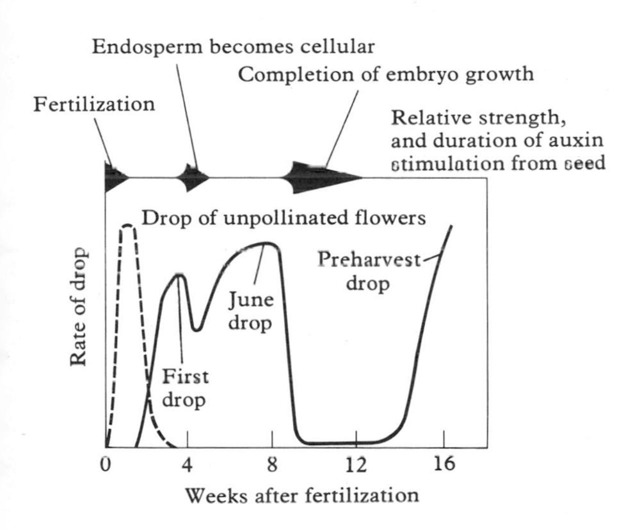

Figure 8-10 Relationship of internal auxin levels in seed and rate of fruit drop of the apple. [After Luckwill, 1953].

strawberry, a member of the rose family, to which also belong blackberry, raspberry, apple, pear, and stone fruits. This remarkable evidence suggests that the physiological and hormonal requirements for ovule and fruit growth are similar for a wide variety of taxonomic groups.

Final Set

After petal fall, the "first drop" occurs—this is a general heavy shedding of undeveloped fruitlets of most species. These are presumed to be mostly from unfertilized flowers. After initial fruit set, there may be two or three more waves of fruit shedding, after which final set is established. The selective shedding of some fruits and the retention of others is, at present, not understood. Since the natural hormone ethylene is known to cause abscission of flowers and fruits, it probably plays a role in this process.

PARTHENOCARPY Fruit set and development without pollination (vegetative parthenocarpy) are found in Oriental persimmon and some cultivars of pear and fig. Stimulative parthenocarpy (pollination but not fertilization) occurs in some varieties of grape (Weaver, 1972). Parthenocarpic cultivars are found to have a higher auxin content at blossom time than do seeded cultivars.

Westwood et al. (1968) found that Anjou pear trees that had produced seeded fruit the previous year would set nearly a full crop of seedless fruit (without pollination) the current year, but would set only half a crop of seedless fruit 2 years after the seeded crop. Woody spurs of trees that had produced a crop of seeded fruit contained more than three times as much auxin (IAA) as spurs from trees that had borne only seedless fruit. It took 4 years to fully deplete this "setting" factor produced by the previous seeded fruit. After such depletion, a fall spray of synthetic auxin (2,4,5-TP) acted much like a seeded crop and induced 60 percent of a crop of seedless fruit to set the following year.

HORMONE RELATIONS A primary site of auxin synthesis in plants is the subapical region of growing shoots; a primary site of GA synthesis is in immature leaves; cytokinins are produced primarily in the roots and transported to the tops via the xylar sap. But there seem also to be a number of other local sites of hormone synthesis in the plant. One such site is the young fertilized ovule and fruit, and the local balance thus generated is important to fruit set, even though the total hormonal balance in the plant also appears to affect set. It seems clear that (1) either auxin, GA, or cytokinins directly causes fruit set or (2) as was shown for tomato, the developing embryo produces GA, which stimulates the production of auxin, the latter causing fruit set. This does not, however, explain why applied auxin does not induce fruit set in a number of species. Perhaps each species requires a specific combination of hormones for set.

In some fruits, embryo abortion occurs after initial set, and the fruit subsequently drop. In Italian prune this occurs in midsummer, when the fruit is about half-grown; however, a spray of 10 to 20 ppm of the auxin 2,4,5-TP at 2 weeks after the start of pit hardening will prevent fruit drop and stimulate normal fruit development (Kochan et al., 1962; Proebsting and Mills, 1961).

Plant hormones may act synergistically in affecting fruit set. Dennis (1973) reports that GA-auxin synergisms were found for cherry, plum, and blueberry. With mango and apple, maximum effects occurred when auxin, GA, and cytokinin were used together. The mechanism or mechanisms of fruit set are unknown, but GA, auxin, cytokinin, ethylene, and ABA seem to have roles in the process.

SEED DEVELOPMENT

Hormones produced in the seed affect fruit growth, fruit set, and general hormone balance in the plant. Stone fruits will not set and grow without normal embryo development. Pome fruits may set without seeds or with few seeds, but growth of the flesh cells exterior to a seedless carpel is reduced. As seeds develop following fertilization, they produce GA, which in turn triggers the production of IAA (auxin), which is related to fruit set. Fruit set and growth of unpollinated ovules treated with GA are similar to fruit set and growth of seeded ovules (Bukovac, 1963). GA production by young pear and apple seeds apparently inhibits floral initiation for next year's crop, causing biennial bearing.

Seed Maturation

Growth of the peach seed and fruit is related to auxin produced in the seed is shown in Figure 8-11. There is periodicity of both auxin synthesis and growth, but no cause-and-effect relationship is evident. During pit hardening, both the flesh and kernel tissues grow

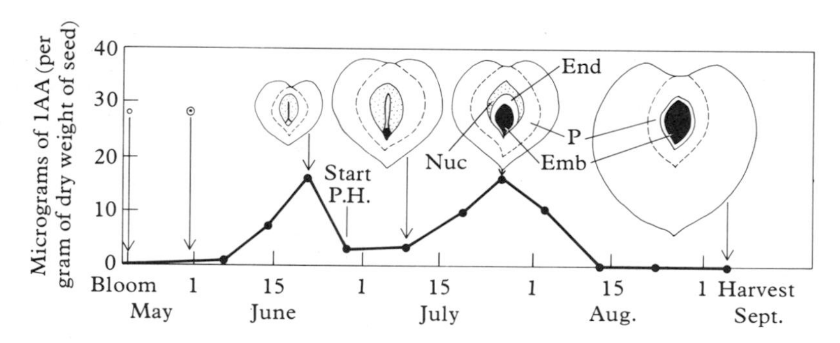

Figure 8-11 Changes in concentration of the growth hormone IAA (auxin), expressed in micrograms per gram of dry weight of the seed of Halehaven peach. Diagrammatic fruits above the graph show the relative sizes of fruit, seed, and seed components. In the seed, the stippled areas indicate the nucellus **(NUC)**, the open areas the endosperm **(END)**, and the solid areas the embryo **(EMB).** The broken line indicates the outer limits of the pit **(P).** Pit hardening **(P. H.)** begins just before July 1. [After Powell and Pratt, 1960]

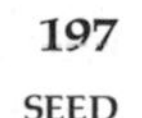

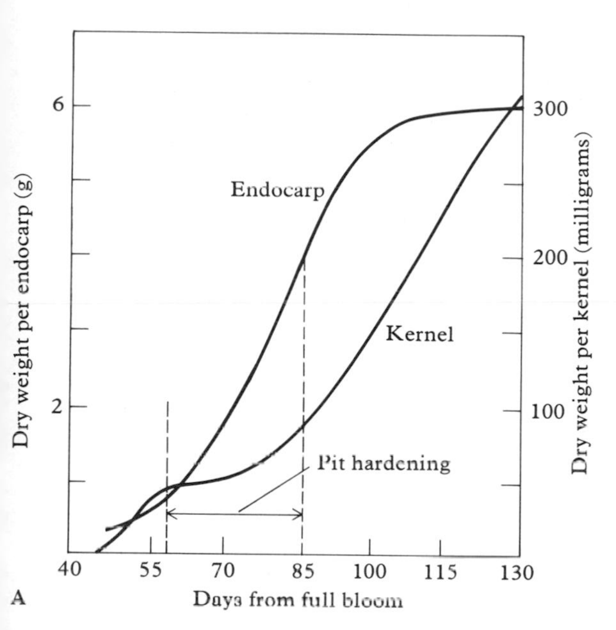

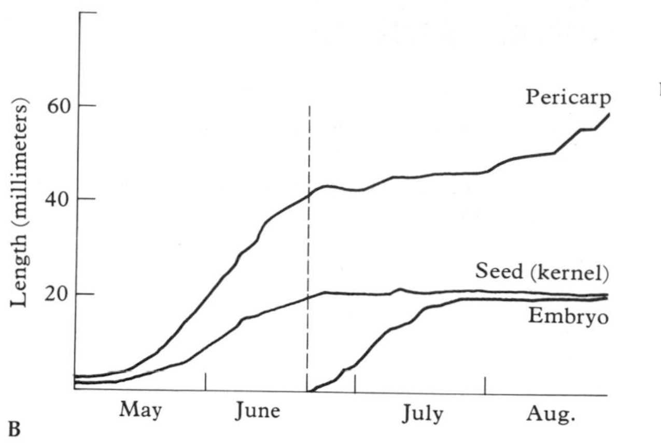

Figure 8-12 **(A)** Seasonal growth of different portions of Elberta peach seed, based on increase in dry weight. Kernel growth was retarded (as was flesh growth) during pit hardening, at which time the pit (endocarp) grew at its most rapid rate. Compare these measurements of growth with measurements of length **(B)**. Measurement of length alone indicates that the kernel has completed its growth by the beginning of pit hardening, but it actually has achieved only about 15 percent of its ultimate dry weight. [Graph A after Batjer and Westwood, 1958a; graph B based on data of Tukey, 1933]

at reduced rates, while the stony endocarp increases rapidly in dry weight (Fig. 8-12A). The growth of the seed in length indicates that growth stops at the beginning of pit hardening (Fig. 8-12B). Yet growth as measured by dry weight shows that it continues to grow until harvest.

Early during seed development, the undifferentiated mass of embryo cells develops into three well-defined structures: the young shoot (epicotyl), the primordial root (hypocotyl), and the cotyledons. Accumulation of stored food in the cotyledons and endosperm signals maturity of the seed. The period of "filling" of the seed is marked by rapid increase in dry weight, although physical size does not change. As embryo enlargement ceases, the seed usually matures as the fruit ripens. The mature dry seed is dormant and may require chilling to bring it out of rest (Fig. 8-13).

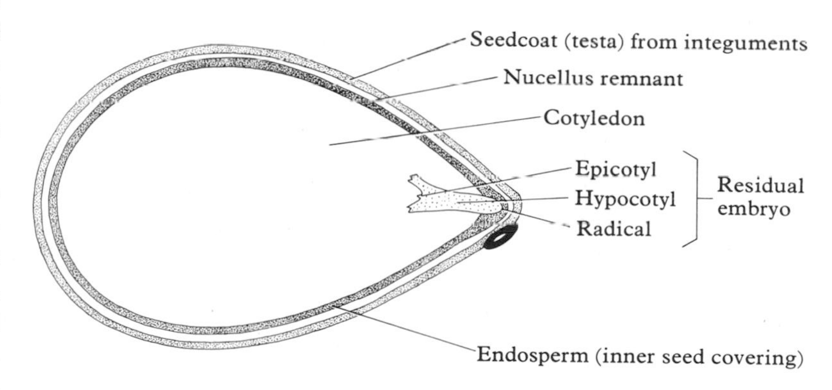

Figure 8-13 Median longitudinal section of a typical pome fruit seed at maturity. The two cotyledons make up most of the seed; the endosperm is reduced to a thin inner seed covering, the residual embryo (considered apart from the cotyledons) is a very small portion of the total. The testa and the residual embryo contain the growth inhibitor ABA which tends to prevent germination.

GENERAL REFERENCES

Crane, J. C. 1969. The role of hormones in fruit set and development. *HortScience* 4:108–111.

Dennis, F. G. Jr. 1973. Physiological control of fruit set and development with growth regulators. *Acta Horticulturae* 34:251–257.

Esau, K. 1953. *Plant anatomy,* Chapters 19 and 20. Wiley, New York.

Griggs, W. H. 1953. *Pollination requirements of fruits and nuts.* Calif. Agr. Ext. Serv. Cir. 424.

Melnick, V. L. M.; L. Holm; and E. Struckmeyer. 1964. Physiological studies on fruit development by means of ovule transplantation *in vivo. Science* 145:609–611.

9

Fruit Growth and Thinning

SEASONAL GROWTH

After pollination, fertilization, and fruit set, the fruit is still very small, and many factors can influence its subsequent growth rate and its ultimate size. Growth can be measured by increases in volume, dry weight, or fresh weight. In this chapter, different kinds of seasonal growth curves will be described, as will some of the factors influencing growth. An understanding of fruit growth and factors affecting it are important to an understanding of the use of fertilizers, pruning, growth regulators, fruit thinning, and size prediction. Each of these topics will be discussed in some detail.

The period of cell division for different fruits varies greatly. Cell division in *Ribes* and *Rubus* are reported to cease at anthesis, while in tart cherry it ceases about 2 weeks after anthesis. The period from anthesis for plum and peach is about 4 weeks, for apple 4 to 5 weeks, and for pear 7 to 9 weeks. Some fruits, such as avocado and strawberry, continue cell division up to maturity. In all fruits, some cell division in the area of the epidermis continues much longer than in the main part of the flesh (Fig. 9-1).

Some time during the cell-division period, cell enlargement begins and proceeds at a rapid rate. At blossom time, intercellular air spaces are absent or very small. Concurrent with cell enlargement, air spaces increase to a maximum for the species in question, then remain relatively constant for the remainder of the season. Early in the cell-enlargement phase, vacuoles form in the cells and increase in size as the cells enlarge, ultimately occupying most of the space in the center of cells. The vacuole is separated from the cell cytoplasm by a semipermeable membrane, through which must pass water and other substances making up the vacuolar (cell) sap. The sap contains

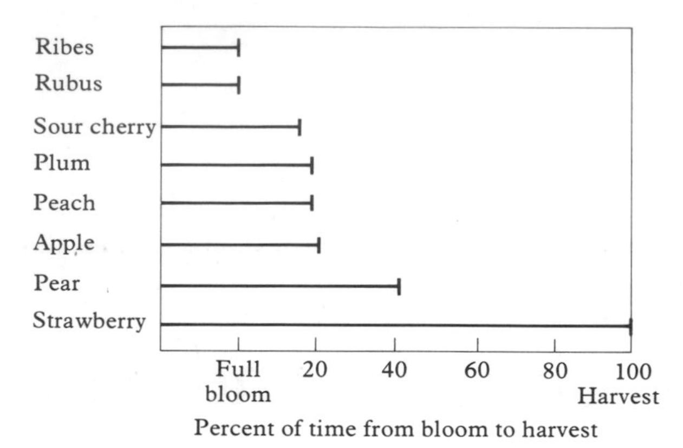

Figure 9-1 Cell division periods for the flesh of different fruit species, as a percent of the total time required for fruit development and maturation.

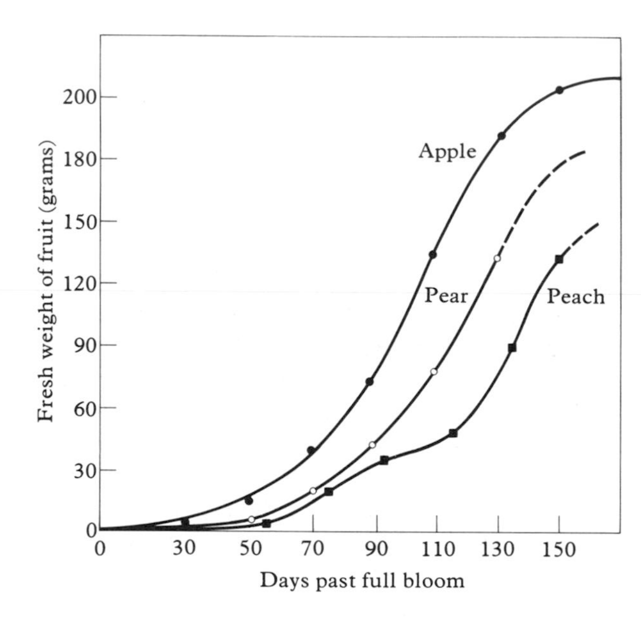

Figure 9-2 Seasonal growth curves for apple, pear, and peach. Apple has a typical S-shaped curve. The growth curve of pear is similar to that of apple except that it does not show the slow growth period at the end (because pears are picked green mature). Peach exhibits a double-sigmoidal growth pattern because of a period of slow growth during pit hardening.

such things as sugars and organic acids, and in the region of the epidermis, contains red and blue pigments, which give these colors to fruit.

The combined growth resulting from cell division, cell enlargement, and air space formation results in a general sigmoidal (S-shaped) curve when fruit volume or weight is plotted as a function of time (Fig. 9-2). This is true of walnut, filbert, pecan, apple, and strawberry. The complete S-shape does not occur with fruits such as pear, which are picked mature-green and are ripened off the tree. If left on the tree to ripen, pear growth would be sigmoidal. Some fruits such as stone fruits (Fig. 9-2), fig, currant, pistachio, and seeded grape have a double sigmoidal growth curve. The first slow growth period of stone fruits coincides with the period of pit hardening, during which lignification of the endocarp (stone) proceeds rapidly while mesocarp (flesh) and seed (kernel) growth is suppressed. Near the end of pit hardening, flesh cells enlarge rapidly until the fruit is ripe, after which growth slows down and finally ceases.

As indicated above, growth of walnut (Fig. 9-3), filbert (Fig. 9-4), and pecan is sigmoidal. Filbert, however, is special because of the long delay between pollination, which takes place in January, and ovary development and fertilization several months later. This growth pattern is described by Thompson (1967) as follows:

After pollination (January), the ovary primordium (or future nut shell) grows very slowly, attaining only about 10 percent of its size in 4 months. In March the ovules (or future kernel) also begin to develop slowly. In mid-May the ovary begins to grow very rapidly and attains 90 percent of its growth in the next 5–6 weeks. During the middle of this rapid growth period (the middle of June) the ovary, which is 8–10 mm in diameter, finally becomes a fully mature organ; that is, the two ovules each contain an egg. One of these eggs is fertilized by the sperm which has been waiting in the tip of a pollen tube located near the top of the ovary since January. The egg in the other ovule is generally not fertilized and the ovule shrivels away. By the first of July, when the nut shell is practically full size (20–24 mm in diameter), the one fertilized egg has developed into a small embryo in practically all of the nuts. At this time there is a maximum amount of white, pithy "fill" tissue. During July the embryo, or kernel, grows very rapidly, pushing back the "fill" tissue so that by early August it occupies almost the entire space within the shell.

The growth of the nut during June is more rapid than for any other fruit or nut of which we have knowledge.

Many factors, such as crop density, heat stress, soil moisture, effective leaf surface, growth-regulator sprays, and so forth, affect the growth rate of fruits. These will be discussed in detail in subsequent sections.

CELL SIZE AND NUMBER

A great deal of work has been done on factors affecting the size and number of cells in apple fruit. Such factors are economically important because they determine fruit size and, to some extent, storage behavior. Bain and Robertson (1951) in Australia reported that large apples had more but not larger cells than small fruit from the same tree. Also, large fruit from light-cropping trees always had larger cells than did smaller fruit from heavy-cropping trees and sometimes contained fewer cells than small fruit from heavy-cropping trees. Fruit thinned with dinitro-o-cresol (DNOC) at full bloom had more but not larger cells than unsprayed controls; thinning with naphthaleneacetic acid (NAA) resulted in neither more nor larger cells. Letham (1961) found that N but not NPK fertilizer resulted in apple fruit with fewer and larger cells than controls—and that these fruit were predisposed to storage disorders. Denne (1960) in England found that heavy prebloom thinning of apples resulted in larger fruit, in part because the cells were larger but mostly because there were more cells per fruit.

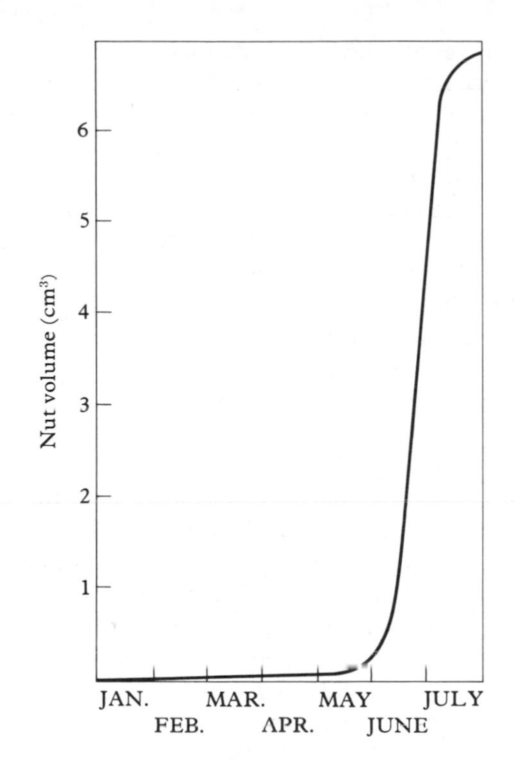

Figure 9-4 Seasonal growth of filbert nut, showing extremely rapid growth during June. [Based on data of M. M. Thompson]

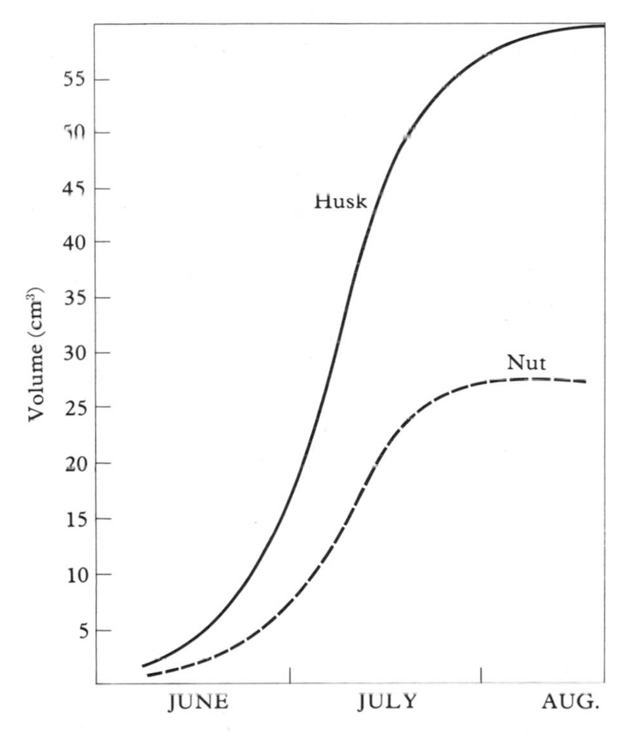

Figure 9-3 Seasonal growth of Franquette walnut in Oregon. [Based on data of H. B. Lagerstedt]

The results cited are complicated by the fact that cell size is larger and cell number less in Australian-grown fruit than those of the same cultivar grown in Washington State. Because of the apparent differences in fruit development between the two areas, Westwood et al. (1967) studied the effects of several factors on fruit development and cell size and number under Pacific Northwest conditions.

Fruit and cell measurements were made in two climatic areas of the Pacific Northwest. Fruit density, resulting from differences in intercellular air space, was greater in small than large fruit of all cultivars. Fruit from light-blooming trees were larger because they contained more cells—and some contained larger cells—than fruit from heavy-blooming trees. Regardless of treatment, small fruit usually contained fewer and smaller cells than large fruit. Chemical thinning resulted in larger fruit, the increase in size being due mostly to more cells per fruit. Chemical thinning after the cell-division period resulted in fruit with more cells than did hand-thinning, indicating that the chemicals selectively remove the small, weak fruits. Unselective early hand thinning, however, resulted in larger fruit, the increase in size being due mostly to larger cells. Center-bloom fruit (those from the terminal flower of the inflorescence) of Delicious had larger but not more cells than side-bloom fruit; Golden Delicious center-bloom fruit had both more and larger cells than side-bloom fruit.

Fruit from early hand-thinned leaders of Delicious trees were compared with those thinned later (Table 9-1). Early, heavy hand thinning (one week after full bloom) resulted in slightly more cells in large fruit than did late, light thinning. This response to early thinning was not found with the small fruit. The small differences found between early-heavy and late-light hand thinning indicate that favorable conditions for cell division were present in both situations.

In Australia, fruit from light-blooming "off year" trees have fewer cells than fruit from heavy-blooming trees, but the opposite is true in the U.S. Very light-blooming Delicious trees in Washington have slightly larger but many more cells per fruit than those with heavy bloom (Table 9-1). Apparently, some limiting factor exists under Australian conditions for cell division in the "off year." A climatic factor, winter chilling, may affect cell division. If a deciduous tree is fully chilled, cell divisions will be maximal if other conditions are favorable for growth. Cell division is limited with suboptimal or no chilling. In Australia, deciduous fruits are grown in the same locales as citrus, and wherever these two kinds of fruit are found together, the winter climate is usually not adequate to satisfy completely the chilling requirement of most deciduous trees. This climatic condition might interact with internal factors induced by the heavy "on year" crop to result in low cell numbers in "off years." However, typical Jonathan fruit grown in Oregon have far more cells (64×10^6) than even "on year" fruit of that cultivar grown in Australia, where "on year" fruit have 46×10^6 cells (and "off year" fruit have 38×10^6) Martin et al. (1964).

Cell number is increased by late chemical thinning of Golden Delicious. NAA, NAAm, and Sevin were applied 18 to 19 days past full bloom; their ultimate thinning action was not complete until after termination of the general cell-division period, 3 to 4 weeks after bloom. Thus chemical thinners are strongly *selective* in their action, that is, they selectively remove the weak, small fruit which contain fewer cells. This may explain why random hand thinning is not as effective as chemicals in sizing fruit. The competition that normally would cause weak, unthinned fruit to drop is diminished by early random hand thinning, thus permitting small, weak fruit to set, while more cell division is stimulated in the remaining fruit.

The results of Westwood et al. (1967) generally agree with the Australian reports on two observations:

1. Large fruit usually have more cells than small ones from the same tree.

2. Early thinning usually stimulates cell division and sometimes cell enlargement. This effect is more pronounced on heavy-setting cultivars.

Westwood et al. (1967) results contrast with the Australian reports in three areas. The Americans found that:

1. Light-blooming trees have fruit with as many or more cells than those from heavy-cropping trees, in contrast to Australian fruit, which have fewer cells per fruit when the bloom is light. This disparity indi-

cates a fundamental difference in climate or culture between the two areas.

2. Thinning with NAA causes Golden Delicious fruit ultimately to have more cells than controls, an effect not found with Delicious or with the other cultivars that were investigated. This probably results from selective thinning rather than from stimulation of cell division.

3. Northwest Delicious and Jonathan apples characteristically contain more cells than the same cultivars in Australia, while Australian fruit typically have larger cells and thus are more susceptible to storage-breakdown disorders associated with large cells. In the U.S. tests, cell number was as high as 121×10^6 and cell size ranged from 199 to 376 $\times 10^{-5}$ cubic millimeters. In Australia, cell number was as high as 80×10^6, while maximum cell size was 600×10^{-5} cubic millimeters. These differences in cell number and cell size probably are related to a difference in climate between the areas.

Table 9-1 Effects of early and late thinning fruit size and density and on cell size and number of Delicious apple.

Early Season Size Group and Thinning Treatment	Fruit Wt. (g)	Fruit Density (S.G.)	Cell Volume ($mm^3 \times 10^5$)	Cells per Fruit ($\times 10^{-6}$)
Small				
Early heavy	149	.856	295	46.6
Late heavy	160	.855	295	50.1
Late light	147	.856	279	49.7
Av. small	152	.856	290	48.8
Large				
Early heavy	233	.827	294	73.1
Late heavy	234	.833	319	68.2
Late light	212	.835	316	62.2
Av. large	226	.832	310	67.8
Very Large				
No thinning, 1959 (light bloom)	430	.802	344	115.6

SOURCE: Westwood et al., 1967.

Cell size and number combine differently to give various sizes of fruit with diverse quality and storage behavior. For example, cell size may be influenced by a number of compensating or interacting factors, some of which tend to increase size at the same time other factors may be acting to decrease it:

Factors That Tend to Increase Cell Size	*Factors That Tend to Decrease Cell Size*
Few cells per fruit	Many cells per fruit
Light bloom and set	Heavy bloom and set
Adequate soil moisture	Inadequate soil moisture
Strong fruiting spurs	Weak fruiting spurs
Center-bloom fruits	Side-bloom fruits
Excess nitrogen	Low nitrogen
High leaf-fruit ratio	Low leaf-fruit ratio
Late-season thinning	Early thinning (increases cell number)
Healthy leaves	Chlorotic or diseased leaves
Excessive chemical or hand thinning	Moderate or no thinning

Some of both sets of factors can act upon a fruit simultaneously. The proper balance of these factors would be expected to produce fruit of optimum size with the desired cell size and number. It appears desirable to grow fruit with relatively many cells of medium size rather than with fewer cells of large size.

Bain's (1961) work with Bartlett pear growth was in many ways similar to that of apple. She found two stages of growth, that is, (1) the slow-enlargement stage during cell division, lasting 42 to 56 days past bloom, and (2) the rapid enlargement stage that follows. Generally, the first stage showed relatively little physiological change but rapid morphological change, while the second stage showed more rapid physiological changes. As will be discussed later, however, temperatures during the whole cell-division phase have a profound effect upon time of maturity and on the quality of the fruit at maturity. Pears, apples, peaches, and possibly other fruits grown at relatively high temperatures during cell division (4 to 8 weeks past bloom) mature their fruits in fewer days than those grown at lower post-bloom temperatures.

FRUIT SHAPE

The shape of fruits can be important economically. Pear cultivars are known by their characteristic shapes, and anything that alters their shapes may reduce their economic value. For example, GA sprays induce parthenocarpic development of some cultivars, but the resultant shape is too elongate for the trade. With apples, buyers want elongated Delicious and Golden Delicious but prefer flattened (oblate) Jonathan and McIntosh.

The apple fruit is classified as a pome; in pomes the fleshy portion arises from the fusion of pericarp and extracarpellary tissues. The two interpretations of the origin of the vegetative portion of the flesh are that it derives (*a*) from the torus (flower receptacle) upon which the calyx, petals, and stamens are born and which encloses the carpels (Kraus, 1913), or (*b*) from the floral tube, which has become fleshy (MacDaniels, 1940). The latter "appendicular" interpretation is generally accepted, although no recent work has been done on either theory. As will be noted later, the distal half of the apple fruit responds to chemical and environmental stimuli and thus determines shape more than does the proximal (stem) end.

The most convenient way of expressing fruit shape is by the ratio of the longitudinal *length* to the transverse *diameter* (L/D ratio). This is a nondestructive measurement and permits comparing shape of very small fruit in early season with large ones later on. L/D ratio may be thought of as *relative* fruit length: the higher the value, the more elongated is the fruit. All fruits are relatively long early in the season, with the L/D declining and finally leveling off before harvest (Fig. 9-5). More detailed changes in apple shape are shown in Figure 9-6.

Some cultivars, such as Delicious, have differently shaped fruits, depending upon growing conditions. As shown in Figure 9-7, fruit grown in a climate with warm clear days and cool nights are more conic-elongate than those grown in hot days and warm nights (Westwood and Burkhart, 1968). Shaw (1914) concluded that fruit shape is determined within 16 days after bloom and that a high temperature during that period results in oblate fruits. He noted that fruits on the upper south portion of the trees he studied were larger and more oblate than those on the opposite side. While temperatures just after bloom do appear to be important, Westwood (1962) showed that changes in shape (the L/D ratio) occur for 60 to 100 days

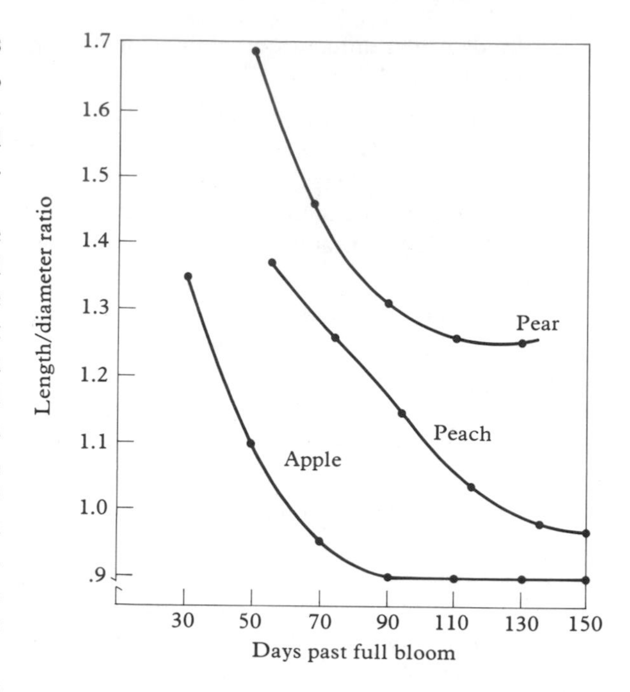

Figure 9-5 Seasonal changes in fruit shape (length/diameter ratio) of Delicious apple, Bartlett pear, and Elberta peach. [After Westwood, 1962]

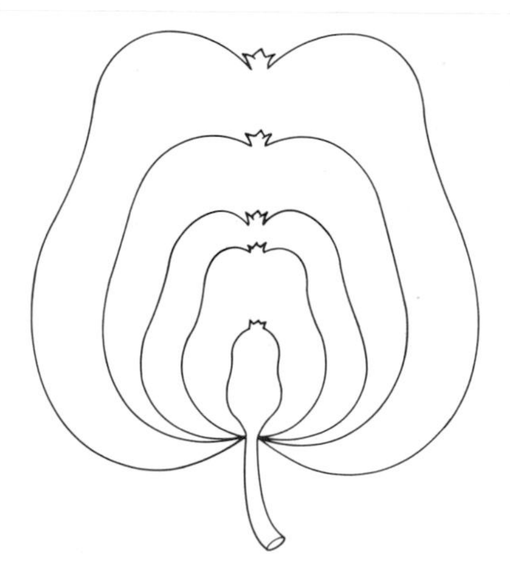

Figure 9-6 Median longitudinal tracings of Delicious apples at five times during the season, showing size and shape changes with time.

A

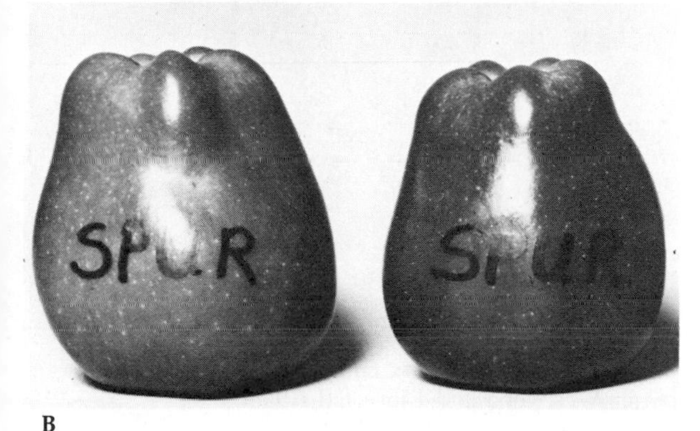

B

Figure 9-7 Shape of Starkrimson Delicious apple **(A)** from the hot climate of western Colorado and **(B)** from the cool climate of western Oregon. In both photos the fruit at left is a center-bloom fruit and the one at right a side-bloom fruit.

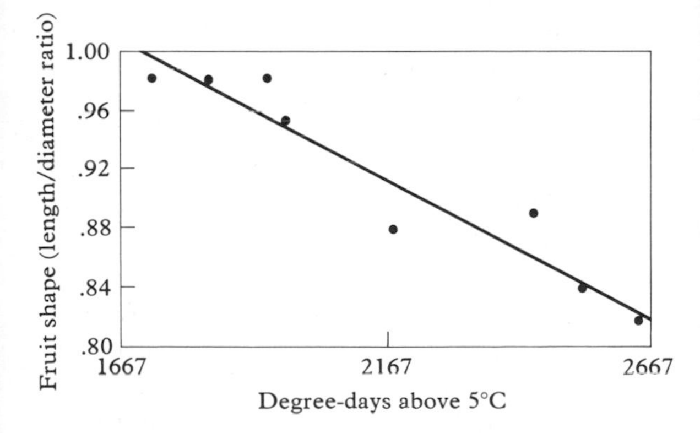

Figure 9-8 Graph showing the influence of growing temperature (degree-days above 5°C) from full bloom to harvest on the shape of Delicious apple.

after bloom (Fig. 9-5). Fruit length was found to be negatively correlated with the number of heat units (degree-days) above 5°C during the entire season (Fig. 9-8), and not just those immediately after bloom. (Degree-days are explained in the Glossary.)

Nonclimatic factors shown to induce elongated fruit were vigorous rootstocks (rather than dwarf), heavy thinning (or light bloom) resulting in a light crop, and center position (rather than side position) on the cluster (Westwood and Blaney, 1963). Also, treatment with GA and some cytokinins (Williams and Stahly, 1969) resulted in more elongated fruit (See Figs. 9-9, 9-10, and 9-11). Recent work indicates that kinetin and IAA have little influence upon shape but that 2,4-D causes small-sized oblate fruits (Fig. 9-12). GA_4 and GA_7 appear to be more effective for fruit elongation than GA_3, but all three occur natu-

Figure 9-9 Delicious apples grown in a hot climate were flat (oblate) shaped (*bottom*) compared to those sprayed with a cytokinin three days after petal fall (*top*). [From Martin, Brown, and Nelson, 1970]

Figure 9-10 Effect of localized application of GA (*arrows*) on growth of **(A)** Wealthy apple and **(B)** Japanese pear. Untreated controls are at left in each case. [A: From Bukovac and Nakagawa, 1968; B: Courtesy of M. J. Bukovac; from Nakagawa, Bukovac, Hirata, and Kurooka, 1968]

Figure 9-11 Effect of a 30 ppm pre-bloom spray of GA_3 on shape of Rome Beauty apple 40 days past full bloom. The main effect was increased growth of the tissues at the calyx end. Control fruit is at left. [After Westwood and Bjornstad, 1968b]

rally in young apple fruits. Effects of GA and cytokinin appear to be additive. Hand thinning at various times between bloom and harvest results in more elongated fruit than controls. Fruit from trees thinned early are more elongate and larger than ones thinned at 60 or 90 days after bloom (Fig. 9-13).

Physical removal or chemical damage to the calyx end of apples (Fig. 9-14) greatly reduces L/D ratio and the effect of GA on fruit elongation, and GA does not cause fruit elongation if applied in lanolin to the *stem* end of the fruit. R. L. Stebbins found that, when different floral parts were cultured *in vitro,* GA stimulated growth only in calyx tissue of apricot. These data suggest that the nonovarian portion of the apple is a fusion of tissues derived from *two* sources. Perhaps the distal portion is derived from calyx bases and the proximal from the pedicel. Variations in shape appear to result from variations in the development of the distal half of the fruit. Pre-bloom sprays of GA to whole trees of Golden Delicious and Rome apples (Westwood and Bjornstad, 1968b) resulted in growth mainly of the calyx end (Fig. 9-11).

When relative fruit length is graphed as a function of time between bloom and harvest, the curve produced is a hyperbola. The L/D ratio of Delicious 20 days after bloom is about 1.40 and declines to as low as .85 at harvest. Fruit of the same weight at harvest but of different shapes tend to have the same number of cells. Elongate fruit have isodiametric cortical cells,

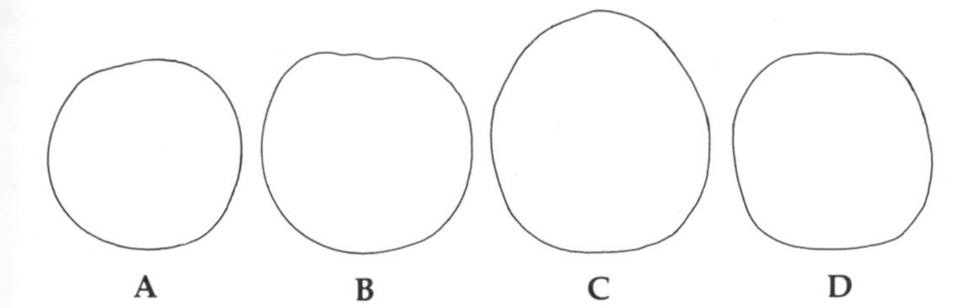

Figure 9-12 Tracings of four Golden Delicious apples, including three that received local hormone treatment on the calyx end five weeks after full bloom. Note the effects on relative size and shape at harvest. **(A)** Fruit treated with 2,4-D, smaller and more oblate than untreated fruit. **(B)** Fruit treated with 1AA, similar to untreated fruit. **(C)** Fruit treated with GA_3, larger and more elongate than untreated fruit. **(D)** Untreated control.

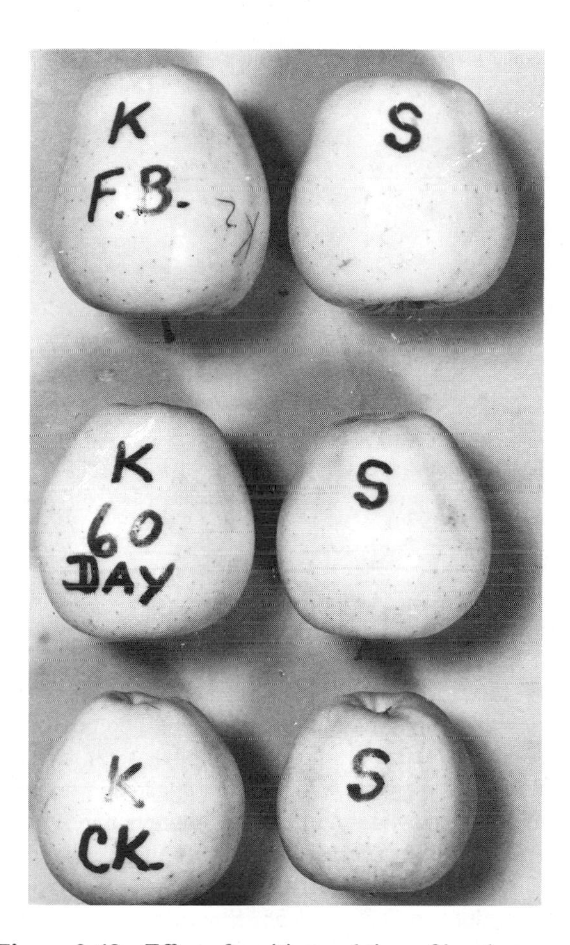

Figure 9-13 Effect of position and time of hand thinning on ultimate size and shape of Golden Delicious apple. *Top:* Thinned at full bloom. *Center:* Thinned 60 days AFB. *Bottom:* Unthinned control. King (center-bloom) fruits are on the left in each case, and side-bloom fruits are on the right. King fruit and early thinning resulted in the largest and most elongated fruit.

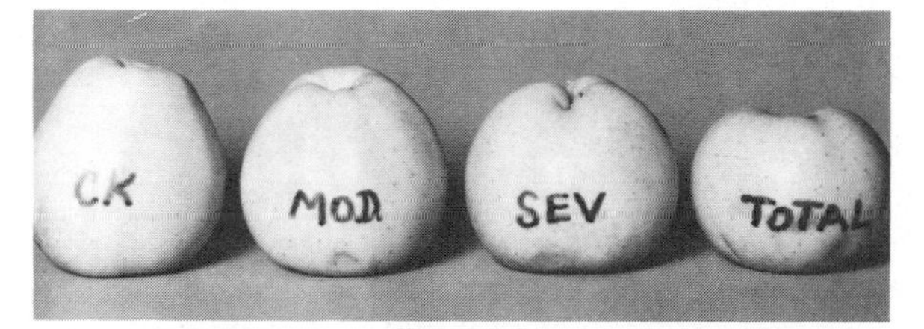

Figure 9-14 Effect of injury or removal of the calyx cup at full bloom on shape of Golden Delicious apple. *Left to right:* untreated control, moderate injury, severe injury, and total calyx-cup removal. Both size and shape clearly depend upon the extent of calyx-cup development.

while oblate fruit have ellipsoidal cells, whose major axis lies at right angles to the longitudinal axis of the fruit. The balance of hormones appears to determine fruit shape by affecting the direction of cell enlargement rather than the total extent of enlargement. An understanding of this phenomenon would enhance our general knowledge of cell growth.

Growth of fruits may be expressed in other ways as well. If, for example, x and y denote measures of any two kinds of growth—for instance, length and diameter—and their specific growth rates are proportional, then the growth is said to be allometric. The following relationships hold for allometric growth:

$$y = \mathrm{b}x^{\mathrm{k}} \quad \text{or} \quad \log y - \log \mathrm{b} + \mathrm{k} \log x$$

where x and y are any two growth measurements, and b and k are constants. Skene (1966) showed that the small changes in shape that occur during growth (as seen in Fig. 9-6) can be studied in this way, using log-log paper to plot the two variables.

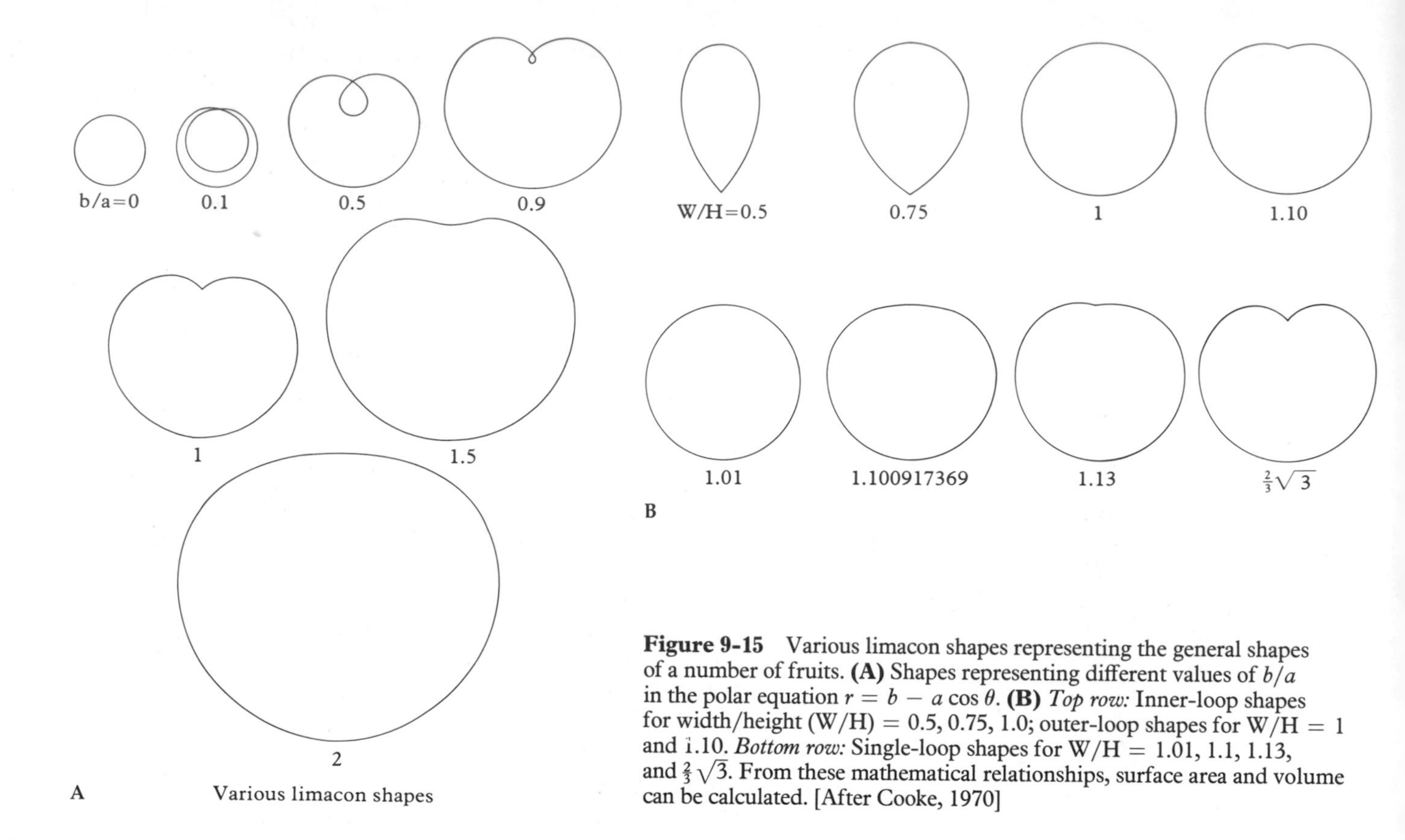

Figure 9-15 Various limacon shapes representing the general shapes of a number of fruits. **(A)** Shapes representing different values of b/a in the polar equation $r = b - a \cos \theta$. **(B)** *Top row:* Inner-loop shapes for width/height (W/H) = 0.5, 0.75, 1.0; outer-loop shapes for W/H = 1 and 1.10. *Bottom row:* Single-loop shapes for W/H = 1.01, 1.1, 1.13, and $\frac{2}{3}\sqrt{3}$. From these mathematical relationships, surface area and volume can be calculated. [After Cooke, 1970]

Cooke (1970) described peach fruit shape as a limacon (Fig. 9-15). The limacon is described by the polar equation:

$$r = b - a \cos \theta$$

where a = diameter which is perpendicular to the axis of symmetry and b = length. As shown in Figure 9-15, shape changes with different values of b/a. When $b/a = 0$, then the limacon degenerates into a circle with diameter a.

FRUIT DENSITY

Specific gravity (weight per unit volume) varies both in ultimate magnitude and in rate of seasonal change among apple, pear, and peach fruits (Fig. 9-16), but generally all types show a decrease from early season to maturity. The first date of sampling was near or after the time the cell-division period ended, when apples and pears weighed 1 to 2 grams, and peaches 2 to 5 grams. Skene (1966) points out that apples contain 86 to 89 percent water, whose specific gravity = 1, and that the water content does not change greatly during growth. Sucrose, the main sugar in apples, has a specific gravity of 1.59, and cellulose has a specific gravity of 1.50 to 1.55. Since most of the dry weight of the fruit is sucrose and cellulose, the average specific gravity for the two would be about 1.55. Thus, the calculated value for apple *cells* is 1.07 to 1.06, depending on water content, and remains so throughout the season. Changes in the specific gravity of fruits during growth must then be due to increases in intercellular and carpellary air spaces; specific

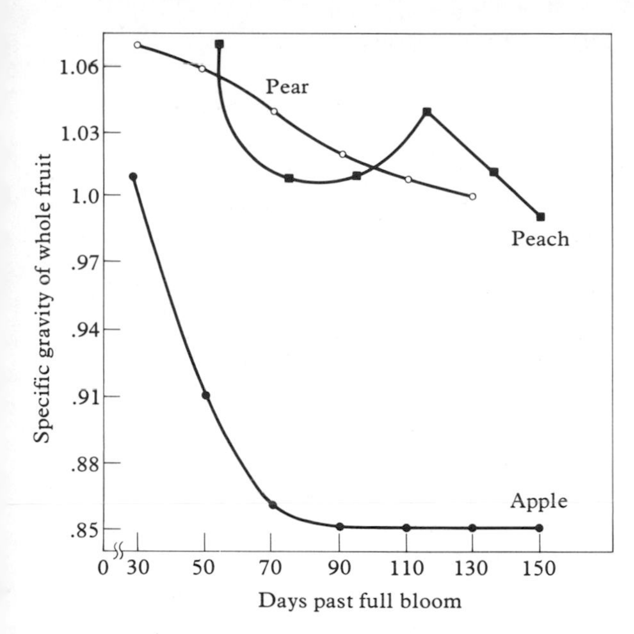

Figure 9-16 Seasonal changes in fruit density (specific gravity) for peach, pear, and apple. Fruit density reflects the extent of air spaces, the amount of lignification (stone cells), and the density of the fruit cells. [Based on data of Westwood, 1962]

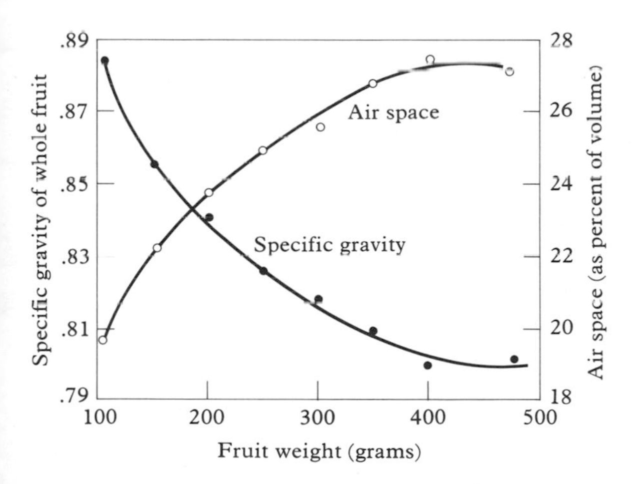

Figure 9-17 The relationship of fruit weight at harvest to specific gravity and intercellular air space in Delicious apple. Smaller fruit are more dense and thus weigh more per box than larger ones. [After Westwood, Batjer, and Billingsley, 1967]

gravity can thus be plotted as the reciprocal of the percent air space of the fruit (Fig. 9-17). It is true of all kinds of fruit that smaller ones are more dense than larger ones, both during the season and at harvest (Table 9-2).

When different portions of the fruit were compared, the core of the apple was more dense than the flesh, even though the carpellary air space was included. This probably reflects the closer packing of small cells at the core, with less intercellular air space or less water in that tissue. Pears, on the other hand, had flesh more dense than the core. This, along with the generally high S.G. of the whole fruit, indicates the greater density of stone cells of the flesh of pear.

Seasonal density of peach shows an irregular pattern (Fig. 9-16): there is an initial decline during post-bloom, an increase during pit hardening, and another decline during final swell. Starting at the beginning of pit hardening, flesh, whole seed, stone,

Table 9-2 Density (specific gravity) changes in whole fruits of apple, pear, and peach during the growing season.

Kind of Fruit and Variety	Days After Full Bloom							
	30	50	70	90	110	130	150	170
Apple								
Jonathan	1.00	.91	.82	.81	.79	.79	.78	—
Delicious	1.02	.91	.86	.85	.85	.85	.85	—
Golden Delicious	1.04	.92	.84	.82	.81	.81	.81	
Rome Beauty	.99	.89	.86	.84	.84	.84	.83	—
Yellow Newtown	1.01	.94	.90	.89	.88	.88	.87	.87
Winesap	1.04	.92	.89	.89	.88	.88	.87	.87
Pear								
Bartlett	1.07	1.06	1.04	1.02	1.01	1.01	—	—
Anjou	1.06	1.05	1.02	1.01	1.00	1.00	1.00	—
Comice	1.08	1.06	1.05	1.04	1.02	1.01	1.01	—
Peach								
Early East	—	1.05	.99	.95	.95	—	—	—
Dixigem	—	1.06	.98	.99	.98	.97	—	—
Redhaven	—	1.02	.99	1.01	1.00	.97	—	—
Fairhaven	—	1.04	1.00	1.01	1.00	.99	—	—
Early Elberta	—	1.03	1.00	1.02	1.03	1.02	.99	—
J. H. Hale	—	1.05	.99	1.00	1.01	.99	.99	—
Sullivan Elberta	—	1.07	1.01	1.01	1.04	1.01	.99	—

SOURCE: Westwood, 1962.

Table 9-3 Differences in specific gravity between tissues of Sullivan Elberta peach from the beginning of pit hardening to harvest.

Kind of Tissue	Days After Full Bloom[1]						
	75	95	115	125	135	145	155
Whole fruit	1.010	1.013	1.036	1.023	1.011	1.000	.990
Flesh	1.009	1.004	.994	.989	.984	.979	.974
Whole seed	1.022	1.079	1.152	1.188	1.221	1.240	1.246
Stone	1.063	1.134	1.206	1.242	1.272	1.281	1.282
Kernel	1.067	1.057	1.048	1.043	1.038	1.033	1.028

SOURCE: Westwood, 1962.
[1]The slow growth period of pit hardening was from 75 to 115 days after full bloom.

and kernel (along with whole fruit), densities of Sullivan Elberta peach were determined at intervals until harvest (Table 9-3). The slow-growth period during pit hardening was complete and the final swell had begun at about 115 days past full bloom. Flesh and kernel specific gravity declined moderately between early pit hardening and harvest; the specific gravity of whole seed and stone increased rapidly from 75 through 135 days after full bloom. The increase in specific gravity of the stone is due to lignification of cell walls and accounts for the increase found in whole fruit during pit hardening. It was previously shown that endocarp tissue increases in dry weight rapidly during the second stage. The extent to which the seed influences the specific gravity of the whole fruit is determined not only by whole-seed density but also by the weight of the seed relative to flesh weight for a given date. Thus it is during pit hardening (when stone weight increases relatively faster than flesh weight) that the specific gravity of the seed has its greatest influence on a whole fruit's specific gravity. During the pit-hardening period, the seed comprises about 25 percent of the total weight of the fruit, while early in the final swell (125 days after full bloom) this value drops to 14 percent. At harvest the seed comprises only 6 percent of the total fruit weight. These data explain why the specifiic gravity of whole fruits of most cultivars increases during pit hardening and then declines during final swell. Early East fruit do not have a slow-growth period during pit hardening, which explains why they do not increase in specific gravity during that period.

FRUIT THINNING

Under optimum conditions most fruits will set more fruit than needed for a full crop. Fruit thinning is done to reduce limb breakage, to increase fruit size, to improve color and quality, and to stimulate floral initiation for next year's crop.

Increasing the leaf/fruit ratio by removing some of the fruit causes the remaining fruit to be larger, but not in direct proportion to the increase in the number of leaves per fruit (Figs. 9-18, 9-19). This means something of a yield reduction but an improvement of fruit size. In general, about twenty to forty leaves per fruit are required to give a proper balance between fruit size and yield. The optimum number of leaves per fruit is somewhat lower, however, for compact mutants and dwarf trees, whose leaves are more efficient than those of standard trees. The leaves of dwarf trees are more efficient because they are exposed to direct sunlight for more hours of the day than those

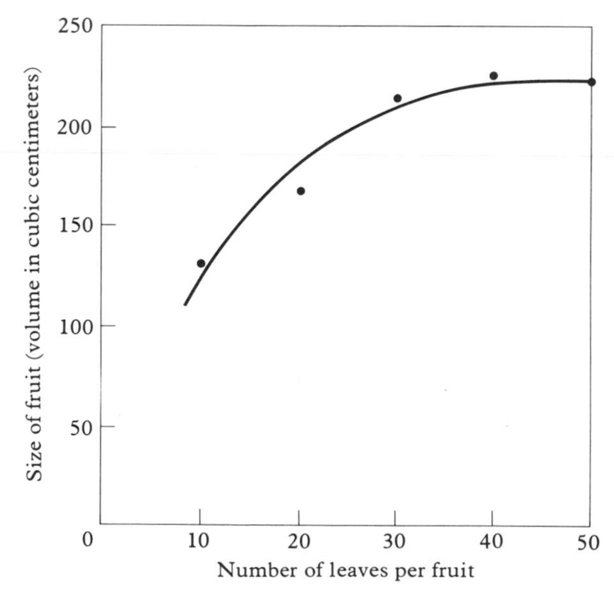

Figure 9-18 Relationship between the number of leaves per fruit and the harvest size of the fruit. At low leaf/fruit ratios, the fruit are small, poorly colored, and do not mature properly. At high leaf/fruit ratios, fruit may develop bitter pit, and yield is reduced. [Based on data of Magness and Overly, 1929]

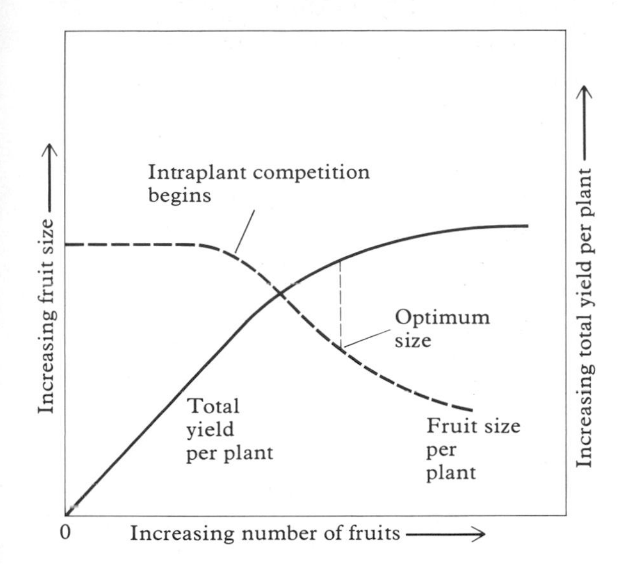

Figure 9-19 Idealized relationship between fruit size and yield. The total yield per tree is slightly reduced by thinning popular large-fruited cultivars of apples and peaches—however, the value of unthinned fruit may be economically worthless. Some reduction in total yield is desirable in order to keep alternate-bearing cultivars producing on an annual basis. [After Jules Janick, *Horticultural Science,* 2nd ed. W. H. Freeman and Company. Copyright © 1972.]

on standard trees. Aldrich (1936) found that pear fruit fed by shoot leaves grew better than those fed by spur leaves. Pruned trees thus have some advantage because they have relatively more shoot leaves than do unpruned trees. But when attempts are made to do enough pruning to eliminate the need for fruit thinning, the usual result is too much reduction in yield.

As pointed out previously, early thinning helps stimulate floral initiation for the next year's crop on cultivars that tend to be biennial. Specifically, thinning removes some of the young embryos, which would produce flowering inhibitors. Apples and to some extent other fruits and nuts tend to have a light crop the season following a heavy crop. Biennial bearing in apple, pear, and filbert is due to lack of floral initiation in the "on year." Walnut and pecan are influenced in some other way because initiation of female flowers occurs just prior to anthesis rather than in the previous year. Pistachio alternate bearing is due to still a third mechanism, that of the dropping of flower buds after they have initiated (Crane, 1971). Apples initiate flower buds early, so thinning must be done within 40 days of full bloom to result in good return bloom. Pears initiate flowers somewhat later, so for them thinning can aid in return bloom if done up to 60 days past full bloom.

If thinning is done before the end of the fruit-cell division period, it may stimulate more cells and hence potentially larger fruit. Still later thinning causes an increase in fruit growth by cell enlargement. The later thinning is done, the less effective it is in increasing fruit size. The degree of thinning depends upon many factors, such as the desired size for a particular market, the amount of pruning done previously, the amount of initial set, and the relative economic penalty for not thinning.

Thinning Methods

The three general methods of thinning are hand, mechanical, and chemical. Hand thinning is simply pulling or breaking off flowers or fruit with the fingers. Historically, fruits were thinned to a predetermined spacing, but in recent years it was found that size thinning is better. Size thinning is the selective removal of small, weak fruits regardless of spacing, but with the same general consideration given to the degree of thinning desired. If size thinning is done, however, one should take care not to retain fruits so close together that they will push each other off the twig as growth occurs. This is mainly a problem with naturally large fruits having short stems.

Mechanical thinning may be done in several ways. At or slightly past bloom, a direct blast of high pressure water from a hand-operated spray rig can thin effectively with an experienced operator. Another method is the use of a stiff-bristled brush to "sweep" off some of the fruits when they are still quite small. The third method is by the use of a power tree shaker of the type used to mechanically harvest fruits and nuts. The shaker head is attached to the trunk of the tree and energy is applied under careful control by the operator. Skill is required to prevent overthinning. Two disadvantages of this method are (1) that it selectively removes the larger fruit because they attain greater directional momentum than small ones during the vibrational action, and (2) that it removes more fruit from the stiff areas of the tree. Also, a varying percentage of attached fruit drop later as a result of injury during shaking.

CHEMICAL THINNING During the 1930's, chemical agents were sought that would remove all of the fruit from trees, because there was little market for the fruit during those Depression years, and it eliminated the need to apply certain pesticide sprays. Of the materials tested, tar-oil distillate was the most effective in removing flowers at the tight-cluster stage. In 1939 it was found that dinitro-o-cyclohexylphenol was promising as a thinning agent. At the same time, sodium dinitro-o-cresylate (DNOC) was found to prevent pollination when applied to stigmas and showed promise as a blossom thinner.

The 1940's were spent exploring the possibilities of dinitro compounds as chemical thinners for tree fruits. Phenol forms seemed to cause such problems as russetting or misshapen fruit. Tests with both acid and sodium DNOC, however, showed similar satisfactory results when used at the proper concentration on apples, on stone fruits and, to a limited degree, on pears. The best time to spray peaches and apricots is a day or two before full bloom; the best time to spray apples and pears is from full bloom to a day or two past full bloom.

From 1958 to 1968 several new post-bloom thinners were introduced (Batjer and Westwood, 1960;

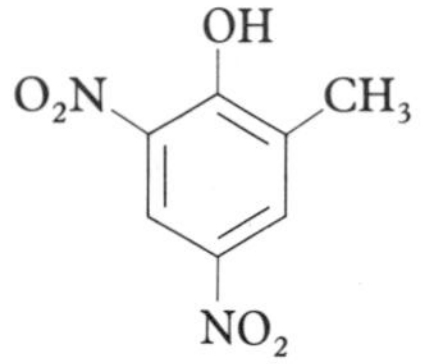

4,6-Dinitro-ortho-Cresol
(DNOC)

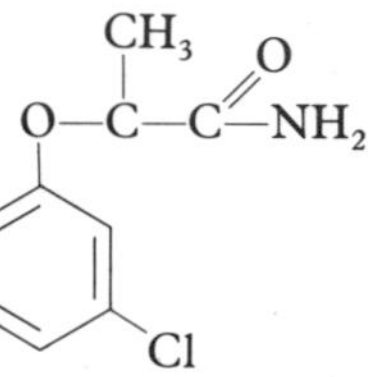

2-(3 Chlorophenoxy)-propionamide
(3 CPA)

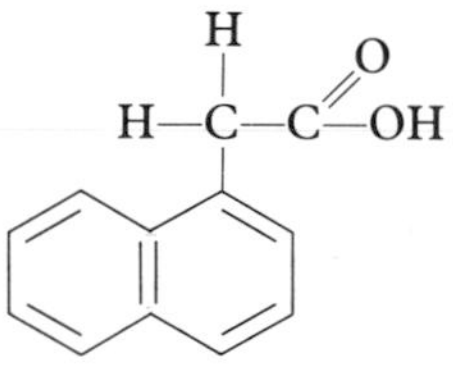

1-Naphthaleneacetic Acid
(NAA)

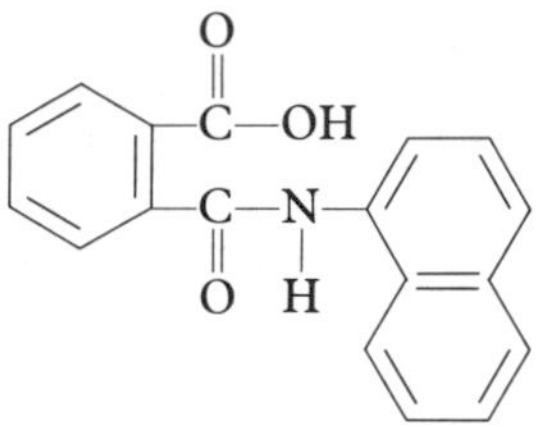

N-1-Naphthyl Phthalamic Acid
(NPA)

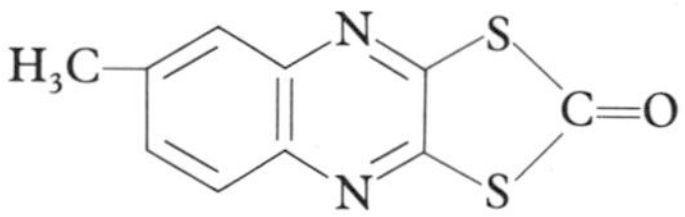

6 Methyl 2,3-quinoxalinedithiol cyclic carbonate
(Morestan®)

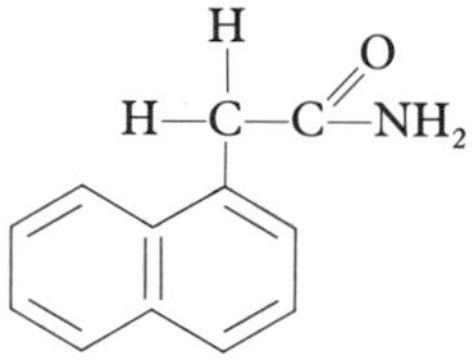

1-Naphthaleneacetamide
(NAAm)

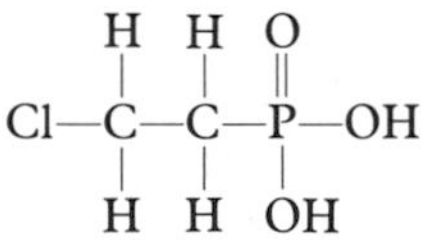

2 Chloroethyl Phosphonic Acid
(ethephon)

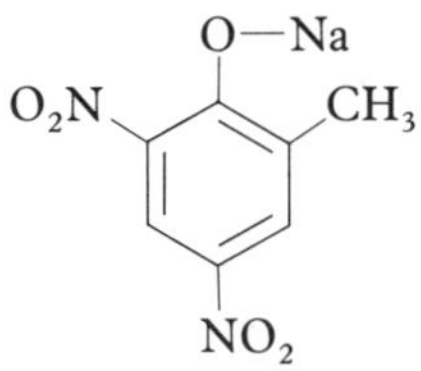

Sodium 4,6-Dinitro-ortho-Cresylate
(Na-DNOC)

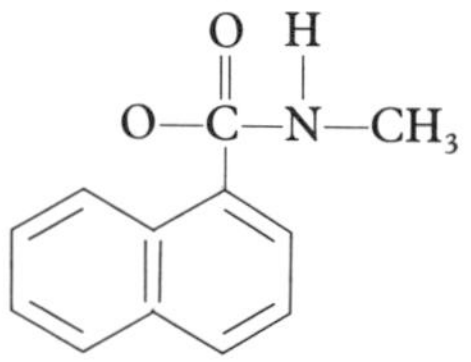

1-Naphthyl N-methyl carbamate
(Sevin®)

Figure 9-20 Structures and names of the most commonly used chemical thinning agents.

Westwood, 1965). These included a group of related methyl carbamates of the general structure:

$$\mathrm{R{-}O{-}\overset{\overset{\displaystyle O}{\|}}{C}{-}\overset{\overset{\displaystyle H}{|}}{N}{-}CH_3}$$

in which R stands for an alkyl or aryl group. Names and structures of common thinning chemicals are shown in Figure 9-20. The best known of these is 1-naphthyl N-methyl carbamate (Sevin), which is quite effective at 20 to 30 days past full bloom on apples, but is ineffective on pears and stone fruits. All of the carbamate thinners are insecticides and should be used with care. Sevin is especially lethal to honey bees, so blossoms on the weeds of the orchard floor as well as late blossoms on the trees should be eliminated or be past petal-fall before using it. Other post-bloom thinners developed during the 1960's are 6-methyl 2,3-quin-oxalinedithiol cyclic carbonate (Morestan*) and 2-chloroethyl phosphonic acid (ethephon). Morestan is effective on apples and ethephon on apples and stone fruits.

During the 1950's and 1960's, several post-bloom thinners were studied for peaches, apricots, and plums, among which were N-1-naphthyl phthalamic acid (NPA), 3-chlorophenoxy-alpha-propionamide (3-CPA) and ethephon, the latter of which shows the most promise.

Recently NAA and NAAm have been tested for use on Bartlett pear. NAA is used at 15 ppm 15 to 21 days past petal fall. NAAm is effective at 25 to 50 ppm 3 to 8 days past petal fall. Table 9-4 gives a general guide to timing and rates of various thinning agents. At best, however, this guide can be used only in conjunction with locally developed programs that make adjustments for local weather and varietal peculiarities.

The advantages of chemical thinning over hand or mechanical thinning are: reduced thinning costs, better fruit size and quality, and better return bloom on biennial cultivars. Possible disadvantages are: the hazard of frost after early sprays, overthinning in some cases, some foliage injury, and variable results with different tree age and vigor. Chemical thinning

*Registered trademark.

Table 9-4 Chemical thinning of pome and stone fruits.

Chemical	Timing[1]	Common Name	Rate of Application[2]	
			Self-sterile Cultivars	Self-fertile Cultivars
		Apple		
DNOC	F.B.	Elgetol 20% DN-dry 40%	237–317 ml 114–150 g	473–630 ml 227–304 g
NAA	15–25 days A.F.B.	NAA	10 ppm[3]	15–20 ppm
NAAm	15–25 days A.F.B.	Amid-thin	30 ppm	50 ppm
NMC	20–35 days A.F.B.	Sevin 50%	227 g	454 g
MQCC	20–30 days A.F.B.	Morestan 25%	227 g	340 g
CEPA	20–30 days A.F.B.	Ethephon	100 ppm	200 ppm
		Peach, Plum, Apricot		
DNOC	60–75% F.B.	Elgetol 20%	473 ml	710–950 ml
3-CPA	7–10 mm seed length	Fruitone CPA	200 ppm	300 ppm
CEPA	30–40 days A.F.B. 8–10 mm seed length	Ethephon	20–200 ppm	20–300 ppm
		Pear		
NAA	21 days A.P.F.	NAA	10–15 ppm	———
NAAm	3–8 days A.P.F.	Amid-thin	25 ppm	———

PRECAUTIONS: Do not spray Sevin or Morestan when there is open bloom and bees in the orchard. Do not spray very young trees or self-sterile cultivars if more than two tree spaces from a pollinizer. Do not use surfactants or wetting agents. Use only chemicals with a current registration for the crop in question. Consult with local extension agent before using for the first time.

[1] F.B. = Full Bloom; A.F.B. = After Full Bloom; A.P.F. = After Petal Fall.

[2] Measurements in milliliters and grams are per 378 liters (100 gallons) water. Rates are based upon dilute spray to the point of drip.

[3] 1 part per million (ppm) = 1 milligram per liter of water.

often varies greatly without apparent reason; this lack of consistency is a constant problem in thinning practice. The following factors tend to increase or decrease the degree of thinning obtained with chemicals:

Increased Thinning	*Decreased Thinning*
Young trees	Mature trees
Rain	Dry weather
High humidity	Low humidity
High maximum temperature	Lower maximums
Frosty nights	No frost
Soft spray water	Hard spray water
Slow drying conditions	Fast drying
High chemical concentration	Low concentration
Very low vigor	Moderate vigor
Close spacing of trees	Wide spacing
Light pruning	Heavy pruning
Heavy bloom	Light bloom
Poor pollination	Good pollination
Addition of wetting agent	No wetting agent
Previous heavy crop	Previous light crop

Use of chemical thinning for the first time should be on a trial basis—and then only on the advice of the local agricultural extension agent or local experiment station.

The thinning effect of DNOC is produced mostly by the direct killing of pollen and pistils, the slowing of pollen tube growth, and by killing the petals of unopened flowers (which prevents exposure of the stigma to pollination). There is also some effect on metabolism, but this seems to be a minor factor in thinning. NAA, NAAm, NPA, and 3-CPA appear to alter the auxin balance of the system, but the specific biochemical action of these materials is not known. In some cases, embryo development is arrested, then fruits drop later. Endosperm development during cytokinesis (cell-forming stage) is sensitive to NAA (Lombard and Mitchell, 1962; Leuty and Bukovac, 1968). Sevin appears to aggregate in the vascular strands of the fruit, where it prevents movement of essential components of growth into the fruit (Williams and Batjer, 1964). Ethephon's action is by the release of ethylene into the tissues, which stimulates the process of abscission during the post-bloom period. The mode of action of Morestan is still unknown.

Fruit Size Prediction

Early season prediction of harvest fruit size is useful in many ways. It allows the grower to purchase packages of the proper size—for example, cup inserts or tray packs. Perhaps the biggest advantage, however, is that it allows supplemental thinning to adjust the fruit to desired harvest sizes. For example, canning peaches should be 6 centimeters in diameter or larger to be number 1 canners. If early season prediction is for an average size of 5 centimeters, then extra thinning would be indicated in order to raise the actual average size to the desired size.

The principle upon which all size prediction is based is the fact that a fruit that is *relatively small* early in the season will be *relatively small* at harvest. This may be because it has fewer cells, is on a poor twig, is in a shaded portion of the tree, etc. Even if all other fruits are thinned from around a small fruit, it still will not attain the harvest size of a relatively larger fruit where much less thinning was done.

Early set of fruit is often erratic; on chemically thinned trees, fruit may be unevenly distributed along the branches. Under such conditions, supplemental space-thinning by hand is not desirable, because to

Table 9-5 Apple size prediction table, showing the relation between size at various times during the season and size at harvest.

Red and Golden Delicious (cm diameter)												
Days Past Full Bloom												Box Size at Harvest
40	50	60	70	80	90	100	110	120	130	140	150	
2.69	3.30	3.8	4.3	4.8	5.1	5.4	5.7	5.9	6.1	6.2	6.3	163
2.77	3.38	3.9	4.4	4.9	5.2	5.6	5.8	6.0	6.2	6.4	6.5	150
2.84	3.43	4.0	4.5	5.0	5.3	5.7	6.0	6.2	6.4	6.5	6.7	138
2.90	3.53	4.1	4.6	5.1	5.4	5.8	6.1	6.4	6.6	6.7	6.8	125
2.95	3.63	4.2	4.7	5.2	5.6	6.0	6.3	6.6	6.8	6.9	7.1	113
3.02	3.73	4.3	4.9	5.3	5.8	6.2	6.5	6.8	7.0	7.2	7.4	100
3.10	3.81	4.4	5.0	5.5	6.0	6.4	6.8	7.1	7.3	7.5	7.6	88
3.28	3.99	4.6	5.3	5.9	6.4	6.8	7.2	7.6	7.8	8.0	8.2	72

SOURCE: Batjer et al., 1957.
NOTE: To use prediction tables—

1. Determine the number of days from full bloom to the date of sampling.
2. Obtain the average diameter of a random sample of fruit from the orchard in question.
3. Look in the table and find the diameter (at the proper number of days past full bloom) which corresponds to the average diameter of your sample.
4. The predicted harvest size is read from the far right column on the same line as the sample diameter.

Table 9-6 The relation of fruit weight at two reference dates to the harvest size of Sullivan's Early Elberta, Elberta, and J. H. Hale peaches.

First Reference Sample Weight[1]	Second Reference Sample Weight[2]	Predicted Harvest Size[3] (cm dia.)
(kg per 200 fruits)		
2.7	3.6	5.1
3.2	4.5	5.3
3.6	5.0	5.5
4.1	5.9	5.7
4.5	6.4	5.9
5.0	7.3	6.1
5.4	7.7	6.4
5.9	8.2	6.5
6.4	9.1	6.7
6.8	9.5	6.8
7.3	10.0	6.9
7.7	11.0	7.1
8.2	11.4	7.2
8.6	11.8	7.3
9.1	12.7	7.4
9.5	13.2	7.6
10.0	13.6	7.7
10.4	14.5	7.8
10.9	15.0	7.9
11.4	15.4	8.0
11.8	16.3	8.0
12.2	16.8	8.1
12.7	17.7	8.2
13.2	18.2	8.3
13.6	18.6	8.4
14.1	19.1	8.5
14.5	20.0	8.6
15.0	20.4	8.7
15.4	20.9	8.8

[1]14 days after the beginning of pit hardening (about July 1).
[2]25 days after First Reference.
[3]2.54 centimeters = 1 inch.

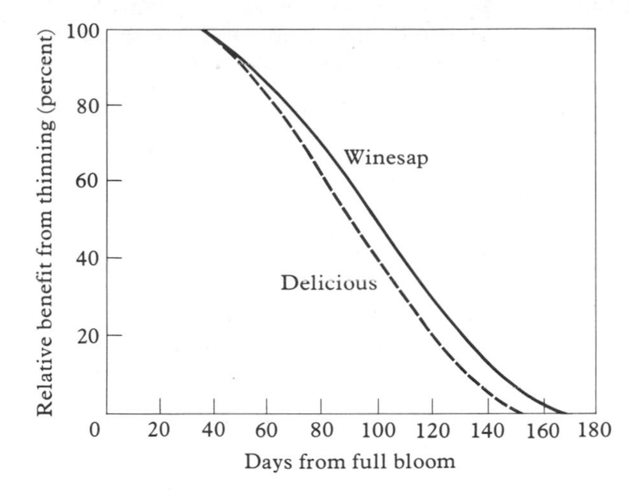

Figure 9-21 Calculated percent benefit (based on 100 percent at 35 days AFB) of hand thinning apples at different times during the season. At 75 and 95 days, the actual benefit found in tests exceeded the theoretical benefit. [After Batjer et al., 1957]

space the remaining fruit out to, say 15 centimeters, probably would result in overthinning. In some situations, no further thinning should be done, but such situations are difficult to evaluate. There is also a human factor. There is the marked tendency when the set is light for the hand thinner to overthin and thus reduce yield unduly. On the other hand, thinners tend to underthin trees with a heavy fruit set. Development of size-prediction charts has provided growers with an early means of determining the degree of hand thinning that is needed (Tables 9-5, 9-6, and 9-7).

APPLE The theoretical benefit of hand thinning apples at different times during the season is shown in Figure 9-21. But according to tests in irrigated orchards (Table 9-8), thinning as late as 95 days past bloom increases size of the remaining fruit more than that calculated from the theoretical curve. This might not be true in nonirrigated orchards or under different climatic and cultural conditions. Size prediction tables likewise might require adjustment for use in some areas.

Size thinning can be done to advantage when one wishes to increase harvest size over that predicted, because when relatively small fruit are thinned off, the remaining larger fruit not only have more leaves to feed them but also have greater potential for growth than the small fruit. Fortunately, chemical thinning selectively removes the smaller, weak fruit in much the same way that a skilled hand thinner would do size thinning.

PEACH Even if the initial set of peaches is heavy, hand thinning must be delayed until ultimate set is established. This coincides with the beginning of the

Table 9-7 Size Prediction Table for Bartlett Pear, in Oregon, showing the relationships between size (diameters in centimeters) at various days after full bloom and size at harvest.

Days Past Full Bloom																	
60	65	70	75	80	85	90	95	100	105	110	115	118	120	125	130	135	140
Preharvest Diameter (cm)												Early Harvest Diameter (cm)		Midharvest Diameter (cm)		Late Harvest Diameter (cm)	
2.6	2.8	3.0	3.2	3.4	3.6	3.9	4.1	4.3	4.6	4.8	5.0	5.1	5.2	5.4	5.6	5.7	5.9
2.8	3.0	3.2	3.4	3.6	3.9	4.1	4.4	4.6	4.9	5.1	5.3	5.4	5.5	5.7	5.9	6.0	6.2
3.0	3.2	3.4	3.7	3.9	4.1	4.4	4.6	4.9	5.1	5.4	5.6	5.7	5.8	6.0	6.2	6.4	6.6
3.1	3.4	3.6	3.9	4.1	4.3	4.6	4.9	5.1	5.4	5.7	5.9	6.0	6.1	6.3	6.5	6.7	6.9
3.3	3.6	3.8	4.0	4.3	4.5	4.8	5.1	5.4	5.7	5.9	6.2	6.4	6.4	6.6	6.9	7.1	7.3
3.5	3.7	4.0	4.2	4.5	4.7	5.0	5.3	5.6	5.9	6.2	6.5	6.6	6.7	7.0	7.2	7.4	7.7
3.7	4.0	4.2	4.5	4.7	5.0	5.3	5.6	5.9	6.2	6.5	6.8	6.9	7.0	7.3	7.5	7.8	8.0

SOURCE: Compiled by Don Berry, Extension Agent, and P. B. Lombard, Horticulturist, 1970.
NOTE: To predict the percentage of fruit in each size class at harvest—

1. Choose the proper vertical line corresponding to the days past full bloom.
2. Measure 10 of the fruit on each of 10 trees—sampling at random from four sides of each tree.
3. For each fruit diameter, place a tally mark nearest diameter class.
4. After sampling 100 fruit, the number of tally marks per diameter class is equal to percentage of fruit in the corresponding classes at harvest.

Table 9-8 A comparison of calculated and actual benefits to size of Winesap and Delicious apple fruit from thinning at various periods from full bloom.

Cultivar and Kind of Value	Benefit from Thinning at Indicated Period from Full Bloom (%)[1]		
	55 days	75 days	95 days
Winesap			
Calculated value	89	75	56
Actual value	85	73	65
Delicious			
Calculated value	87	68	47
Actual value	88	80	63

SOURCE: Batjer et al., 1957.
[1]The thinning treatment at 35 days from full bloom was taken to represent 100% benefit in fruit size from thinning.

slow-growth period, which, for Elberta, is 70 days past full bloom and 14 days past the beginning of pit hardening (Fig. 9-22). This is the "First Reference" date for predicting harvest size. Hand thinning, if any, is done after taking samples for size prediction. The prediction chart (Table 9-6) may be used for trees that were chemically thinned as well as those that were not, because early selective thinning by chemicals results in faster growth to First Reference so that the sample will reflect the larger size potential. Second Reference is set at 25 days after the first, at the end of pit hardening but before the final swell.

Several factors can affect final fruit size, the most important of which are the amount of thinning and the time of thinning. The earlier thinning is done and the heavier the thinning, the larger the harvest size will be. Other factors—such as soil moisture, fertilizer, and pruning—may affect fruit size if they limit fruit growth. Harvest size is directly influenced by the amount of thinning, and total yield is reduced by excessive thinning (Table 9-9). However, reduction of yield is not directly proportional to the degree of thinning because fruit size increases with heavier thinning. Thinning in one test was done at First Reference (Fig. 9-22), and the data show clearly that the amount of thinning done at that time influenced harvest size.

Here is a stepwise procedure for the practical use of the size-prediction chart (Table 9-6):

1. Determine the start of pit hardening as explained previously (usually about June 15), then add 14 days to get the First Reference date.

2. At First Reference, obtain the weight in kilo-

Table 9-9 Effect of different amounts of thinning on harvest size and yield of Elberta peaches.

Amount of Thinning	Fruit Left On (%)	Number	Yield (m tons per ha)	Relative Yield (% of Light)	Harvest Size (cm dia.)	Fruit Volume Increase (%)
Light	72	1400	50.9	(100)	6.38	0
Medium	57	1200	46.4	93	6.53	8
Heavy	39	820	35.9	72	6.78	21

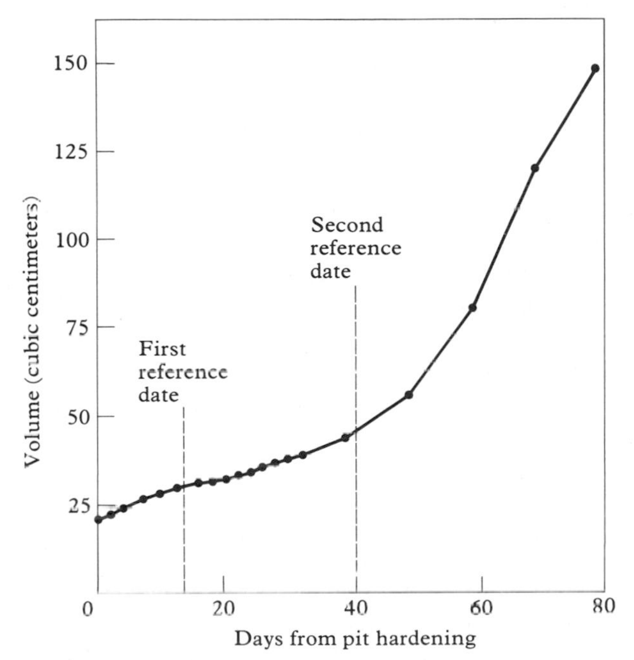

Figure 9-22 Elberta-peach growth pattern from the start of pit hardening, showing first and second reference dates during pit hardening, at which times the prediction chart is used for estimating final size. Pit hardening begins about 60 days after full bloom. [After Batjer and Westwood, 1958b]

grams of a 200-fruit sample (20 fruit per tree from each of 10 trees). If the sample is not taken on the exact reference date, adjust the weight by adding 114 grams for each day ahead of First Reference that the sample is taken; if the sample is taken after First Reference, subtract 114 grams for each day. Find in Table 9-6 the weight (under First Reference) corresponding to your sample weight, and read directly on the same line the predicted average harvest size. For example, if your 200-fruit sample weighs 5.9 kilograms, the average harvest size is predicted to be 6.5 centimeters in diameter. If prediction for a large orchard is desired, take one sample for each 2 or 3 hectares.

3. On the basis of the predicted harvest size, decide whether light, moderate, or heavy thinning should be done to obtain the desired size. Keep in mind that the sizes shown in Table 9-9 are given only to show the relative effect of different amounts of thinning. The actual size obtainable in a specific orchard will depend upon climate; soil type and depth; irrigation; fertilizer practice; pruning; and fruit set. Fruit size at First Reference results from an integration of all of these factors up to that time. Since the prediction chart is based on "average" commercial thinning at First Reference, final size may be smaller or larger, according to the amount of thinning done. Any factor—such as frost, chemical thinning, or mechanical blossom thinning—that reduces initial fruit set will tend to increase the size at reference date. When overthinning is done before the reference date, the actual harvest size will likely be larger than predicted.

4. At Second Reference (25 days after First Reference), about a month after the general hand thinning is complete, take another 200-fruit sample and weigh as described in Step 1. Find the sample weight in Table 9-6 (under Second Reference) and read the harvest size predicted by the sample. This prediction may or may not be the same as for the same orchard at First Reference, depending upon the amount of

thinning done and other factors. From this second prediction, decide whether supplemental hand thinning is needed to obtain the desired harvest size. Thinning at Second Reference is not as effective as early thinning, but it can be worthwhile, especially if selective size thinning is done to eliminate small fruits that will never make size.

The use of the prediction table is for average size only. Under standard orchard conditions, a predicted average size at harvest of 6.6 centimeters diameter would mean that 30 percent of the fruit would be between 6.3 and 6.6 centimeters in diameter, and 15 percent of them would finish between 6.0 and 6.3 centimeters in diameter.

If it is desirable to know the size above which 90 percent of the fruit will finish, then merely subtract 0.5 centimeter from the average harvest diameter predicted at either First or Second Reference date. As an example, suppose a 200-fruit First Reference sample of J. H. Hales weighed 9.5 kilograms. The predicted average harvest size would be 7.6 centimeters in diameter, and 90 percent of the crop would be likely to finish larger than 7.1 centimeters in diameter. This procedure may be used for both Elbertas and J. H. Hales of any average size.

The distributions given in Table 9-6 are based on orchards where the usual space thinning was done. It can be assumed that size thinning would narrow the range of harvest sizes, so that a larger percentage of the crop would finish within 0.3 centimeter of the average size.

Each grower should perform his thinning according to the predicted size he obtains from Table 9-6, in relation to the specific fruit size and tonnage he desires. Some tonnage must be sacrificed in order to obtain fruit larger than predicted for a given size at a reference date. However, this may be feasible under certain market conditions. During a year when J. H. Hales less than 7.3 centimeters in diameter brought $45 per ton and those larger than that brought $116 per ton, a grower might do much extra thinning (and thereby reduce tonnage considerably), in order to obtain the greatest net return for his crop. It is up to each grower to evaluate his own situation and use the prediction and distribution tables to decide how best to thin for his particular set of circumstances.

The prediction chart is intended only as a general guide to aid in fruit thinning. Since there is much orchard-to-orchard and season-to-season variability, all of which may affect fruit size, each orchardist must determine how closely his orchard conforms to the average orchard indicated in the table.

PEAR The prediction chart for pear (Table 9-7) was developed in much the same way as that for apples (Williams et al., 1969). But because pears are picked at a firm "green mature" stage for proper storage and ripening, they are growing at a much faster rate at harvest than are most apple cultivars. For this reason, thinning of pears can be done near harvest and without losing the benefit of a rapid growth rate following thinning. In some areas, Bartlett pears are size-picked several times (instead of being picked in a "once-over" harvest); hence, the table indicates growth during early, mid-season, and late harvest. An early partial picking constitutes late thinning and results in faster growth of the fruit remaining on the tree. Volume calculations made from the diameters given in Table 9-7 show that both small and large fruit increase about 55 percent between early harvest at 118 days and late harvest at 140 days. This means a 10 percent increase in tonnage every 4 days. However, the smallest fruit on the tree gain only 38 cubic centimeters in volume during harvest, while large ones gain 99 cubic centimeters. For this reason, it is advantageous to do size thinning early in the season if the prediction chart indicates that harvest size will be small. But, as is true of other fruits, thinning will reduce total yield.

GENERAL REFERENCES

Bain, J. M., and R. N. Robertson, 1951. The physiology of growth of apple fruits I. *Australian J. Sci. Res.* 4: 75–91.

Batjer, L. P., and H. D. Billingsley. 1964. *Apple thinning with chemical sprays.* Wash. Agr. Exp. Sta. Bull. 651.

Davis, L. D., and M. M. Davis. 1948. Size in canning peaches: The relation between the diameter of cling peaches early in the season and at harvest. *Proc. Am. Soc. Hort. Sci.* 51:225–230.

Leopold, A. C., and P. E. Kriedemann. 1975. *Plant growth and development.* 2nd ed. McGraw-Hill, New York.

Martin, D. and T. L. Lewis. 1952. Physiology of growth in apple fruits. III. Cell characteristics and respiratory activity of light and heavy crop trees. *Australian J. Sci. Res.* 5:315–327.

Martin, D.; T. L. Lewis; and J. Cerny. 1954. The physiology of growth in apple fruits. VIII. Between tree variation in cell physiology in relation to disorder incidence. *Australian J. Biol. Sci.* 7:211–220.

Robertson, R. N., and J. F. Turner. 1951. The physiology of growth of apple fruits. II. Respiratory and other metabolic activities as functions of cell number and cell size in fruit development. *Australian J. Sci. Res.* 4:92–107.

Westwood, M. N. 1962. Seasonal changes in specific gravity and shape of apple, pear, and peach fruits. *Proc. Am. Soc. Hort. Sci.* 80:90–96.

10

Plant Efficiency: Growth and Yield Measurements

The ultimate aim of the fruit grower is to convert the energy of the sun into the maximum yield per hectare of high-quality fruit with a minimum of input costs. Estimates of plant growth and efficiency can be made in various ways in order to aid in finding out where solar energy goes after entering the system. Measurements include shoot length; fresh or dry weight per unit time; leaf area as related to land surface (leaf-area index, LAI); leaf efficiency; trunk, stem, or limb size in cross-sectional area; plant volume, fruit size, shape and density; and yield per unit of plant size or per unit area of land surface. Combinations of these measurements in plantings not only can indicate the efficiency of the system but can also identify potential limiting factors, which, when corrected, can improve yield efficiency.

VEGETATIVE GROWTH

Depending on the situation, length of shoots may or may not be a good measure of actual growth. If shoots are affected by shading or by injury to the leaves, they will be long and spindly and not as high in dry weight as shorter shoots not so affected. However, the measurement of shoot length is nondestructive and can be used when the more direct measure of fresh or dry weight would destroy the shoot and prevent further observation.

Leaf area can be estimated in several ways, such as by using a planimeter on tracings, by measuring the reduction of airflow through a screen covered with leaves, and by measuring fresh or dry weight. Perhaps the easiest nondestructive method is that of multiplying leaf length by the width. This L × W product bears a linear relationship to actual leaf area for a number of species (Ackley et al., 1958). Such products can be converted to actual area by finding (or by area tests) the constant for converting the L × W product to area for the cultivar in question. For example, Delicious apple leaf area can be estimated by multiplying the L × W product by .71.

Photosynthetic efficiency of leaves can be estimated by the increase in dry weight between 5 A.M. and 2 P.M. on the same day. The amount of such dry-weight increase has been shown to be a good index of net photosynthesis. The best method is to tag the leaves to be used, then to take a one-square-centimeter plug from each leaf with a leaf punch at 5 A.M.; another one-square-centimeter plug is taken from the opposite half of each leaf blade at 2 P.M. The difference in dry weight between the two samplings is determined after drying at 70°C to constant weights. Foliage-efficiency tests are useful in showing the effects of light, temperature, soil moisture, fertilizers, insect infestations, disease infection, or chemical sprays.

There has long been a need for some simple measurement that accurately reflects total tree size. The simplest field measurement is trunk diameter or circumference. Early workers showed that such measurements were related to tree weight as well as to fruit yield. Waring (1920) reported that, in orchard experiments, yields based on a unit of trunk circumference were more meaningful than yield per tree. This was due to variations in tree size within a given test. Heinicke (1921) found that tree growth resulting in a doubling of trunk circumference resulted in a 7.3-fold increase in weight of young apple trees. Sudds and Anthony (1928) also found both yield and tree growth to be related to trunk circumference, but the correlation was much better if the circumference was squared or cubed. Murray (1972) reported a constant and predictable correlation between the volume of a shoot system and the cross-sectional area of the stem supporting it. More recently, workers in England have attempted to elucidate this relationship under various conditions of culture, rootstock, and cultivar. Pearce (1952) expressed the relationship by the equation $W = AG^b$, where W = tree weight, G = trunk girth, b = some power of girth, and A = a constant. He showed that b was quite constant within a given trial. When he used this equation on other workers' data, he obtained values of b ranging from 2.6 to 2.87. While these calculated relationships between trunk circumference and tree weight seem reliable, they are awkward to compute and result in unnecessary work.

Increase in volume or weight of a cylinder or stem plotted as a function of circumference or diameter gives a curved line. The volume of a cylinder equals its cross-sectional area times its length, so that cross-sectional area bears a linear rather than curvilinear relationship to both volume and weight. Thus, the simple conversion of trunk girth to cross-sectional area should serve to estimate tree weight better than a more complex equation. This would provide a simple linear measure by which to estimate bearing surface per acre in orchards of different ages or with different tree spacings. Simply squaring the trunk diameter also makes it linearly related to tree weight but does not permit calculating yield efficiency—which is fruit weight per unit of trunk cross-section.

Recently, trunk cross-sectional area was found to bear a linear relationship to total above-ground weight of apple, peach, and cherry trees. Thus the use of a simple caliper for trunk diameter measurements, converted to square centimeters of cross-section, can be used to estimate the potential bearing surface of any orchard tree as long as it has not been pruned heavily to prevent crowding. The estimate is best on young uncrowded trees that have been trained rather than severely pruned. This relationship permits the calculation of yield efficiency as fruit weight per square centimeter of trunk cross-sectional area. Estimates were made of maximum bearing surface potential per hectare (as square centimeters of total trunk area) for several kinds of tree fruits and nuts (Table 10-1).

A lever dendrometer, developed at the University of Idaho by Verner (1962), measures the radial growth of a tree trunk during periods of one or more days (Fig. 10-1). This new instrument has been used suc-

Table 10-1 Estimated maximum bearing surface per hectare (as trunk cross-section equivalent) for several tree crops.

Cultivar	Maximum Trunk Area per ha[1]	
	Standard (cm^2)	Dwarf (cm^2)
Apple, Delicious	77,600	50,000
Apple, Golden Delicious	95,900	56,600
Apple, Gravenstein	121,300	92,200
Apple, Rome Beauty	101,550	57,800
Apple, Jonathan	83,800	45,500
Apple, Yellow Newtown	84,750	63,750
Pear, Bartlett	111,200	69,450
Pear, Anjou	123,550	81,550
Pear, Bosc	118,600	76,600
Pear, Comice	160,600	119,600
Pear, Seckel	111,200	69,700
Pear, Eldorado	116,150	74,150
Montmorency cherry	72,650	—
Sweet cherry	86,000	—
Italian prune	93,900	—
Peach	79,100	—
English walnut	86,000	—
Barcelona filbert[2]	91,800	—

SOURCE: Westwood and Roberts, 1970.
[1] See Table 10-2 for diameter-to-cross-section conversion table.
[2] Filbert data supplied by Dr. H. B. Lagerstedt, ARS-USDA.

cessfully in Idaho in scheduling orchard irrigation and in monitoring the growth of various forest-tree species. Unlike the dial-gauge dendrometers previously available, the new lever dendrometer provides readings that are not affected by daily trunk shrinkage. The Idaho instrument always shows maximum radius attained by the trunk since the previous reading.

A direct-recording dendrometer is shown in Figure 10-2. This unit provides a measurement of radial growth similar to that provided by the other instruments, but it has the advantage of providing a continuous record of growth and shrinkage over both short and long periods. It also has the advantage of obtaining much data with very little time required of the operator. One disadvantage is that it is expensive, and thus relatively few trees can be measured at one time.

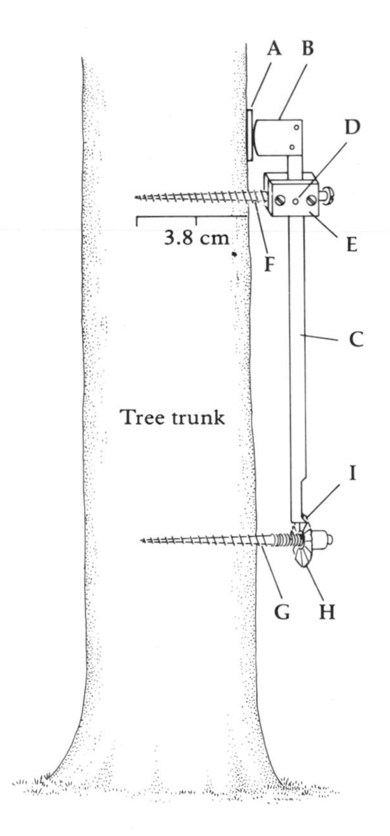

Figure 10-1 Idaho lever dendrometer. As described by Verner (1962), **A** is a small piece of light-gauge zinc fastened to the bark with waterproof glue and provides a firm surface against which to place bar **B**. **C** is a lever pivoting on metal pin **D**. Slight tension is provided on the lever by screws near point of attachment D, holding the lever in the position to which it has been moved by the pressure of plate A (a result of radial growth). The lever and pivot are mounted on metal block **E**, which has a hole drilled through it vertically to receive wood screw **F**, to which the block is made fast by a knurled set screw (not shown). Screw F, screwed into the trunk to a depth of about 3.8 cm, supports the fulcrum. Screw G is screwed in to the same depth. **G** carries a threaded collar with a flange, **H**, which is calibrated on its margin in units of one-fifth and one twenty-fifth of its circumference; this flange serves as a dial in measuring growth. Screws F and G extend far enough into woody tissue so they are not affected by growth. When a reading is made, the pointed projection **I** on the end of the lever makes contact with the undersurface of the dial; a slight lateral movement of the lever indicates the moment of contact. One complete turn of the dial equals .013 cm of radial growth. The smallest subdivision of the dial represents .0005 cm radial growth. Tree growth moves A outward, presses on B, moving point I inward. Thus the growth reading is done by screwing the dial H until it touches I. [After Verner, 1962]

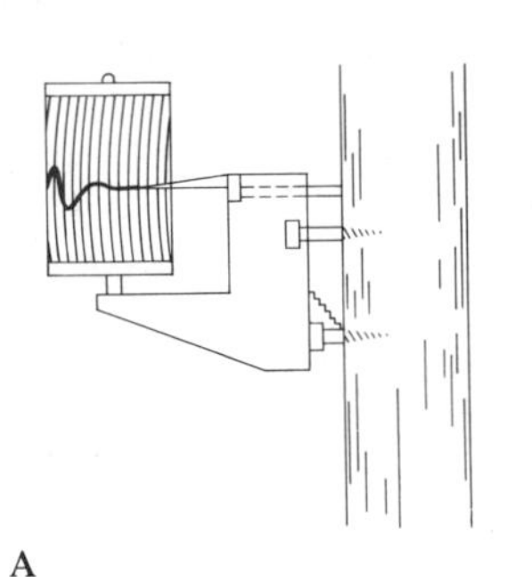

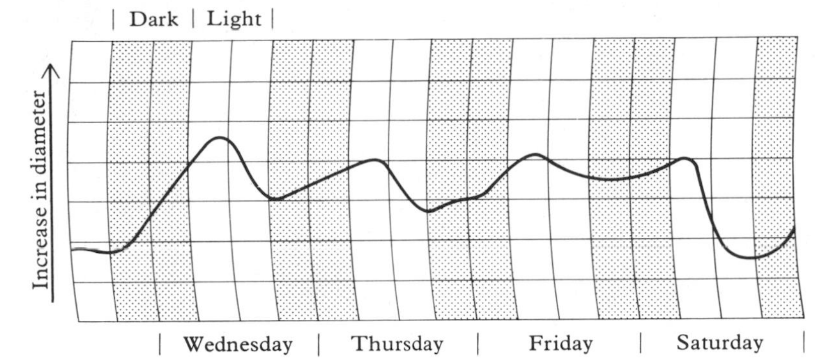

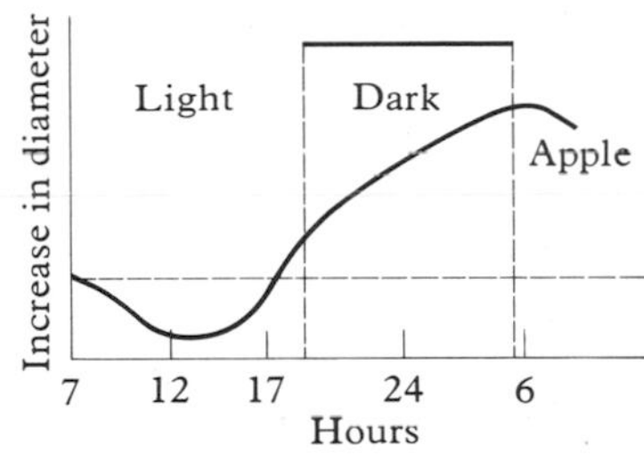

Figure 10-2 **(A)** Direct-recording dendrometer, showing trunk shrinkage during daylight and growth at night. **(B)** Similar diurnal shrinkage in light and growth at night occurs with fruit, as measured by an electronic transducer. [A: After Martin A. Zimmermann "How Sap Moves in Trees." Copyright © 1963 by Scientific American, Inc. All rights reserved; B: After Tukey, 1964a]

Total above-ground tree volume can be calculated from height and spread measurements as follows:

1. For a tree that is taller than it is wide (prolate spheroid), Volume $= \frac{4}{3}\pi ab^2$.

2. For a tree which is wider than tall (oblate spheroid), Volume $= \frac{4}{3}\pi a^2 b$, where $\pi = 3.1416$

$$a = \tfrac{1}{2} \text{ the major axis}$$
$$b = \tfrac{1}{2} \text{ the minor axis}$$

Height and spread of trees may be measured directly or estimated with the use of a sighting device described by Ford (1960). Readings from calibrated scales are made when the device is held at arm's length while standing a specific distance from the tree to be measured (Fig. 10-3). The volume estimates described above may be useful with trees that have been overpruned or that have trunk and frame stocks of different types. With such trees, the estimate of bearing surface by trunk diameter and cross-sectional measurement cannot be used.

FRUIT GROWTH

The data of Table 10-2 are compiled so that simple linear measurements of fruits can be easily converted to volume. Fruit growth is a function of volume (or weight) increase. Volume in cubic centimeters is listed for a large number of diameters. Most fruits approximate a sphere but may deviate some from it. If more exact volume is desired, the fruit in question can be checked for volume (by water displacement), and a correction factor can be applied to the values listed, permitting the use of routine diameters to obtain volume.

Diameter by itself is generally a poor measure of fruit growth because it is not linearly related to volume or weight. (For example, 1 centimeter of diameter growth on a 2-centimeter fruit represents an increase in volume of about 10 cubic centimeters, but 1 cm diameter growth on an 8-centimeter fruit represents an increase in volume of more than 110 cubic centimeters.) However, when the diameter of a fruit is sufficiently correlated with a stage of development, it can be a useful indicator. Thus, the diameter of apples

Figure 10-3 Sighting device for measuring height and spread of trees. **(A)** Sighting to the top of a tree a given distance away. **(B)** Sighting the width of the tree with the instrument oriented horizontally. (For details, see Ford, 1960).

at a given time after bloom helps predict harvest size, and the length of peach seeds indicates the best time to apply post-bloom chemical thinners. Such measures are useful because they correlate with a particular development stage.

To obtain an accurate measure of the seasonal growth of a particular fruit, the best procedure is to tag a number of fruit on the tree and then measure the diameter or circumference as they grow—later converting to volume. If fresh or dry weights are desired, pick nontagged fruits of the *same* diameter as the average of the tagged fruit. Fruits such as peach have a different diameter at the cheek than at the suture, so this must be considered in calculating volume. For such fruits, a circumference is better because it integrates the two diameters.

Some outward dimensions (of nuts and of pits of stone fruits) are fixed long before the ultimate dry weight is attained. Peach stones attain maximum size (in length and diameter) early in the pit-hardening period, although dry weight and density continue to increase for several weeks. This type of growth results from the lignification of the secondary walls of the stone cells.

The following formulae were used in computing the values in Table 10-2:

$$\text{Inches} = \text{centimeters} \times .3937$$

$$\text{Circum.} = \text{diameter} \times \pi = \text{diameter} \times 3.1416$$

$$\text{Area} = \pi r^2 = \tfrac{1}{4}\ \pi d^2 = .7854 d^2$$

$$\text{Volume of sphere} = \tfrac{4}{3}\ \pi r^3 = \tfrac{1}{6}\ \pi d^3 = .5236 d^3$$

Exact volume of fruit or other plant parts may be obtained by direct volume displacement or by weighing fruits under water as follows:

1. Weigh a container of water, allowing enough space for fruit submersion.

2. Submerse fruit while the container is still on the scales. To avoid air bubbles on the fruit, which cause erroneous readings, put a few drops of a wetting agent or detergent in the water to reduce surface tension. Keep fruit from touching the sides or bottom of the container by holding it under water with a dissecting needle held steady by a ringstand and clamp.

Table 10-2 Conversion of diameter or circumference measurements to trunk cross-sectional area or to volume of fruits (calculated as a sphere).

Diameter		Circum-ference (cm)	Trunk or Branch Cross-sectional area (cm²)	Volume of Fruit as Sphere (cm³)	Diameter		Circum-ference (cm)	Trunk or Branch Cross-sectional area (cm²)
(cm)	(in)				(cm)	(in)		
0.5	.197	1.57	.1963	.0654	25.5	10.04	80.1	511
1.0	.394	3.14	.7854	.5236	26.0	10.24	81.7	531
1.5	.591	4.71	1.767	1.767	26.5	10.43	83.3	552
2.0	.787	6.28	3.142	4.189	27.0	10.63	84.8	573
2.5	.984	7.85	4.909	8.179	27.5	10.83	86.4	594
3.0	1.18	9.42	7.07	14.14	28.0	11.02	88.0	616
3.5	1.38	11.00	9.62	22.45	28.5	11.22	89.5	638
4.0	1.57	12.6	12.57	33.51	29.0	11.42	91.1	661
4.5	1.77	14.1	15.90	47.71	29.5	11.61	92.7	683
5.0	1.97	15.7	19.63	65.45	30.0	11.81	94.2	707
5.5	2.16	17.3	23.76	87.13	30.5	12.01	95.8	731
6.0	2.36	18.8	28.27	113.1	31.0	12.20	97.4	755
6.5	2.56	20.4	33.18	143.8	31.5	12.40	99.0	779
7.0	2.76	22.0	38.48	179.6	32.0	12.60	100.5	804
7.5	2.95	23.6	44.18	220.9	32.5	12.80	102.1	830
8.0	3.15	25.1	50.27	268.1	33.0	12.99	103.7	855
8.5	3.35	26.7	56.75	321.5	33.5	13.19	105.2	881
9.0	3.54	28.2	63.62	381.7	34.0	13.38	106.8	908
9.5	3.74	29.8	70.88	448.9	34.5	13.58	108.4	935
10.0	3.94	31.4	78.5	524	35.0	13.78	110.0	962
10.5	4.13	33.0	86.6	606	35.5	13.98	111.5	990
11.0	4.33	34.5	95.0	697	36.0	14.17	113.1	1018
11.5	4.53	36.1	103.9	796	36.5	14.37	114.7	1046
12.0	4.72	37.7	113.1	905	37.0	14.57	116.2	1075
12.5	4.92	39.3	122.7	1023	37.5	14.76	117.8	1104
13.0	5.12	40.8	132.7	1150	38.0	14.96	119.4	1134
13.5	5.31	42.5	143.1	1288	38.5	15.16	121.0	1164
14.0	5.51	44.0	153.9	1437	39.0	15.35	122.5	1195
14.5	5.71	45.6	165.1	1596	39.5	15.55	124.1	1225
15.0	5.91	47.1	176.7	1767	40.0	15.75	125.7	1257
15.5	6.10	48.7	188.7	1950	40.5	15.94	127.2	1288
16.0	6.30	50.2	201.1	2145	41.0	16.14	128.8	1320
16.5	6.50	51.8	213.8	2352	41.5	16.34	130.4	1353
17.0	6.69	53.4	227.0	2572	42.0	16.54	131.9	1385
17.5	6.89	55.0	240.5	2806	42.5	16.73	133.5	1419
18.0	7.09	56.5	254.5	3054	43.0	16.93	135.1	1452
18.5	7.28	58.1	268.8	3315	43.5	17.13	136.7	1486
19.0	7.48	59.7	283.5	3591	44.0	17.32	138.2	1521
19.5	7.68	61.3	298.6	3882	44.5	17.52	139.8	1555
20.0	7.87	62.8	314		45.0	17.72	141.4	1590
20.5	8.07	64.4	330		45.5	17.91	142.9	1626
21.0	8.27	66.0	346		46.0	18.11	144.5	1662
21.5	8.46	67.5	363		46.5	18.31	146.1	1698
22.0	8.66	69.1	380		47.0	18.50	147.7	1735
22.5	8.86	70.7	398		47.5	18.70	149.2	1772
23.0	9.06	72.3	415		48.0	18.90	150.8	1810
23.5	9.25	73.8	434		48.5	19.09	152.4	1847
24.0	9.45	75.4	452		49.0	19.29	153.9	1886
24.5	9.65	77.0	471		49.5	19.49	155.5	1924
25.0	9.84	78.5	491		50.0	19.68	157.1	1963

3. Read the weight of the container plus the water plus the submersed fruit.

4. The difference in grams between the two weights is equal to the volume of the fruit in cubic centimeters.

If fruit density is desired, simply divide fruit weight by volume to get the specific gravity.

Diurnal fluctuations in fruit growth have been reported for pome and stone fruits by L. D. Tukey (1959). Using a sensitive electronic device, he found that fruits shrink during the morning hours, recover in the afternoon, and grow mostly at night (Fig. 10-2). This growth pattern results from water loss during transpiration; thus shrinkage is greatest when the stomata are wide open and leaves are pulling water from the fruit. Such a sensitive measuring device as L. D. Tukey (1964) describes would be most useful in obtaining very small growth increments during short periods of time.

The exact shape of fruit may be found by tracing on paper median longitudinal sections of fruit. But a simple nondestructive measure of shape that can be used regardless of fruit size is the polar length/transverse diameter ratio (L/D ratio). This L/D ratio has been used to trace shape changes of fruit from petal fall to maturity. Both pome and stone fruits begin with high L/D ratios, which decline hyperbolically during the season (Fig. 9-3). Measurements are usually made with a vernier caliper.

Fruit cells are relatively large; their size is easily measured with an ocular micrometer mounted in a compound microscope. Hand sections of fruit tissue are mounted in water on standard glass slides and need not be stained. The cells to be measured are oriented by using a microscope fitted with a stage manipulator. The scale of the micrometer in the ocular lens is in arbitrary units and must be calibrated by comparison with a stage micrometer marked in known units. Once calibration is complete, the objective lens is not changed to one of a different power. Once the shape of the cells being studied is determined, their volumes can be calculated as prolate or oblate spheroids in the same way that volume of whole trees is calculated.

CROP DENSITY

Bloom

Bloom density and set are usually estimated from an examination of three or more uniform limb units per tree, evenly spaced around the tree. Very low, weak, or shaded limbs should be avoided. The position and type of limb units selected are arbitrary, but one should be consistent on all trees used.

The best time to get flower bud counts is during the late bud-swell stage or not later than the late-pink stage. It is hard to make accurate counts after the flowers start to open. Begin counting at the basal end of the limb (the starting point can be marked by a fingernail on the bark) and systematically count flower buds toward the tip, taking each side branch as it comes. Go over the limb a second time and count only leaf buds this time. Write both the flower-bud and the leaf-bud counts on a durable manila tag and tie the tag tightly to the limb where the mark was made, so that later when you come back to make fruit-set counts, you will know where to start. Some fruits have single flower buds; others have several flowers per bud. The individual flowers are counted on peach and apricot, but on apple, pear, cherry, and prune the clusters are counted (each cluster comes from a single flower bud).

Fruits such as apple and pear bear most of their flowers on 2-year spurs, but sometimes fruit buds will develop on 1-year lateral buds on the terminal shoots. In general, these lateral flowers do not set well, so in getting limb counts, it is usually advisable to rub off these lateral flowers and count only the spur bloom. Bloom density as percentage of total buds is:

$$\text{Percent bloom} = \frac{\text{flower buds}}{(\text{leaf buds} + \text{flower buds})} \times 100$$

Set

By going back to the same limb units used for the bloom-density counts, set can be obtained by simply counting the number of fruit (after the final drop) on each tagged limb. Set of pome fruits is usually expressed as number of fruits setting per 100 *flower clusters;* set of stone fruits is expressed as the number setting per 100 *flowers,* thus:

$$\text{Pome fruit percent set} = \frac{\text{number of fruit set}}{\text{number of flower clusters}} \times 100$$

$$\text{Stone fruit percent set} = \frac{\text{number of fruits set}}{\text{number of flowers}} \times 100$$

Another method of obtaining fruit set (or more properly, crop density) is the counting of fruit on limbs of known size. Limb size is determined by converting calipered basal diameter to square centimeters of cross-sectional area (from Table 10-2). Crop density is simply the number of fruit setting per square centimeter of limb size. This measure of crop load reflects the combined influence of original bloom density and fruit set.

Yield Estimates

Yields are measured by direct weighing or by counting the number of level field boxes. If the latter method is used, several field boxes of fruit should be weighed at the start to determine the average weight per box so that the yield in boxes can be converted to kilograms if desired. Different kinds of fruit have different densities; for example, a field lug that holds 17.2 kilograms of apples will hold 20.9 kilograms of pears. When bulk bins are used, measure the height of the fruit in the bin and convert to kilograms or metric tons, or weigh the bins on a tractor-mounted metric scale.

It is often desirable to obtain fruit size as well as yield, since yield is a function of both set and final size. To get a good random sample of fruit for size determination, simply take a few fruit from each box after the tree is picked. Either fill a level box in this manner or carry platform scales along and count the number of fruit per box or per a given weight. The diameter of the fruit may also be measured, using a vernier caliper (calibrated in centimeters or decimal fractions of inches).

Yield Efficiency

Tree yields by themselves may be used to compare treatments with a control, but to be really meaningful, yields should be converted to yield efficiency and to amount per hectare, based on tree spacing. The usual method of obtaining plant efficiency is to calculate grams or kilograms of fruit per square centimeter of cross-sectional area of the trunk. This estimates the efficiency of the bearing surface. Yield per hectare (ha) estimates the efficiency of both bearing surface and land surface (Figs. 10-4, 10-5).

$$\text{Yield efficiency (tree unit)} = \frac{\text{kg yield}}{\text{cm}^2 \text{ trunk area}}$$

$$\text{Yield efficiency (ha basis)} = \text{tree yield} \times \text{no. trees per ha}$$

Data of Table 10-1 indicate maximum potential bearing surface but not what maximum yield per hectare to expect. Based upon a wide variety of experiences and upon the fruit size and quality required for

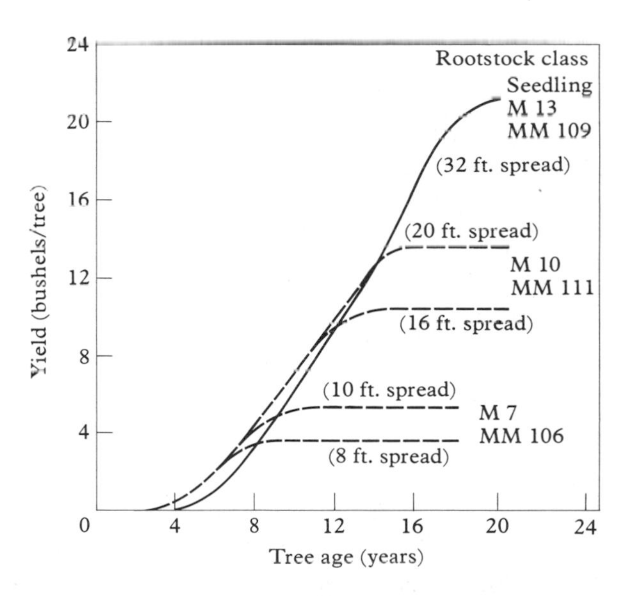

Figure 10-4 Generalized yield curve for apples on various vigorous and semidwarfing stocks. [After J. C. Cain, "Tree spacing in relation to orchard production efficiency," N.Y. Agr. Exp. Sta. Res. Circ. 15. 1969.]

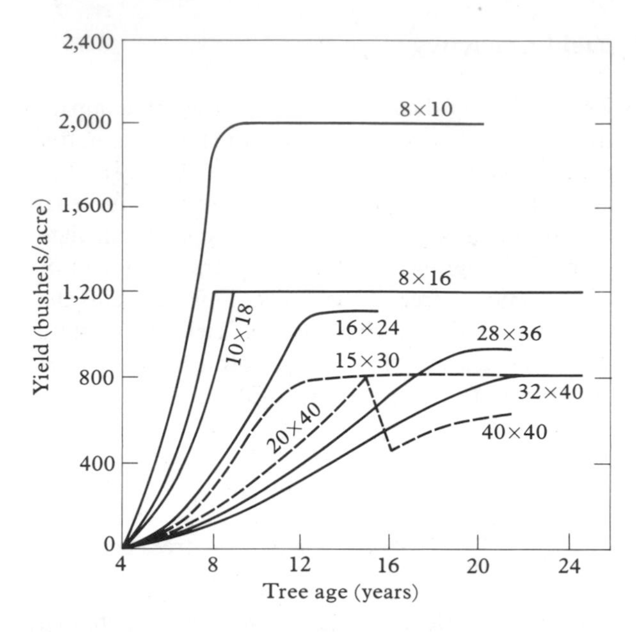

Figure 10-5 Calculated yield per acre at maximum density for different-sized trees at different ages. Planting densities are shown in feet. [After J. C. Cain, "Tree spacing in relation to orchard production efficiency," N.Y. Agr. Exp. Sta. Res. Circ. 15. 1969.]

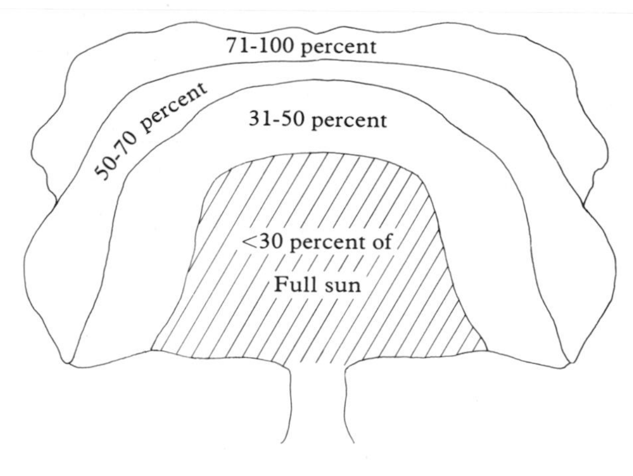

Figure 10-6 Average light distribution, as a percentage of full sunlight, in standard-sized Delicious apple trees. Shaded interior produces inferior fruit of poor color. [After Looney, 1968]

economic marketing, the following are estimated maximum yields for a number of crops:

Fruit	*Maximum m tons per ha*
Apple	112
Pear	90
Peach	56
Prune	45
Sour cherry	27
Sweet cherry	22
Filbert	4.5
Walnut	9.0
Strawberry, everbearing	67
Strawberry, spring bearing	34
Blackberry	25
Red raspberry	16
Black raspberry	13
Blueberry	22
Cranberry	56
Grape	45

If the yields of a given planting are lower than those listed above, and if there is adequate bearing surface, then one or more factors are adversely affecting flowering, fruit set, fruit growth, or some combination of the three. The maximum yields of a number of fruits in California are somewhat higher than those listed above; efficiency is in part related to the amount of sunlight reaching the leaves (Fig. 10-6); leaves receiving less than 30 percent of full sun do not produce good fruit.

GENERAL REFERENCES

Chaplin, M. H.; M. N. Westwood; and R. L. Stebbins. 1973. Estimates of tree size, bearing surface, and yield of tree fruits and nuts. *Ann. Proc. Oregon Hort. Soc.* 64: 94-98.

Leopold, A. C. 1964. *Plant growth and development.* McGraw-Hill, New York.

Steward, F. C. 1968. *Growth and organization in plants.* Addison-Wesley, Reading, Massachusetts.

Westwood, M. N. 1969. Tree size control as it relates to high density orchard systems. *Proc. Wash. State Hort. Assoc.* 65:92-94.

11

Crop Maturity

The proper maturation of the crop and its specific maturity at harvest are important to success in fruit growing. As will be shown, optimal maturity varies with intended use—e.g., fresh market, canning, freezing, drying. Thus, it is important to know beforehand what the fruit will be used for. Only then can criteria be developed to define or establish proper harvest maturity. Before discussing maturation, maturity indices, and quality, definitions of terms will be given. One should distinguish clearly between maturity and ripening. "Mature" means ready to harvest, while "ripe" means ready to eat or process. The two conditions sometimes coincide. Details of ripening are considered in Chapter 13, Post-Harvest, Storage, and Nutritional Value.

DEFINITIONS

Maturation: The processes by which the fruit develops from the immature to the mature state. The term "maturation" is variously applied to:

1. The entire course of fruit development.
2. Only the period of development just preceding senescence.
3. The time between the final stage of growth and the beginning of ripening.

The third definition is the one generally accepted by horticulturists, but the beginning and end of maturation are not as clearly defined as some other stages, such as anthesis, fertilization, or cell division.

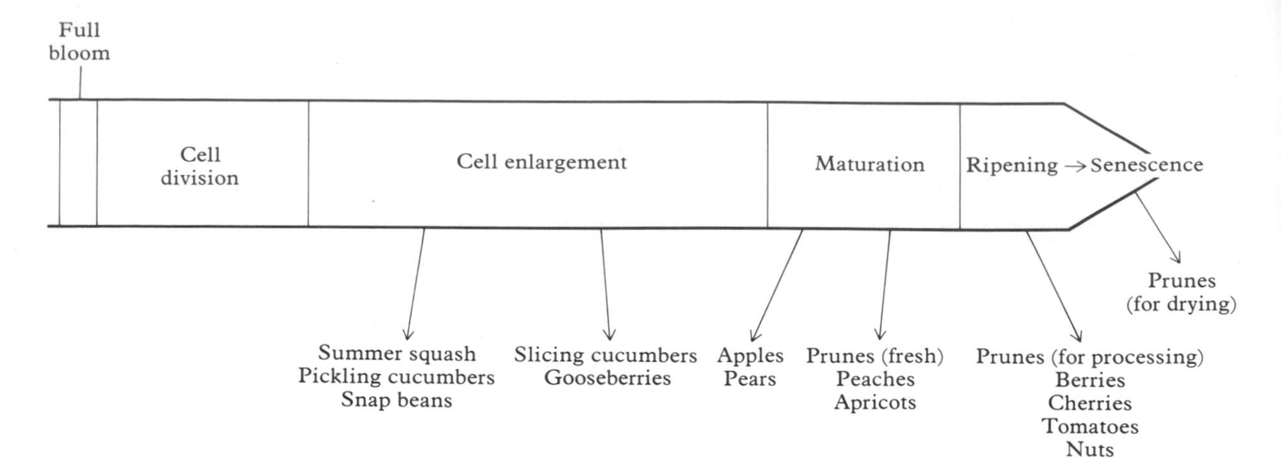

Figure 11-1 Horticultural maturity in relation to physiological development and maturity for several fruits, nuts, and vegetables.

Maturity: The end point of maturation—but with two distinct interpretations:

1. *Physiological maturity* is the attainment, after full development, of the stage just prior to the start of ripening.

2. *Horticultural maturity* is the stage at which growth or development is optimum for a particular use. This use must be specified before maturity criteria can be established. The crop may be at an optimum point of development for consumption or for processing at harvest, or it may ripen to acceptable quality after harvest (and possible storage). Transportation and handling requirements affect these maturity standards. Some types of horticultural maturity are shown in Figure 11-1, and three important classes are listed here:

a. *Harvested physiologically immature:* Green cucumbers, green tomatoes, summer squash, gooseberries, and cherries for brining.

b. *Harvested firm mature but ripened later:* Pears, winter apples, fresh prunes, apricots, and peaches.

c. *Harvested when ripe:* Berries, cherries, slicing tomatoes, nuts, prunes for canning or drying, fruits for roadside markets.

MATURATION

Maturation comprises physical, biochemical, and physiological changes. Physical changes include a decrease in firmness; changes in texture; a decrease in skin chlorophyll and an increase in carotenes and xanthophylls (the change from green to yellow ground color), and an increase in anthocyanins (which contribute a red or blue overcolor). Internal chemical and physiological changes include a decrease in starch (in some fruits); an increase in sugars, soluble solids, and soluble pectin; a decrease in acidity; and, for some, a decrease in respiratory activity.

Changes in composition during maturation (Fig. 11-2) follow a specific pattern, determined by genetic makeup but modified by environment; environmental effects result in seasonal variations. Substances that increase during maturation are reducing sugars, sucrose, carotenes (in some fruits), xanthophylls, anthocyanins, citric acid (in some pears), starch (in bananas), protein, ethyl alcohol, soluble pectin, and esters. These changes are accompanied by an increasing capacity for anaerobic respiration, an increasing sensitivity to ethylene, and increasing vacuolation of cells. Substances that decrease during maturation are starch (in pome fruits) and malic acid (in most pome

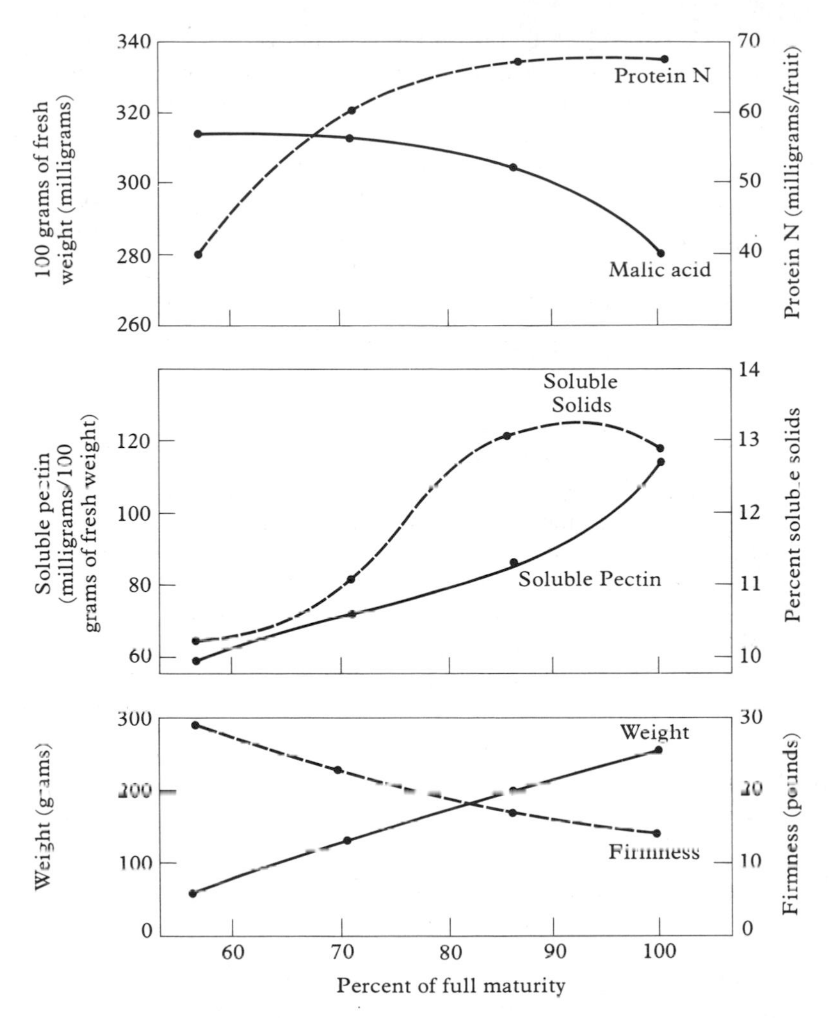

Figure 11-2 Changes in weight, firmness, and chemical composition during maturation of Anjou pears. In common with many other fruits, maturity results in a decline in organic acids and firmness and an increase in weight, soluble solids, and soluble pectins. [After Wang, Mellenthin, and Hansen, 1972]

fruits). Flesh firmness and respiration decrease; plastids degenerate; there is a decrease in capacity to withstand physical and physiological stress and a decrease in resistance to pathogens.

INDICES OF MATURITY

The factors usually measured to specifically indicate horticultural maturity of fruits are flesh firmness; skin color, flesh color, sugar content, content of soluble solids, total content of acids, chlorophyll content and carotene content. Also used are the number of days from full bloom (Ryall et al., 1941) and heat-unit accumulation during specific periods of the growing season. (For an explanation of the latter, see "Degree-days" in the Glossary.) Approximate harvest periods for several crops are given in Figure 11-3.

Pear Maturity

Pears are harvested firm mature, are stored in this condition, and then are ripened before fresh consumption or processing. The pressure test for firmness at harvest is the best single index of maturity (see Table

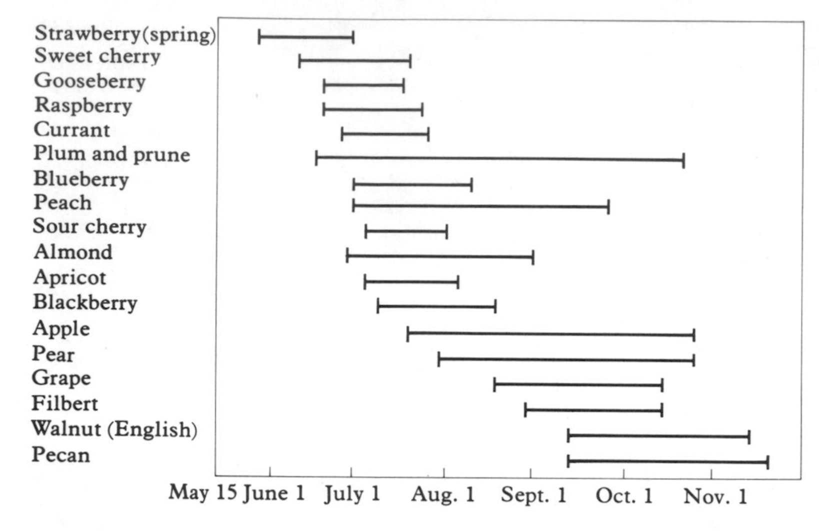

Figure 11-3 Range of ripening dates for several cultivars of fruits and nuts grown in mid- rather than early- or late-season areas. Some crops, however, such as almond and pecan are grown only in the more southern latitudes.

Table 11-1 Flesh firmness as a maturity index for pears as measured by the U.S. tester with a .8-cm[1] head on peeled cheeks.

Cultivar	Firmness Range	
	Pounds	kg
Williams (Bartlett)	23–17	10–7.7
Howell	17–13	7.7–6.8
Bosc	15–13	6.8–5.9
Packham's Triumph	15–13	6.8–5.9
Anjou	15–13	6.8–5.9
Easter	15–13	6.8–5.9
Eldorado	15–13	6.8–5.9
Winter Nelis	15–12	6.8–5.4
Conference	14.5–11	6.6–5.0
Forelle	14–12	6.4–5.4
Clairgeau	14–11	6.4–5.0
Glou Morceau	14–11	6.4–5.0
Kieffer	13.5–12	6.1–5.4
Seckel	13–11	5.9–5.0
Flemish Beauty	13–10	5.9–4.5
Passe Crassane	13–10	5.9–4.5
Comice	12–10	5.4–4.5
Hardy	11–9	5.0–4.1
Angouleme	11–8	5.0–3.6

SOURCE: USDA Circ. 627 (Haller, 1941).
[1] .8 cm diam. head = $\frac{5}{16}$ inches.

11-1). Optimum firmness will vary, however, in different climates and with different rootstocks. Fruit grown in a hot climate or on Oriental rootstocks are firmer at optimum maturity than are other fruit. Soluble solids (mostly sugars) may be used, but they are more variable according to season, growing conditions, variety, and crop load. Heat-unit accumulation for 5 to 9 weeks following bloom is inversely correlated with the number of days from full bloom to harvest maturity. This can be worked out for each cultivar. For example, Anjou pear maturity varied from 130 days past full bloom when early season degree-days (above 6.1°C daily mean) were 528, to 155 days when degree-days were only 278 (Mellenthin, 1966). Bartlett was mature at 105 days when the degree-days 5 weeks after bloom were 333 (above 4.5°C daily mean) but required 130 days when the post-bloom degree-days were 167 (Fig. 11-4). Also, cool weather below 10°C during the preharvest period causes Bartlett pear to mature and ripen prematurely and thus shortens the number of days from full bloom to harvest (Wang et al., 1971). The most reliable index for pear maturity is flesh firmness,

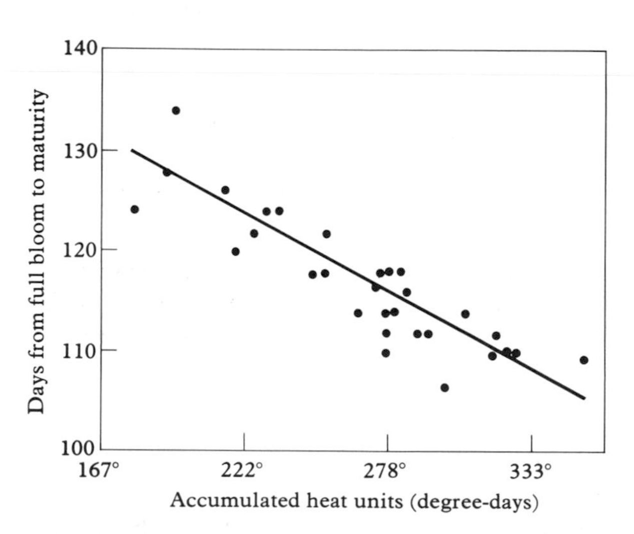

Figure 11-4 The relationship of the number of accumulated heat units during the first 36 days after full bloom to the days required from full bloom to mature Bartlett pear fruit. The warmer the post-bloom period, the shorter the time to maturity. [After Lombard, Cordy, and Hansen, 1971]

but it should be used in conjunction with heat units during the post-bloom period (and the preharvest period for Bartlett).

Apple Maturity

Flesh firmness is not a good index for early harvest of apples because it is not closely related to maturity. However, firmness tests have value for later harvest and during storage, particularly when they are used with other indices.

The number of days from full bloom to harvest is a fairly good index of apple maturity, especially when used with the observed change from green to white-yellow ground color and the attainment of a minimum amount of soluble solids in the juice (about 11 percent for Delicious). Also, it is an index that can be adjusted for a given area in the following manner (Fisher, 1962; Blanpied, 1964): If the average date of full bloom has been established over several seasons, one can predict maturity in a given year by subtracting 1 day for each 3 days that bloom is later than average or adding 1 day for each 3 days that bloom is earlier than average. For example, if Delicious apple on average is mature at 150 days with a bloom date of April 25, and if full bloom for a given year is April 22, fruit will be mature in about 151 days; if full bloom is April 28, fruit will be mature in about 149 days; and so forth.

The flesh of apples should be white or cream colored rather than greenish at harvest, and it should be somewhat sweet to the taste rather than starchy. But for long cold storage, apples should not be fully eating ripe at harvest, as this leads to various types of early physiological breakdown and loss.

Prune and Plum Maturity

Optimum maturity for prunes depends upon whether they are to be eaten fresh, canned, or dried. Those to be eaten locally can be picked more mature than those destined for distant fresh markets. Prunes to be stored for later shipment or canning are picked at a less mature stage than those to be canned or dried directly after harvest. Fruit picked before they reach full eating maturity must be ripened properly to be of good quality when eaten or canned (Gerhardt and English, 1945). Obviously, ultimate quality depends upon post-harvest storage and ripening conditions as well as maturity and condition at harvest.

As is true of other fruits, prune cultivars have a characteristic number of days from full bloom to maturity. In California, French prune requires an average of 158 days to mature, with a range of 145 to 165 days during a 14-year period. In western Oregon, the Italian prune requires about 150 days to mature and has varied between 146 and 156 days during 8 years. This can be used as a general harvest guide. At about 140 days past full bloom, preliminary tests can be started to determine maturity more precisely. Other indices of maturity are percent soluble solids, flesh firmness, flesh and skin color, total acidity, and soluble solids/acids ratio.

Prunes grow at a daily rate of 0.5 to 1.0 percent during the preharvest period. From the earliest time the fruit may be picked for fresh shipment until they are tree ripe, they will gain 20 to 25 percent in weight. Thus, while it may be necessary at times to pick at minimal maturity, some loss in tonnage will result. Since prunes do not store carbohydrates as starch (as does the apple), there is no increase in soluble solids (sugars) during off-the-tree ripening.

For storage and distant shipment, Italian prunes are picked at less than the eating ripe stage but mature enough to ripen to acceptable quality. In preliminary tests, Hartman (1926) found that flesh firmness was a better indicator of maturity than were sugars, solids, or acids of the fruit. Considerable variability in firmness was found, however, between orchards and between seasons. Tucker and Verner (1932) in Idaho reported that flesh firmness was less variable than sugar content in estimating maturity. At that time the sugar and soluble solids were being measured by a hydrometer, which was subject to considerable error and was much less accurate than modern refractometers. Later Gerhardt et al. (1943) reported that the soluble solids/acids ratio was the best index, but percent soluble solids and skin color were also useful tests. The solids/acids ratio between 12 and 15 was satisfactory and soluble solids were best at 14 to 16 percent. A report from Idaho (Verner, 1962) showed that firmness declined and soluble solids increased linearly during maturation. But Gerhardt and Schomer (1955), working with Early Italian prune, found that firmness and soluble solids were not dependable as maturity indices. They found that flesh color change and solids/acids ratio were best. Yet Fisher (1940) in British Columbia reported that percent soluble solids in Italian prune, as measured by a refractometer, was the best index for fresh shipment. He found that 17 percent soluble solids was the minimum maturity level for acceptable quality. The several differences reported here may have resulted from different measuring devices or techniques but may also have been due to real differences in other factors. Climate may cause seasonal fluctuations in both firmness and chemical composition (Hansen, 1959). The data in Table 11-2 indicate some general firmness readings of several plum cultivars picked for fresh shipment.

Prunes for drying are harvested at a somewhat more mature stage than those used for canning. Sugar is added to the canned prune to bring it to the desired sweetness, while the desired sugar content for dried prunes is obtained by allowing them to ripen longer on the tree. Claypool et al. (1962) reported that the change in flesh color and firmness of French prune were the best indices of maturity. They found that soluble solids, while related to maturity, fluctuated with crop load and with climate. Roberts (1964) later reported that maturity could be found by using both solids and firmness tests. The best maturity for drying was with 25 to 35 percent soluble solids and 1 to 2 pounds flesh firmness (as measured by a standard Magness-Taylor tester with a plunger (.8 cm diameter) pressed into a peeled surface of the cheek of the fruit). Baker and Brooks (1944) reported that days from full bloom was a fair index of maturity, but it varied from the usual 158 days by about 10 days in some exceptional seasons. Hansen (1959) reported that Italian prunes for drying were at optimum maturity when solids were up to 20 percent, solids/acid ratio was 22, and the flesh had turned amber (1.025 micrograms of carotene per 100 grams). He concluded that the best single field test of maturity for drying was best denoted by the attainment of maximum soluble solids, that is, when successive samples show no increase. Beyond that time, the fruit tended to become overripe, showing stem-end shrivel, tan or brown flesh, and loss of acidity. Fruit softening as measured by the Magness-Taylor tester (with skin on) also was correlated with ripening, but that index varied according to seasonal climate and weather during the preharvest maturation period. The major differences in firmness were between years rather than between orchards during the same year.

Table 11-2 Upper and lower limits of firmness (.8-cm head, U.S. tester) for plums and prunes for distant shipping without precooling.

Cultivar	Firmness Range	
	Pounds	kg
Beauty	13–9	5.9–4.1
Burbank	20–14	9.1–6.4
Climax	18–13	8.2–5.9
Diamond	20–15	9.1–6.8
Duarte	15–11	6.8–5.0
Giant	16–11	7.3–5.0
Santa Rosa	18–12	8.2–5.4
Wickson	15–12	6.8–5.4
Formosa	13–9	5.9–4.1
President	16–11	7.3–5.0
Italian	14–10	6.4–4.5

SOURCE: USDA Circ. 627 (Haller, 1941).

The change of flesh color in the prune from yellow-green to amber appears to be a good visual measure of maturity. Other indices related to maturity are flesh softening (firmness), sugar buildup, red-to-blue skin color, decline in acidity, and the solids/acids ratio. Of these the flesh color and solids/acids ratio seem to be the most reliable in all climates and orchards. Yet the rate of nitrogen fertilizer may alter both flesh color and firmness. Excessive nitrogen tends to result in greener and softer flesh. The effects of rootstocks on maturity indices are as yet unknown, although it is known that rootstocks may alter firmness. The biggest problem with both pressure tests (firmness) and soluble solids is that they tend to vary from year to year and from orchard to orchard too much to permit the use of single values either for minimal or for optimal maturity. Some combination of these indices should be used rather than any single index. One index with merit is the soluble-solids/firmness ratio. During maturation, solids are increasing while firmness is decreasing. To standardize this test, a single uniform pressure test would have to be used. In the past, several different instruments were used—some were applied to a peeled cheek and others to unpeeled fruit. Some modification of the tester now in use would permit more accurate readings in the optimum maturity range. This ratio might also be combined with a specific acid or a specific soluble-solids level for each cultivar.

Peach Maturity

The main indices for peach are ground color change from green to straw color, flesh firmness, and days from full bloom (Rood, 1957). As is true of other fruits, warm weather during the post-bloom period reduces the number of days required to reach harvest maturity (Batjer and Martin, 1965). Table 11-3 shows firmness values suggested for distant shipment of several peach cultivars.

Cherry Maturity

Increases in soluble solids (sugars) and increases in fruit color have been considered the best indices of maturity, but recent evidence indicates stem:fruit removal force may be better (Richardson et al., 1975). They found that the attachment between stem and fruit loosens as the fruit becomes mature and the reduced force required to remove the fruit from the stem can be quantified and used as a maturity index. Also fruit-acid level is important in sour cherries. As indicated above, optimum maturity varies with intended use. Cherries for brining are picked prematurely, before color and the amount of soluble solids have developed enough for the fresh market. Cherries picked for distant markets are picked earlier than those intended for local markets or for canning. Levels of soluble solids and color are established locally for each cultivar and each use.

Table 11-3 Upper and lower limits of firmness (.8-cm head) for distant shipment of peaches.

Cultivar	Firmness Range		Remarks
	Pounds	kg	
Belle	14–12.5	6.4–5.7	Pared
Carmen	12–9	5.4–4.1	Pared
Early Crawford	20–14	9.1–6.4	Pared
Early Elberta	18–14	8.2–6.4	Unpared
Elberta	19–14	8.6–6.4	Unpared
J. H. Hale	20–17	9.1–7.7	Unpared
Phillips Cling	12–8.8	5.4–4.0	

Source: USDA Circ. 627 (Haller, 1941).

Maturity of Nuts

Filberts are mature when they are shed from the husk in September, October, or November. But although they are fully mature when they drop to the ground, they still must "cure" by losing moisture. Walnuts are mature 1–4 weeks before hull dehiscence (Fig. 11-5). Oils are formed in walnut by the time the packing tissue is brown—they are mature at that time. They are artificially dried to 3–4-percent moisture immediately after harvest in warm air tunnels at 38–43°C for

Figure 11-5 Hull dehiscence of walnut, which occurs one to four weeks after the nuts are mature. [Courtesy of G. C. Martin]

walnut and 35–40°C for filbert to a final moisture content of 8–10 percent. Almonds are mature when they are loose enough to be knocked and hulled. Delaying harvest until the hulls are quite dry does not impair almond quality, but the longer the delay, the greater the threat from the navel orangeworm. Pecans are mature when the husks open from around the nuts. Mature nuts do not fall all at one time, so mechanical shakers are used to bring them down. The curing of mature nuts involves the drying away of some moisture along with increases in fatty acids and changes in flavor components.

Maturity of Small Fruits

Grape maturity is best indicated by the balance between soluble solids (sugars) and acids. Picking too early results in levels of acids that are too high and in reduced yield. Picking too late results in high sugars and low acids. The optimum sugar/acids index varies with the intended use of the grapes—whether for juice, jelly, fresh or wine; it also is different for different cultivars (Figs. 11-6 and 11-7).

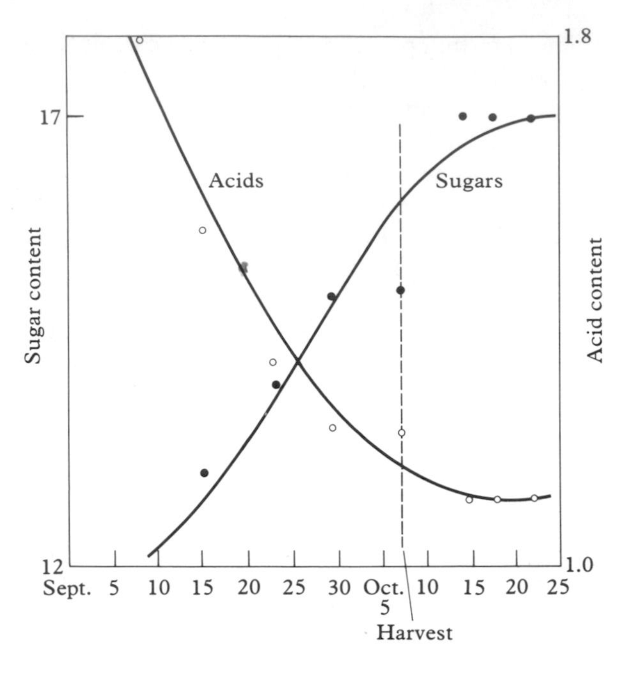

Figure 11-6 Changes in sugars and acids of Concord grape during maturation. [Based on data from Nelson Shaulis, N.Y. Agr. Exp. Sta. Bull. 805]

Strawberry maturity for fresh shipment occurs when the fruit just attains all-over red color or has only a small area of white and is firm. Even one day later the fruit may be too soft to ship. For immediate processing, however, the fruit is picked more mature and thus attains higher soluble solids, better overall quality, and larger size. Figure 11-8 indicates harvest dates for the important areas of production in the U.S.

Raspberries and blackberries are ready to pick when they separate readily from the stem. Raspberries should be fully colored and should not crumble. Blackberries are mature when they are fully colored, when the tips of the druplets are entirely filled, and when the fruit is loose on the stem. Boysen and Youngberries do not develop their highest quality until they are fully ripe and well colored.

Gooseberries for pie are picked at the firm green stage, at which time they are too sour to eat fresh. They sunburn easily after harvest and so should not be left uncovered in the field.

When they are to be used for jelly, currants are picked slightly immature to impart a bit more tartness to the flavor.

Blueberries are mature when they have reached full blue color (reddish in some cultivars) at the stem end and are loose on the pedicels. Because they hang well on the bush when ripe, they need be picked only once a week over a period of 6 or 7 weeks.

Cranberries are harvested in most areas during September and October. Partly ripe berries will color up in storage at 13° to 16°C if well ventilated but do not color well at 0.5° to 1.0°C or under poor ventilation. Also, chilling injury occurs at such low temperatures, (Lutz and Hardenburg, 1968). Both sugars and acids decline during storage at 2° to 5°C, but the desirable tart flavor is maintained if the berries remain sound. Mature berries have about 2.35 percent acids and 3 to 6 percent sugars. The best quality is achieved with high sugar content.

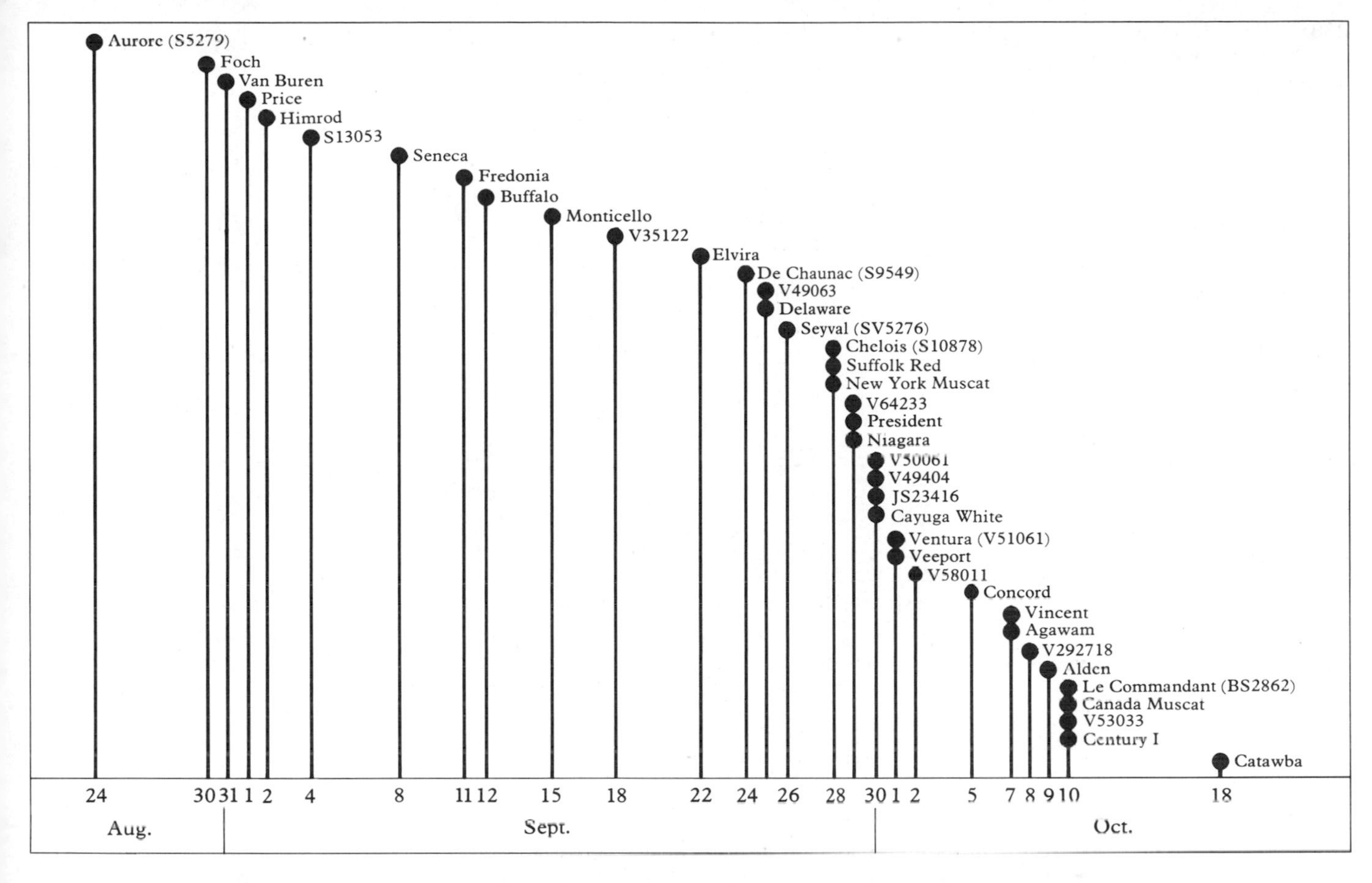

Figure 11-7 Harvest dates for grapes in Ontario. Harvest in Ontario is a month or more later than in warmer areas. [After Bradt, Hutchinson, Ricketson, and Tehrani, 1974]

Fig Maturity

The fig is like many other fruits in that its growth and maturity is hastened markedly by treatment with ethephon, which releases ethylene in the tissues (Fig. 11-9). Sprays with auxins such as 2,4,5-TP also hasten fig maturity by stimulating endogenous ethylene synthesis. Ethylene appears to be a primary ripening hormone in plants. It will be discussed more fully in the section on ripening in Chapter 13.

FRUIT COLOR

As fruits mature they take on red, blue, purple, or yellow colors characteristic of the cultivar; at the same time, most cultivars begin to lose their green undercolor. These pigment changes are indications of both quality and stage of maturity.

Some pigment changes require direct light on the skin of the fruit, and others do not. Red, blue, and purple colors of fruits result from synthesis of antho-

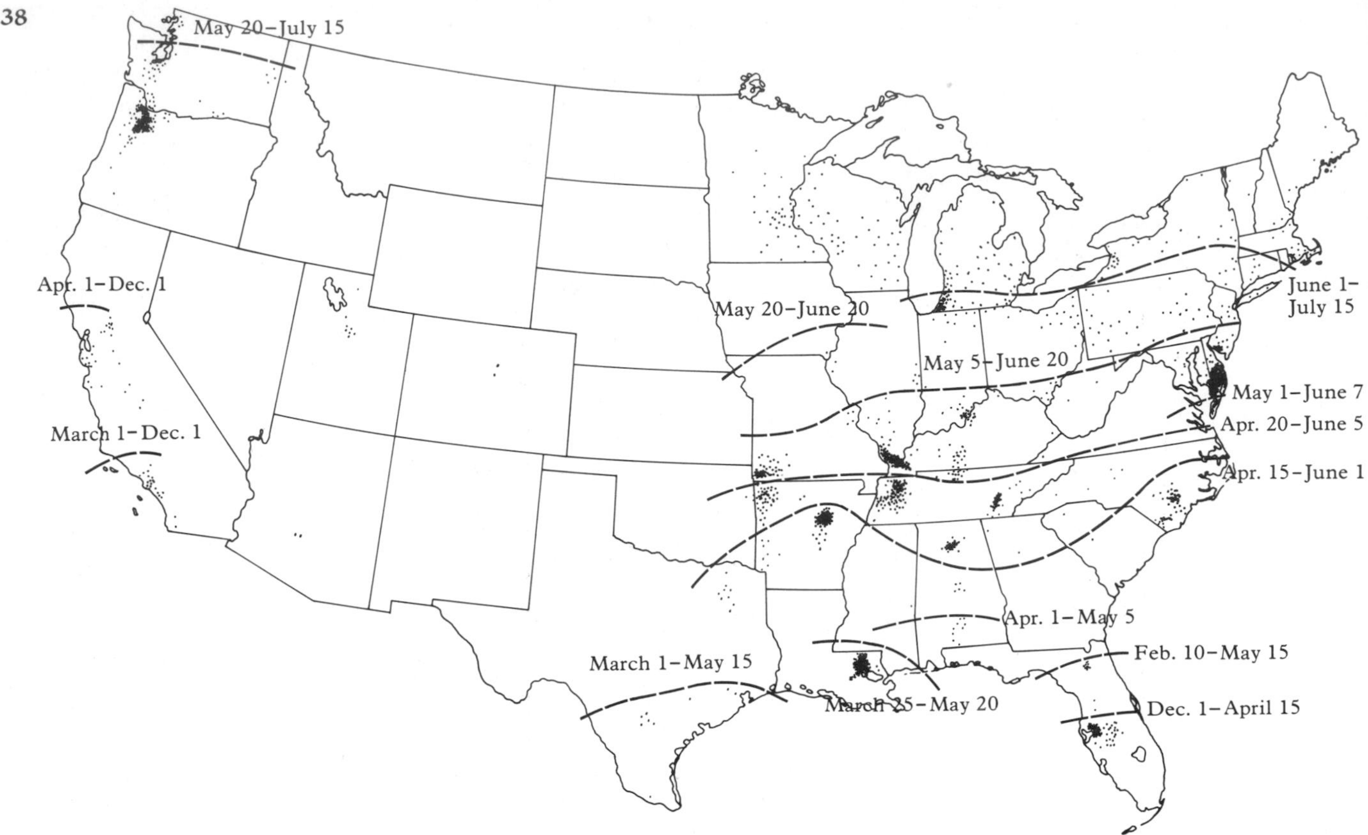

Figure 11-8 Map shows the location of the principal commercial strawberry-producing regions, the approximate ripening time in each region, and the northward progression of the strawberry season. [After USDA]

cyanin pigment. In apple, pear, peach, nectarine, and apricot, direct light to the fruit is needed for red color development (Fig. 11-10). Other fruits, such as grape, cherry, plum, prune, raspberry, and blueberry, do not require direct light for anthocyanin synthesis. Yellow pigment does not require direct light to develop.

Red pigmentation is affected by various cultural and environmental factors such as pruning, thinning, fertilization, temperature, and light. Factors that result in a high level of carbohydrates in the fruit during the preharvest period tend to increase anthocyanin pigments. Excess nitrogen and improper pruning can reduce color by depleting carbohydrates and by preventing proper light distribution in the tree. Although individual leaves produce maximum photosynthate at light intensities below that of full sun, the whole tree produces more dry weight in full sun than in lower light. Leaves within the canopy receiving less than 30 percent of full sunlight are ineffective. Thus any factor such as clouds, rain, haze, fog, or smoke, which attenuates incoming radiation can reduce total photosynthate production in orchard trees. Emphasis here will be on those factors affecting light-requiring pigment synthesis.

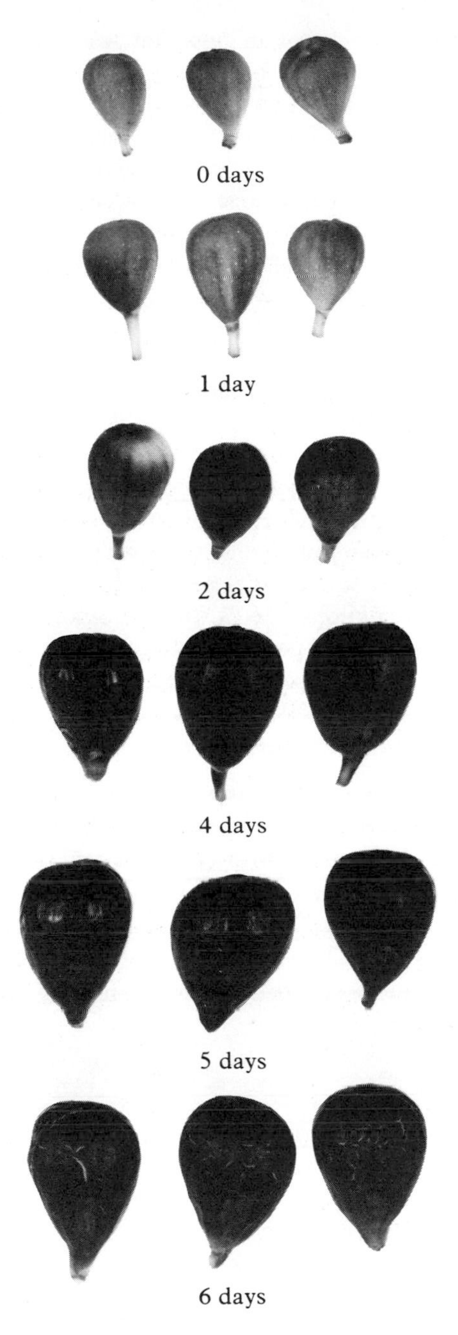

Figure 11-9 Fig maturity and ripening occur normally on the tree but can be greatly accelerated by ethylene treatment. Ethylene exposure in days is shown, and each was photographed one week after the start of treatment. [After Maxie and Crane, 1968]

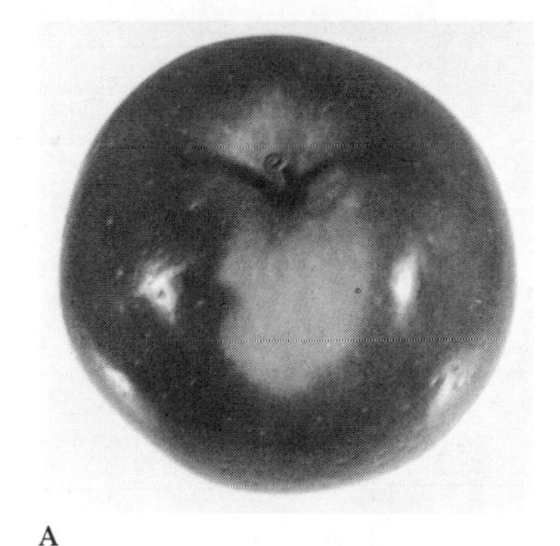

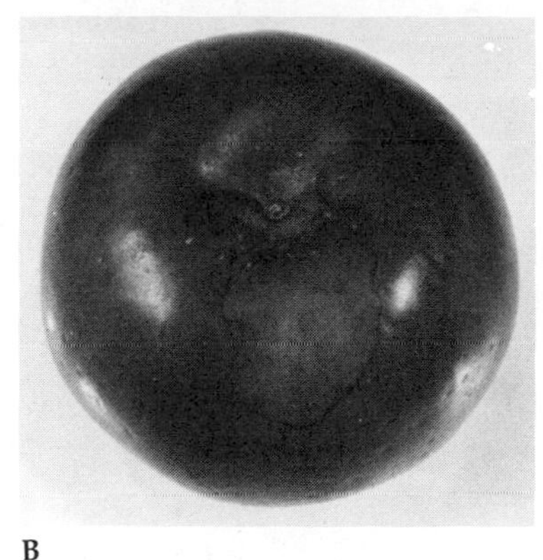

Figure 11-10 Effect of direct light on red pigment in skin of Rome Beauty apple. **(A)** Unpigmented area resulted from being covered by a leaf. No pigment developed during six months of cold storage in the dark. **(B)** The same fruit after daily illumination with low-intensity fluorescent light for one week. The entire area became red, though not so intense as that illuminated with sunlight.

Consumer Preference

In one area of the U.S., buyers preferred red eating apples over yellow or green ones but did not show this preference for cooking apples (Table 11-4). In this test, juiciness, blemish-free skin finish, size, firmness, and internal condition were qualities rated higher than fruit color by consumers who bought apples. However, an earlier test indicated that red apples were preferred by consumers: more bags of red than green Rome Beauty apples were sold, even when the price

Table 11-4 Color preferences of consumers who bought apples, as the percent of those responding.

Fruit Color	Type of Apple	
	Eating	Cooking
Deep red	43	19
Bright red	41	23
Yellow	19	6
Part red and part green	14	25
Green	5	23

SOURCE: Meyers and Weidenhamer, 1966.

Table11-5 Sales of Rome Beauty apples during 4 weeks as related to fruit color and price.

Fruit Color	Price per 2.3 kg[1] Bag	Volume Sales (bags)	Sales Value
Green	49¢	393	$193
Red	49¢	743	$364
Red	55¢	363	$200
Green	55¢	227	$125

SOURCE: Evans and March, 1957.
[1]2.3 kg = 5 pounds.

for red apples was higher (Table 11-5). Consumers in different parts of the world may differ in their preferences, depending upon local customs and the nature of the main cultivars grown. In general, because of consumer preference, highly colored fruit from a normally red cultivar are deemed to be of better quality than poorly colored ones from the same cultivar, but yellow and green cultivars such as Golden Delicious, Yellow Newtown, Mutsu, and Granny Smith are being well received by consumers.

Pigment Synthesis

The red pigment of apple is idaein (cyanidin-3-galactoside), as reported by Sando (1937). The generalized steps for synthesis within the fruit cell vacuoles are shown in Figure 11-11. Siegelman and Hendricks (1958) reported that idaein synthesis in isolated apple skin was dependent upon two light-requiring phases, the first of which was an induction phase in which no pigment was formed, followed by a production phase. Both phases had action spectra near a wavelength of 650 nanometers. The researchers theorized that the photoreceptor is a flavo-protein similar to Acyl Coenzyme A dehydrogenase. The need for red light (650 nm) in these tests is inconsistent with other data which indicate that ultraviolet light is most important for the pigment synthesis.

Influence of Environment

Because of the importance of light in pigment synthesis, it is appropriate to discuss the quality and intensity of light as influenced by environmental factors. Solar radiation just beyond the earth's atmosphere is estimated to be 2 calories per square centimeter per second (145 kilolux, or 13,500 foot candles). Of this total radiation, 7.5 percent is ultraviolet, 41 percent is visible, and 51.5 percent is infrared. According to Platt and Griffiths (1965), light is modified as it passes through the atmosphere in the following ways:

1. It is scattered by air molecules and by molecules smaller than the wavelength of the radiation.
2. It (especially ultraviolet radiation) is absorbed by atmospheric gases (O_2, O_3, CO_2, H_2O).
3. It is scattered and absorbed by cloud masses.
4. It is scattered and reflected by particles of sizes equal to or greater than the wavelength of the incident light.

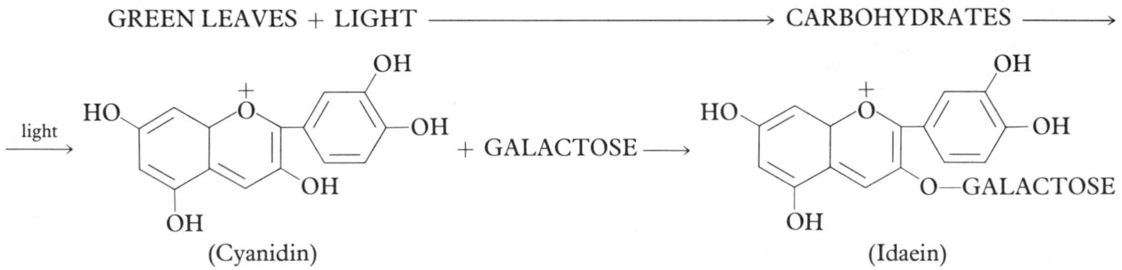

GREEN LEAVES + LIGHT ⟶ CARBOHYDRATES ⟶ (light) ⟶ Cyanidin + GALACTOSE ⟶ Idaein

HO, OH, OH, OH, OH (Cyanidin)

HO, OH, OH, O—GALACTOSE, OH (Idaein)

Figure 11-11 Synthesis of the red pigment idaein in the skin of apple fruit. Carbohydrates are produced first; then, cyanidin is produced in the presence of light; finally, the sugar galactose is added chemically to complete the synthesis.

Of the 43 percent of the incoming light that reaches the earth's surface, 25 percent is direct light and 18 percent is diffuse sky radiation. Both are important in the pigmentation of fruit.

Ultraviolet and blue light are more effective in producing anthocyanin pigments in fruits with a light requirement than are the longer wavelengths. Magness (1928) found that low-intensity ultraviolet light for one hour per day increased the color of Jonathan apples. He also found that ordinary window glass, which reduces greatly the shorter wavelength radiations and degrades them to longer wavelengths, reduced pigment synthesis in detached fruits (Table 11-6). The window glass changed both the intensity and the quality of light; both factors probably affected color. Schrader and Marth (1931) reported that the intensity of light reaching the fruit was directly related to pigmentation of apples on the tree (Table 11-7). In their study, only the fruit were shaded, so that photosynthesis by leaves was held constant in all treatments. This work was done during two seasons, in which the fruit were bagged with cheese cloth or muslin from June or July until harvest. The data were consistent both for cultivar and season, and indicate the need for adequate exposure of the fruit to light for good pigmentation. The experiment also showed that the exposure of chlorophyll in the fruit skin to full light resulted in larger fruit than did exposure to attenuated light. Fruit receiving only 39 percent full sun were 20 percent smaller at harvest than controls (Table 11-7). This finding is consonant with the common observation that, under orchard conditions, fruit produced in the shaded portions of a tree have poorer color and are smaller than those borne at the periphery (Heinicke, 1966).

The recent work by Siegleman and Hendricks (1958) showing a photocontrol for anthocyanin in apples at a wavelength of about 650 nanometers indicates the need for red light in at least one step in the synthesis. It is unlikely, however, that red light is ever limiting under field conditions because previously cited work indicates that ultraviolet and blue light are of primary importance in the field. Apparently enough red light is available in all orchard situations.

Table 11-6 Color development of detached apples under glass and in direct sunlight.

Cultivar	Solid Red Color (%)			
	At the Start		After 12 Days	
	Under Glass	Full Sunlight	Under Glass	Full Sunlight
Jonathan	7	9	44	96
Delicious	5	5	17	74
Rome Beauty	2	2	9	36

SOURCE: Magness, 1928.

Table 11-7 Effect of light intensity to fruit on the development of red pigment in nine apple cultivars.

Fruit Measurement	Amount of Light Reaching Fruit (as % of full sun)			
	100%	81%	61%	39%
Diameter (cm)	7.1	7.0	6.9	6.6
Volume (cm³)	187	180	172	150
Size (as % of control)	(100)	96	92	80
Percent visual red color	57	28	10	1

SOURCE: Schrader and Marth, 1931.

ATMOSPHERIC CONDITIONS To obtain data on reductions in radiation due to smoke, clouds, and rain, P. B. Lombard (Medford Experiment Station) obtained radiation readings and observations on atmospheric conditions from the weather bureau there (Table 11-8). Radiation was measured with a recording Pyrheliometer, and equivalency in visible

Table 11-8 Effect of smoke, clouds, and rain on incident solar radiation in Medford, Oregon.

Condition	Radiation (% of ETR[1])	Percent of Clear Day	Visibility (km)
Clear or slightly cloudy	73	100	16–40
Light smoke	69	95	9.5–13
Clouds—no smoke	65	90	13–48
Clouds—some smoke	64	89	6.5–11
Rain—no smoke	45	62	8–40
Heavy smoke	42	58	5
Smoke and rain	32	44	5–8

[1]Extra-Terrestrial Radiation.

light is not known. These data serve only to show the general relationship between sky condition and radiation. They indicate that the poorest conditions are created by heavy smoke, rain, or smoke and rain. Recent work by Heinicke (1964) on light distribution in apple trees indicates that leaves are ineffective when they receive less than 30 percent of full sunlight and that light diminishes rapidly from the periphery to the center of a tree. Thus, if incoming light were considerably less than full sunlight because of smoke, dust, clouds, or rain, a correspondingly larger proportion of the interior leaves of a tree would receive less than the critical level of 30 percent full sun.

In dry air, the scattering of light is inversely proportional to the fourth power of the wavelength of the ray, so the blue end of the spectrum is scattered much more than is the red. This is the reason a clear sky appears blue. This is also the reason that diffuse sky radiation is important for fruit pigmentation. Depending upon local atmospheric conditions and latitude, diffuse sky radiation accounts for 25 to 54 percent (in summer) and 19 to 73 percent (in winter) of the total short wavelength radiations received. Generally, the most highly colored fruits at maturity also are of highest quality.

TEMPERATURE Under field conditions, apples color best in climates with clear bright days and cool nights during the preharvest period. The benefit of cool nights appears to be an indirect effect on coloring by reducing respiration loss of carbohydrates, because Magness (1928) showed that cold night temperatures on detached fruit did not enhance pigmentation.

CULTURE Pruning, tillage, fruit thinning, fertilizer use, and pest control probably affect fruit color directly to the extent that they influence effective leaf area, leaf/fruit ratio, carbohydrate level, and the degree of fruit shading before harvest. Magness (1928) found a close relationship between leaf/fruit ratio, sugar content, and red color of Delicious apple (Table 11-9). Fruit with a low sugar content and low leaf/fruit ratio failed to color well even when exposed to optimum sunlight. Heinicke (1964) showed that the best apple color was produced when fruits and leaves were exposed to 70 percent or more of full sun, while adequate color was produced when they were exposed to between 40 and 70 percent full sun. Below 40 percent full sun, fruit developed inadequate color. Below 50 percent they were small in size. Soluble solids were directly related to light exposure.

Table 11-9 Effect of fruit thinning on sugars and red color in Delicious apple.

Fruit Thinning (leaves per fruit)	Total Sugars (%)	Red Color (%)
10	9.8	23
20	11.2	26
30	11.6	42
50	13.2	51
75	14.5	58

SOURCE: Magness, 1928.

Excess nitrogen in the tree increases growth, decreases available carbohydrates, and thus reduces color. Benson et al. (1957), however, found that although high levels of nitrogen delayed apple maturity, if the fruit were left on the tree until mature, then color was satisfactory. This may have been due to cooler night temperatures and more favorable carbohydrate levels resulting from the later maturity. Recent work (unpublished) indicates that foliar sprays of urea later than early summer very drastically reduce both red and yellow color in apples. This seems to be a direct inhibitory effect of urea and not merely the effect of increasing the nitrogen level.

Proper thinning-out pruning can increase the amount of light getting into trees, but severe pruning stimulates excess growth that ultimately reduces light inside the canopy (Table 11-10).

Genetic Potential and Fruit Color

As a point of interest, at least one apple cultivar (Beacon) is able to develop red pigment in the dark after harvest. This could prove valuable in a breeding program. This trait is similar to that of prune plums, which gain a considerable amount of purple skin color in dark storage.

Table 11-10 Effect of pruning method on fruit color and light inside the canopy of Elberta peach trees.

Pruning Method	Light in tree (in August as % of full sun)	Red Blush Color (%)	Yellow Undercolor (green-to-yellow) Rating[5]
Corrective[1]	60	54	3.6
Thinning out[2]	60	58	3.7
Conventional[3]	55	42	3.3
Severe[4]	35	27	3.2

SOURCE: From Westwood and Gerber, 1958.
[1]Corrective = Slight pruning of broken, crossing branches.
[2]Thinning out = Moderate pruning by thinning out shoots.
[3]Conventional = Same as thinning out, plus heading back remaining shoots by half.
[4]Severe = Removal of 50–75% of new shoots, plus severe heading back of those remaining.
[5]2 = Green; 3 = Green-yellow; 4 = Amber-yellow.

During the last 30 years a great many genetic mutants have been selected that are more highly colored than the original cultivars, under similar growing conditions. Use of such mutants permits harvest at maturity without sacrificing optimum color. Often, the original cultivars were harvested at an overmature stage because they were left on the tree too long in the hope of getting better color.

GENERAL REFERENCES

Allen, F. W. 1932. Physical and chemical changes in the ripening of deciduous fruits. *Hilgardia* 6:381–441.

Brown, D. S. 1952. Climate in relation to deciduous fruit production in California. V. The use of temperature records to predict the time of harvest of apricots. *Proc. Am. Soc. Hort. Sci.* 60:197–203.

Fisher, D. V. 1962. Heat units and number of days required to mature some pome and stone fruits in various areas of North America. *Proc. Am. Soc. Hort. Sci.* 80:114–124.

Gerhardt, F., and D. F. Allmendinger. 1946. The influence of naphthaleneacetic acid spray on the maturity and storage physiology of apples, pears, and sweet cherries. *J. Agr. Res.* 73:189–206.

Haller, M. H. 1941. *Fruit pressure testers and their practical application.* USDA, Circ. 627.

Magness, J. R., and G. F. Taylor. 1925. *An improved type of pressure tester for the determination of fruit maturity.* USDA, Circ. 350.

Platt, R. B., and J. F. Griffiths. 1965. *Environmental measurement and interpretation.* Reinhold Pub. Corp., New York. Pp. 57–91.

Tukey, H. B. 1942. Time interval between full bloom and fruit maturity for several varieties of pears, apples, peaches, and cherries. *Proc. Am. Soc. Hort. Sci.* 40: 133–140.

12

Harvest

The main objective of a harvest is to obtain the maximum saleable crop with the least possible loss. This requires that the crop attain full size and maturity without unnecessary preharvest drop. A natural consequence of maturity, however, is abscission, the separation of the fruit or nut from the plant. Preharvest drop is not a problem with small fruits, such as raspberry, blueberry and cherry, but it can result in serious losses with large fruits, such as apple or pear. With the recent advent of mechanical harvesting, it has become advantageous to apply chemicals that loosen the crop quickly and uniformly so that machine harvest can take place at a specified time following the spray. With hand harvest, however, sprays are applied to prevent loosening and drop.

PREHARVEST TREATMENT

Preventing Pome-Fruit Drop

During maturation, ethylene is generated in the fruit, initiating both ripening and the process of abscission, which causes drop. Synthetic auxin sprays help delay abscission. Abscission takes place when changes occur to weaken the walls of the layer of cells connecting the fruit and the stem or woody spur. These changes occur by enzymatic action as the fruit matures. Cells in the abscission zone are small parenchyma cells with relatively little lignification of cell walls. These cells are held together by the middle lamella, containing calcium and magnesium pectate residues, which are strongly bound to the cell-wall structure. Pectinic acid chains are bound together by the cross-linking action of these two divalent metals. Specific enzymatic action replaces calcium in the structure with methyl groups (which are monovalent and do not furnish cross-linkage) and cause the middle lamella and part of the cell walls to dissolve, so that the cells are no longer cemented together.

The process of separation has more than one phase, the first being methylation, in which the action of the enzyme, pectin methylesterase (PME) decreases binding by decreasing cross-linkages. PME is found in higher concentrations near the abscission zone than elsewhere in the stem. Finally, another enzyme (polygalacturonidase) breaks up the long chain molecules, resulting in separation. At this point, the fruit is held on the tree only by the vascular strands of the conducting system, which are easily broken. Wittenbach and Bukovac (1974, 1975) showed that peroxidase activity is stronger in the abscission zone of cherry than in fruit, which would keep IAA at a low level. They also found that ethephon treatment during stage III hastened abscission of sour cherry, although ethylene levels in the fruit were not correlated with loosening. Probably both peroxidase and ethylene levels aid abscission if the timing and concentration are right.

The onset of pome-fruit maturity is preceded by production of ethylene in the tissues, which induces ripening of fruit and hastens abscission. Although these two functions of ethylene may be related, they are not the same, because the interactions of hormones and ethylene differ in the two situations. For example, 2,4-D treatment retards ethylene-induced abscission but actually causes an increase in ethylene production and subsequent ripening of pears (Hansen, 1946; Dewey and Uota, 1953). Insofar as abscission is concerned, ethylene apparently activates the enzymes that dissolve the middle lamella. NAA and 2,4-D, on the other hand, probably directly inhibit one or more enzyme systems and thus prevent abscission. It should be stressed that processes resulting in abscission are not reversible. Once the pectates have been dissolved and the fruit is loose and ready to drop, hormone sprays will not help. They must be applied before such processes are underway. And, even though the spray prevents dropping, it does not slow down the normal maturation of the fruit. Thus fruit is sometimes left on the tree too long and becomes overmature.

All of the synthetic regulators used to prevent preharvest drop are thought to function in the same way in preventing abscission, but some of the side effects are different. It was shown as early as 1939 that synthetic hormone sprays could delay abscission several days to several weeks in apples (Gardner, et al., 1939). Of the many hormones first tried, NAA and its derivatives were most effective. Recently other effective materials have been marketed, namely 2,4-D and 2,4,5-TP. Of these, 2,4,5-TP is superior because of its longer lasting action.

In general, trees should have healthy green leaves at the time of application because most of the hormone enters the plant through the leaves. Also, the spray should thoroughly cover all parts of the tree. Batjer and Thompson (1948) showed that the application of NAA to selected apple spurs prevented drop of fruits on those spurs but had practically no effect on adjacent untreated spurs. They also found that treating spur leaves was much more effective than treating the fruit.

Temperatures below 24°C during spraying produce less response than those of 24°C and above; however, a relatively cool period several days after the spray may prolong the effective period. Other weather factors may alter the leaf cuticle and thus affect hormone uptake. Trees preconditioned in high relative humidity or in red (rather than blue) light—simulating the light on cloudy days—appear to have thinner, less waxy cuticles, thus permitting more rapid penetration of externally applied hormones. Bukovac showed that the time of day at which sprays are applied seems to have no effect on hormone uptake. Since leaf stomata are open in the forenoon but closed in the afternoon, cuticular rather than stomatal entry is indicated. Re-wetting of the leaves by rain within three days of spraying tends to increase the uptake, as does relatively slow drying induced by high humidity after the spray.

A number of spray additives also influence the amount of hormone absorbed by the trees. Wetting agents such as Tween-20 and Spray Modifier increase absorption, while plastic emulsions such as Plyac or AC-33 reduce uptake. When NAA is mixed with summer oil (at 125 milliliters per 100 liters) stop-drop action is increased, but injury to foliage also is increased. Sprays to which lime or calcium have been added are much less effective than the straight hormone in water. Hard water particularly reduces NAA uptake. Adding hormones to wettable-powder pesticide sprays may also reduce stop-drop response (Westwood and Batjer, 1960).

Other factors influencing spray effectiveness are tree condition, timing, and spray concentration. Weak trees in poor vigor may not respond at all. Sprays applied too early tend to increase fruit ripening and may not last the full harvest season. Repeat sprays usually are not detrimental but also are not too effective. Within the range of effective concentrations, effectiveness increases as concentration increases. But one should never exceed the suggested concentration (or its equivalent per hectare rate) because injury to both tree and fruit may result. Spraying should begin about 7 days before earliest estimated picking maturity. The hormone NAA takes effect in apples and pears within 3 or 4 days and is effective for 3 or 4 weeks depending upon weather and tree condition; 2,4,5-TP takes effect in apples in 7 to 12 days and lasts 5 to 7 weeks. The following concentrations are suggested:

Chemical	Pears*		Apples	
	ppm	g/ha	ppm	g/ha
2,4-D	3	—	—	—
NAA	5	60	10	119
2,4,5-TP	—	—	20	—

*Bartlett and Bosc respond to treatment. Anjou is much less responsive.

Preventing Prune Drop

Italian and Early Italian prune tend to undergo considerable midsummer fruit drop. This appears to be caused by embryo abortion following normal pollination and early development. A spray of 5 to 20 ppm 2,4,5-TP two weeks after the beginning of pit hardening (determined by cutting thin slices across the tip of the fruit about June 1) will nearly eliminate this "blue drop." In some cases, yield has been doubled by the spray. In western Oregon, this spray is usually applied in mid-June for best results. It should be applied when less than a full crop of fruit has set initially. When oversetting occurs, the blue drop may provide some desirable fruit thinning.

Side Effects of Hormone Sprays

The most common side effects of hormones are twig and shoot injury and accelerated ripening or core breakdown of the fruit. Apples are much less sensitive than pears, but both will show these effects (*a*) if concentrations higher than those suggested are used, (*b*) if sprays are applied too early, or (*c*) if weather is hot. At 5 ppm, NAA does not hasten ripening of Bartlett pear, but at 10 ppm it increases both ripening and core breakdown. Likewise, 2,4-D at 2 to 3 ppm is satisfactory, but 5 ppm is too much. Bartlett is especially sensitive to 2,4,5-TP. A spray of 10 ppm in the fall results in early ripening and calyx-end breakdown of fruit the following year. This effect of hormones on ripening is reported to be worse on trees whose nitrogen levels are too high. The general effect of growth regulators on ripening appears to result from ethylene stimulation in the fruit. With 2,4,5-TP on prunes, dieback of shoots in the top of the tree and bud killing inside the canopy are sometimes reported, and fruit quality is occasionally reduced. These side effects can be diminished by using a spray of lower concentration than the usual 20 ppm.

Because of the stability of most synthetic growth regulators in trees, misuse can result in a buildup in the tissue, giving an undesirable carryover from year to year. In one case where 2,4-D weed killer drifted into a Bartlett pear orchard in early summer, severe fruit breakdown and shoot dieback occurred within a few weeks, and the following year the crop again was a total loss although the symptoms were milder. In this orchard it was 4 years before the symptoms disappeared even though no further hormone was applied. For this reason the recommended dosages should not be exceeded. Dosages are established to give the maximum beneficial effect while minimizing known adverse side effects.

Chemical Looseners

Tests on various fruits have shown salicylic acid, ascorbic acid, or iodoacetic acid to be active in promoting fruit abscission. At times, however, leaves are removed rather than fruit. Recently, a new class of chemicals known as 2-haloethane phosphonic acids was made available. The commercial product (ethephon) contains about 16 percent 2 chloroethane phosphonic acid, 36 percent mono-2-chloroethyl ester of phosphonic acid, and 13 percent 2-chloroethane phosphonic anyhydride. These compounds affect abscission and other processes primarily by the release of ethylene directly in plant tissues. Many of the effects of ethephon can be duplicated by fumigation with ethylene gas. Abscission induced by ethephon promises to be of value in post-bloom chemical thinning, chemical loosening for mechanical harvesting, uniform early ripening of fruits (Unrath, 1973), and defoliation of nursery stock to aid early digging.

Studies in Michigan indicated that ethephon is more effective than other chemicals in reducing the fruit removal force (FRF) needed to separate Montmorency fruit from their stems 10 to 15 days after spraying. Loosening occurred at the fruit end of the stem rather than at the spur. Napoleon cherries in Oregon were sprayed with ethephon and pull tests made 8 days later (Table 12-1). Both concentrations significantly reduced the force needed to separate the fruit from the stem. Here also loosening occurred entirely at the fruit:stem abscission zone. Treated fruit appeared to be slightly more mature than controls.

Early post-bloom fruit drop occurs by abscission at the stem:spur zone, while at maturity all abscission is at the fruit:stem zone (Table 12-2). The same situation seems to hold for other stone fruits. Thus, the use of ethephon to loosen stone fruits for mechanical harvest will result in a high proportion of fruits removed without stems. Although this is desirable for canning, it prevents "stem-on" harvest for the premium-grade brining cherry.

Table 12-1 Effect of ethephon sprays 8 days before harvest on fruit maturity and stem loosening of Napoleon cherry.

Ethephon Concentration (ppm)	Firmness Test (00-type durometer)	Soluble Solids (%)	Fruit Removal Force	
			Fruit:Stem (g)	Stem:Spur[1] (g)
0	75	18.4	583	1117
250	74	19.4	306	1044
500	71	19.2	312	1131

[1]These separated with part of the spur remaining on the stem and thus did not form an abscission layer at that point.

Table 12-2 Abscission of Napoleon cherries 40 days after full bloom and at maturity as related to the two abscission sites.

Stage of Fruit Development	Place of Separation		
	Stem:Spur (%)	Fruit:Stem (%)	Neither Zone[1] (%)
Immature (40 days A.F.B.)			
Sound fruit	20	20	60
Discolored fruit	90	0	10
Mature	0	100	0

SOURCE: Westwood, 1968.
[1]The only possible abscission zones are at the stem:spur or the fruit:stem connection points.

Ethephon treatment causes the husks of pecans and walnuts to loosen and split off. This effect is found both on the tree and off. Recently, it was learned that ethephon also matures the filbert husk, thus accelerating release of the nut, which is held by the husk for some time after the nut is mature. Sprays of 750–1000 ppm ethephon to filberts the third week in August causes about 90 percent nut drop in 3 weeks. This then permits early pickup harvesters to get the crop in ahead of rainy weather (Lagerstedt, 1972a).

HAND HARVEST

The correct way to pick fruits is by lifting up with a slight twisting motion rather than pulling down or straight away from the spur. Pulling down results in many fruits being removed without stems, and it is more difficult to pick the fruit (Fig. 12-1). Proper

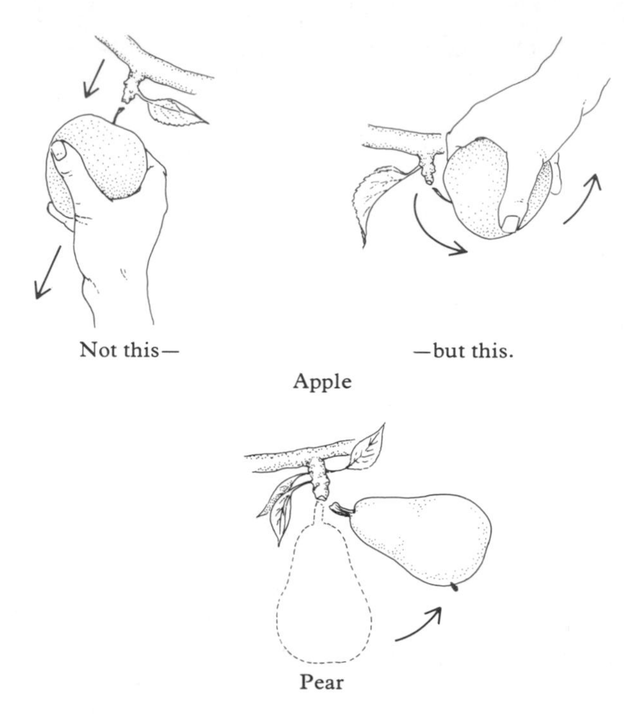

Figure 12-1 Hand picking of apples and pears should be a lifting-twisting motion rather than a straightaway pull. This principle holds generally for many other species as well. [After Stebbins and Bluhm, 1975]

picking generally requires careful handling at every step to prevent bruising:

1. Grasp the fruit gently so as not to cause finger marks.

2. Place the fruit gently in the picking bag or container. Never drop it in.

3. Empty fruit from the picking bag or bucket with an easy rolling motion to prevent fruit from bumping against one another.

Proper picking and handling will prevent stem pulls, skin punctures, and bruising—all of which can ruin a good fruit. The softer fruits require more careful handling than the firmer ones to avoid bruising, but firm fruit tend to get skin punctures more readily than softer types.

MECHANICAL HARVEST

Shake-and-Catch System

This system employs a tree or limb shaker and a catching frame or canvas under the tree. With prune cultivars that ripen fully on the tree, the shake-and-catch system is satisfactory. Italian prunes for the fresh market might be harvested with this system, provided trees are well pruned and good equipment is used. It is the standard method of harvest for drying or canning.

Tree shaking for harvest—particularly with walnuts, almonds, and prunes—dates back to the 1930's in California. Today, it is also used with peaches and apricots. In the eastern states and to a lesser extent in the West, apples for processing are being harvested with shakers. Experiments with harvesting Napoleon sweet cherries for brining have been conducted in several states. However, New York workers report that preliminary results were not promising. Fruit removal with Windsor, their principal cultivar, was 62 percent at the most advanced maturity, and increased shaking force removed 80 percent, but a large number of spurs were also removed. Results in Michigan were better. Almost all Montmorency cherries there are now harvested by this method. Also orchardists in the Stockton, California area began commercial mechanical harvest of sweet cherries for brining in 1965, following successful tests started in 1961. Tests on mechanical harvesting of Napoleon cherries in The Dalles, Oregon began in 1966, but only for brining cherries.

SPECIAL PRUNING Trees in the Stockton area are top-worked on Mahaleb frames and pruned to be comparatively stiff and upright, but those in the Oregon test were unpruned, limber and willowy. Pruning for stiffness would result in much less fruit injury and greater removal.

The distance from the ground to the lowest limb on the trunk is referred to as the "height of head." Most catching frames can now seal around the trunk at or near the ground, and the height of head need be no more than 30 to 60 centimeters. However, trunk shakers work much better when they can be operated at heights of 60 to 90 centimeters above the ground. A limb shaker can be used on the trunk when the trees are young. For shaking then, a height of head of 60 to 90 centimeters is desirable.

For limb shaking, a tree should have three or four main scaffold branches. Secondary scaffold branches should not arise closer than 45 to 60 centimeters from the trunk. The limbs should be as upright as possible. This allows the shaker operator: (1) to make only three or four attachments, (2) to avoid bark injury (horizontal limbs are more easily skinned), and (3) the operator can see the place to clamp easily, thus increasing his speed. Trees should be pruned for stiffness to increase shaking efficiency.

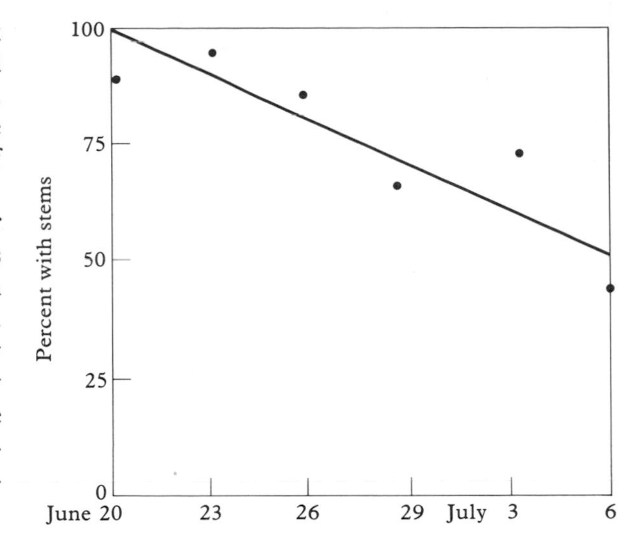

Figure 12-2 Percent of machine-harvested sweet cherries having their stems attached. [After Stebbins, 1967]

SHAKING CHERRIES In a study by Stebbins (1967) on harvesting Napoleon cherries for brining (see Chapter 13), the Roberts catching frame was used with a Wagco limb shaker mounted on the catching frame. The frame is tractor-powered and is a one-piece unit with permanently stretched canvas on the conveyor side and collapsible wings which fold around the tree on the other side. Machine-harvested fruit was brined in the field immediately after harvest in a standard brine of sulfur dioxide and lime in water.

Figures 12-2 to 12-5 (from Stebbins, 1967) indicate the results of these tests. The average percent of cherries machine-harvested with stems attached decreased as fruits matured during the harvest period (Fig. 12-2). Although machine harvest caused slightly more injury (Fig. 12-3), this did not markedly affect grade after brining. The increase in brown color on the last three picking dates was the primary reason for loss of grade on those dates (Fig. 12-4). The percentage of hand-picked fruit in the best grades (1 and 2) changed very little, but that of machine-harvested fruit gradually decreased as the fruit became more mature (Fig. 12-5).

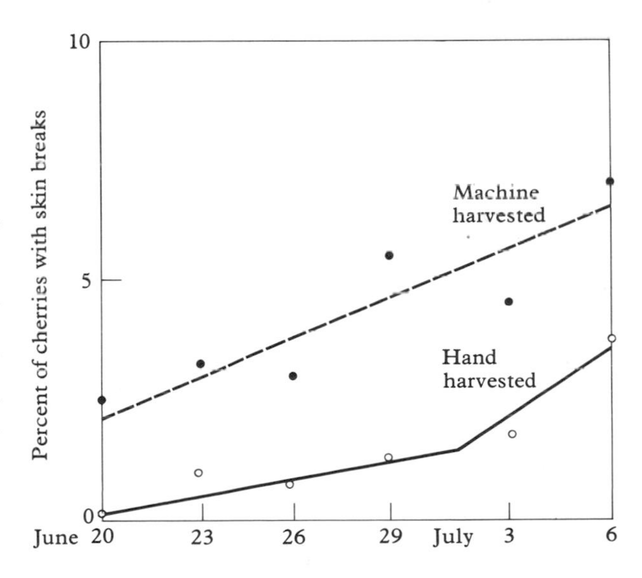

Figure 12-3 Percent of sweet cherries with skin breaks due to mechanical injury after hand or machine harvest. [After Stebbins, 1967]

Loss in grade due to mechanical harvesting can be held down by minimizing (*a*) fruit damage in the orchard and (*b*) time elapsed from shaking to brining. Brown discoloration, the main source of grade reduc-

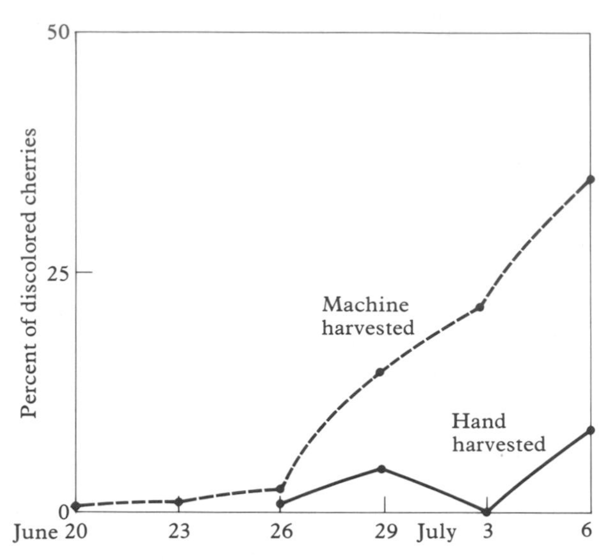

Figure 12-4 Percent of sweet cherries with enough brown discoloration to reduce grade after hand or machine harvest. [After Stebbins, 1967]

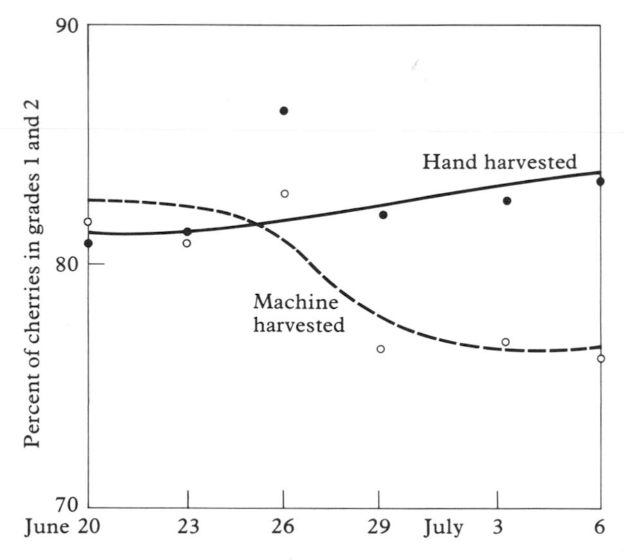

Figure 12-5 Percent of sweet cherries in combination grades 1 and 2 when brined after hand or machine harvest. [After Stebbins, 1967]

tion, results from the oxidation of bruised portions of the fruit. Immediate brining in the orchard reduces oxidative browning.

TYPES OF SHAKERS Basically two types of tree shakers are available: trunk shakers and limb shakers. Trunk shakers have the advantage of a single hookup, with the frame and shaker system operated by one man, and are easy to operate. But many trees are not trained for trunk shaking, and there is thus a greater power requirement. Limb shakers are adapted to more tree shapes and require less power, but the multiple hookup is slower and requires more labor and more operator skill. Shaker power requirements vary with the kind of fruit and the tree size. Prunes are easily removed with a low-powered shaker. Walnut and pecan trees are much larger and require a more powerful shaker. Long-boomed limb shakers are most often used on walnuts because power requirements are less when shaking limbs fairly high in the tree. Most trunk shakers provide an orbital shaking pattern. The size of weights and pulleys in the shaking head are increased to handle larger trees.

The design of the clamp is important, particularly when much shaking force is to be applied. A C-clamp is much preferable to the old pincers-type clamp for limb shakers (Fig. 12-6). The newer clamp designs have larger pads and are more flexible so that they do less injury to trunks or limbs. Trunk damage is minimal with well-designed clamps, good closure-pressure regulation, well-pruned trees, and careful operation. Originally, most trunk shakers had insufficient padding of clamps to prevent damage even with skilled operation. Recent designs of trunk shakers include pillows filled with air, sand, or ground walnut shells at the pressure points inside the clamp, with a rubber cover which slips against the pillow and prevents bark damage.

CATCHING FRAMES Catching frames cost between $3000 and $40,000. Within this range many features are offered. The cheapest type of "frame" is simply a conveyor with a canvas that can be rolled out under the tree (Fig. 12-7). Hydraulic power is supplied by the tractor. The canvas is attached to ropes which go

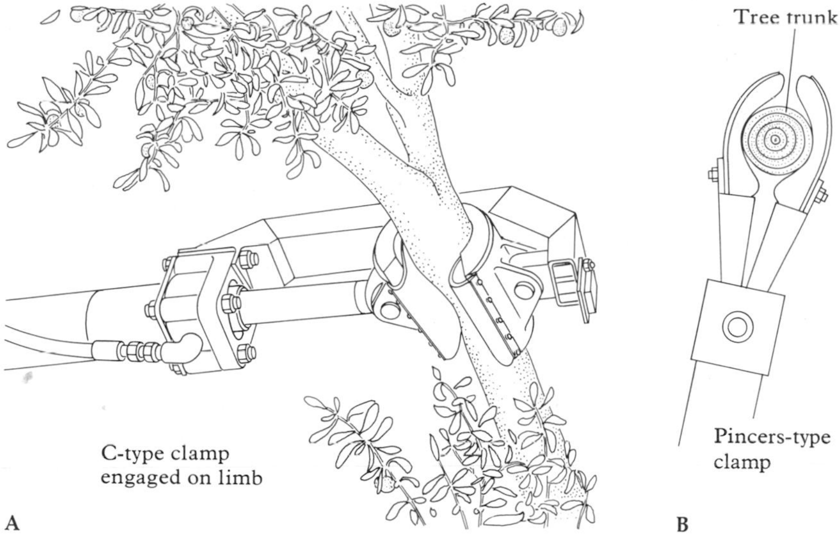

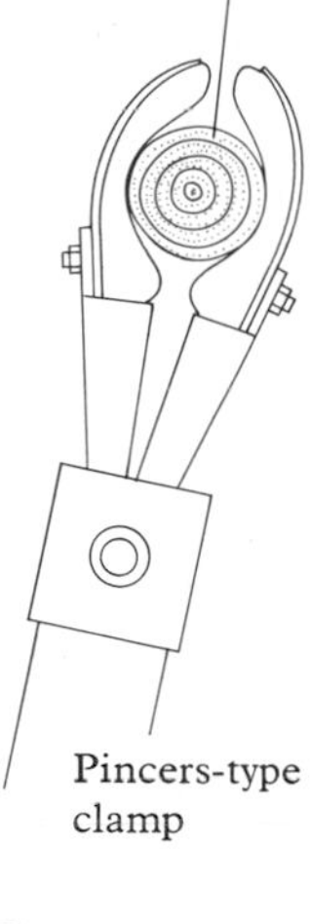

Figure 12-6 Two general types of clamps used to attach to trunks for harvest shaking. **(A)** The preferred C-clamp engaged on a limb. **(B)** The less-desirable pincers-type clamp engaged on a tree trunk. [A: After Fridley and Adrian, 1966]

Figure 12-7 Roll-out canvas in place to receive the fruit shaken from the tree. Reel at right draws fruit to conveyor, which in turn delivers it to bulk container. [Courtesy of R. B. Fridley, University of California, Davis]

to a roll-up bar, which draws the fruit that fall on the canvas onto the conveyor. A helpful modification includes a pan that hinges down to the ground on one side. The fruit is lifted onto the conveyor by the pan rather than coming up in the canvas where it is damaged. A shaker can be mounted on the conveyor portion and powered hydraulically by the tractor.

Stretched canvas frames cost more originally but save considerable labor. Some employ two separate, self-powered halves with a trunk shaker mounted. One-piece frames are available either self-powered or towed. A special Saran-type material for catching frames is found to reduce sour-cherry "scalding" in Michigan. Design features are changing every year and are too numerous to mention here. One model is shown in Figure 12-8.

Pick-Up Harvest Systems

With a pick-up system, the crop is either shaken to the ground or allowed to fall and then is windrowed and picked up with a pick-up machine. The main steps of this operation include: (1) ground preparation, (2) shaking, (3) sweeping into rows (windrowing), (4) pick-up, and (5) cleaning. Some cleaning of fruit with air jets can be done during pick-up; additional cleaning outside the orchard is usually required. This system is not suitable for most fruits for fresh market but is used for walnuts, pecans, filberts, almonds, pistachio, and cider apples in Europe (Fig. 12-9). Contamination with soil organisms, such as those producing aflotoxins, and with colon bacteria are problems with this method of harvest.

Figure 12-8 Catching frame around a cherry tree being harvested with a trunk shaker. [Courtesy of R. L. Stebbins]

For pick-up operations, one must be able to drive under the trees for ground preparation and windrowing operations. For filberts, herbicide-treated strips in the tree row make cross-travel unnecessary. Weed control by flail mowing rather than cultivation has become popular since the adoption of the herbicide-treated strip. Flail mowing greatly reduces the time required for preparation of the orchard floor for harvest. Because windrowing (as shown in Fig. 12-9) is done along the tree row, pollinizers which cannot be mixed with the main cultivar should be planted in separate rows so they can easily be kept segregated at harvest.

Standards of evenness and firmness of ground must be met for pick-up operations to be successful. Small undulations, dirt clods, or soft ground result in loss of yield or a dirty product. Mole, gopher, and mouse tunneling, and earthworm castings, are problems in pick-up operations. Usually, about twice as many earthworms live in the top few inches of untilled soil as in tilled soil. Some soil types are so sandy or silty that they cannot be rolled firm enough for pick-up operations. Other soils contain enough clay that they cannot be worked fine enough. Fortunately, the majority of soils can be worked into a condition suitable for mechanical pick-up. With filberts, the blank nuts fall before the good ones, but blank pistachio nuts fall later. In prunes, some of the summer drop may still be on the ground. The orchard must be flailed, dragged, or floated and rolled one last time before harvest to force the blanks or summer drop down into the soil—and out of reach of the pick-up equipment. Most prune orchards are now harvested by trunk shaker and catching frame, so that orchard floor preparation is no longer needed.

Leaves that fall prior to harvest are a major problem in pick-up operations. With prunes, walnuts, and filberts, leaf fall is earlier in orchards deficient in potassium. With prunes and walnuts, boron deficiency causes early leaf fall. A heavy application of nitrogen can induce a deficiency of boron or potassium, thus increasing leaf fall. Heavier nitrogen fertilization, with maintenance of sufficient boron and potassium usually keeps the leaves on the trees longer. Drought or excessively deep cultivation, as well as injury from insects such as mites or aphids, increases early leaf fall. Many leaves can be separated from the crop before pick-up operations begin. The usual method is to use a fan to blow them out of the orchard or into an already-harvested row. This is widely practiced with walnuts in California but will work for other crops provided the leaves are not too wet. A simple blower can be made with a wheel-barrow chassis, a 4 or 5 horsepower air-cooled gasoline engine, and a small airplane propellor. At present in California the blower is mounted on the pick-up machine. Before the pick-up operation, nuts are windrowed into one or two rows from about .6 to 1.2 meters wide between tree rows. They may be windrowed again for a final clean-up of late-falling nuts and those missed the first time. Rains will bring the last filberts down but will also create muddy conditions for the pick-up operation. Either a helicopter or a ground-level blower will help in removal of the last of the crop so that harvest can begin before the fall rains begin. A little rain before harvest can be helpful in reducing dust. (See the earlier discussion of the use of ethephon to cause early nut drop.)

Sweepers for nuts vary in size and cost. Growers with large acreages generally favor the riding sweepers over the walk-behind kind. Small power sweepers for working close to trees are available. Most growers have leaf-blowers attached to the sweepers, but some still use hand rakes to remove the nuts from around the tree trunks before sweeping. Mud collects on the canvas draper in front of the sweeper but is no problem when a polyethylene plastic draper is used. There

A

B

Figure 12-9 **(A)** Sweeper, which places fallen filbert nuts in windrows. **(B)** Pickup filbert harvester, which uses an airstream to eliminate leaves and husks from the nuts. [A: Courtesy of H. B. Lagerstedt; B: Courtesy of R. L. Stebbins]

are at least two different basic designs for sweepers: (1) the reel type and (2) the kind with revolving bars with metal tines. The reel type usually has rubber or fabric paddles.

Walnuts, almonds, pecans, and pistachios are shaken from the trees before pick-up harvesting. Walnuts and almonds come down in the hulls during early harvest and must be run through a hulling machine. Mud is the main problem with walnut harvest in western Washington and Oregon because the nuts mature so late. For mechanical harvesting, there is considerable advantage to planting earlier-maturing cultivars. The nuts can be shaken from the trees when the hulls are easily separated from the nuts but before many of the nuts have fallen from the tree. If Franquette is harvested too early, the shells after hulling will have a reddish cast. When the nuts can be hulled without showing red-shell, harvest can commence (provided a huller is available). Earlier harvest means fewer leaves to cope with. If walnuts are allowed to stay on the ground for more than a few days, particularly in wet weather, the kernel color will darken from the phenols in the pellicle, reducing the grade. Basically, mechanical harvest is about the same for walnut, pecan, filbert, almond, and pistachio. Large boom-type limb shakers are used for walnut and pecan because of the large size of the trees. With the smaller trees of almond and pistachio, trunk shakers are adequate.

PICK-UP EQUIPMENT There are three basic types of pick-up machines: (1) the reel-type, (2) the belt and paddle-type, and (3) suction machines. With some harvesters, the paddles travel about twice as fast as the belt. All types of harvesters can be used, but the reel-type is the most commonly used. Since prunes cannot be windrowed without excessive damage, the pick-up attachment is flanked on either side by rubber augers that move the fruit in front of the belts. At present, most of the prunes are harvested by shake-and-catch equipment. The suction-type machines have not been successful for filberts, but have been used for apples in Europe. The reel-type harvester is about twice as fast as the belt-and-chain type. If the ground is not well leveled, the reel-type harvester may leave more nuts than the belt-type, since the reels cover a wider area than the belts. The belt-type harvester can be designed so as to follow the contour of the ground very closely, but nuts must be placed in a narrower windrow than for the reel-type. Both machines tend to pick up everything, including leaves, twigs, and nuts. Mud is more of a problem with the chain-and-paddle type harvester. Some problems with pick-up operations include: (1) getting around in tight places with bulky equipment, (2) leaves and mud in machine, and (3) grass and chickweed preventing good pick up.

CLEANING EQUIPMENT ON HARVESTERS Perhaps the greatest problem in mechanical harvesting is getting a clean product. Processors and buyers complain about "clean away" or "wash away" and usually penalize the grower when it reaches a certain point. Probably one of the cheapest places to clean the product is in the field. With prunes, all that is required is an adequate leaf blower on the conveyor that loads the bin.

Nuts present a more difficult cleaning problem. Nuts must be separate from clods, hulls, blank nuts, leaves, rocks, and twigs. Cleaning devices include: (1) blowers or suction cleaners, (2) revolving rollers, (3) a revolving drum or "squirrel cage," and (4) revolving tines or brushes. One of the best ways to separate leaves and nuts is by blowing as they fall from the end of a conveyor into a bin or another cleaning device. Hulling cylinders have been developed for removing hulls. The squirrel cage helps remove leaves, clods, and small sticks but is too slow for many operations. Corn-husker rollers in the bottom of a cleaning bin pull leaves and twigs down between them and send the nuts to the rear. They work better when the material is either completely dry or when it is kept moist by application of water. Separation of wet leaves and hulls from walnuts with rollers has been successful.

All of this cleaning equipment must operate rapidly enough to keep ahead of the pick-up section—otherwise, either the whole operation is slowed down or the nuts are not thoroughly cleaned. Problems encountered in the cleaning process include: (1) hard clods the same size as nuts, (2) mud-clogged conveyors, (3) wet leaves that won't blow away, (4) sticks and trash that jam equipment, and (5) the slowness of the whole process. Partial cleaning in the field followed by more thorough cleaning after harvest may offer a more economical overall solution than attempts to do the whole job in the orchard.

Bulk Handling

With rapid mechanical harvesting, the old methods of handling—which employ small, hand-carried containers—will not suffice. There is no point in having harvesting and cleaning equipment capable of harvesting more than a ton of nuts or up to 8 tons of prunes per hour if excessive labor is required to load and transport the product from the orchard. Handling must be mechanized also. The orchard bin, the picking platform, and the bulk trailer provide some answers to this problem (Fig. 12-10). When bins are carried on the harvester or are towed behind, a fork-lift unit is required in the orchard to move the bins out to orchard's edge and to load them onto a truck or trailer. Straddle trailers greatly reduce loading and unloading time. They are available in various sizes at various prices. One man pulling a straddle trailer can load, transport, and unload eight 1.2 × 1.2 meter bins by himself. The trailer can be pulled by a jeep or an ordinary pickup truck. The straddle trailer requires hydraulic power to lift the bins from the 10 × 10 centimeter planks on which they are stacked. A fork lift can be attached to almost any orchard tractor. Bulk trailers towed behind the harvester might be more practical for nut crops, since smaller quantities are harvested per hour and no fork lifts are required. Of course, the length of the haul and the conditions in the orchard greatly influence the type of equipment that will be required.

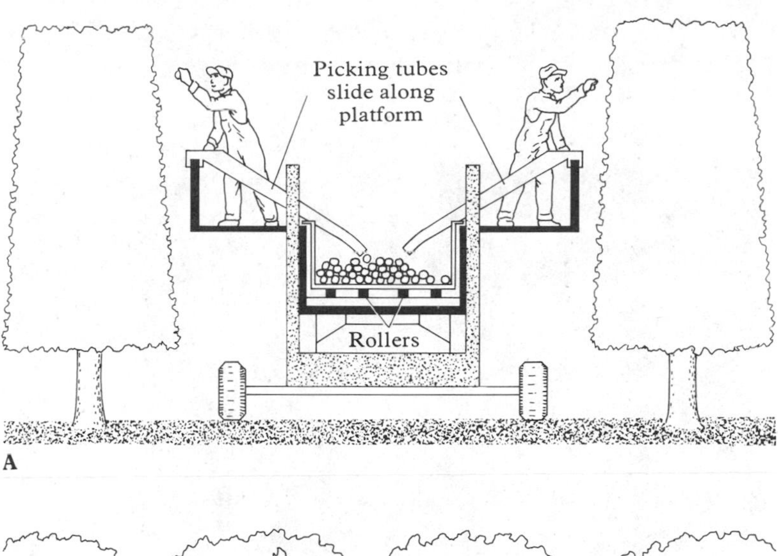

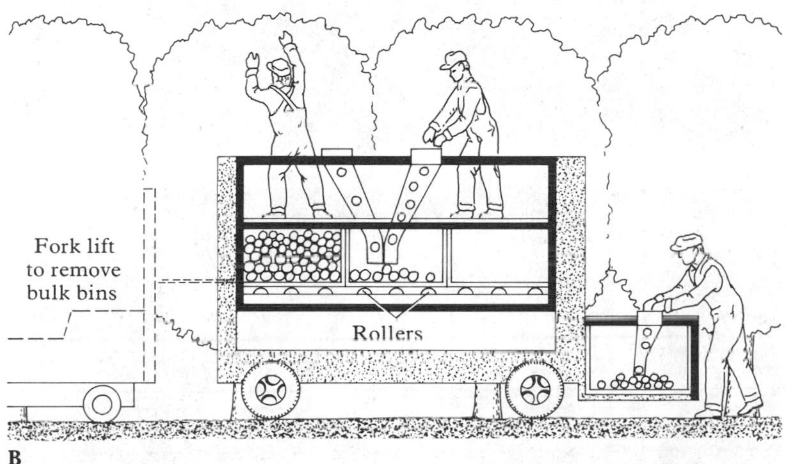

Figure 12-10 **(A, B)** Two cross-sectional views of a harvest platform with bulk containers. The platform is used in hand harvesting fruit from treewalls. The fruit on the lower half of the treewall are picked from the ground, and the remainder from a single-level platform. Picking tubes replace the picking bag to permit greater freedom of movement for the picker. Elimination of both ladders and picking bags increases picker efficiency.

Problems With Equipment

Growers who have converted to mechanical harvesting have found that shutdowns are costly. When one depends on a machine for harvest in a short period of time, any mechanical failure is serious. Air-cooled engines don't last long under dusty conditions. For this and other reasons, more and more equipment manufacturers are making use of hydraulic motors—driven by the power take-off from the main engine—

A

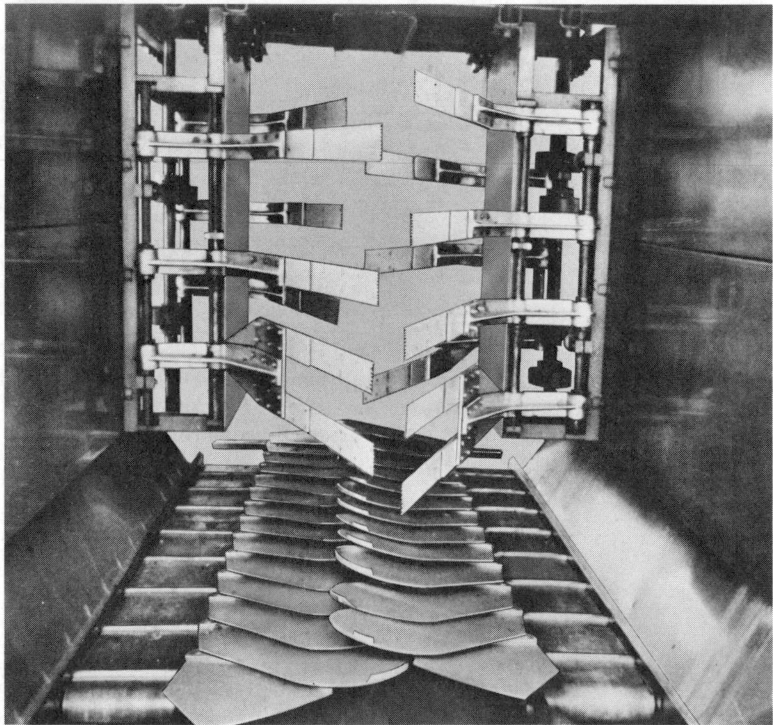

B

to operate fans, lifts, and other equipment. Hydraulic motors are permanently sealed and free of dust. They can be run with power from a single, large, well-protected engine, such as a tractor.

Here are a few other important points: Drive-belt slippage and wear is another common problem; use of parallel belts provides extra reliability. Modern sealed bearings are dust-free and require little maintenance. Mud in fan housings overloads the fan and reduces efficiency—it is advisable to carry a water supply for cleaning purposes. A well-pruned orchard will normally have fewer sticks to catch in machinery. In the first year or two of mechanical shaking, the stick and leaf problems are likely to be much more severe than in later years.

Over-Row Harvesters

The continuously moving over-row harvester is used for harvesting grapes, blueberries, raspberries, blackberries, and might also be used for dwarf tree fruits. The harvester straddles the row, and fruit is dislodged by two sets of vibrating tines mounted on rotating drums, whose speed is synchronized with the machine. Likewise, the catching belts at the bottom move along both sides of the row and catch the fruit as it falls (Figs. 12-11 and 12-12). Berries that naturally loosen as they are mature, such as blackberry and raspberry, are often more uniform in maturity with mechanical harvest than when harvested by hand. Hand pickers tend to pick any berries which are the color of mature berries, even though they are not yet loose on the stem. Grapes are machine-harvested for processing but not for the fresh market. Clusters tend to break up and the berries tend to be damaged by crushing in the bulk containers, necessitating immediate processing.

Figure 12-11 **(A)** General view of the over-row mechanical harvester for grapes. **(B)** Rear view of the harvester, showing harvesting arms, overlapping movable plates for straddling posts and trunks, and collector conveyers. [Courtesy of Chisholm-Ryder Co., Inc., Niagara Falls, New York]

A

B

Figure 12-12 Over-row harvester for brambles. **(A)** Side view showing the rotating drums with vibrating fingers to dislodge the berries **(B)** Harvester in operation. [Courtesy F. J. Lawrence, USDA]

Combine Harvesters

Strawberry harvesters strip or clip the vines in a single harvest, so cultivars and cultural practices are selected to concentrate the maturity period. Maximum yield of good fruit is obtained when the once-over harvest is done during the second week of the ripening season. The basic unit is either a mower that removes the entire top of the vine or a set of mechanical fingers which comb the fruit from the plant. In both types, the second stage separates the fruit from leaves and debris and the third stage conveys the partially cleaned fruit to a container (Fig. 12-13).

Multilevel Harvesters

Large fruits, such as apple, pear, and peach sustain a great deal of damage when harvested with conventional shake-and-catch systems by striking tree branches or by banging against other fruit on the catching frame. Multilevel catching arms are now being used to reduce such damage (Figs. 12-14 and 12-15). These units are experimental but similar systems probably will be developed as commercial units.

Cranberry Harvesters

In addition to hand-scoop harvesting, various types of machines are used for cranberry (Figs. 12-16 and 12-17). The dry harvester rakes or combs the berries from the vine and deposits them on an inclined conveyor belt and hence into a container behind the machine. With flood harvesting, a reel-type machine moves through the bog, beating the berries off. They then float to the surface and are moved by the wind to

A

B

Figure 12-13 Mechanical strawberry harvester. **(A)** General view of the harvester. **(B)** Close-up of the unit that clips off both leaves and berries and conveys them into the machine, where an airstream separates the leaves and other debris from the fruit. Because this is a once-over harvester, and because it is a "clipper" rather than a "stripper," it is desirable to develop cultivars that concentrate fruit maturity into a short period and that bear fruit on strong upright stalks. [Courtesy Oregon State Extension Service]

A

B

Figure 12-14 Multiple-inflated-tube pear harvester with trunk shaker, adapted for narrow treewalls. **(A)** Overall front view, with solid collection surface (folded) at bottom right. **(B)** Harvesting arms inserted into a pear tree. When trunk is shaken, the fruit separate and roll down the inclined, inflated tubes. [From Fridley, 1974]

Figure 12-15 Rear view of a mechanical apple harvester with bin filler and unloader. Note the multilevel catching frame at left. Second operator's platform is at right of bin, and driver sits low on the left and ahead of the bin. [From Millier Rehkugler, Pellerin, Throop, and Bradley, 1975]

Figure 12-16 Cranberry dry harvesters. **(A)** Hand scoop used to comb off the berries. **(B)** Darlington dry picker. [Courtesy Oregon State Extension Service]

A

B

A

B

Figure 12-17 **(A)** Reel-type wet harvester, which beats the fruit from the vines. **(B)** The fruit are less dense than water and float to the surface where they are collected at one point and loaded in a truck with a conveyor belt. [Courtesy Oregon State Extension Service]

one corner, where a conveyor lifts them out of the water and into bulk containers. Harvesting with scoops or mechanical fingers is usually done in a single direction to minimize damage to the vines.

GENERAL REFERENCES

Am. Fruit Grower. 1969. Where we stand on mechanical harvesting of strawberries. 89:13-51.

Crandall, P.C., and J.E. George. 1967. Here's a harvest of mechanical caneberry harvesters. *Am. Fruit Grower* 87: 20-38.

Fridley, R.B., and P.A. Adrian. 1966. *Mechanical harvesting equipment for deciduous tree fruits.* Calif. Agr. Exp. Sta. Bull. 825.

Hedden, S.L.; H.P. Gaston; and J.H. Levin. 1959. Harvesting blueberries mechanically. *Michigan Quart. Bull.* 42:24.

Levin, J.H. 1969. Mechanical harvesting of food. *Science* 166:968-974.

Richardson, D.G.; D. Kirk; and R. Cain. 1975. Brining cherries mechanical harvesting experiments. *Proc. Oregon Hort. Soc.* 66:18-22.

13

Post-Harvest, Storage and Nutritional Value

POST-HARVEST PHYSIOLOGY

Harvested fruits and nuts are living things, using O_2 and stored substrates while giving off CO_2 in the process of respiration. The rate of respiration varies with stage of maturity, kind of fruit, temperature, chemical treatment, and composition of surrounding atmosphere. Before, as well as after harvest, fruits are undergoing a complex series of enzymatically controlled biochemical reactions, such as conversion of starch to sugars, changes in the form of sugar, use of sugars to respiration, decrease in organic acids, changes in pectic compounds, and production of volatile compounds.

Ripening

Ripening is the transformation of physiologically mature fruit from an unfavorable state of firmness, texture, color, flavor, and aroma to a more favorable state for consumption (Dilley, 1969). In many species ripening occurs before harvest—this is true of berries, stone fruits, nuts, figs, and grapes. In other species, such as pear, quince, late apples, avocado, and persimmon, ripening takes place largely or entirely after harvest.

A fruit is said to be physiologically mature when ripening can occur. It proceeds under the control of natural hormones whose stimuli bring about a

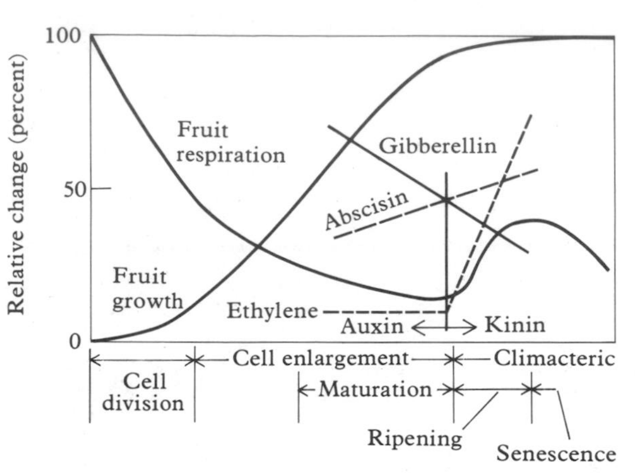

Figure 13-1 Hypothetical scheme for the hormonal control of fruit ripening via regulation of the endogenous ethylene concentration. The vertical line intersected by the gibberellin and abscission lines marks the onset of the ripening process. Prior to the intersection, gibberellic acid is dominant to abscisic acid. In this circumstance, a low rate of ethylene synthesis is maintained, and ripening is delayed. Beyond the intersection, abscisic acid is dominant to gibberellic acid; this results in a higher rate of ethylene synthesis, causing accumulation of an active level that induces ripening. Auxin is considered to enhance ripening by shortening the time required to reach the GA-ABA intersection, while kinin is considered to have the opposite effect. [After Dilley, 1969]

sequence of biochemical and physical changes—in aggregate called "ripening." One possible model for the natural ripening sequence is shown in Figure 13-1. Treatment with certain growth regulators can trigger the ripening of normally "immature" fruits.

ETHYLENE This plant hormone can be considered the principal ripening hormone for fruits, although ripening may not be controlled by ethylene alone. Ethylene without exception promotes ripening of physiologically receptive tissue (Abeles, 1973). When ethylene is removed from fruit by ventilation or hypobaric (low-pressure) storage, ripening is usually inhibited or delayed. As fruits age, they become increasingly sensitive to ethylene and will respond faster and at a lower concentration than when they were at a less mature stage. An increase of ethylene usually is followed by a rise in respiration in fruits, such as apple, that have a climacteric (sharp rise in respiration just before ripening), but exceptions to this rule are reported. Likewise, an increase in ethylene in nonclimacteric fruits, such as citrus, usually does not cause an increase in respiration, yet exceptions have been noted. Thus, there appears to be no uniform and predictable relationship between ethylene production and respiration (Abeles, 1973). Fruit still attached to the plant ripen more slowly than similar but detached fruit. It is now known that ethylene synthesis occurs at a more rapid rate after the fruit are detached. Attached fruit are under the influence of plant leaves, which apparently produce an ethylene inhibitor.

AUXINS When fruit have reached a certain physiological stage, treatment with an auxin (2,4-D; 2,4,5-TP; etc.) seems to stimulate maturation and ripening through stimulation of ethylene production. Early ripening by such treatment has been achieved with fig, apricot, pear, tomato, banana, and others. Treatment with ethylene or an ethylene generator, such as ethephon (CEPA), also will stimulate ripening.

GIBBERELLINS AND KININS GA generally delays ripening and reduces respiration; thus GA and ethylene seem to have opposing effects. Kinins tend to oppose some of the ripening reactions, but act more as anti-senescent agents than as general anti-ripening hormones.

ABSCISIC ACID ABA is a natural growth regulator that in many cases acts as a GA antagonist. Dilley (1969) suggests that as GA declines during maturation, ABA increases to a point at which the effective GA level is too low to inhibit the action of ethylene—thus ripening is initiated (Fig. 13-1).

Wang et al. (1972) recently showed that softening of pears was stimulated by a level of ethylene that was too low to cause the climacteric rise in respiration, showing that the climacteric rise is not essential to the softening process. Ripening is now considered to be an active process involving metabolic energy, hormones, enzymes, changes in membrane permeability, protein synthesis, changes in pectic substances, and production of volatile esters. As fruit mature, they become more sensitive to ethylene, so that a lower concentration will stimulate the ripening process. But higher amounts of ethylene applied early also can stimulate ripening.

RIPENING REACTIONS Ripening brings about physical as well as chemical changes. Softening results from changes in the pectic substances binding cells together, making them less firmly cemented. Sorbitol (D-glucitol) in some fruits is converted to fructose, the sweetest of the common sugars. Starch is converted to sugars, increasing the sweetness further. Stone fruits do not produce starch, thus much of their increase in sweetness comes from conversion of sorbitol to fructose and from the reduction in acid content, which reduces the sour taste present in unripe fruits. Polyphenolic compounds (tannins) decline during ripening, removing the astringent or bitter taste that they cause. Skin and flesh color often change during ripening, changing from greenish to yellow or orange. The red skin color of apples, pears, peaches, and apricots does not change after harvest, unless the fruit is exposed to direct light, but the blue or red color of plums and some other fruits increases during ripening in storage, even in the dark (Allen, 1932).

Many of the distinctive flavors of fruits come from the volatile esters produced during ripening. Each fruit type has a different set of flavor esters that, in the aggregate, we recognize as "apple" or "peach" flavor. These volatile constituents may be reduced or lost during long storage, so that freshly picked ripe fruit usually have a richer flavor than stored ones. Apples and particularly pears, however, if picked firm mature for long storage, do not at that time contain some of the flavor components produced by ripe fruit.

Pears may require special ripening for best quality. They ripen best and with more aromatic flavors at 18.3°–21.1°C than at either lower or higher temperatures. In fact, some cultivars will not ripen if the temperature is above 24°C. Since pears are stored in the firm mature state and require special ripening conditions for good quality, some shippers have preconditioned their fruit before sending it to the retailer. This is done by beginning the ripening process, then placing them in refrigerated cars before softening is advanced enough that bruising in handling would be a problem. At the retail market, the fruit are still firm but ripen normally under a variety of conditions to good eating quality. Once a fruit is eating ripe, its shelf life is very short but may be prolonged by refrigerated storage.

COLD STORAGE

Temperature profoundly affects respiration rate and thus the storage life of the fruit. Generally fruit stored at 15.6°C respires at a much higher rate than if stored at 0°C. In regard to heat evolution, fruit stored at 0°C releases only 10 to 20 percent as much heat as when it is stored at 15.6°C. Thus, low-temperature storage prolongs the effective life of the fruit many times over that of storage at room temperature. Increasing the CO_2 and decreasing the O_2 levels of the storage atmosphere also markedly reduce respiration and prolong storage life. This is of particular benefit to fruits that are injured by −1° to 0°C storage and must be stored at higher temperatures. The rate of ripening reactions is influenced by temperature and type of atmosphere. For example, picked apples left standing in the orchard at 21°C will ripen as much in three days as they would in a month at −1°C. Cold storage also reduces moisture loss and the growth of decay-producing organisms, which often infect fruits. Delays in cooling harvested fruit to −1° to 0°C storage temperatures often result in shorter useful storage life and increase certain types of physiologic disorders.

Specific Crops

The optimum storage temperature and the storage life of fruits varies with species and cultivar (Table 13-1). For best results, fruits should be harvested at

Table 13-1 Recommended storage conditions, storage life, heat evolution, and physical characteristics of deciduous fruits.

Commodity	Storage Temp. (°C)		Relative Humidity (%)	Approx. Storage Period (days)	Highest Freezing Point (°C)	Water Content (%)	Specific Heat[1] (k cal/kg/°C)	Heat evolved (k cal/m Ton/day)[1] when stored at:		
	Low	High						0°C	4°–5°C	20°–21°C
Apple	−1	4	90	90–240	−1.50	84.1	.87	139–250	306–444	1028–2139
Apricot	−.6	0	90	7–14	−1.06	85.4	.88	——	500–2306	3667–7639
Avocado	4	13	85–90	14–28	−0.28	65.4	.72	——	1222–1833	4500–21194
Berries										
Blackberry	−.6	0	90–95	2–3	−0.83	84.8	.88	1083–1194	1917–2500	9528–11778
Blueberry	−.6	0	90–95	14	−1.28	82.3	.86	139–639	556–750	3167–5333
Cranberry	2	4	90–95	60–120	−0.89	87.4	.90	167–194	250–278	667–1111
Currant	−.6	0	90–95	7–14	−1.00	84.7	.88			
Dewberry	−.6	0	90–95	2–3	−1.28	84.5	.88			
Elderberry	−.6	0	90–95	7–14	—	79.8	.84			
Gooseberry	−.6	0	90–95	14–28	−1.11	88.9	.91	417–528	750–833	
Loganberry	−.6	0	90–95	2–3	−1.28	83.0	.86			
Raspberry	−.6	0	90–95	2–3	−1.11	80.6	.85	1083–1528	1889–2361	
Strawberry	0		90–95	5–7	−0.78	89.9	.92	750–1083	1000–2028	6250–11972
Cherry (sour)	0		90–95	3–7	−1.67	83.7	.87	361–806	778–806	2389–3056
Cherry (sweet)	−1	−.6	90–95	14–21	−1.78	80.4	.84	250–333	583–861	1722–1944
Fig	−.6	0	85–90	7–10	−2.44	78.0	.82	——	667–806	3472–5806
Grape (vinifera)	−1	−.6	90–95	90–180	−2.17	81.6	.85	83–139	194–361	
Grape (American)	−.6	0	85	14–56	−1.28	81.9	.86	167	333	2000
Nectarine	−.6	0	90	14–28	−0.89	81.8	.85			
Peach	−.6	0	90	14–28	−0.94	89.1	.91	250–389	389–556	3611–6250
Pear	−2	−.6	90–95	60–210	−.156	82.7	.86	194–417	306–611	1833–4278
Persimmon (D. kaki)	−1		90	90–120	−2.17	78.2	.83		361	1222–1472
Plum and prune	−.6	0	90–95	14–28	−0.83	85.7	.89	111–194	250–556	1028–1583
Pomegranate	0		90	14–28	−3.00	82.3	.86			
Quince	−.6	0	90	60–90	−2.00	85.3	.88			

SOURCE: Lutz and Hardenburg, 1968.
[1]Specific heat; 1 BTU/lb./°F = 0.999 k cal/kg/°C.
Heat evolved; 1 BTU/ton of fruit/day = 0.27778 k cal/m ton/day.

the correct maturity and placed immediately in cold storage at high relative humidity. Some fruits are injured at −1° to 0°C storage even though they are not frozen. McIntosh and, in some instances, Jonathan and Yellow Newtown apples develop an internal browning (brown core) when stored at −1°C. This disorder is avoided by storage at 2°–4°C. Of course, respiration is increased and storage life decreased at this higher temperature. For this reason these cultivars are best stored in a controlled atmosphere (CA) of low O_2 (3 percent) and high CO_2 (5 percent) to prolong storage life and maintain quality. The main objective is to harvest the fruit and place it in optimum storage conditions for maximum storage life when it is neither undermature nor overmature and at the same time retains the ability to ripen at full quality. Quality after storage and proper ripening can never be better than quality when placed in storage. Poor-quality fruit placed in storage will be poorer quality when brought out. The best storage merely preserves the original quality.

FRUITS During and after storage, fruits continue to ripen. If not consumed or processed, they will become overripe; quality will deteriorate rapidly, terminating in the death of the cells. Two kinds of fruit are recognized because of their different ripening behavior. Nonclimacteric fruits, such as berries and cherries, have no appreciable lag time between maturation and ripening—there is a smooth flow of physiological processes with no clear distinction between the two. Climacteric fruits, such as apples and pears, have a steady-state period of some time between maturity and the sharp rise in respiration (the climacteric), which marks the beginning of ripening. Most climacteric fruits will ripen on the tree if left long enough, but better quality and longer storage life are obtained by harvesting before the climacteric. The avocado is an exception, however, in that both the climacteric and ripening do not occur as long as the fruit remains attached to the tree. Pears are peculiar too, because some cultivars (Anjou, Bosc, Comice, Eldorado, Packham's Triumph), when picked at a mature stage, will not ripen until stored at a low temperature (−1° to 2°C) for several weeks. Treatment with ethylene can cause ripening without cold storage, and studies show that with storage there is a gradual buildup of endogenous ethylene in the tissues. Apparently, cold storage brings about ethylene synthesis and makes the tissue sensitive to lower concentrations of ethylene. Storage also seems to activate or stimulate synthesis of enzymes required for normal ripening processes.

NUTS In shell nuts are unlike fruits because, even though they are alive, respiration and metabolic activity is much lower at the low moisture content of the dried nuts. Filberts are dried to an in-shell moisture content of 7–8 percent, and may be stored 14 months at 21°C. At −1° to 2°C and 65 percent relative humidity (RH) they can be stored for two years, and at −3° to −4°C for 4 years. Pecans and walnuts have a high oil content (60–70 percent) and thus are subject to rancidity during storage. Off flavors of rancid nuts are caused by oxidation or hydrolysis of the fats and oils in the nut to the free fatty acids. The first rancidity of unshelled nuts can be detected after 4 months at 21°C and after 2 years at 1°. Storage at this lower temperature is best at 75–80 percent RH and 3.5–6.0 percent moisture. Most nuts when fully dried can be stored for very long periods at −18°C without injury or loss of quality.

Chestnuts are the exception in that they are not dried before storage. They contain about 45 percent carbohydrates, 5 percent oil, and 50 percent moisture at harvest, and can be stored at 0° to 2°C for about 6 months. The main problems are mold and spoilage in storage due to the high moisture content, but if the nuts are dried, the kernels become hard and inedible.

Storage Systems

Common cold-storage units are built of several construction materials, but all units must be well insulated to keep outside heat from getting into the storage (Fig. 13-2). Insulation must have a low heat conductivity and also should be moisture resistant, easy to install, inexpensive, and should not contain volatile substances that might adversely affect the fruit. Some good insulating materials are corkboard, fiberglass, rock wool, and vermiculite—all of which pass .58 or less kilocalories per square meter per hour per centimeter of thickness per one-degree Celsius temperature difference between the outside and inside of the unit (Table 13-2). (This is equivalent to .3 $BTU/ft^2/hr/inch\ thickness/°F$.) A vapor barrier must be installed between the outside wall and the insulation to prevent moisture from moving in from the warm outside air and condensing in the insulation. Moisture here would reduce the insulating value and might cause some materials to deteriorate or rot. The vapor barrier should be made of vapor-proof building paper, sealed at the joints with odorless asphalt.

The refrigeration unit is simply a heat pump that draws energy out of the storage room and discharges it outside. It consists of two main parts: a condenser, and a cooling (evaporator) coil (Fig. 13-3). These two units are connected by pipes containing the refrigerant. During operation, the refrigerant (which is a gas at ordinary pressure) enters the condenser as a gas. It is compressed and condensed to a liquid, then cooled. It then returns by pipe to the cooling coil inside the storage room where it is released into the coil through an expansion valve. The release of pressure causes

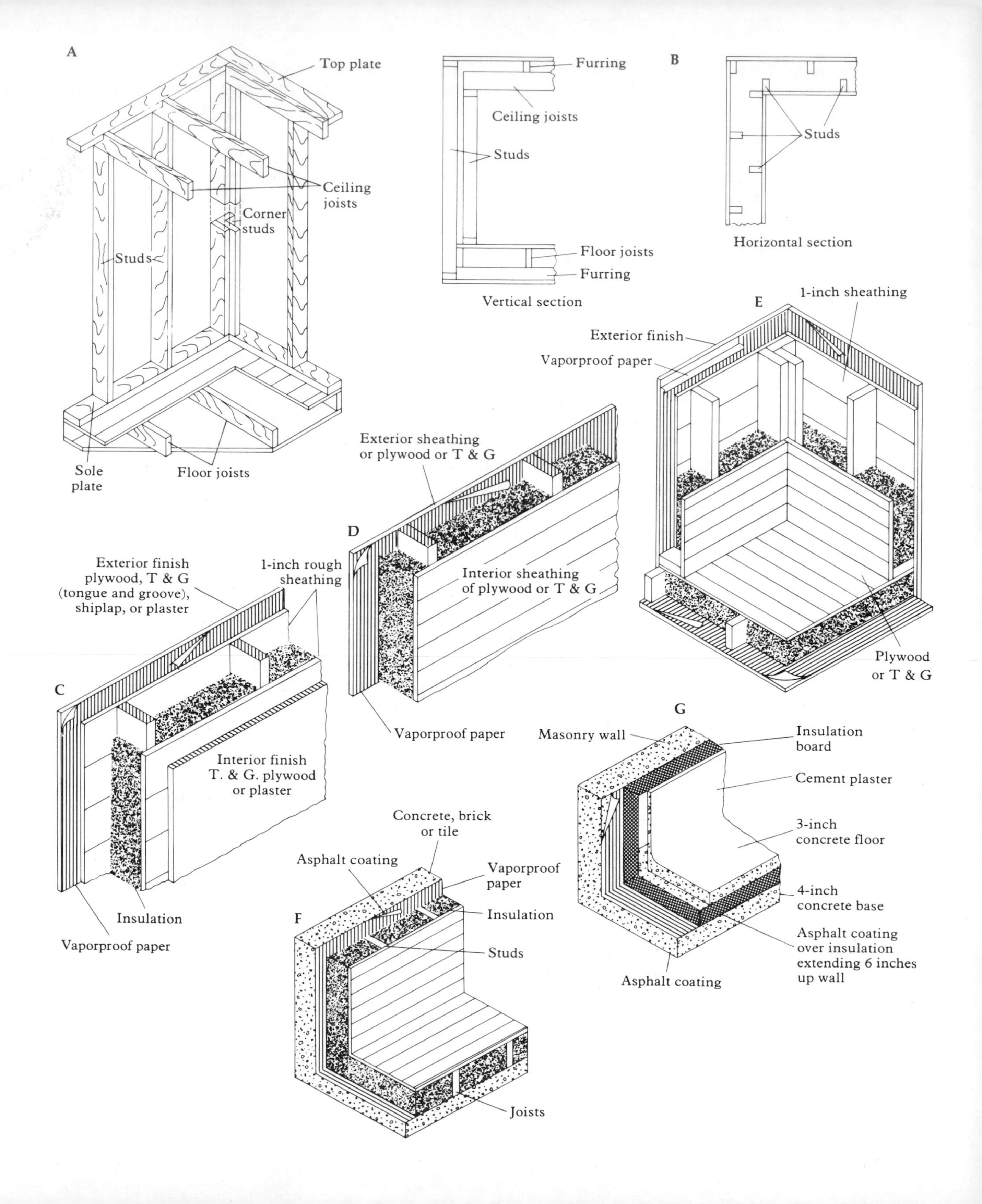

A
Top plate
Ceiling joists
Corner studs
Studs
Sole plate
Floor joists
Furring
Ceiling joists
Studs
Floor joists
Furring
Vertical section
B
Studs
Horizontal section
E
1-inch sheathing
Exterior finish
Vaporproof paper
Plywood or T & G
D
Exterior sheathing or plywood or T & G
Interior sheathing of plywood or T & G
Vaporproof paper
C
Exterior finish plywood, T & G (tongue and groove), shiplap, or plaster
1-inch rough sheathing
Interior finish T. & G. plywood or plaster
Insulation
Vaporproof paper
G
Masonry wall
Insulation board
Cement plaster
3-inch concrete floor
4-inch concrete base
Asphalt coating over insulation extending 6 inches up wall
Asphalt coating
F
Concrete, brick or tile
Asphalt coating
Vaporproof paper
Insulation
Studs
Joists

Figure 13-2 Details of different types of cold-storage construction showing important aspects of insulation and vapor-barrier placement. The following descriptions are taken from the original figure by Tavernetti (1948): **(A)** Solid-stud and joist framing. This is the simplest and commonest type of framing for insulation up to 6 inches thick. **(B)** Staggered-stud and crossed-joist framing, commonly used for insulation that is over 6 inches thick. With this type, practically all the wall, floor, and ceiling spaces have some insulating material within them. **(C)** Wall construction with double sheathing on both exterior and interior. **(D)** Wall construction with solid studs and single sheathing on both sides. **(E)** Wall and floor construction using solid studs with double exterior sheathing and single interior sheathing. **(F)** Wall and floor construction using concrete exterior and wood interior. **(G)** Wall and floor construction using concrete and board-type insulation. With both types of construction, the inside surface of the concrete should be sealed with asphalt before the vapor barrier is installed. [After Tavernetti, 1948]

Table 13-2 Conductivity of some insulating and structural materials.

Material	Conductivity[1]	Thickness (cm) to Equal 1 cm of Corkboard
Balsam wool	.52	.90
Balsa wood	.67	1.16
Brick	9.61	16.66
Corkboard	.58	1.00
Cork (granulated)	.60	1.04
Celotex	.63	1.09
Concrete	15.38	26.67
Cotton	.52	.90
Firtex	.63	1.09
Foamglass	.81	1.40
Fiberglass	.52	.90
Insulite	.63	1.09
Mineral wool	.52	.90
Rock wool	.52	.90
Rock cork	.65	1.13
Redwood bark fiber	.50	.87
Sawdust	.79	1.37
Shavings	.79	1.37
Vermiculite (expanded)	.54	.94
Wood	1.44	2.50

SOURCE: Tavernetti, 1948.
[1]Kilocalories per square meter per hour per centimeter of thickness per degree Celsius temperature difference between outside and inside.

the liquid refrigerant to evaporate into a gas. This requires that it absorb a large amount of heat energy to go from a liquid to a gas (heat of vaporization). The required heat is drawn from the room, causing it to become cold. The heat trapped in the gas refrigerant is then conveyed by pipe out of the room and back to the condenser in a closed system, where the cycle is repeated continuously until the desired temperature is attained. The motor that drives the compressor is turned on and off by a thermostat set at the desired temperature. Many refrigerants can be used, but the most common are NH_3, Freon 12 (CCl_2F_2), CH_3Cl, and SO_2. Fans are used to circulate the air inside the storage room so that an even temperature is maintained throughout. Cold spots may be found, however, where fruit might be frozen. Such cold or warm spots can be avoided by using baffles to change the direction of air movement in the room. Also, stacking patterns and container design are important in assuring uniform storage temperature.

CONTROLLED ATMOSPHERE Controlled-atmosphere (CA) storage requires that the storage room be airtight, so that a specific ratio of O_2 and CO_2 can be maintained. Ordinary air contains 78 percent N_2, 21 percent O_2, and .03 percent CO_2. The atmosphere in which apples are stored usually contains only 3 percent O_2 and up to 5 percent CO_2. As fruit is loaded into the CA room and sealed, fruit respiration uses O_2 and releases CO_2 and H_2O. Respiration is slowed down both by the lower O_2 level and the higher CO_2 level. Of particular note is the fact that high levels of CO_2 are inhibitory to ethylene action, and ethylene is necessary for ripening. In order to speed the adjustment of O_2 and CO_2 levels, nitrogen gas may be pumped into the unit, and O_2 reduced and CO_2 in-

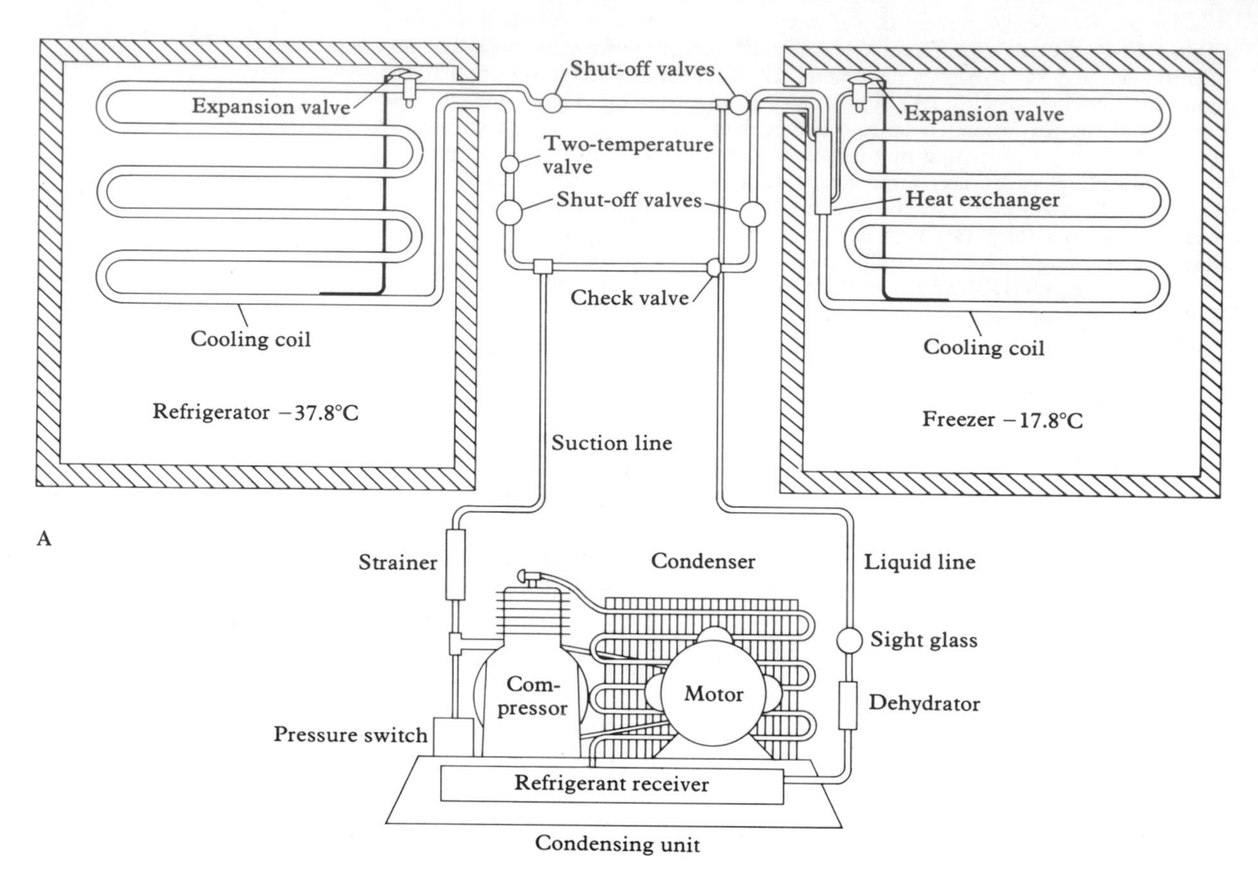

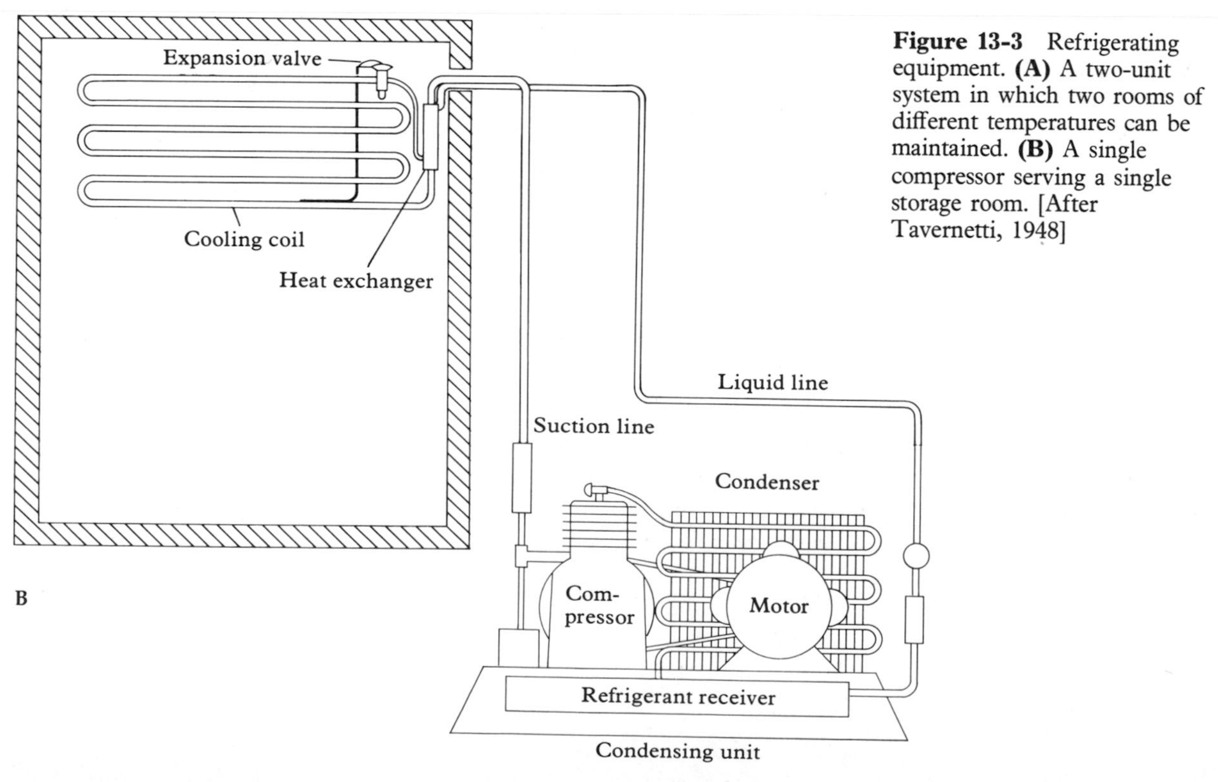

Figure 13-3 Refrigerating equipment. **(A)** A two-unit system in which two rooms of different temperatures can be maintained. **(B)** A single compressor serving a single storage room. [After Tavernetti, 1948]

creased by the use of special gas-fired burners in the unit. When the O_2 reaches the desired level, it is maintained there by controlled ventilation. Levels of CO_2 that are too high can be toxic to fruit, so the excess is removed by circulating the air through a sodium hydroxide solution or over sacks of lime which traps some of the CO_2. This type of storage is of greatest value in storing apples such as McIntosh, which cannot be stored at temperatures below 2.2°C, and in extending the life of winter pears (Bosc, Comice, and Anjou), which tend to lose the ability to ripen after long periods in regular cold storage. Cultivars such as Delicious apple, which can be stored at −1° to −0.5°C, have very low respiration rates and thus show very little difference when stored in CA at the same temperature (Olsen and Schomer, 1964). Larger differences reported by commercial operators are probably due to the fact that better-quality fruit is usually placed in CA. A recent innovation to prolong the storage and shelf life of strawberries and raspberries is to subject them to a high charge of CO_2 for a short time as they are being placed in cold storage prior to air shipment to market. This high initial charge of CO_2 is being used experimentally on apples and pears to reduce respiration during the cool-down period in storage.

STORAGE PROBLEMS The greatest demand on refrigeration equipment is during the load-in period at harvest. At first, the principal problem is removing field heat, after which the problem becomes keeping the temperature as uniform as possible. The refrigeration load is relatively light after field heat is removed —it consists of the fruit's heat of respiration, the heat entering through insulation and open doors, and the heat from power equipment and workmen in the room.

The time required to cool fruit to a desired temperature depends on the kind of fruit, the amount of field heat, and the cooling method, as shown in Figure 13-4. As shown, the most effective methods of rapid cooling are hydrocooling, forced-air cooling, and vacuum cooling. Arrangement of fruit in the storage room to permit circulation of air through the stacks also is important in rapid cool-down. Much of the potential storage life of fruit can be lost by rapid deterioration if cooling is too slow at the start (Mitchell et al., 1972). The necessary capacity of the refrigerating unit depends upon the kind of fruit to be stored, the size of the storage, and the efficiency of the system. Refrigeration capacity is measured in tons—that is, the amount of cooling done by one short ton of melting ice at 0°C during 24 hours. This is equal to 72,576 kilocalories of heat absorbed per day or 3024 kilocalories per hour (12,000 BTU/hour). Capacity for loading-in is considerably higher than for maintenance of a given temperature. For example, the refrigeration capacity required for the load-in period for pears coming in at the rate of 13.6 metric tons of fruit per day (300 metric tons total in 22 days) is about 14.4 tons of refrigeration. After the field heat is removed and the storage temperature is down to −1.1°C, this same amount of fruit can be maintained in storage with only about 3.4 tons of refrigeration (Lutz and Hardenburg, 1968). Because of their higher heats of respiration, such fruits as cherry, strawberry, raspberry, and blackberry require more refrigeration capacity (Table 13-1). Also, the heat from building leakage and other sources must be added to obtain the total refrigeration requirement for a given crop. Because of the reduction of refrigeration requirement after the initial pull-down, at least two compressors should be installed, one with about one-third of the total capacity needed. After the fruit has been cooled and the outside air temperature gets colder as winter approaches, the smaller compressor should be adequate to maintain storage temperature. Having two units with each storage also permits overhaul and maintenance without costly shutdowns.

QUALITY

Criteria

The foregoing data on storage and ripening are simply the ingredients used to develop and maintain acceptable quality as judged by the consumer. As with matu-

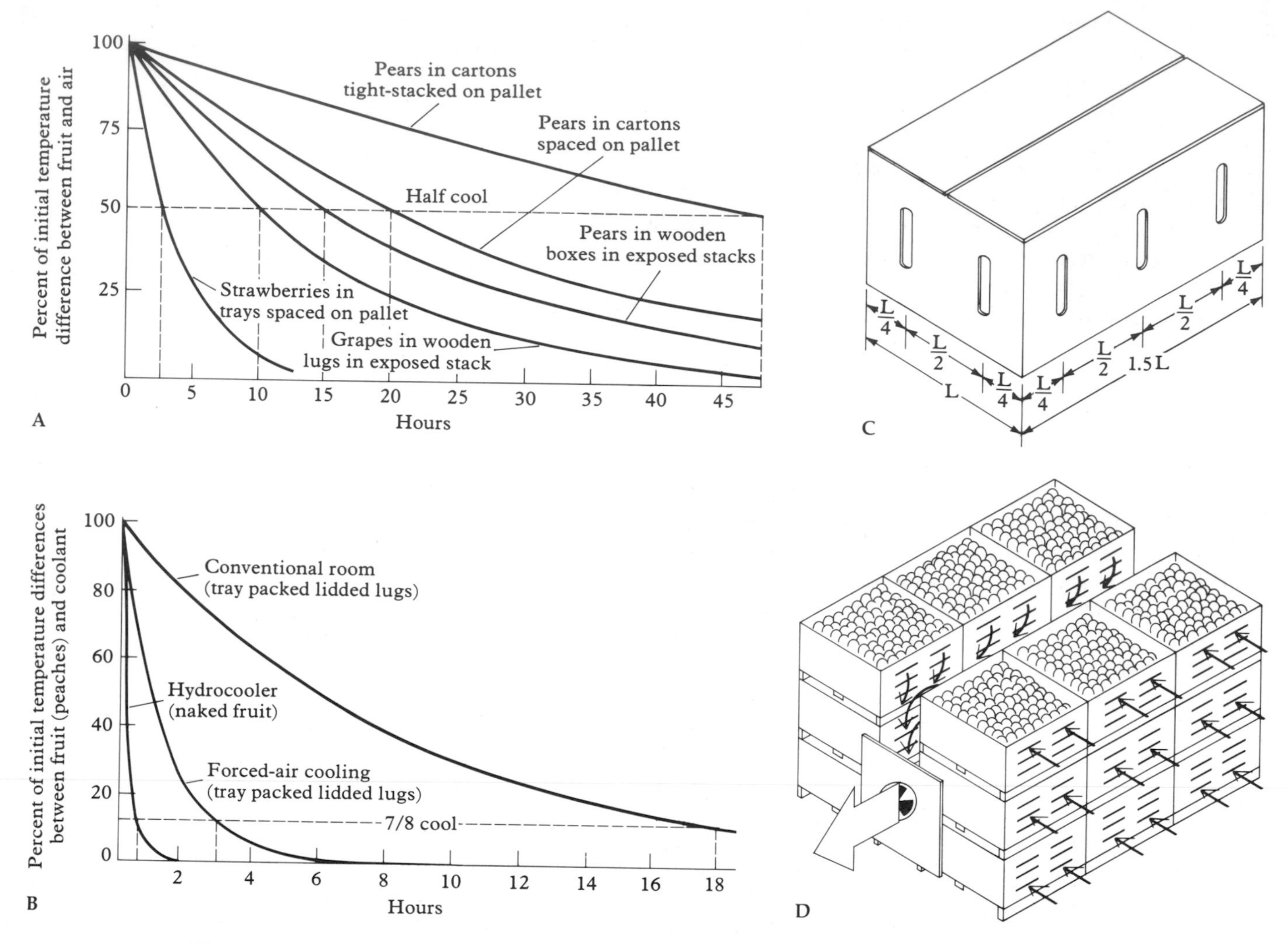

Figure 13-4 Methods of cooling fruit as it comes into storage from the field. **(A)** With average initial fruit-pulp temperatures, the half-cooling time ranged from 2 hours to 48 hours, depending upon the type of fruit and the kind of package. **(B)** A comparison of the speed of cooling peaches by different methods. **(C)** Recommended corrugated container venting pattern for room cooling. With dimensions of L × 15.5 L, and with vents centered at a distance of L/4 from corners and L/2 apart, vents will be aligned when containers are cross-stacked as shown in **D.** Air flow will then, with proper end baffles, be forced to pass through the containers and around the fruit as the exhaust fan pulls air out of the center chamber. [After Mitchell, Guillou, and Parsons, 1972]

rity, quality is based upon intended use. Fresh-fruit quality is affected by culture, climate, harvest date, and storage; in addition, it is affected by handling, transport to market, retail handling, and its acceptance by the ultimate consumer. Quality of fruit intended for processing may be based upon other criteria. In particular, the condition at the moment of processing and the unique characteristics of cultivars and species determine the quality for canning, freezing, drying, concentrating, or fermenting. The main factors relating to quality are sugar and organic-acid content; color; firmness; texture; juiciness; flavor; nutritional value; absence of diseases, disorders, or insects; and general appearance.

Quality Measurements

Chemical constituents such as pectins, vitamins, sugars, total solids, and acids are measured by routine laboratory instruments and chemical tests. Volatile esters (flavor essences) and other gases in the fruit, such as ethylene, CO_2, and O_2, are measured by the gas chromatograph and gas analyzer.

Physical measurements of quality factors include various color measures of skin and flesh with both reflectance and light-transmission meters and charts, devices to measure fresh or processed texture of fruit, and firmness (Fig. 13-5). Several special devices are available for the different fruits and nuts.

Finally, quality is measured organoleptically—that is by the use of the human senses of taste and smell. Special statistical methods have been used to get unbiased comparisons of selected fruit samples with a standard cultivar. Rating sheets for these tests are carefully prepared and samples are coded to eliminate possible bias. Organoleptic tests are the final tests done at the stage where the product, either fresh or processed, is normally consumed. Such tests are frequently performed by trained experts and are correlated with known consumer tastes.

Physiological Disorders

POME FRUITS Physiological disorders (see Fig. 13-6) of fresh fruit in contrast to diseases caused by pathogens may result from some adverse environmental condition either before or after harvest, and may be associated with some cultivars and not others, because of genetic predisposition. Some disorders, such as scald and shrivel of apples, are characteristic of immature fruit, while such disorders as internal breakdown or water core occur in overmature fruit. Some disorders affect only certain tissues, while others affect the entire fruit. Some are visible at harvest—for example, water core and bitter pit—while others occur only after a period of storage. With

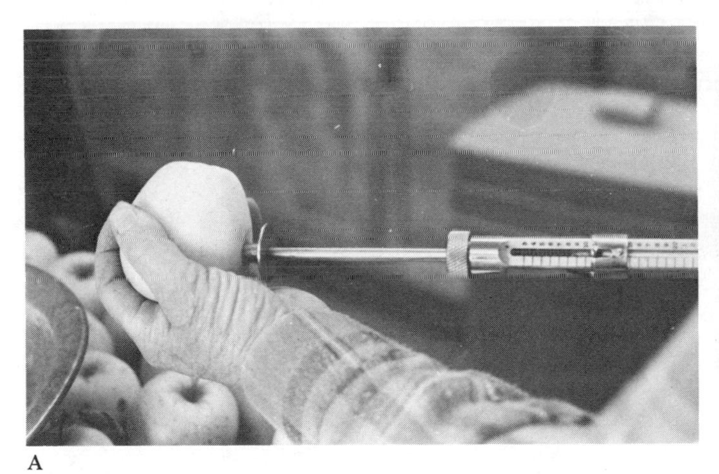

A

B

Figure 13-5 Common maturity tests. **(A)** The Magness-Taylor (U.S.) tester measures flesh firmness. **(B)** The hand refractometer is an easy-to-use instrument that measures soluble solids (mostly sugars) of the fruit juice. [Courtesy of David Newlyn]

Figure 13-6 Two physiological disorders of stored pears. **(A)** Storage scald, and **(B)** friction discoloration. Three pathogen-induced diseases: **(C)** Bullseye rot of apple. **(D)** Rhizopus rot of apple. **(E)** Grey mold (*left side of fruit*) and blue mold (*right side*). Losses in storage may nullify an efficient production program. [A, B: Courtesy of D. G. Richardson; C, D, E: Courtesy of D. L. Coyier]

some disorders, predisposition occurs during the growing season, but appears only after storage.

Hansen (1961) has summarized the effects of climate on physiological disorders of pome fruits. Disorders associated with high-temperature growing seasons are cork spot and mealy breakdown of pear, and bitter pit and water core of apple. Those associated with low temperatures during growth are scald of Anjou pear, premature ripening of Bartlett pear, and internal browning and brown core of apple. Although data are not available on the critical growing temperatures that predispose fruit to many disorders, wide variations from long-term mean temperatures in an area indicate the probability of certain disorders. Hot climates have mean temperatures near the critical temperature for high-temperature disorders. Similarly, in districts where the climate is considered cool for a given crop, low-temperature disorders are more likely to occur because the mean is closer to the critical temperature inducing low temperature problems. For example, Newtown apple grown in the cool climate of Watsonville, California frequently develops internal browning (Overholser et al., 1923). This disorder occurs sometimes in the moderate climate of Hood River, Oregon, but is unknown in the hot climate of Albemarle County, Virginia. Similarly, premature ripening of Bartlett pear occurs only in cool climates. It is now known that only a few hours chilling at 5°–10°C during the month before harvest can induce premature ripening (Wang et al., 1971). Sprays of GA_3 or SADH tended to prevent abnormal ethylene production and also prevented premature ripening. The duration of a period of critical temperature may be brief. The predisposition to brown core of Bosc pear can take place within a week during the final stage of maturity (Hansen, 1961). Bitter pit of apple and cork spot of pear (Fig. 13-7) are disorders associated with hot, dry seasons as well as high leaf/fruit ratios. They appear to be caused by undue moisture stress, which contributes to calcium deficiency in the fruit. Aldrich et al. (1940) reported that moisture stress in the pear tree was increased by a relatively high evaporating power of the air (caused by a combination of high temperature and low humidity), even when there was sufficient soil moisture. Sprays of the growth regulant TIBA, which increases bitter

Figure 13-7 **(A)** Bitter pit of apple and **(B)** cork spot of pear are similar physiological disorders associated with low calcium levels in the fruit. They usually are visible at harvest, but mild forms may not be apparent until after a period of storage.

A

B

pit, also inhibits calcium translocation within the tree. Currently, the use of foliar sprays and post-harvest dips of calcium solutions show promise in reducing both bitter pit and cork spot. Methods of accurately predicting disorders related to climate would be useful, but ultimately we need cultural and chemical tools which can prevent such problems, since weather modification is usually more difficult. One possible method of modifying microclimate, however, is with over-plant sprinklers to reduce the effects of a heat wave.

Some disorders occur in the storage unit as a result of storage temperatures that are too low. Most apples are best stored at −1°–0°C, but some cultivars develop disorders at such temperatures (Lutz and Hardenburg, 1968). Jonathan apple from some areas develops soft scald at 0°C, McIntosh at 0°C develops brown core, and Yellow Newtown from cool districts develops much internal browning at 0°C. Such cultivars are stored at 2°–4°C, and they further benefit from controlled-atmosphere storage, which slows respiration and prolongs storage life.

STONE FRUITS A number of disorders of different cultivars of plum and prune seem to be related in a general way to internal browning of Italian prunes. Conditions attributed to heat waves prior to harvest (and which can be reproduced by applied heat after harvest) are described as drought spot, Kelsey spot (Proebsting, 1936), heat spot, or heat injury similar to pit burn on apricot. Most of these conditions start as a visible spot or clearing of tissue, the area becoming brown or discolored and soft as it progresses. Mature fruit is more susceptible than immature fruit, and the internal oxygen level may be depressed as a result of high respiration. Maxie and Claypool (1957) reported that prunes kept in a high concentration of oxygen after harvest did not develop the injury even at high temperatures, but in a low concentration of oxygen they developed injury at 20°C. They suggested that this type of disorder is caused by depletion of oxygen from a high rate of respiration. This in turn results in abnormal (anaerobic) respiration deep in the tissue, leading to tissue breakdown, browning, and death. Verner et al. (1962) describe internal browning as the breakdown of the flesh next to the pit. The term "pit burn" used by California workers appears to be the same disorder. In Idaho it is usually found as a post-harvest and post-storage condition but has been seen at harvest. An incipient stage seems to be a translucent area of flesh next to the pit, followed by browning and tissue breakdown from the pit cavity outward into the flesh (Fig. 13-8). Typically no symptoms are seen at harvest or during 2 or 3 weeks of cold storage, but the disorder develops rapidly at room temperature as the fruit is placed in retail markets.

Mutant strains of early-maturing Italian prunes are more susceptible to internal browning than the standard Italian prune. Early strains such as Richards Early Italian, Demaris, and Milton Early Italian were found to be similar in maturity, quality (sugar/acids ratio, etc.), and degree of internal browning. Early Italian strains are poorer in canning quality and more susceptible to internal browning than Italian (Fogle et al., 1955). It is generally known that fruit of Early Italian strains do not store well, usually contain less soluble solids than Italian, and often break down on the tree before or at harvest. Firmness appears to be related to the incidence of breakdown. The softest prunes at any given time always are more susceptible to browning than are firmer fruit. Some rootstocks induce more internal browning, apparently by accelerating maturity.

A complicating factor related to later browning is "chilling" injury sustained by fruit in cold storage. Early Italian prune stored at temperatures above 0°C and below about 7°C develop more internal browning than fruit stored at 0°C or slightly below. Walter Kochan in Idaho (personal communication) also finds that similar above-freezing storage temperatures increase browning of Italian prune. Apparently the temperature range just above 0°C allows certain enzyme activity in the tissue that predisposes the fruit to browning when ripened later.

Both early and standard Italian respond to chemical sprays. GA, urea, and cycloheximide thiosemicarbazone (CTS) reduce browning when applied about one month before harvest. However, sprays of 2,4,5-TP,

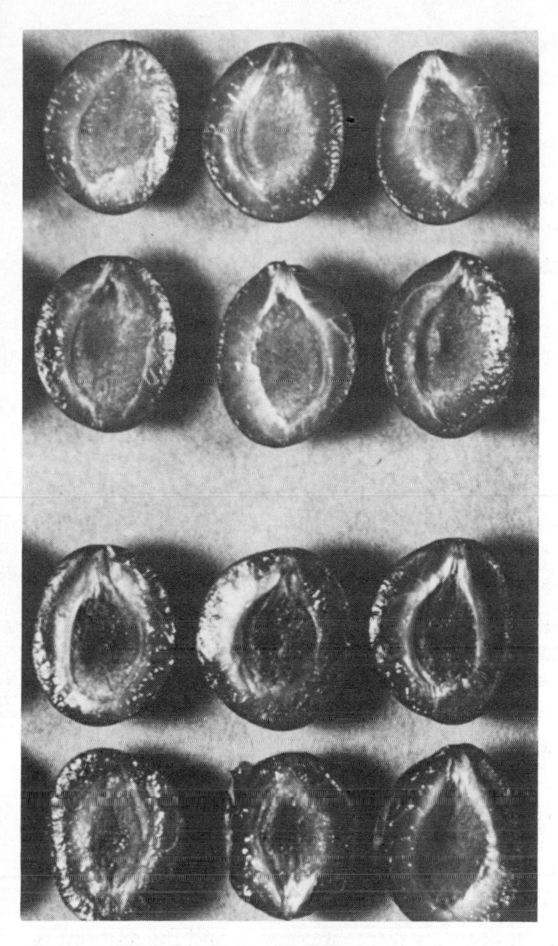

Figure 13-8 In the lower two rows, various degrees of internal browning of Italian prune can be seen. The fruit in the upper two rows are healthy and show no internal browning. [From Verner, Kochan, Loney, Moore, and Kamal, 1962a]

if applied too early (less than 250 degree-days above 5°C base after full bloom), caused more internal browning than occurred on controls.

Two physical factors also affect browning severity. The compression of fruit increases browning—apparently tissue injury releases enzymes responsible for the browning reactions. Also, fruit with stems show more browning than those with no stems. Stem-off fruit likely has a more direct port of entry for oxygen into the pit cavity, which would tend to prevent anaerobic respiration in the internal tissue.

To summarize, internal browning can be reduced by certain preharvest sprays, by picking the fruit while still firm, by preventing undue compression (bruising) and by storing the fruit at −1° to 0°C rather than slightly higher temperatures. Also, prevention of high rates of fruit respiration during preharvest hot spells, if possible, might reduce browning. Many of the cited factors relating to internal browning appear to be associated with a degree of advanced maturity. Normal fruit develop internal browning as a consequence of overmaturity and senescence. With very susceptible cultivars, such as Early Italian, several favorable treatments and conditions possibly might be combined to bring the incidence of internal browning down to acceptable levels.

Gum spot of prunes is closely related to leaf curl and is thus probably caused by moisture stress in the fruit. Trees with little crop have significantly more gum spot than do those with average or heavy crops. Both fruit gumming and leaf curl appear to be related to low cropping and to internal water loss. The leaves lose water to the atmosphere and in turn draw water from the fruit during stress, resulting in local tissue breakdown. Any such tissue damage in prune appears to cause gumming. On shallow, dry soil, curl and gumming can occur even with a full crop, while light-crop trees in deep, moist soil may show very little disorder. Proebsting and Fogle (1957) reported that Early Italian showed more gum spot than Italian. In fact, they showed that leaf curl, leaf spot, partial defoliation, and gum spot all can be caused on Early Italian simply by reduced crop density. The appearance of these symptoms is associated with a lower moisture content of the foliage, suggesting that the general problem is one of internal water balance.

Stem-end shrivel of prunes prior to harvest is a poorly understood disorder. Workers in British Columbia found shrivel to be associated with cool seasons. But in Pennsylvania it was considered to be related to fruit shrinkage during stress periods. Once again this is related to the ability of leaves to remove water from the fruit. In Washington, shrivel was reported to be associated with dry soil and moisture stress in the leaves. Also, in one instance, it was associated with trees on *Prunus besseyi,* a rather weak, dwarf rootstock. Stem-end shrivel is generally thought to be a weather problem because during some seasons it has been widespread through all of the fruit areas of the Northwest. Even though this points to weather, the disorder has not been defined well enough to establish a cause and effect relationship. Walter Kochan of Idaho indicated that rain during the preharvest period is the principal cause of stem-end shrivel. He produced it by preharvest over-tree sprinkling. Experience in Oregon confirms this, but as yet the physiological explanation is lacking.

Solution pocket of brined cherries has increased during the past 15 years. The disorder is manifest by subepidermal splits in the flesh (caused by movement of water into the cells prior to their being killed by the SO_2 solution), which fill with brine solution and with ruptured cell contents. Pockets may occur anywhere in the fruit but are commonly at the suture. Affected fruits sometimes are not firm enough to pass through a pitting machine without being torn, increasing cullage and lowering grade. Sweet cherries are brined in a solution of sulfur dioxide and lime rather than the usual salt-brine method used on other crops. Cameron and Westwood (1968) showed that the incidence of solution pocket is related to the sugar content of the fruit and to the degree of turgidity at the time fruit is placed in the brine. Fruit harvested early in the season have fewer solution pockets than fruit picked late in the season. The cooler the temperature during harvest, the greater the incidence of solution pocket. In fruit with a sugar content above 18 percent of soluble solids, the percentage of solution pockets is reduced when picked later in the day when fruit were less turgid. In tests, a prebrining treatment with sugar solutions or alcohol to draw water from the fruit resulted in a marked reduction in solution pockets. Rapid killing of fruits either with 50 percent dimethyl sulfoxide (DMSO) or 100 percent ethanol prevented the formation of solution pockets. On the basis of these data, it should be possible to reduce the number of solution pockets by reducing the water content of the fruit. This is accomplished by picking after transpiration shrinkage, either late in the day or when temperatures are high. Allowing picked fruit to sit in the sun also reduces the number of solution pockets but cullage increases due to scald and fruit rots. It is apparent that solution pockets are related to a rapid increase in turgor pressure from the movement of water into living cells with high sugar content, prior to their being killed by the brine solution. Any procedures, either in the orchard or in processing, that reduce either the sugar or water content of the fruit will tend to decrease the percentage of solution pockets. Also, increasing the concentration of the toxic agent in the brining solution would reduce the disorder.

Rain cracking of sweet cherries occurs mostly during the harvest period when the fruit is mature or nearly so and has been wet with rain for some time. Primary cause of cracking is the absorption of water directly through the skin of the fruit and not through the root system (Verner and Blodgett, 1931). The absorption of water by the cherry fruit is an osmotic process and occurs at the point of contact between the skin and a drop of water. Both weight and volume of fruit increase upon immersion in water; the rate of water absorption is proportional to the osmotic concentration of the juice; when immersed in sugar solutions, the fruit's volume increases, and the amount of cracking varies inversely with the solution concentration. However, there is little or no correlation between the percentage of soluble solids in thirteen different cultivars and their tendency to crack. For a given cultivar, however, higher soluble solids result in more cracking. The injuries from cracking may vary from short breaks, which are apparently only skin deep, to larger ruptures extending most of the length of the fruit and into the underlying flesh. The severity of cracking depends upon a number of factors, chief of which are the cultivar (Zielinski, 1964), the stage of maturity (percent soluble solids), and the duration of contact with liquid water. The order of cracking susceptibility of eight cultivars was found to be as follows: Bing > Napoleon > Lambert >

Emperor Francis > Giant > Schmidt > Yellow Spanish > Montmorency, which did not crack (Tucker, 1934). In New York, it was reported that Bing cracks badly, followed in order by Lambert, Giant, Gil Peck, and Geant d'Hedelfingen. The cultivars Lamida, Ebony, and Spalding, developed at the Idaho station, are slightly more resistant to cracking than Bing and Lambert. Also the cultivars Sue, Van, Sam, and Corum are less susceptible than Napoleon.

Losses due to cracking may be minimized by proper harvesting practices. The fruit should be picked as soon as it reaches maturity. When it rains, immediate harvesting is desirable, since the longer cherries remain on the tree, the greater the cracking injury. Some attention has been given to the removal of water from the surface of the fruit following rains. Shaking water from leaves and fruit is suggested and the use of a helicopter flown at tree-top-height to blow off the water. From the long-range viewpoint, breeding programs under way ultimately may produce desirable crack-resistant cherries.

Bullock (1952) reported that cracking may be reduced as much as 60 percent if the fruits are sprayed with 1 ppm NAA 10 days after shuck fall or 30 days before harvest. Timing and concentration of the spray are important factors in the effectiveness of the treatment, which also may be modified by the size of the crop. He also reported that preharvest sprays of Ca^{++} and Al^{+++} ions reduced cracking. Ackley (1962) reported that preharvest sprays of GA_3 show some promise in reducing cracking. In one series of four commercial blocks, NAA reduced rain cracking from 32 to 11 percent with the Bing variety. The use of NAA 30 days before harvest, followed by a lime spray (28 kg per hectare) at 8 to 14 days before harvest, reduced water cracking more than either treatment alone (Westwood and Bjornstad, 1970b). This indicates that the physiological effects of the two chemicals are different, so that they become additive when applied to the same trees. Lime sprays usually reduce cracking but leave an undesirable residue on fruit harvested for fresh market. This residue may be removed by a weak solution of acetic acid used as a post-harvest wash. Preharvest lime sprays generally result in fewer solution pockets of brined cherries and a lower percentage of poor grade fruit than unsprayed controls.

Fruit pitting of sweet cherry is a condition in which areas near the surface of the fruit become sunken, forming dimples or pits (Couey, 1971). Pitting may occur before or after harvest, and there are at least three different sources: from feeding by sucking insects such as the soldier bug, from bruising during handling, and from physiologic injury—the origin of which has not been determined. The insect injury is of two types. If it occurs early, the affected fruit can easily be detected and removed during sorting because it is greatly distorted and the pits are deep. But if the injury occurs only a few days before harvest, little damage may be evident at that time. Later, the injured tissue surrounding a puncture will collapse, forming a pit with a bit of white or brown tissue at the bottom. The insect often feeds in an arc at several sites close together so that eventually several pits overlap making one large pit with irregular margins. Pits caused by bruising are usually round, and the underlying tissue is discolored. These lesions may be classed as pits or as bruises by the inspector. They occur mostly on the sides or tips of the fruit. The third type of pitting is usually irregular in outline, lacks discoloration, and does not have the white or brown puncture mark in the pit. It is usually found on the shoulder of the fruit. Some evidence suggests that it is caused by adverse low-temperature stress during post-harvest cooling, but the low incidence of the disorder some years indicates that it is in some way related to growing conditions. The precise conditions have not been identified, but the disorder is similar in some respects to internal browning of prune. Stanley Porritt in British Columbia recently found that storage at 0°C or lower resulted in more pitting than storage above 0°C and that hydro-cooling of fruit increases the amount of pitting. Also, there is some evidence that high temperature and moisture stress during fruit ripening on the tree increases pitting.

Peaches stored at 0°C for more than about 4 weeks tend to have dry or mealy flesh and may show marked browning around the pit.

NUTS Shell perforation of walnuts is observed mostly at the tip of the nut but may also be found along the suture and sometimes at the base. Sometimes only thin spots in the shell occur rather than complete holes. All of the causes of this disorder are not known, but one cause is damage by the common aphid. Apparently many factors which prevent normal shell development can cause perforations. Sunburn at critical stages or nuts infected with bacterial blight may result in thin spots or holes. Below-normal summer temperatures also seem to cause perforations.

Poor seal is the lack of a tight seal between two halves of the walnut along the suture. The condition is more prevalent in large than small nuts. It usually develops after nut fall and may be caused by leaving nuts on the ground for long periods before harvest. Drying that is too rapid, uneven drying, or overdrying of nuts seems to cause some poor seal. Also, drought possibly contributes to the disorder. To avoid it, nuts are harvested as soon as possible and dried evenly at no higher than 43°C.

Dark kernels or kernels with pepper spots are caused by high growing temperatures of 38°C or above for walnuts. This is more of a problem in California than in Oregon. Kernel shrivel of walnuts, however, is a condition caused by lower than optimum growing temperatures and is more common in Oregon than California. Temperatures of 27°–32°C are needed during the several weeks preceding harvest to obtain well-filled kernels with a high oil content.

Blank nuts of filbert may result from lack of fertilization (parthenocarpy) or from embryo abortion at various stages of development. Some pollinizer cultivars tend to be related to a higher percentage of blanks than others. Nutritional imbalance and soil-moisture stress are thought to cause some degree of blank nuts.

Brown stain of filbert (Fig. 13-9) is a disorder that occurs some years but not others. It appears as a visible brown streak along the nut shell just after the first of July. The brown tissue starts in the shell tissue between the vascular strands running along the polar axis of the nut, then enlarges to include a wider area of the shell. It appears to be a physiological rather than a pathological disorder. In two seasons when brown stain was widespread, a heat wave followed by a rainy period occurred just before the disorder was seen. As yet the exact cause has not been found.

Figure 13-9 General appearance of brown stain, a physiological disorder of filbert. (Courtesy of H. B. Lagerstedt, USDA)

SMALL FRUITS Grapes at harvest may have a proportion of "shot" berries (millerandage) in which small, seedless berries persist along with large normal ones. This may be characteristic of some cultivars—for example, Muscat—but it also can be caused by zinc deficiency. Zinc deficiency also causes the shedding or shelling of berries (coulure) soon after bloom. Uneven ripening of grapes is a condition in which some berries of a cluster are green while others are mature. This condition may be caused by abortive or too few seeds in the green berries. It also may be caused by excessive soil moisture late in the season. Treatment with GA_3 two weeks before harvest tends to result in more uniform color and maturity. Fumigation of grapes with SO_2 in storage is done to reduce infection by molds. However, if too much SO_2 enters the fruit through lenticels and injuries, localized tissue damage and bleaching occurs.

Crumbly berry is a condition of red raspberry in which the druplets tend to separate and the fruit falls apart. It is caused by too few druplets setting per flower. This can happen as a result of poor weather or virus infection of the plants.

"Red-berry" disorder of blackberries in the West is caused by the microscopic blackberry mite that feeds around the core and between the druplets, resulting in red druplets that never turn black. The mite is controlled by early-season sprays of lime sulfur.

Cranberries are injured (scalded) by preharvest high temperatures following a cool, cloudy period. This can be prevented by sprinkling before the temperature reaches 27°C. Also, a form of physiological breakdown occurs in mature cranberries in which pectic substances in the cell walls are degraded, causing softening. Doughty et al. (1967) found that both breakdown and the respiration rate of fruit were reduced if harvest was two weeks earlier than commercial maturity. Cranberries are injured at storage temperatures below 2°C, so they are stored at 2°–4°C.

Nutritional Value

While it is not within the scope of this book to consider in detail the nutritional quality of fruits and nuts, it is worthwhile to summarize the main factors relating to the ultimate value of these crops in human nutrition. Too often it seems, the general opinion is encountered that fruits and nuts are luxury items; they are nice to include in the diet but they really aren't necessary to high-quality nutrition. Perhaps that kind of thinking is the reason why, in 1965, Americans ate diets containing less calcium, vitamin A, and vitamin C than they did 10 years earlier (Dean, 1970). Fruits generally are low in calories but are rich sources of vitamins A and C, and iron. Strawberry and persimmon are higher in vitamin C than citrus, and black currant is four times higher. Fruits are moderately good sources of thiamin, niacin, and calcium. They are high in potassium and low in sodium and thus usually an excellent food for people on salt-free diets (Table 13-3). Nuts are energy-rich foods with moderate levels of carbohydrates and protein. In addition they are rich sources of the vitamins thiamin, riboflavin, and niacin and the minerals calcium, phosphorus, iron, and potassium (Table 13-3).

Does an apple a day really keep the doctor away? A study done at Michigan State University indicated that eating two apples per day resulted in better health and better performance than not eating apples—but the reason for this was not apparent (Dedolph et al., 1961). Recently, however, research at Rutgers University (Fisher et al., 1964, 1965) indicated that fruit pectins have a remarkable influence on animal physiology. Pectins tend to trap cholesterol, preventing its deposit in the linings of blood vessels. This results in a lowering of blood pressure and a reduction in the symptoms of atherosclerosis. Also it has been widely reported that people who lack fiber in their diets run a much greater risk of getting cancer of the colon than those whose diet contains fiber. Unpeeled fruits are good sources of dietary fiber. Finally, it has been known for many years that fruits, such as prunes and pears, contain significant amounts of the simple sugar fructose, which is a natural laxative. Thus, in addition to their vitamin and mineral content, fruits are for several reasons a good buy and an investment in good health.

In spite of the benefits to health, the rate of fruit consumption has been declining for many years. Advertising and promotion of fresh fruits have not reversed this trend. Using apples as an example, consumption rates for different countries are shown in Table 13-4. Annual consumption of 7.7 kilograms per capita in the U.S. today is only a small fraction of the 28 kilograms per capita consumption for the U.S. from 1910–1915. Perhaps advertising and promotion should use the basic nutritional and healthful qualities of these crops to promote them rather than relying on catchy slogans and cosmetic appeal.

Table 13-3 Nutritional Composition of Deciduous Fruits and Nuts (per 100 grams edible portion).

Food Item[1]	Water (%)	Calories[2]	Protein (g)	Fat (g)	Carbohydrates (g)
Almond, dried	4.7	598	18.6	54.2	19.5
Apple, fresh	84.8	56	.2	.6	14.1
Apricot	85.3	51	1.0	.2	12.8
Apricot, dried, sulfured	25.0	260	5.0	.5	66.5
Blackberry	84.5	58	1.2	.9	12.9
Blueberry	83.2	62	.7	.5	15.3
Butternut	3.8	629	23.7	61.2	8.4
Cherry, sour, red	83.7	58	1.2	.3	14.3
Cherry, sweet	80.4	70	1.3	.3	17.4
Chestnut, fresh	52.5	194	2.9	1.5	42.1
Chestnut, dried	8.4	377	6.7	4.1	78.6
Cranberry	87.9	46	.4	.7	10.8
Currant, black, European	84.2	54	1.7	.1	13.1
Currant, red	85.7	50	1.4	.2	12.1
Elderberry	79.8	72	2.6	(.5)	16.4
Fig	77.5	80	1.2	.3	20.3
Fig, dried	23.0	274	4.3	1.3	69.1
Filbert (Hazelnut)	5.8	634	12.6	62.4	16.7
Gooseberry	88.9	39	.8	.2	9.7
Gooseberry, canned, water pack	92.5	26	.5	.1	6.6
Grape, American type	81.6	69	1.3	1.0	15.7
Grape, European type	81.4	67	.6	.3	17.3
Hickory nut	3.3	673	13.2	68.7	12.8
Jujube (Chinese date)	70.2	105	1.2	.2	27.6
Jujube, dried	19.7	287	3.7	1.1	73.6
Loganberry	83.0	62	1.0	.6	14.9
Nectarine	81.8	64	.6	t	17.1
Papaw	76.6	85	5.2	.9	16.8
Peach	89.1	38	.6	.1	9.7
Peach, light syrup	84.1	58	.4	.1	15.1
Peach, frozen, sweetened, not thawed	76.5	88	.4	.1	22.6
Pear	83.2	61	.7	.4	15.3
Pear, canned in light syrup	83.8	61	.2	.2	15.6
Pecan	3.4	687	9.2	71.2	14.6
Persimmon, Japanese	78.6	77	.7	.4	19.7
Persimmon, Native	64.4	127	.8	.4	33.5
Piñon (Pinenut)	3.1	635	13.0	60.5	20.5
Pistachio Nut	5.3	594	19.3	53.7	19.0
Plum, Damson	81.6	66	.5	t	17.8
Plum, Japanese and hybrid	86.6	48	.5	.2	12.3
Plum, Prune type	78.7	75	.8	.2	19.7
Pomegranate (pulp)	82.3	63	.5	.3	16.4
Prune, dried	28.0	255	2.1	.6	67.4
Quince	83.8	57	.4	.1	15.3
Raisin	18.0	289	2.5	.2	77.4
Raspberry, black	80.8	73	1.5	1.4	15.7
Raspberry, red	84.2	57	1.2	.5	13.6
Strawberry	89.9	37	.7	.5	8.4
Strawberry, frozen, sweetened	71.3	109	.5	.2	27.8
Walnut, black	3.1	628	20.5	59.3	14.8
Walnut, Persian or English	3.5	651	14.8	64.0	15.8

SOURCE: Watt and Merrill, 1963.
KEY: Numbers in parentheses denote values imputed—usually from another form of the food or from a similar food. Dashes denote lack of reliable data for a constituent believed to be present in measurable amount.
t = "trace" of constituent present. I. U. = International units.

Vitamins					Minerals				
Vit. A (I. U.)	Thiamin, B_1 (mg)	Riboflavin, B_2 (mg)	Niacin, B-Vit. (mg)	Ascorbic Acid, Vit. C (mg)	Calcium (mg)	Phosphorus (mg)	Iron (mg)	Sodium (mg)	Potassium (mg)
0	.24	.92	3.5	t	234	504	4.7	4	778
90	.03	.02	.1	7	7	10	.3	1	110
2,700	.03	.04	.6	10	17	23	.5	1	281
10,900	.01	.16	3.3	12	67	108	5.5	26	979
200	.03	.04	.4	21	32	19	.9	1	170
100	(.03)	(.06)	(.5)	14	15	13	1.0	1	81
—	—	—	—	—	—	—	6.8	—	—
1,000	.05	.06	.4	10	22	19	.4	2	191
110	.05	.06	.4	10	22	19	.4	2	191
—	.22	.22	.6	—	27	88	1.7	6	454
—	.32	.38	1.2	—	52	162	3.3	12	875
40	.03	.02	.1	11	14	10	.5	2	82
230	.05	.05	.3	200	60	40	1.1	3	372
120	.04	(.05)	.1	41	32	23	1.0	2	257
600	.07	.06	.5	36	38	28	1.6	—	300
80	.06	.05	.4	2	35	22	.6	2	194
80	.10	.10	.7	(0)	126	77	3.0	34	640
—	.46	—	.9	t	209	337	3.4	2	704
290	—	—	—	33	18	15	.5	1	155
200	—	—	—	11	12	10	.3	1	105
100	(.05)	(.03)	(.3)	4	16	12	.4	3	158
(100)	.05	.03	.3	4	12	20	.4	3	173
—	—	—	—	—	t	360	2.4		
40	.02	.04	.9	69	29	37	.7	3	269
—	—	—	—	13	79	100	1.8	—	531
(200)	(.03)	(.04)	(.4)	24	35	17	1.2	(1)	170
1,650	—	—	—	13	4	24	.5	6	294
—	—	—	—	—	—	—	—	—	—
1,330	.02	.05	1.0	7	9	19	.5	1	202
440	.01	.03	.6	3	4	13	.3	2	133
650	.01	.04	.7	40	4	13	.5	2	124
20	.02	.04	.1	4	8	11	.3	2	130
t	.01	.02	.1	1	5	7	.2	1	85
130	.86	.13	.9	2	73	289	2.4	t	603
2,710	.03	.02	.1	11	6	26	.3	6	174
	—	—	—	66	27	26	2.5	1	310
30	1.28	.23	4.5	t	12	604	5.2	—	—
230	.67	—	1.4	0	131	500	7.3	—	972
(300)	.08	.03	.5	—	18	17	.5	2	299
250	.03	.03	.5	6	12	18	.5	1	170
300	.03	.03	.5	4	12	18	.5	1	170
t	.03	.03	.3	4	3	8	.3	3	259
1,600	.09	.17	1.6	3	51	79	3.9	8	694
40	.02	.03	.2	15	11	17	.7	4	197
20	.11	.08	.5	1	62	101	3.5	27	763
t	(.03)	(.09)	(.9)	18	30	22	.9	1	199
130	.03	.09	.9	25	22	22	.9	1	168
60	.03	.07	.6	59	21	21	1.0	1	164
30	.02	.06	.5	53	14	17	.7	1	112
300	.22	.11	.7	—	t	570	6.0	3	460
30	.33	.13	.9	2	99	380	3.1	2	450

[1]Unless otherwise stated, the "Food Item" is considered to be in its fresh raw state.
[2]The food calorie (k cal) is equal to 1000 gram calories of heat.

Table 13-4 Annual per capita consumption of apples in several countries.

Country	Amount Consumed (kg)
France	43.9
Germany	39.4
Belgium	28.1
Netherlands	27.7
United Kingdom	12.7
Argentina	9.0
U.S.	7.7
South Africa	3.5

SOURCE: Strydom, 1975.

From Chapter 1 (Table 1-1) we see that world deciduous fruits and nuts coming to market reach nearly 100 million metric tons annually. But because of the very perishable nature of fruits, this figure probably represents only a fraction of the amount consumed. Much fruit is grown locally in noncommercial gardens and is consumed without being counted as part of the world production. Because most of these fruits are eaten fresh, they contain more vitamins than processed fruit. Fruits certainly deserve a more important place in the diets of many countries, particularly in the U.S.

GENERAL REFERENCES

Abeles, F. B. 1973. *Ethylene in plant biology.* Academic Press, New York.

Hansen, E. 1966. Post-harvest physiology of fruits. *Ann. Rev. Plant Physiol.* 17:459–480.

Lutz, J. M., and R. E. Hardenburg. 1968. *The commercial storage of fruits, vegetables, and florist and nursery crops.* USDA, Agr. Handbook 66.

Rose, D. H., and R. C. Wright. 1952. Commodity storage requirements. In USDA *Refrigerating data book,* 4th ed., Chapter 19.

Ryall, A. L., and W. T. Pentzer. 1974. *Handling, transportation, and storage of fruits and vegetables.* The Avi Publishing Co., Westport, Connecticut.

Watt, B. K., and A. L. Merrill. 1963. *Composition of foods.* USDA, Agr. Handbook 8.

14

Growth Regulators

NAMES AND TYPES

Names

Below are listed the common names (which are often abbreviations and trade names) of the principal plant-growth regulators. The first-listed common name or abbreviation is the one most commonly used. The synonyms listed are also found in the literature. Trade names are sometimes also used as common names.

ABA. 3-methyl-5-(1′-hydroxy-4′-oxo-2′,6,6-trimethyl-2′-cyclohexen-1′ yl)-*cis,trans*-2,4-pentadienoic acid. SYNONYMS: abscisic acid; abscisin II; dormin. TYPE: Growth inhibitor.
abscisic acid. *See* ABA
abscisin II. *See* ABA.
Alar. *See* SADH
Amchem 66-329. *See* ethephon.
Amo-1618. Ammonium (5-hydroxycarvacryl) trimethyl chloride piperidine carboxylate. SYNONYMS: 2-isopropyl-4-dimethylamino-5-methylphenyl 1-piperidine-carboxylate-methochloride; ACPC. TYPE: Growth inhibitor.
BA. 6-benzylamino purine. SYNONYMS: benzyladenine; BAP; Verdan. TYPE: cytokinin.
BAP. *See* BA.
benzyladenine. *See* BA.
B-Nine; B-9; B-995. *See* SADH.
BNOA, NOA. β-naphthoxyacetic acid. TYPE: Auxin.
BTP. *See* PBA.
carbaryl. *See* Sevin.
CBBP. *See* Phosfon-D.
CCC. (2-chloroethyl)trimethylammonium chloride. SYNONYMS: chlorocholine chloride; chlormequat; Cycocel. TYPE: Growth inhibitor.
CEPA. *See* ethephon.

chlormequat. *See* CCC.
chlorocholine chloride. *See* CCC.
3-CP. 3-chlorophenoxy-α-propionic acid. TYPE: Auxin.
3-CPA. 3-chorophenoxy-α-propionamide. TYPE: Auxin.
4-CPA. 4-chlorophenoxyacetic acid. SYNONYMS: *p*-chlorophenoxyacetic acid; PCPA. TYPE: Auxin.
Cycocel. *See* CCC.
2,4-D. 2,4-dichlorophenoxyacetic. TYPE: Auxin.
DMSO. dimethylsulfoxide. TYPE: Adjuvant.
DNOC. 4,6-dinitro-*o*-cresol (or sodium 4,6-dimitro-*o*-cresylate). TYPE: Caustic.
dormin. *See* ABA.
Duraset.*N-meta*-tolyl phthalamic acid. SYNONYM: 7R5. TYPE: Auxin.
Elgetol. sodium 4,6-dinitro-*o*-cresylate. *See* DNOC. TYPE: Caustic.
ethephon. (2-chloroethyl)phosphonic acid. SYNONYMS: Ethrel; CEPA; Amchem 66-329. TYPE: Ethylene generator.
Ethrel. *See* ethephon.
FAP. *See* kinetin.
fenoprop. *See* 2,4,5-TP.
GA_3 (gibberellic acid.) 2β,4a,7-trihydroxy-1-methyl-8-methylene-$4a_a$,4bβ,-gibb-3-ene-1_a,10β-dicarboxylic acid, 1,4a-lactone. (Subscripts indicate specific analogues, such as GA_1, GA_2.) TYPE: Gibberellin.
Gibberellin(s). One or more of the known gibberellins (*see* GA_3). General references to exogenous gibberellins may be assumed to refer to GA_3 or KGA_3, commercially available compounds that have equal activity when used on an acid-equivalent basis.
heteroauxin. *See* IAA.
IAA. indoleacetic acid. SYNONYMS: indole-3-acetic acid; 3-indoleacetic acid; indolylacetic acid; heteroauxin. TYPE: Auxin.
IBA. indolebutyric acid. SYNONYM: indole-3-butyric acid. TYPE: Auxin.
2iP. 6-(γ,γ-dimethylallylamino)-purine. TYPE: Cytokinin.
KGA_3. potassium gibberellate (potassium salt of GA_3). TYPE: Gibberellin.
kinetin. 6-furfurylamino purine. SYNONYMS: *N*-furfuryladenine; FAP. TYPE: Cytokinin.
maleic hydrazide. *See* MH.
MH. maleic hydrazide. TYPE: Growth inhibitor.
NAA. naphthaleneacetic acid. SYNONYM: α-naphthaleneacetic acid. TYPE: Auxin.
NAAm. naphthaleneacetamide. SYNONYMS: NAD; NAAmide. TYPE: Auxin.
NAAmide. *See* NAAm.
NAD. *See* NAAm.
naptalam. *See* NPA.
NPA. *N*-1-naphthylphthalamic acid. SYNONYM: naptalam, Alanap. TYPE: Auxin.
PBA. 6-(benzylamino)-9-(2-tetrahydropyranyl)-9*H*-purine. SYNONYM: SD8339, BTP. TYPE: Cytokinin.
PCPA. *See* 4-CPA.
Phosfon. *See* Phosfon-D.
Phosfon-D. 2,4-dichlorobenzyltributylphosphonium chloride. SYNONYMS: Phosfon, CBBP. TYPE: Growth inhibitor.
POA. phenoxyacetic acid. TYPE: Auxin.
PPG. *N*-(6-purinyl) α-phenylglycine. TYPE: Cytokinin.
7R5. *See* Duraset.
SADH. succinic acid; 2,2-dimethylhydrazide. SYNONYMS: *N,N*-dimethylaminosuccinamic acid; Alar; B-Nine; B-9; B-995. TYPE: Growth inhibitor.
SD8339. *See PBA.*
Sevin. 1-naphthyl *N*-methyl carbamate. SYNONYM: carbaryl. TYPE: Abscission inducer.
silvex. *See* 2,4,5-TP.
2,4,5-T. 2,4,5-trichlorophenoxyacetic acid. TYPE: Auxin.
2,4,5-TB. 2,4,5-trichlorophenoxybutyric acid. TYPE: Auxin.
2,3,5,6-TBA. 2,3,5,6-tetrachlorobenzoic acid. TYPE: Auxin.
TIBA. 2,3,5-triiodobenzoic acid. TYPE: Auxin synergist.
2,4,5-TP. 2-(2,4,5-trichlorophenoxy)propionic acid. SYNONYMS: fenoprop; silvex. TYPE: Auxin.
Verdan. *See* BA.
zeatin. 6-(4-hydroxy-3-methyl-2-butenylamino) purine. TYPE: Cytokinin.

Types

The functioning of a plant depends upon specific levels of natural hormones, each in balance with the others. The achievement of specific agricultural objectives, however, may depend upon the proper balance of natural *and applied* growth regulators. This balance changes throughout the growing season—for example, NAA is used during post-bloom to chemically thin apples, but it is used later in the season on the same plant to prevent preharvest fruit drop. Both timing and concentration are critical in achieving these specific responses. Growth regulators, both natural and synthetic, may be divided into five groups, based on differences in their structures and effects: (1) auxins, (2) gibberellins, (3) cytokinins, (4) ethylene and ethylene-generators and (5) growth inhibitors. The different types of growth regulators will be discussed in terms of their function, sites of synthesis, and their interactions. The discussion in this chapter deals with the general effects of growth regulators on plants; more details on their uses in pomology are given in sections dealing with specific uses.

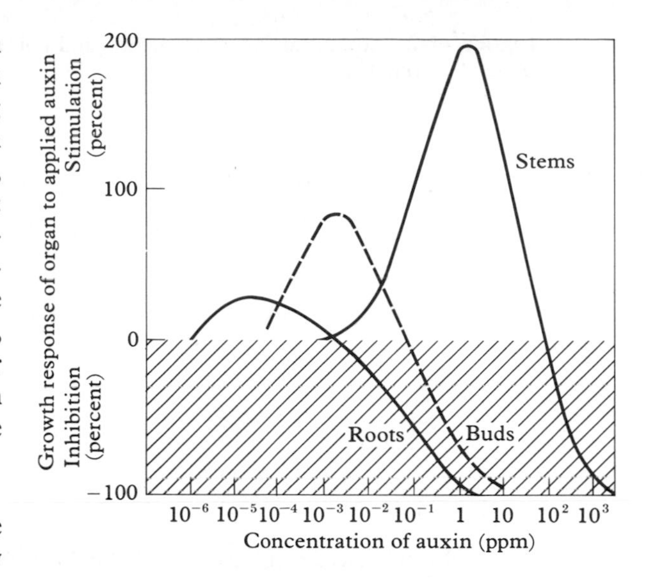

Figure 14-1 The effect of auxin concentration on the growth of roots, buds, and stems. Roots and buds are stimulated at lower concentrations than stems; at 1 ppm, stems are strongly stimulated while roots and buds are completely inhibited. [After K. V. Thimann, *Am. J. Bot.* 24: 407–412, 1937.]

AUXINS The primary natural auxin appears to be indole-3-acetic acid (IAA). Several related naturally occurring indole compounds such as indoleacetaldehyde, indolepyruvic acid, and indoleacetonitrile also are reported as auxins, but their auxin activity is probably due to their conversion to IAA. Synthetic auxins include NAA, BNOA, NAAm, IBA, 3-CPA 2,4-D, 2,4,5-T, and 2,4,5-TP.

IAA is produced mainly in subapical regions of actively growing shoots, in young leaves, and in developing embryos. It controls the rate of cell enlargement by affecting the extensibility of the cell wall. Auxins may induce or retard abscission of young fruitlets or retard abscission of mature fruit. Auxins can stimulate ethylene synthesis in fruits, thus hastening ripening. Auxins such as IBA, NAA, and 2,4-D stimulate rooting of stem cuttings of many species. The downward flow of IAA from shoot tips both stimulates cambial cell division and inhibits lateral bud development. The higher IAA level on the shady side of a shoot tip results in the bending of a stem toward the light (phototropism). Figure 14-1 shows the responses of auxin in different plant parts. Natural and synthetic auxins are shown in Figure 14-2.

Triiodobenzoic acid (TIBA) at 1 to 75 ppm inhibits auxin-mediated growth, but at .01 to .1 ppm it acts as an auxin synergist. Thus it is not an anti-auxin as some suppose, and TIBA-induced auxin inhibition is not reversed by adding more auxin. TIBA in many woody species stimulates floral initiation. Also it interferes with transport of auxin, calcium, and perhaps other substances. This action results in wider branch angles on young trees, and increased bitter pit in some apple fruit.

GIBBERELLINS All gibberellins are natural products of the fungus *Gibberella fujikuroi* or of higher plants. The review by Lang (1970) lists A_1-A_4, A_7, A_9-A_{16}, A_{24}, A_{25} isolated from the fungus and A_1-A_9, A_{13}, A_{17}-A_{23}, A_{26}-A_{29} from higher plants. GA_3, A_4, and A_7 have been found both in immature apple seeds and in grapes, A_1 and A_3 in hazel seed, and GA_{32} in peach and apricot. In 1976, the number of gibberellins identified reached forty-five. The commercially available GA_3 and $GA_{4\text{-}7}$ are extracted from cultures of the fungus. The structures of many gibberellins and related compounds are given in Figures 14-3, 14-4, and 14-5.

Figure 14-2 Structural formulas, names, and abbreviations of some auxins and auxin precursors. [After Weaver, 1972.]

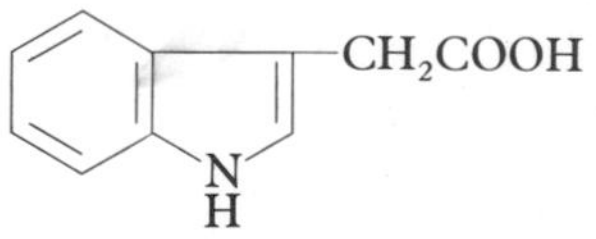

Indoleacetic acid (IAA)

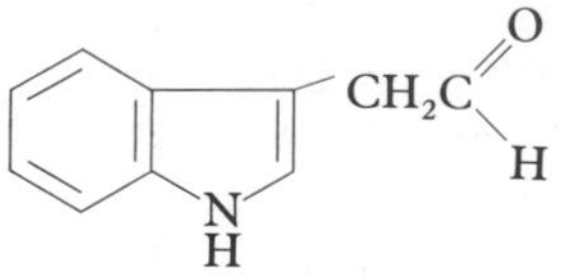

Indoleacetaldehyde (IAAld)

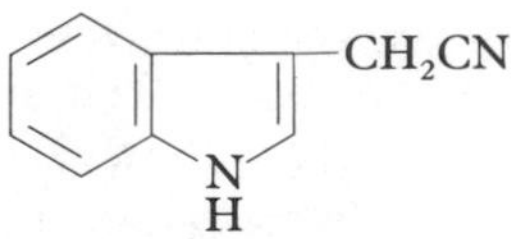

Indoleacetonitrile (IAN)

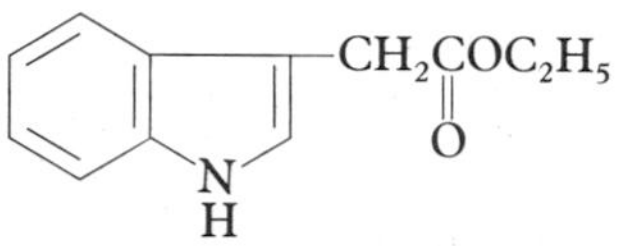

Ethylindoleacetate (IAEt)

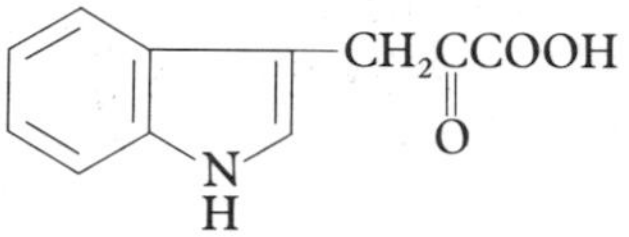

Indolepyruvic acid (IPyA)

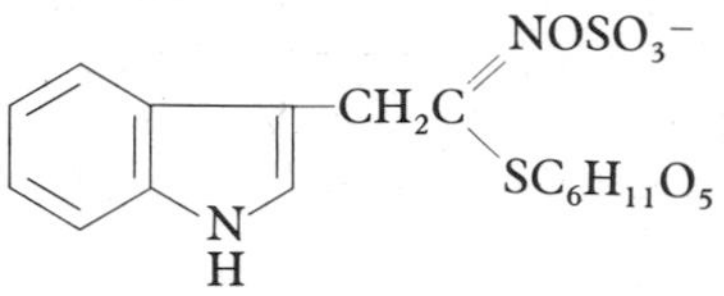

Glucobrassicin

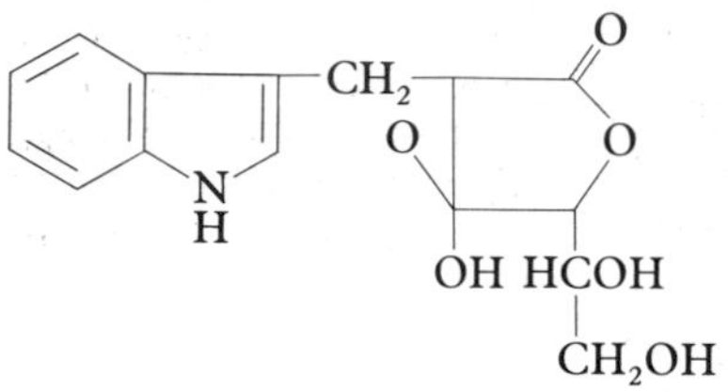

Ascorbigen

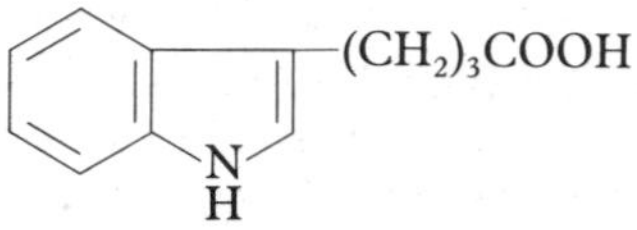

Indolebutyric acid (IBA)

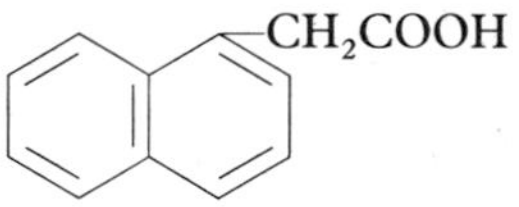

α-Naphthaleneacetic acid (NAA)

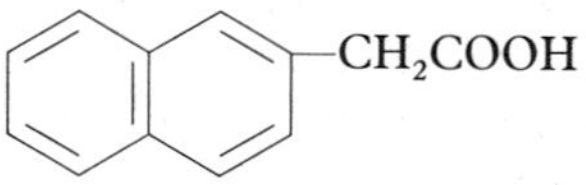

β-Naphthaleneacetic acid

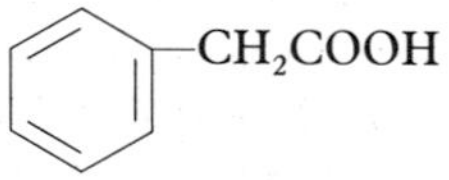

Phenylacetic acid

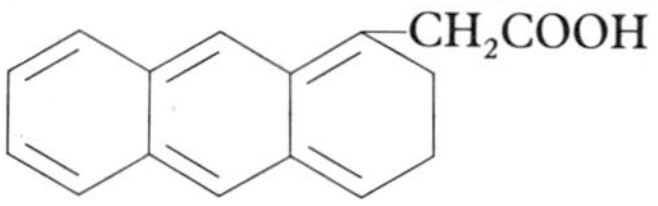

Anthraceneacetic acid

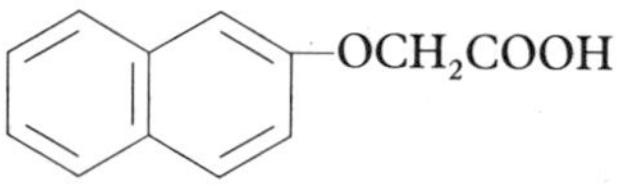

β-Naphthoxyacetic acid (BNOA)

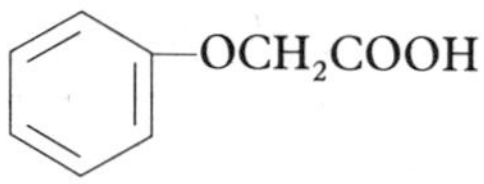

Phenoxyacetic acid (POA)

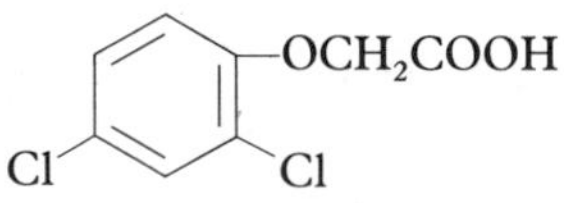

2,4-Dichlorophenoxyacetic acid (2,4-D)

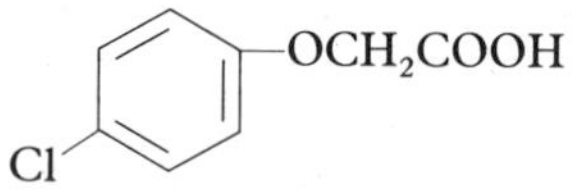

4-Chlorophenoxyacetic acid (4-CPA)

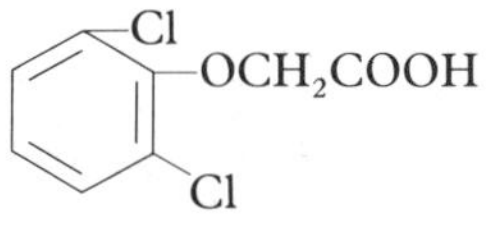

2,6-Dichlorophenoxyacetic acid (2,6-D)

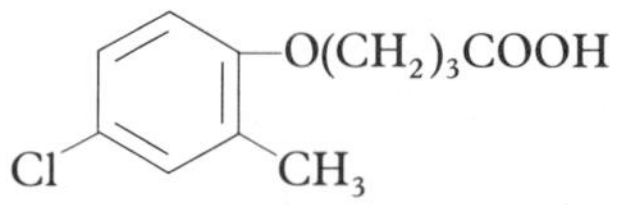

4-[(4-Chloro-*o*-tolyl)oxy]butyric acid (MCPB)

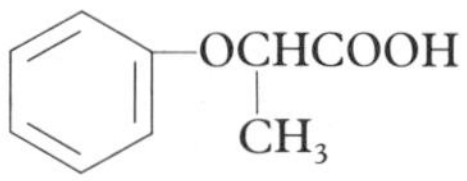

2-Phenoxypropionic acid

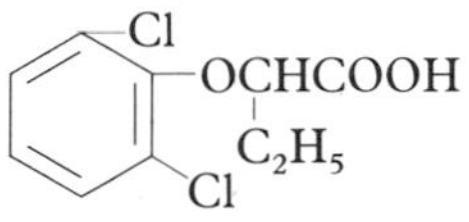

2-(2,6-Dichlorophenoxy)butyric acid

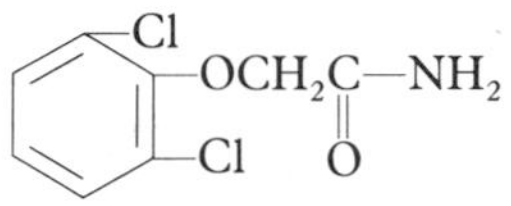

2,6-Dichlorophenoxyacetamide

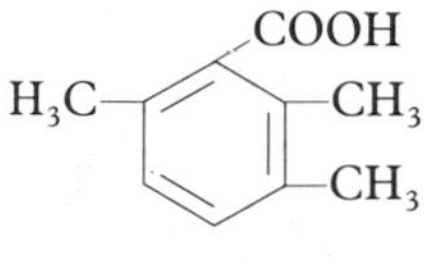

2,3,6-Trimethylbenzoic acid

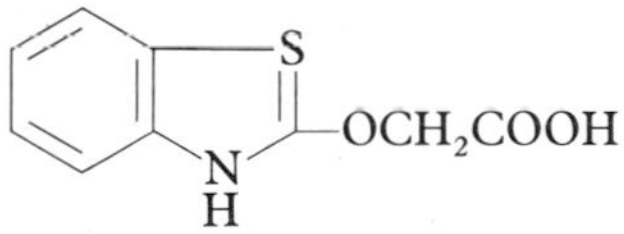

Benzothiazole-2-oxyacetic acid (BOA)

Figure 14-3 Structural formulas of some known gibberellins. Heavy lines and wedges indicate bonds lying above the plane of the ring; broken lines indicate bonds lying below this plane. [After Lang, 1970]

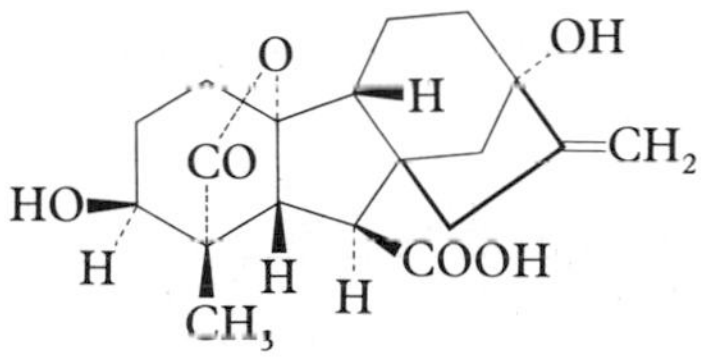

GA$_1$

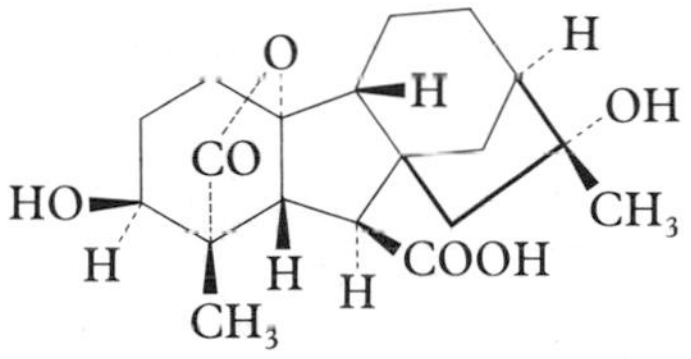

GA$_2$

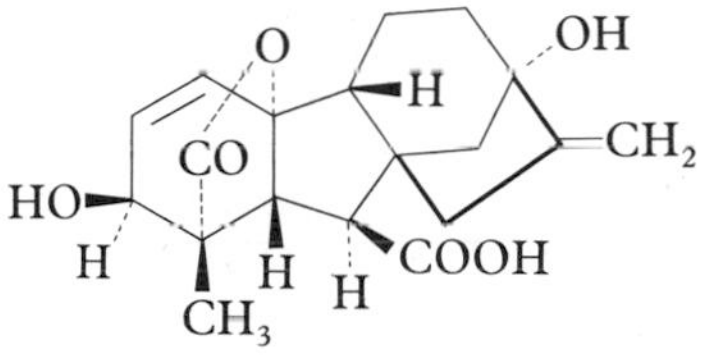

GA$_3$ (Gibberellic acid)

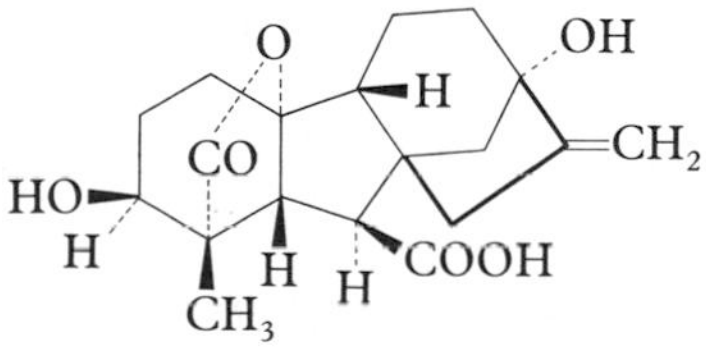

GA$_4$

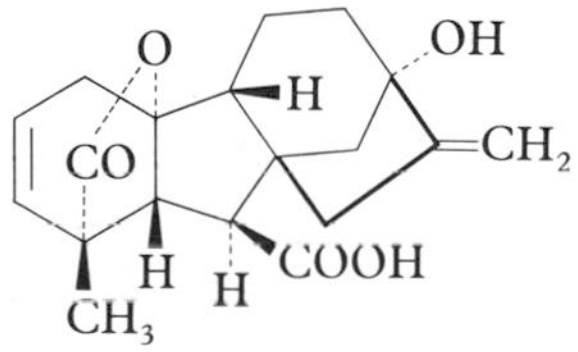

GA$_5$

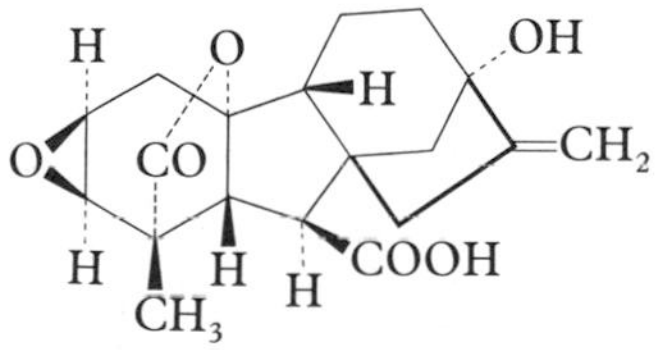

GA$_6$

(continued)

Figure 14-3 (continued)

GA_7

GA_8

GA_9

GA_{10}

GA_{11}

GA_{12}

GA_{13}

GA_{14}

GA_{15}

GA_{16}

GA_{17}

GA_{18} (*Lupinus* gibberellin I)

GA_{19} (bamboo gibberellin)

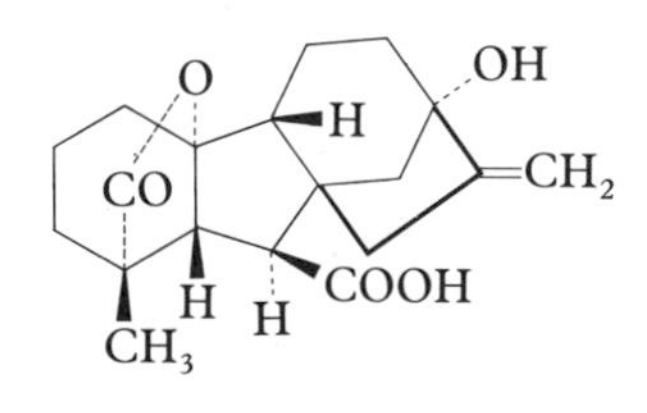

GA_{20} (*Pharbitis* gibberellin I)

GA_{21} (*Canavalia* gibberellin I)

Figure 14-3 (continued)

O OH H CO $=CH_2$ H H COOH CH_2OH

GA_{22} (*Canavalia* gibberellin II)

OHC OH H $=CH_2$ HO H H H H_3C H COOH COOH

GA_{23} (*Lupinus* gibberellin II)

OHC H H $=CH_2$ H_3C H H COOH COOH

GA_{24}

HOOC H H $=CH_2$ H_3C H H COOH COOH

GA_{25}

O H O H HO H CO $=CH_2$ HO H H H COOH CH_3

GA_{26}

O CH_2 H H HO H CO $=CH_2$ HO H H H COOH CH_3

GA_{27}

HOOC OH H $=CH_2$ HO H_3C H H COOH COOH

GA_{28}

O OH HO H CO $=CH_2$ H H COOH CH_3

GA_{29}

OH O H CO H $=CH_2$ HO H COOH CH_3

GA_{30}

OH O H CO H $=CH_2$ H COOH CH_3

GA_{31}

OH O OH CO H $=CH_2$ HO H COOH OH CH_3

GA_{32}

OH OH O H CO H $=CH_2$ O H COOH CH_3

GA_{33}

HO O H CO H $=CH_2$ HO H COOH CH_3

GA_{34}

HO O H CO H $=CH_2$ HO H COOH CH_3

GA_{35}

CHO H H $=CH_2$ HO H_3C H COOH COOH

GA_{36}

O CH_2 H CO H $=CH_2$ HO H COOH CH_3

GA_{37}

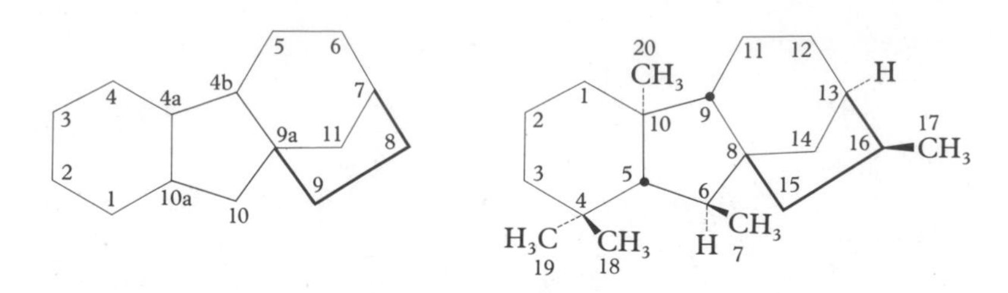

Figure 14-4 Structural formulas of the gibbane molecule *(left)* and *ent*-gibberellane (*right*). Heavy lines and wedges indicate bonds that lie above the plane of the ring system; broken lines indicate bonds that lie below this plane. [After Lang, 1970]

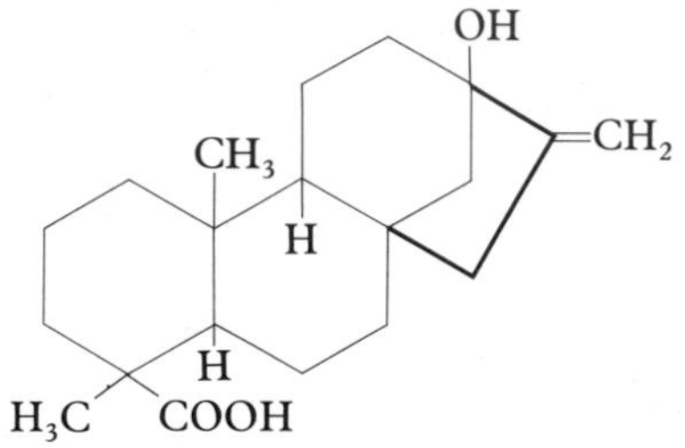

Steviol

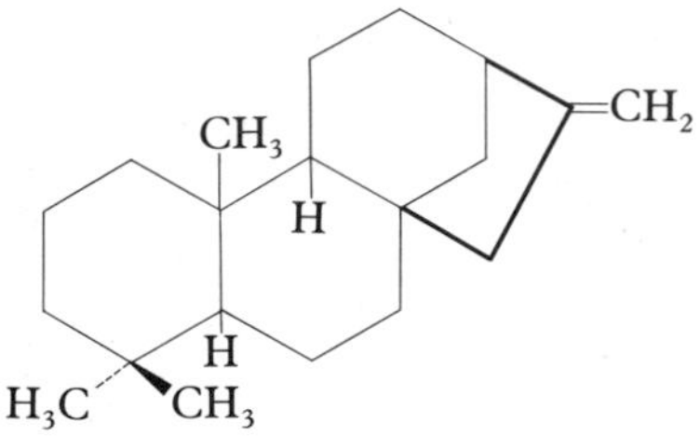

Kaurene

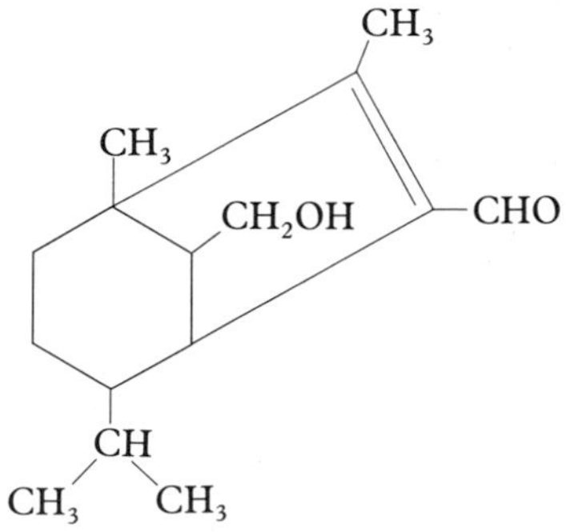

Helminthosporol

Figure 14-5 Structural formulas of some compounds that lack a gibbane skeleton but evidence gibberellin activity. [After Weaver, 1972]

Gibberellins in trees are produced mainly in very young leaves, young embryos, fruits, and in roots. They function in cell elongation, aid in breaking rest of seeds and dormant buds, prevent flower initiation, and seem to work with auxin to prevent abscission of young fruits. GA retards coloring and maturity of sweet cherries and enhances elongation of apple (Fig. 14-6) and pear fruits. In grapes, it improves seedless set, increases berry size, and loosens the clusters by promoting pedicel elongation (Fig. 14-7).

CYTOKININS Several N^6-substituted adenine derivatives are known to promote cell division (cytokinesis), and, since the isolation of kinetin in 1955 by Miller et al., both natural and synthetic cytokinins have been used to affect plant growth (Fig. 14-8). Six known cytokinins are kinetin, zeatin, 2iP, BA, PBA, and PPG. In addition to promoting cell division, cytokinins have been found to regulate nucleic acids (DNA and RNA); to regulate apical dominance and branching; to regulate bud initiation; to enhance seed germination; to influence transport of nutrients and metabolites; to prevent abscission and senescence of flowers, fruits, and leaves; and to inhibit root initiation (Helgeson, 1968 and Letham, 1969).

Primary sites of cytokinin synthesis are roots and young fruit. Its movement in plants appears to be both upward from the roots in xylar sap and downward in phloem from points of application or synthesis in aerial organs. Upward flow of cytokinins in the xylar sap of trees reaches a peak in spring about the time of full bloom, declines to a low level in late summer and remains low throughout the winter. This seasonal pattern is consistent with the known role of cytokinins in fruit set, leaf growth, and control of senescence (Luckwill and Whyte, 1968).

Figure 14-7 Effect of GA sprays on berry size and cluster loosening of Thompson Seedless grape. Control cluster is at left. [Courtesy of R. J. Weaver, University of California]

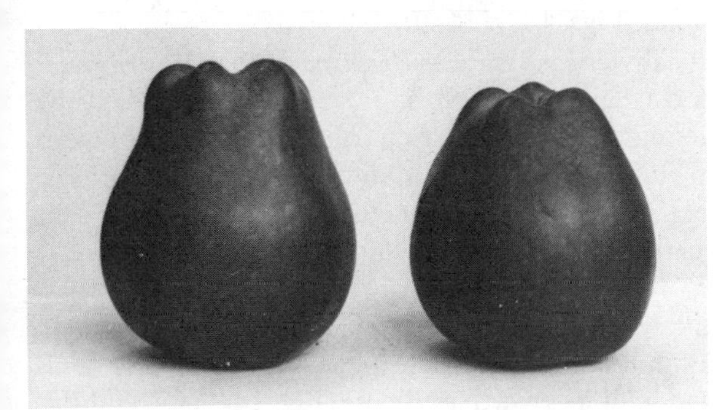

Figure 14-6 *Left.* Effect of GA_{4-7}, as a petal-fall spray, on the shape of Delicious apple. *Right.* The control also has the desirable conic shape, but the calyx lobes are not as well developed.

ETHYLENE Many sites and functions of ethylene as a plant hormone have been discovered in the past 40 years. Kidd and West (1932) suggested that ethylene gas, produced by apples, increased the rate of respiration and hastened ripening. Zimmerman and Hitchcock (1933) found that ethylene and other unsaturated gases stimulated root initiation on cuttings.

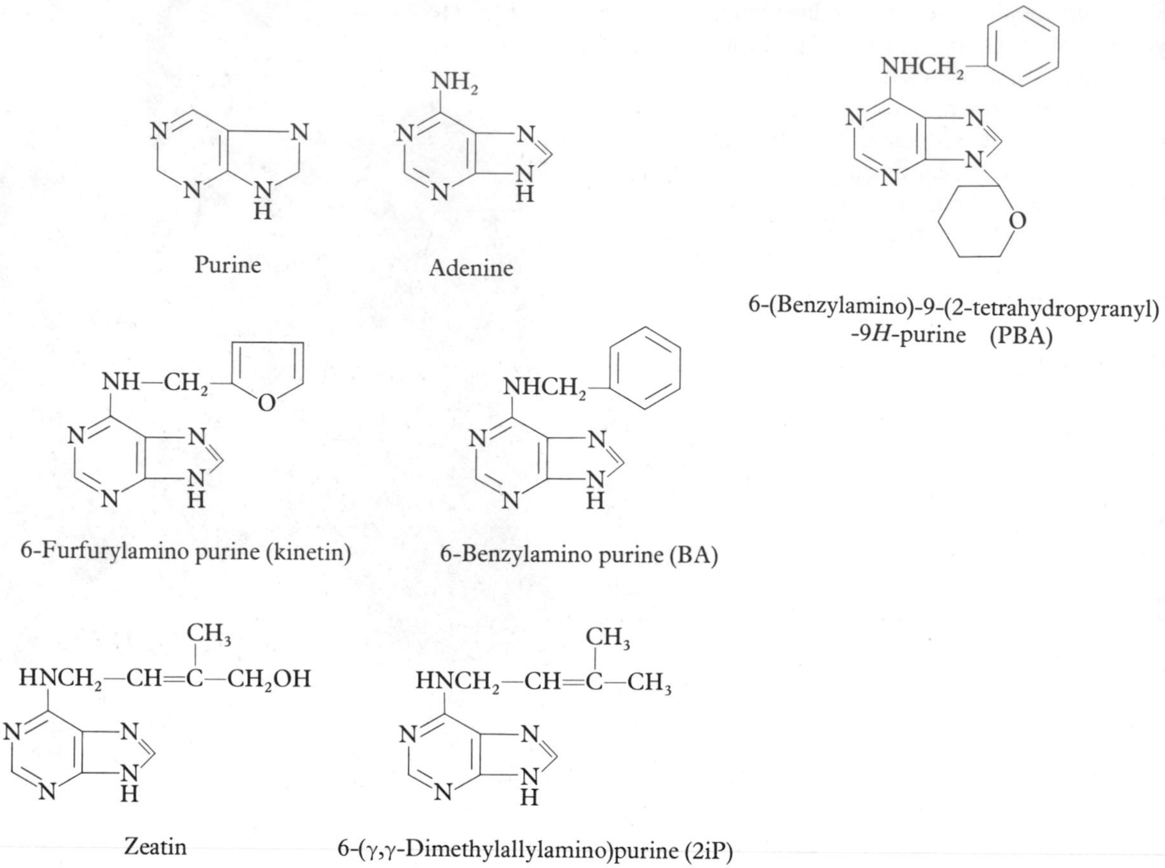

Purine

Adenine

6-(Benzylamino)-9-(2-tetrahydropyranyl)-9*H*-purine (PBA)

6-Furfurylamino purine (kinetin)

6-Benzylamino purine (BA)

Zeatin

6-(γ,γ-Dimethylallylamino)purine (2iP)

Figure 14-8 Structural formulas of purine, adenine, and some cytokinins. Kinetin, BA, and PBA are synthetic; zeatin and 2iP are naturally occurring. [After Weaver, 1972]

Hansen (1946) showed that application of the auxin 2,4-D to pear fruits resulted in an increase in ethylene synthesis, followed by ripening. In recent years, there has been a surge of new interest in ethylene with the introduction of ethylene-generating chemicals, for example, ethephon. When applied to plants, it decomposes to produce ethylene (C_2H_4), phosphate, and HCl. The ethylene thus produced in tissues causes several effects of interest: it hastens ripening and color development; it promotes abscission of leaves, fruits and nuts; it stimulates floral initiation; it breaks rest in buds and seeds; it causes gumming of some species; it inhibits lateral bud development in concert with auxin (Figs. 14-9, 14-10).

Ethylene is synthesized in many parts of the plant, particularly under physical stress. It is produced by young fruits and maturing fruits, but is also produced in injured tissues of all kinds. It seems to promote transverse rather than longitudinal expansion of cells (Galston and Davies, 1969).

GROWTH INHIBITORS ABA is a natural growth inhibitor that plays a role in the rest of buds and seeds and inhibits growth of shoots (Fig. 14-11). Its inhibi-

Figure 14-9 Fruit growth of apricot as related to rate of volume increase of its component parts and to levels of hormones, based on data contained in several works. The levels of hormones bear no relationship to one another but only to time after anthesis. [After Crane, 1969]

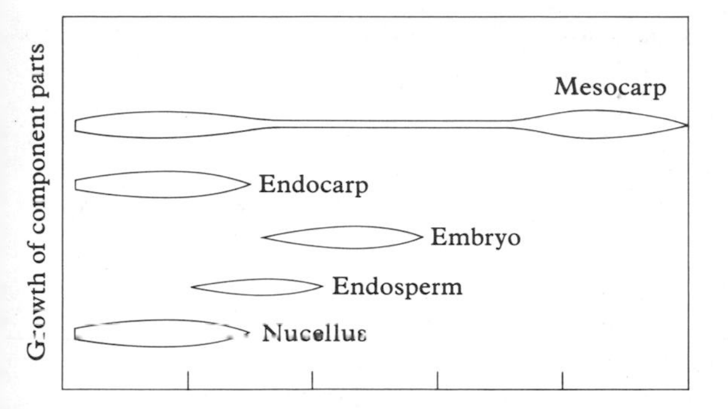

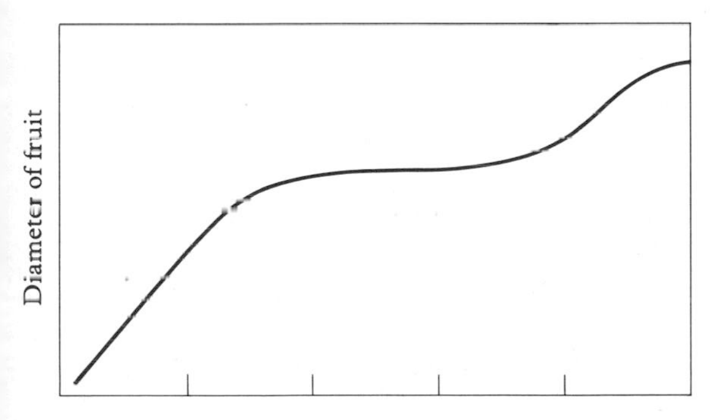

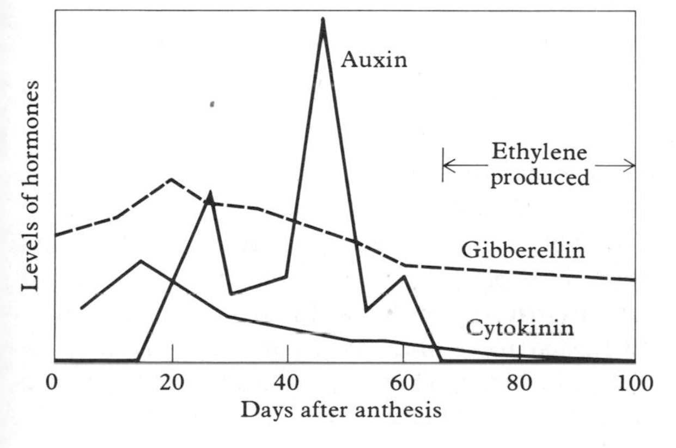

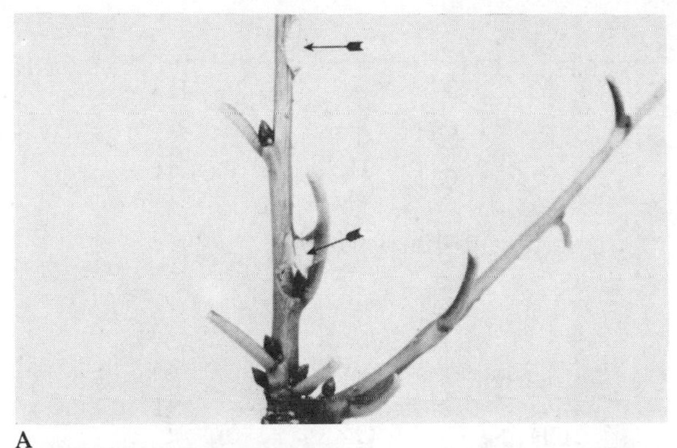
A

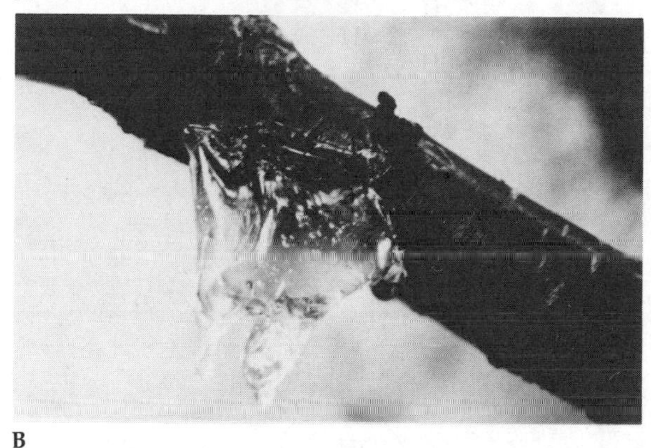
B

Figure 14-10 Gumosis of Montmorency cherry in response to treatment with the ethylene generator ethephon. **(A)** Gum exudate from current shoots (*arrows*). **(B)** Heavy gumming of older branch. [From M. J. Burkovac, *J. Am. Soc Hort. Sci.,* Vol. 94, 1969]

tion of the actions of auxin, GA, and cytokinin appears to be nonspecific in many situations. In some cases, however, it blocks RNA synthesis, specifically inhibiting GA-induced enzyme production. ABA sometimes stimulates fruit and leaf abscission and senescence. The regulatory role of ABA in resting buds or seeds appears linked to the changing levels of GA and other promoters, which rise sharply as chilling is complete and the rest period is broken.

Mature leaves are a primary site of ABA synthesis during the shortening days of late summer, but many

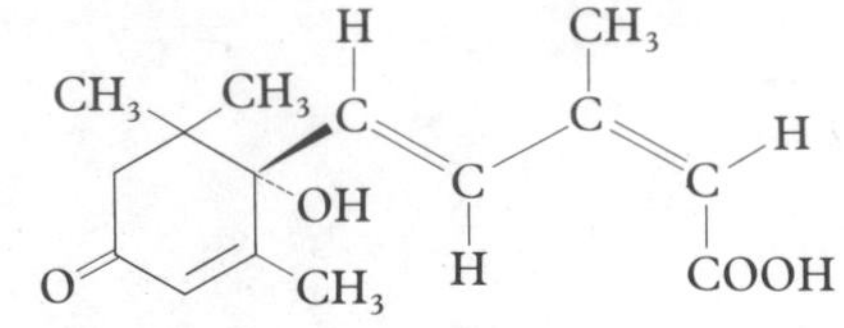

(S)-Abscisic acid (ABA)

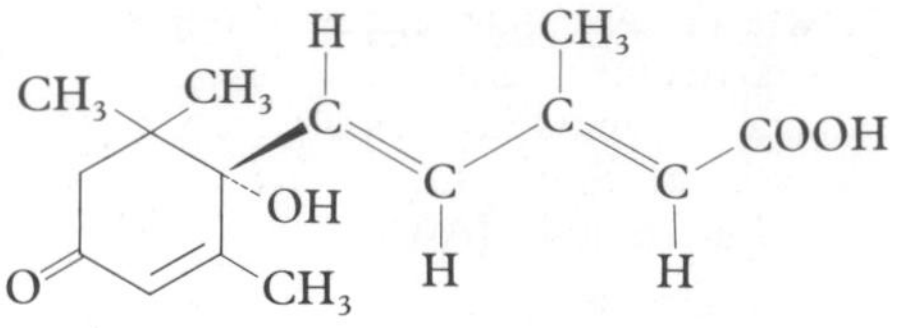

2-*Trans*-abscisic acid (ABA)

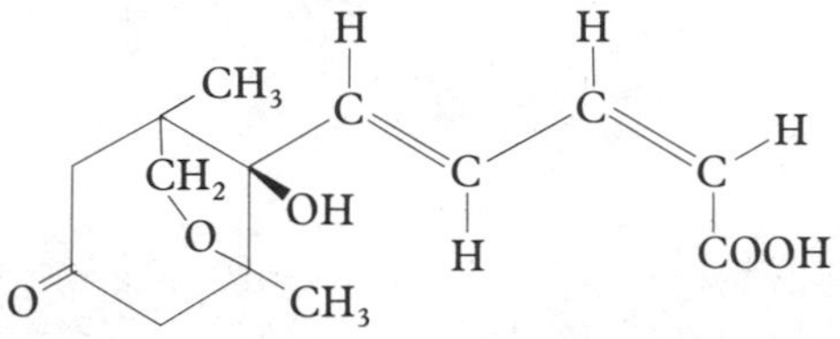

Phaseic acid

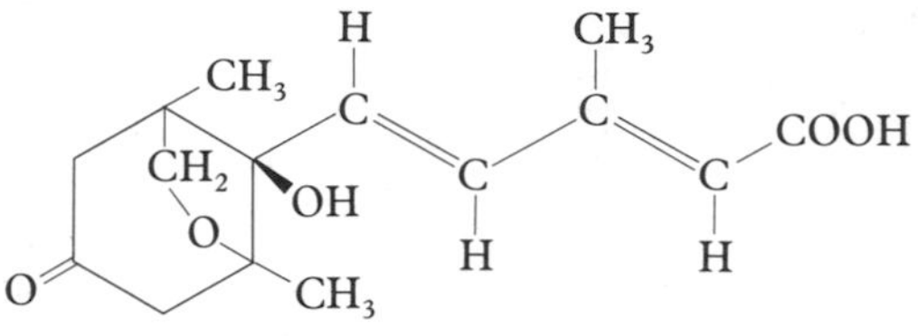

2-*Trans*-phaseic acid

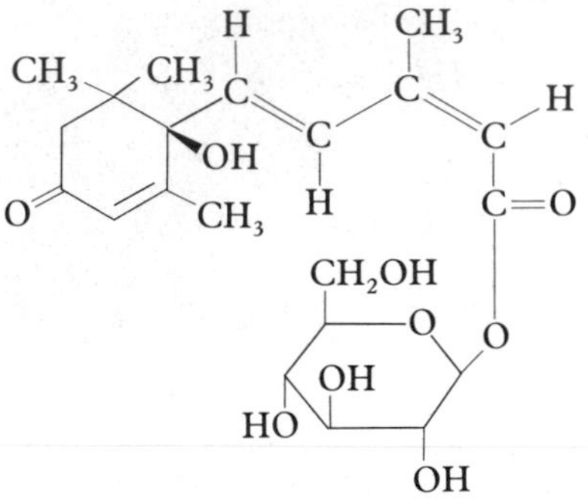

(+)-Abscisyl-β-D-glucopyranoside

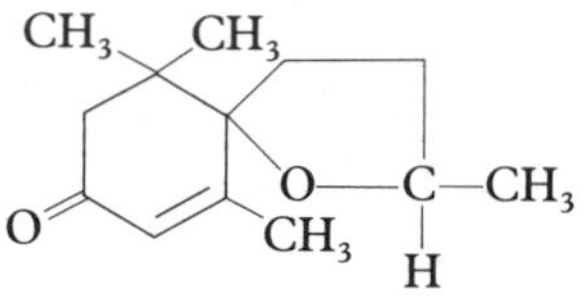

Theaspirone

Figure 14-11 Structural formulas of ABA and some naturally occurring related substances. [After Addicott and Lyon, 1969]

other tissues are known to synthesize ABA. In particular, stems or roots under physical or moisture stress produce ABA. Both immature and mature fruit contain relatively large amounts of ABA.

Several growth retardants, namely CCC, Amo-1618, and Phosfon D (Fig. 14-12) inhibit GA biosynthesis. They increase flower initiation and reduce vegetative growth. The growth retardant SADH also reduces shoot growth and increases flowering (Fig. 14-13), but, in addition, it promotes fruit set in grape and apple, retards ripening, reduces size, and alters shape in apple (Fig. 14-14), yet enhances color and hastens maturity in cherry. In addition to the inhibitors listed above, there are a number of natural phenolics, including derivatives of benzoic and cinnamic acids, lactones of the coumarin group, and flavonoids.

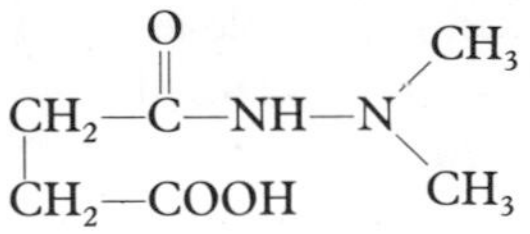

$HOCH_2CH_2NHNH_2$

Succinic acid-2,2-dimethylhydrazide (SADH; B-995; B-9; Alar)

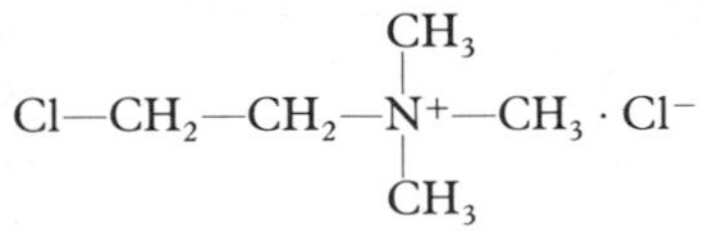

(2-Chloroethyl)trimethyl ammonium chloride (CCC; Cycocel)

β-Hydroxyethylhydrazine (BOH)

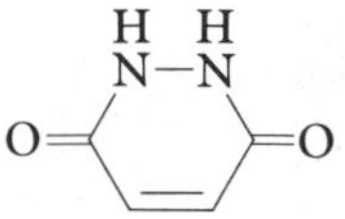

1,2-Dihydro-3,6-pyridazinedione (MH; maleic hydrazide)

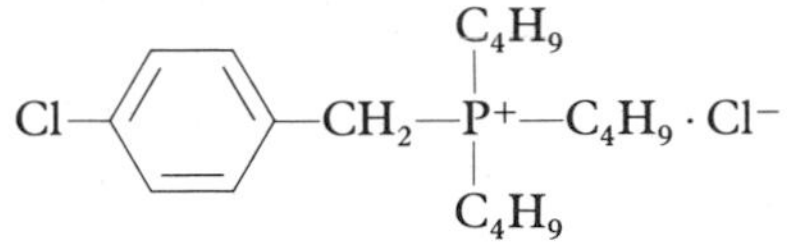

2,4-Dichlorobenzyltributylphosphonium chloride (Phosfon-D)

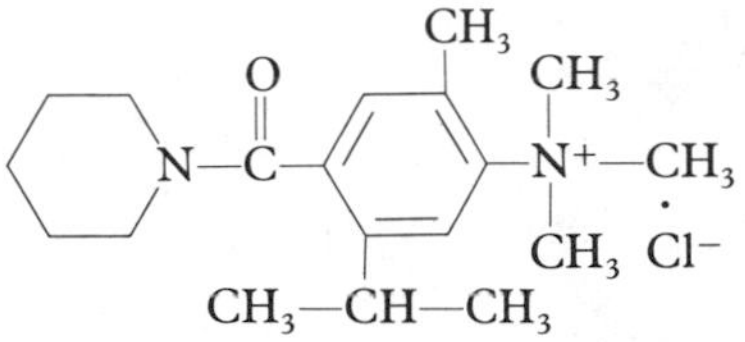

Ammonium (5-hydroxycarvacryl)trimethyl chloride piperidine carboxylate (Amo-1618)

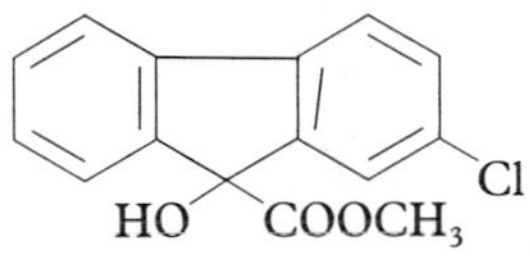

Methyl-2-chloro-9-hydroxyfluorene-9-carboxylate (IT 3456, a morphactin)

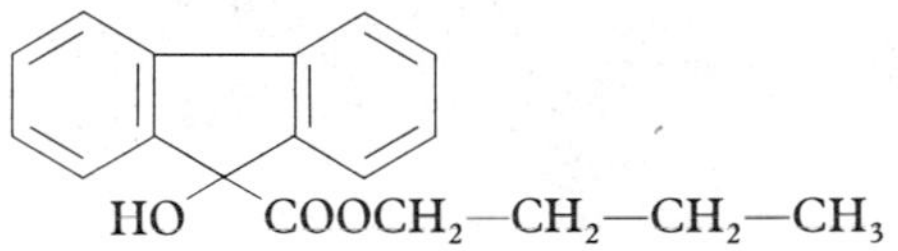

N-Butyl-9-hydroxyfluorene-9-carboxylate (IT 3233, a morphactin)

Figure 14-12 Structural formulas of some synthetic inhibitors. SADH, CCC, Phosfon-D, and Amo-1618 are important plant growth retardants. [After Weaver, 1972]

GROWTH-REGULANT INTERACTIONS

Hormones do not act alone, nor do they have a single function. They affect different organs differently and at least some operate sequentially. The extreme complexity of the functional interrelations among auxins, GA, cytokinins, ABA, and ethylene has prevented a complete understanding of them. But they must be understood as an integrated whole if they are to be understood at all. In general they seem to control the genetic information of the cell in which enzymes are made for specific biochemical reactions, for example: DNA $\rightarrow$ mRNA $\rightarrow$ enzyme.

Only a few specific reactions are known. GA promotes the synthesis of the enzyme α-amylase, which in turn increases starch hydrolysis and thus enhances germination in some seeds. It also seems to mediate the synthesis of tryptophan, a precursor of auxin. Cytokinins appear to form an intimate part of some transfer RNA's and possibly in this way to influence cell division and other growth phenomena. Ethylene promotes nucleic acid and enzyme-protein synthesis, leading to abscission and ripening. ABA, on the other hand, seems to shut down the whole process.

Auxin promotion of growth depends upon the organ and the concentration. Optimum promotive concentrations are low for roots, intermediate for buds, and high for stems. It is now known that so-called auxin-inhibited growth results from auxin-stimulated ethylene production. The critical concen-

A B

Figure 14-13 SADH at 500 ppm 50 days after full bloom increased floral initiation in **(A)** Bosc pear compared with **(B)** control.

Figure 14-14 SADH treatment reduces the size of apple fruit and makes it less elongated. This is the opposite effect of a GA treatment, which results in more elongated fruit. *Left to right:* Control, 1000 ppm, and 2000 ppm SADH. [From Williams, Bartram, and Carpenter, 1970. *HortScience* 5(4):257]

tration of auxin that induces ethylene is different for different tissues. In apical dominance, lateral buds are inhibited by auxin-induced ethylene at the nodes. Cytokinins not only reverse the inhibitory effect of auxin to cause bud growth but also counteract the effect of applied ethylene.

To oversimplify the way hormonal factors probably regulate fruit growth: GA is dominant at the earliest stage; cytokinin is dominant during the cell-division phase; auxin becomes dominant during cell enlargement (vanOverbeek, 1962). These sequential changes in hormone levels occur naturally in fertilized, seeded fruit, but parthenocarpy may be induced by early GA application, which substitutes for that produced by young embryos.

As the fruit begins to mature, another hormonal sequence occurs. A natural or applied increase in auxin results in increased ethylene synthesis, which in turn mediates the production of "ripening" enzymes and stimulates abscission. However, such auxins as 2,4-D or 2,4,5-TP applied to maturing fruit will *prevent* ethylene-stimulated abscission while stimulating ethylene-induced ripening.

Chilling, to remove rest in seeds, appears to reduce ABA and increase GA levels, resulting in germination. Incompletely chilled seeds will often germinate when extra GA is applied to counteract ABA. Because both synthesis and action of one hormone may be related to one or more others, several different treatments or combinations of treatments can achieve the same net result. For example, GA treatment to young fruit results in an increase in auxin synthesis. Thus in some fruits, such as pear, treatment with either GA or auxin induces parthenocarpic set. Another example is the auxin-ethylene relationship in maturing fruit. Auxin treatment induces ethylene synthesis, so that treatment with either will hasten ripening.

Because of the sequential action of plant hormones, timing and concentration are critical. When two or more regulants are to be used, the sequence of their use may also be critical. Because most growth regulants are applied as foliar sprays, factors affecting retention and penetration of sprays by plant cuticles are important to understand. In general, plants developed under high temperature, low humidity, and high light intensity develop thicker, more waxy cuticles than those developed under the opposite conditions. Penetration in the latter case is much greater. Also, post-spray conditions of high humidity (slow drying) and high temperatures result in greater than average penetration. Finally, the addition of wetting agents (surfactants) may either enhance or reduce penetration depending upon the specific regulant and the specific surfactant. Surfactants should not be used unless specified. Table 14-1 indicates the current uses of growth regulants in deciduous fruit and nut production.

Table 14-1 Uses of growth regulants in pomology.

Use	Crop	Concentration or Amount[1]	Timing of Treatment[2]
	Gibberellins (GA)		
Reduce yellows virus effect	Sour cherry	15–25 ppm	10–15 days APF
Delay maturity	Sweet cherry	5–10 ppm	3 wks. before harvest
Larger, firmer fruit			
Reduce rain cracking			
Improve fruit shape and size	Apple	5–25 ppm	First PF
Improve fruit set	Pear (some cultivars)	10–20 ppm	At FB or PF
Prevent premature ripening	Pear (Bartlett)	100 ppm	4 wks. BH
Improve fruit quality	Italian prune	20–50 ppm	4–5 wks. BH
Improve fruit set and size	Blueberry, cranberry	10–50 ppm	FB–PF
Increase fruit and cluster size	Black Corinth grape	2.5–5 ppm	Just AFB
Increase fruit size	Thompson seedless grape	2.5–20 ppm	FB
Loosen clusters	Thompson seedless grape	20–40 ppm	At fruit setting
Induce seedlessness	Delaware grape	100 ppm	Before FB
Increase fruit size, hasten maturity	Delaware grape	100 ppm	At FS
Loosen clusters, reduce rot	Compact clustered grapes	1–10 ppm	2–3 wks. BFB
Increase seed germination	Apple, pear, cherry, filbert	5–100 ppm	Pre-germination
	Auxins (NAA, NAD, IBA, 2,4-D, 2,4,5-TP)		
Chemical thinning	Apple	10–20 ppm NAA	15–25 days AFB
Chemical thinning	Pear	10–15 ppm NAA	15–21 days AFB
Chemical thinning	Apple	20–50 ppm NAD	7–14 days AFB
Increase fruit set	Pear	2–7.5 ppm 2,4,5-TP	After harvest
Prevent preharvest drop	Pear	10 ppm NAA	3 wks. BH
Prevent preharvest drop	Apple	20 ppm NAA	3–4 wks. BH
Prevent preharvest drop	Apple	10 ppm 2,4,5-TP	5–6 wks. BH
Prevent preharvest drop	Apricot, Italian prune	5–20 ppm 2,4,5-TP	2 wks. APH
Reduce rain cracking	Sweet cherry	1 ppm NAA	35 days BH
Rootsprout control	Apple, pear, prune, cherry, filbert	1000 ppm 2,4-D or NAA	Early summer
Increase fruit set	Blackberry	β-NOAA 50–100 ppm	Berries ½ size
Rooting of cuttings	Various species	20–200 ppm IBA soak	Before callusing
Rooting of cuttings	Various species	500–5000 ppm IBA quick dip	Before callusing

[1]These uses and concentrations are not recommendations for use. Each use of a chemical must be in accord with state and federal regulations.
[2]FB = Full Bloom; AFB = After FB; PF = Petal Fall; BH = Before Harvest; ABB = After Bud Break.

(continued)

Table 14-1 (continued)

Use	Crop	Concentration or Amount[1]	Timing of Treatment[2]
	Cytokinins (BA, kinetin)		
Increase branching	Several species	100–200 ppm	Early summer
Increase fruit length	Apple	BA, 25 ppm	FB–10 days AFB
Increase seed germination	Several species	100–500 ppm	1 day soak
	Ethylene generators (ethephon)		
Floral initiation	Many species	100–1000 ppm	Early summer
Chemical training	Italian prune	200–500 ppm	Early summer
Chemical thinning	Peach, plum, apple	20–200 ppm	4–8 wks. AFB
Induce ripening and coloring	Apple, fig	250–500 ppm	1–2 wks. BH
Hasten hull dehiscence	Walnut	400–500 ppm	Early hull crack
Hasten husk splitting	Filbert	900–1000 ppm	First nuts loose
Induce fruit abscission and facilitate harvest	Peach, cherry, plum, pear, apple	500–2000 ppm	10 days BH
Induce fruit abscission and facilitate harvest	Cranberry, blueberry, currant	500–2000 ppm	10 days BH
Induce fruit abscission and facilitate harvest	Grape	250 ppm	2 wks. BH
Delay bud opening	Cherry, plum, peach	200–800 ppm	Early fall
Hasten defoliation	Nursery stock	2000 ppm	Before harvest
Increase seed germination	Several species	100–500 ppm	1 day soak
	Growth retardants (SADH and CCC)		
Increase fruit set	Grape	100–1000 ppm CCC	Foliar spray
Flower initiation	Pear	1000 ppm CCC	40–50 days AFB
Flower initiation	Pear, apple	500–1000 ppm SADH	30–40 days AFB
Growth control	Apple	1000–2000 ppm SADH	30 days AFB
Prevent preharvest drop, improve quality	Apple	1000–2000 ppm SADH	45–60 days AFB
Delay bloom, increase set	Apple	4000 ppm SADH	Fall spray
Advance maturity, color, loosens fruit	Cherry, peach, plum	500–2000 ppm SADH	2–5 wks. AFB
Delay bloom	Almond	2000–4000 ppm SADH	June, Sept, Oct
Increase fruit set	Grape	2000 ppm SADH	Early bloom
Prevent premature ripening	Pear	SADH	30 days BH
Chemical training, branching	Pear	500 ppm SADH	Early summer
	TIBA		
Increase branch angles	Several species	50 ppm	3–4 wks. ABB
Flower initiation	Several species	25 ppm	4–6 wks. AFB

GENERAL REFERENCES

Mitchell, J. W. 1966. Present status and future of plant regulating substances. *Agr. Sci. Rev.* 4:27–36.

Weaver, R. J. 1972. *Plant growth substances in agriculture.* W. H. Freeman and Company, San Francisco.

Wellensiek, S. J. 1972. Growth regulators in fruit production. *Acta Horticulturae* 34:1-507.

Wittwer, S. H. 1971. Growth regulants in agriculture. *Outlook in Agr.* 6:205–217.

15

Dormancy and Plant Hardiness

Growing plants are nonhardy and incapable of becoming hardy, so dormancy during cold winter months is necessary to survival. In their natural habitats, plants are seldom injured by cold because they have evolved adaptive physiological mechanisms that permit them to be dormant during severe winter weather. For example, the leaves of high-latitude (boreal) species sense the shortening of day length in late summer and initiate inhibitor mechanisms that cause the plant to stop growth well before the first heavy frosts of autumn. Another mechanism causes cold acclimation in response to these first nonkilling frosts. Low-latitude species from areas with mild winters tend to continue to grow as long as temperature and soil moisture are favorable and there is no danger of winter-killing. Middle-latitude temperate species have developed yet a third kind of adaptive physiology incorporating some of the characteristics of both high- and low-latitude species. Winters in the middle latitudes may fluctuate between cold and mild temperatures. Species there have developed long rest-period chilling requirements; they will not begin to grow in midwinter even though it may warm up to growing temperatures for periods of several days.

Like other deciduous plants, deciduous fruits and nuts stop growing in the fall; before winter, they drop their leaves and are dormant; in the spring, their growth resumes. This synchrony between plant and environment insures their survival during the cold winter months. However, the protection of cultivated plants against winter injury may present problems not found in natural habitats. Many cultivated species were either bred for specific fruit-quality factors or have been selected and moved to climates other than that in which they evolved. Thus, many domestic forms are not completely adapted to the environment in which they are cultivated. Cultural practices then become important in augmenting the natural ability of the species to survive cold winters—for example, water and fertilizers are manipulated to prevent late-season growth. The choice of cultivar, rootstock, and interstock also is important in some climates. The grower must know when and how to protect flowers, buds, and bark from extreme winter cold or spring frosts. This chapter deals with the physiology of dormancy and rest; acclimation and hardiness; the technical aspects of how freezing kills plants; practical considerations of hardiness; and principles of frost protection.

Because of some confusion in the use of terms relating to dormancy, a brief discussion of terminology and a few definitions will facilitate understanding the sections to follow. "Dormancy" is the general term used to denote the inactive state. There are several kinds of dormancy:

Quiescence. When buds are dormant as a result of external conditions unfavorable to growth (for example, temperature, available water, photoperiod).

Correlative inhibition. When buds are prevented from growing by the inhibitory influence of another plant part (for example, the dormancy of lateral buds due to the dominance of the shoot terminal).

Rest. When buds are dormant because of internal physiological blocks that prevent growth even under ideal external conditions for growth. Chilling temperatures above freezing terminate rest.

Onset of rest. The transition in autumn between quiescence and full rest.

Typically, a plant adapted to the temperate zone has a grand period of growth during the first half of summer, after which growth ceases and terminal buds form. At the beginning of this period, the buds are quiescent and may be forced into growth by such things as pruning, defoliation, irrigation, or nitrogen fertilizer. In autumn, the onset of rest begins. During this period, rest becomes progressively deeper until some time in October, November, or December in the Northern Hemisphere, depending on the species or cultivar. A specific amount of chilling is required to terminate or break rest and restore the bud's ability to expand and grow again. Effective chilling temperatures to terminate rest are between 0°C and about 7°C or perhaps as high as 10°C for some species. Internal inhibitor systems, which regulate rest, seem to be enzymatically altered at these temperatures. Rest often is terminated during winter in both high- and low-latitude species. After that, bud growth will occur whenever favorable temperatures occur. It is during this winter period that, if the temperature warms, buds become active and lose much of the hardiness they had during rest. Hardiness is lost rapidly when growth resumes. At full bloom, the plant lacks hardiness and must be protected from frost to prevent injury.

REST OF BUDS AND SEEDS

Buds

Shortening day length in late summer triggers the cessation of growth in many species. Leaves are the receptors for this short-day response, the actual mechanism being the conversion of the pigment phytochrome from one form to another. But whether or not a species responds to day length, quiescence begins some time between midsummer and late fall. For example, the mature apple tree tends to set terminal buds early, and apricot ceases growth several weeks later (Walker and Seeley, 1973). The transition from quiescence to rest is usually complete by October or November. Leaf-fall normally occurs during this transition.

The periods of onset of rest, rest, and breaking rest are accompanied by changes in growth-regulating hormones and metabolism (Figs. 15-1, 15-2). Research indicates that inhibitors such as ABA tend to increase, while promoters and respiration decrease, as buds go into rest. At the termination of rest, there is a sharp rise of promoters relative to inhibitors, and respiration increases sharply. The chilling required to break rest varies from very little in, say, almond to more than 2000 hours for maximum growth in some grape cultivars (Fig. 15-3). Intermittent warm and

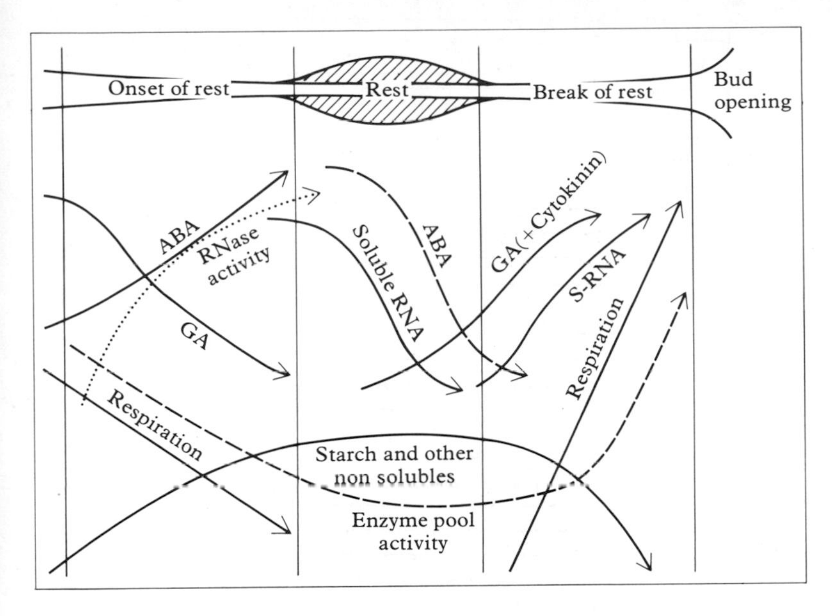

Figure 15-1 Schematic description of metabolic activity in relation to stages of rest. [After Lavee, 1973]

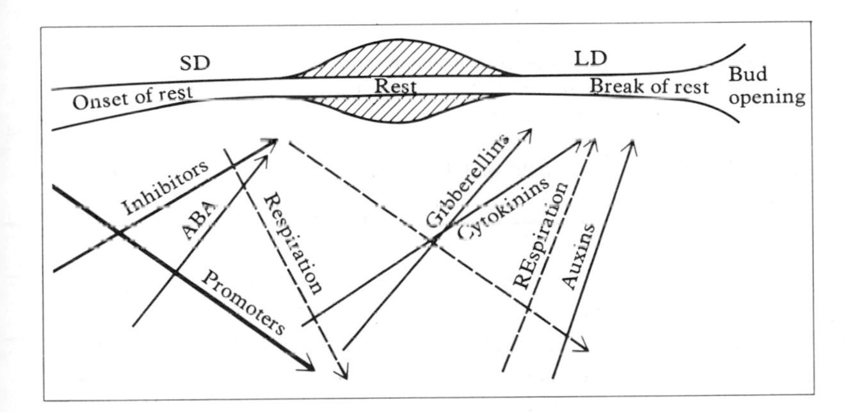

Figure 15-2 Schematic description of the changes of growth-regulating substances in relation to stages of rest. [After Lavee, 1973]

cold periods during winter may require more bud chilling than continuous chilling temperatures (Overcash and Campbell, 1955).

Seeds

In many ways, rest in seed resembles that in buds of the same species (Westwood and Bjornstad, 1968a). A study with seed of fourteen pear species serves as an example of the way seed and tree development are related. The greater ease of experimentation with seed permits their use in elucidating dormancy phenomena. Species from warm winter climates require less chilling (3 to 10°C) than those from colder climates, ranging from 5 days in *Pyrus pashia* to 180 days with *P. pyrifolia.* Temperatures at or below freezing are relatively ineffective in breaking rest.

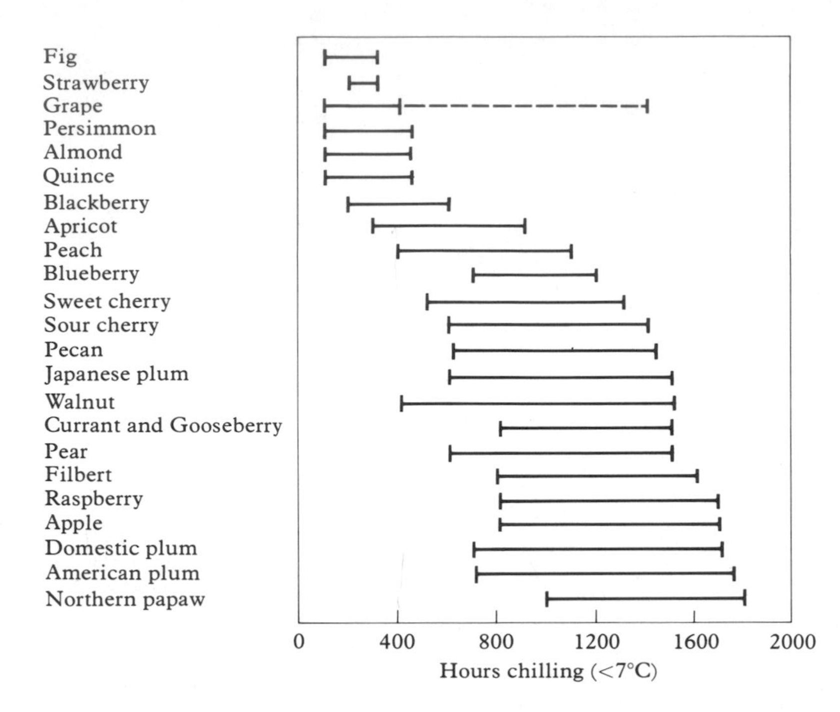

Figure 15-3 Approximate chilling requirements to break winter rest for fruit and nut species. The ranges shown for each species indicate the differences between low- and high-chilling cultivars within the species. Grape will grow with very little chilling but will begin growth much faster after long chilling. [Partially based on data of Chandler, Kimball, Philp, Tufts, and Weldon, 1937]

Drying chilled seed reduces subsequent germination. Seedcoat inhibitors (which can be leached with water) reduce germination and growth of chilled seed. Germination of long-chilling species is enhanced by a pre-chilling soak of GA_3.

There are several natural controls of seed germination in temperate plants, only one of which is removed by chilling. Hard impervious coverings such as the stony endocarp of stone fruits, may impede oxygen and water penetration. All such controls are presumed to have survival value by properly synchronizing plant development with the environment and seasonal sequence in which the species evolved. The control mechanisms in seed fit a logical pattern for protecting the viability of the seed and also insuring the establishment of young seedlings in nature. As fruit mature and drop to the ground, the seed, although moist and mature, do not germinate, because they are in rest and because they have inhibitors in the seedcoats (Strausz, 1969). During winter, the moist seed (either in the fruit or on the soil surface) are chilled and rest broken, but most will fail to germinate if the seedcoat inhibitors are not washed out by rainfall or surface moisture. If there is not enough water to wash out the inhibitors, the soil might be too dry to sustain and establish the new seedling. Thus, the seedcoat inhibitors prevent germination under dry conditions. As the seed dry out, they again go into rest and may establish after the next winter cycle, because they remain viable after drying. This extended period of viability affords a greater chance of survival. The specific chilling requirement for a given seed determines whether it gets enough chilling to germinate in a given winter season. This requirement varies from seed to seed and permits the development by natural selection of physiological types (ecotypes) within a species, each adapted to a somewhat different environment. Chilling reduces the inhibitor ABA in the seed (Fig. 15-4).

Once having established the parameters describing dormancy and germination control in a group of species, one can make selections from species or interspecific hybrids according to predetermined objectives, either for studies in basic physiology or for direct use in agriculture. The chilling requirements of seed are similar to those of buds of the trees grown from the seed, so that chilling behavior of seed will predict the requirements of the trees. For example, if trees are desired which have only a 200-hour chilling requirement, one can collect large numbers of seed and subject them to 200 hours of moist chilling and grow only those seed which germinate. Likewise if a long chilling requirement is desired, one can chill a large lot of seed, and periodically discard those which germinate with less than the desired chilling. In both buds and seeds, rest appears to be controlled by the balance of promoters and inhibitors rather than by one or the other alone.

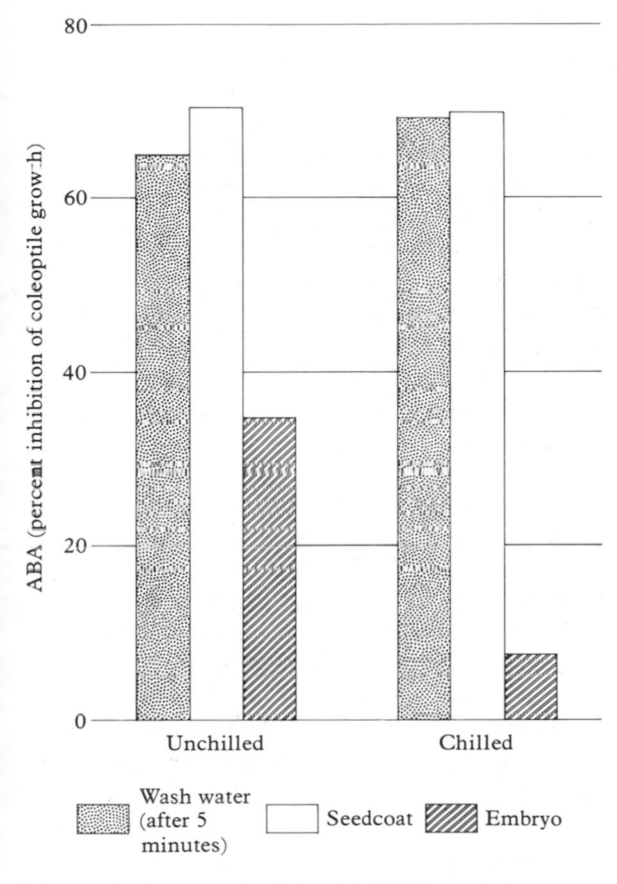

Figure 15-4 The abscisic acid (ABA) level in domestic pear seed is related to washing and chilling. Unchilled embryos have much more ABA than chilled ones, but the amount of ABA in the seedcoat and in the wash water is not affected by chilling. In this controlled test, extracts from pear seeds and wash water were applied to oat coleoptiles; inhibition was reduced only when the chilled-embryo extract was applied. [After Strausz, 1969]

FREEZING INJURY

Kinds of Injury

Low temperature injury is probably the most important factor determining the distribution of plant species on earth. Common types of damage to deciduous fruits are winter sunscald of thin barked species; frost splitting of tree trunks; blackheart of stems; frost heaving of soil and damage to crowns of herbaceous perennials, such as strawberry; freezing of roots; midwinter kill of dormant flower buds; death of cambium in twigs, branches, and trunks; and frost damage to flowers and fruit during spring and fall (Weiser, 1970). Some plants are very resistant to freezing, but many of our most important fruit species have only moderate hardiness. There are two main areas in which attempts are being made to reduce freeze damage. The first is simply to breed more hardy cultivars. The second is to use hardy intermediate framestocks, physical protection, climate modification, and chemical and cultural manipulation (*a*) to slow down growth and induce wood maturity in early autumn, (*b*) to prolong winter rest, and (*c*) to delay growth and dehardening in spring.

How Freezing Kills

Explanations of how freezing kills plants are complicated by the fact that the same plant will vary drastically in hardiness at different seasons, adjacent cells or tissues are not equally hardy at a given moment, and different plant organs do not possess the same degree of hardiness. For example, in midwinter, living xylem ray tissues may be killed at −35°C, but adjacent cambial and phloem cells may survive to −50°C. Flowers in the buds may be uninjured, while the tissues attaching the bud to the branch are killed—yet the reverse often occurs. Also, microclimate influences hardiness, as shown by the fact that roots of a species are much less hardy than stems above ground, yet stems covered with soil lose hardiness while roots exposed to above-ground conditions become hardier.

When woody plants are in active growth, death of tissues occurs almost at the moment of freezing, which, due to a small amount of supercooling, is at about −2° to −8°C. With slow freezing, heat of fusion warms the tissue somewhat, and continued freezing occurs at only a slightly depressed freezing point of −0.3° to −1.0°C, due to solutes in the water. In the fall, as woody plants attain some hardiness, death no longer occurs at the moment of freezing, even though supercooling and the extent of freezing-point depression remain unchanged. In nature, the relatively slow rate of freezing results in the first ice being formed outside the cell protoplasts, where water is purest. Hardy plants can survive extensive amounts of extracellular freezing. But, if freezing is rapid, say 10 degrees per minute, then ice crystals may form suddenly in the protoplasm (this is termed "intracellular freezing"), and death invariably occurs. However, at extremely rapid rates of cooling in the laboratory, noncrystalline ice (vitrification) may occur in the protoplasm, and even nonhardy cells are not killed. Thus, it is the consequences of ice rather than low temperature that kills plants.

Cooling rates rapid enough to cause intracellular freezing are rare in nature, but relatively rapid fluctuations in temperature are common in some climates. Sunscald injury occurs on the south-facing branches or southwest sides of tree trunks, under conditions in which incident radiation warms the tissues during a sunny winter day, followed by rapid cooling when shaded by obstructions or after sunset. Sunscald injury may be due to intracellular freezing or to loss of hardiness by the warm temperature. Since the rate of water exit from cells is critical to whether freezing occurs inside or outside the cells, factors affecting the permeability of membranes to water are important. Membranes of hardy plants are more permeable to water than are nonhardy ones.

Slow freezing injury or death occurs with many cultivated trees, shrubs, and vines at temperatures between −15° and −40°C at cooling rates that occur naturally. Weiser (1970) has summarized a large body of research data relating to such injury and has presented a chronology of events leading to death by slow freezing: Supercooling → Freezing of extracellular water → Rapid propagation of ice through the stem → A rise in tissue temperature (exotherm) from release of heat of fusion → Further cooling after readily available water is frozen → Movement of water from the protoplast out of the cell in response to the external vapor pressure deficit → Continued extracellular freezing of water → Growth of ice crystals → Shrinkage of protoplasts, plasmolysis, and concentration of cell solutes → Slowed migration of water to external ice, as most or all freely mobile water is frozen → Continued temperature reduction until the critical temperature is reached → Granulation of protoplasm → Death (Fig. 15-5). This sequence of events applies to the phloem (bark) of some hardy species. In trees and shrubs native to high-latitude boreal forests, which survive −196°C, all water is frozen. There may be multiple exotherms, but the first and last are the only ones that signal events crucial to survival. Recent evidence has been reviewed (Burke et al., 1976), indicating that for many temperate fruit and nut species deep supercooling occurs as a freezing-avoidance mechanism in flower buds and xylem-ray cells. Depending upon the species, spontaneous nucleation of water occurs at −25° to −40°C, at which point the tissues are killed by intracellular freezing (Fig. 15-6).

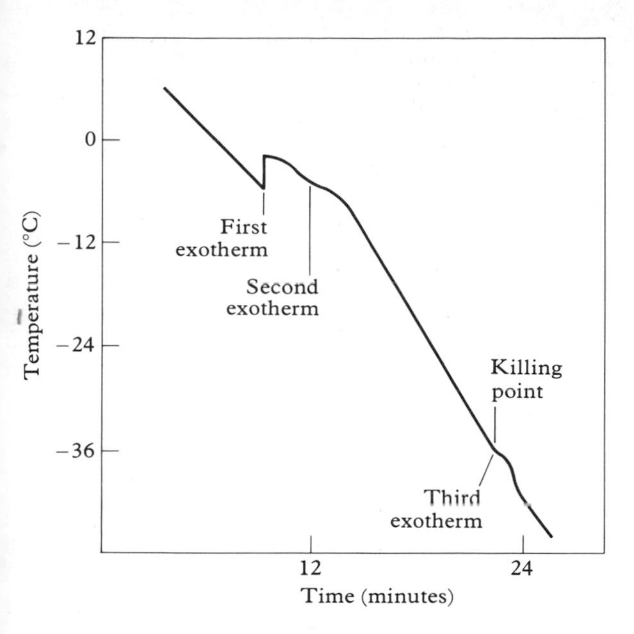

Figure 15-5 A typical freezing curve—a record of tissue temperature during the controlled freezing of an acclimated stem section of a semihardy woody species. Exotherms are the points at which the heat of fusion from the freezing of water in the excised stem detectably raises the sample temperature. This figure shows the cooling curve of a stem that initially supercooled to −6°C before freezing. The third exotherm has been observed to be the killing point of stem tissues in a number of woody species. [After Weiser, 1970]

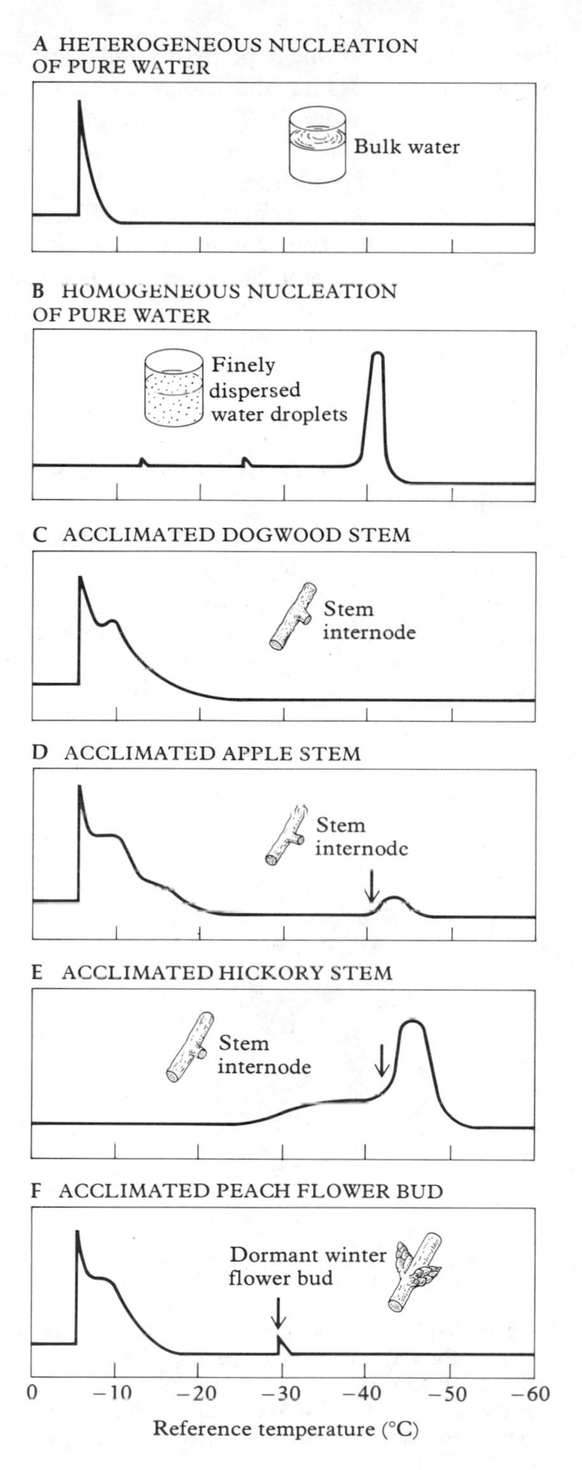

Figure 15-6 Effect of deep supercooling of water as a freezing-avoidance mechanism in temperate species. **(A)** Exotherm of water with little supercooling. **(B)** Exotherm of water at about −40°C with deep supercooling. **(C)** Acclimated dogwood; no supercooling. **(D, E, F)** Exotherms at −30 to −40° for apple, hickory, and peach, indicating deep supercooling. When nucleation occurs, intracellular water is frozen and tissues are killed. [After Burke, Gusta, Quamme, Weiser, and Li, 1976]

One hypothesis (vital-water exotherm) to explain death in the moderately hardy species is suggested by Weiser (1970). He proposes that during freezing a point is reached when all freely mobile water has been withdrawn from cells and frozen extracellularly, leaving only "vital water" in the protoplasm. As temperature continues down, vital water is pulled out of the protoplasm, setting off a chain reaction of denaturation, further vital-water release, and death. Such events, however, do not take place in the very hardy boreal species (Fig. 15-7).

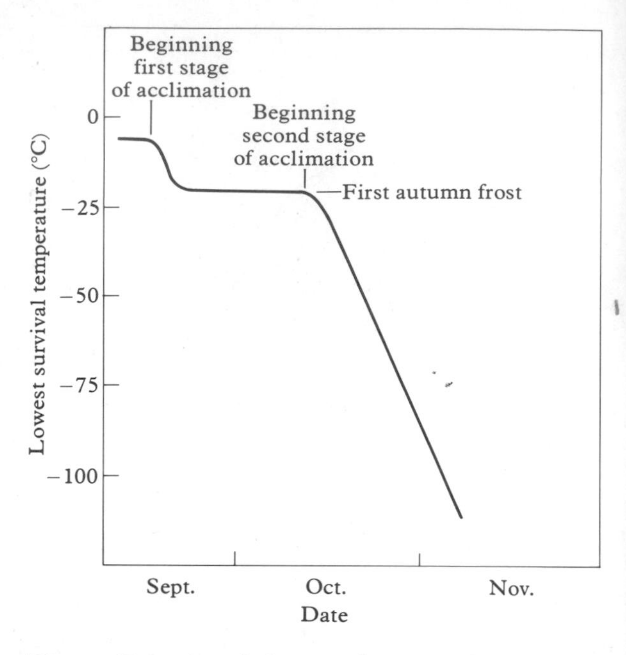

Figure 15-8 A typical seasonal pattern of cold resistance in the living bark of *Cornus stolonifera* stems in Minnesota. In nature, acclimation in this hardy shrub—and in a number of other woody species—proceeds in two distinct stages, as shown. The beginning of the second stage of acclimation characteristically coincides with the first autumn frost. [After Weiser, 1970]

COLD ACCLIMATION

Hardy woody plants undergo a series of changes in late summer and fall that prepare them for colder winter temperatures (Fig. 15-8). In adapted species, the environmental cues are shortening day length and cool nights, followed later by frosts of 0° to −5°C. However, many cultivated plants are either so much altered by breeding and selection, or else the climate is so unlike that of their place of origin, that they do not respond physiologically to these cues and thus do not acclimate properly in some production areas. Thus, one of the principal problems with deciduous plants is late-fall or early-winter freezes for which the plants are not physiologically prepared: freeze damage can be seen in the coastal clone of dogwood in Figure 15-9.

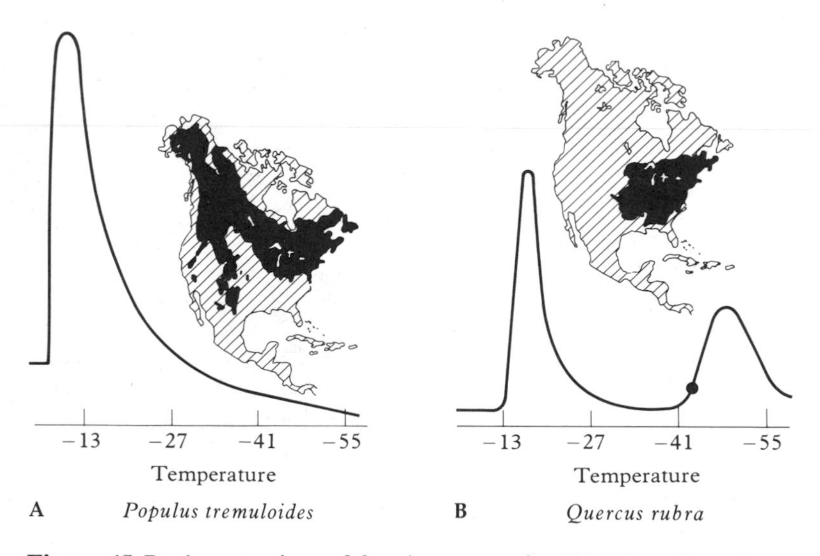

Figure 15-7 A comparison of freezing curves for **(A)** a boreal species with one for **(B)** a temperate species. The former shows no exotherm following the initial one for extracellular water. The latter shows supercooling to about −43°C, then an intracellular freezing exotherm, which causes death at this point. [After Burke, Gusta, Quamme, Weiser, and Li, 1976]

It is now known that many physiological changes take place during acclimation, so it is not merely a passive event brought on by cessation of growth (quiescence or rest). For example, during acclimation of dogwood, a hardy shrub, changes are found in proteins, lipids, tissue hydration, hardiness promoters, carbohydrates, organic acids, free and bound amino acids, and in nucleic acids, which provide basic

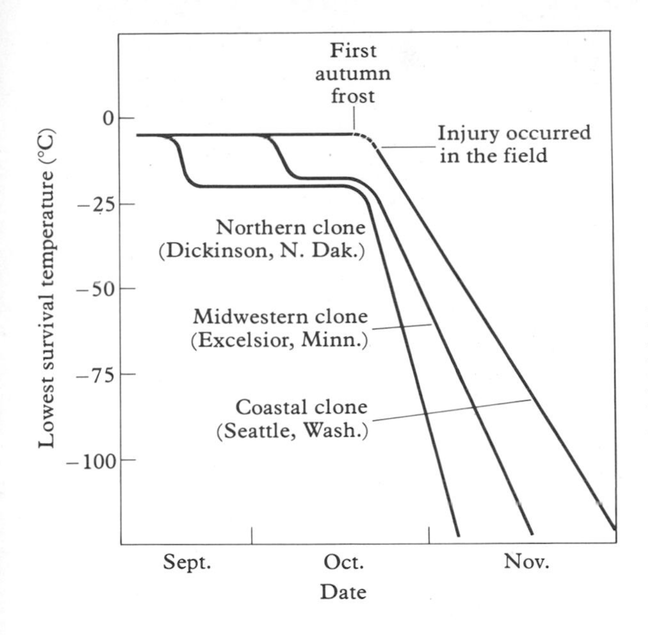

Figure 15-9 A diagrammatic comparison of the typical seasonal patterns of cold resistance in the living bark of three climatic races of *Cornus stolonifera.* The acclimation curves shown are for clones from North Dakota, Minnesota, and Washington grown in the field in Minnesota. Races from regions with mild climates and long growing seasons acclimate later and more slowly than clones from regions with severe climates and short growing seasons. Over twenty-five clones that have been collected from widespread locations in North America became resistant to −196°C by midwinter in Minnesota. [After Weiser, 1970]

information to the cells. In woody plants, these hardening processes do not take place during growth.

The first stage of acclimation of an adapted species is induced by short days, which causes growth to cease. At this stage, the induction of rest is not required—only the cessation of growth. The second stage of acclimation (Figs. 15-8 and 15-9) requires low temperature (frost). Possibly, enzyme systems are activated by frost, which are metabolically active during the warmer days following cold nights. Hardiness is quickly lost with even a few hours of thawing to a more moderate temperature. Known rapid midwinter fluctuations in hardiness seem to be closely related to the temperature the preceding day.

Fuchigami et al. (1971), using grafted, split plants of dogwood—in which part of a plant could be exposed to different light and temperature regimes—studied acclimation in a way that should be of interest to all fruit growers. His major results are:

1. Plants depleted of food reserves cannot acclimate.

2. Leaves respond to the short-day stimulus to start first-stage acclimation.

3. Low temperature inhibits the short-day induced response.

4. On long days, leaves produce a translocatable acclimation inhibitor.

5. On short days, leaves produce a translocatable acclimation promoter.

6. The hardiness promoter moves from leaves to stems through the bark.

7. The promoter from a hardy genotype enhances hardiness of a less hardy type when the two are grafted together.

8. The frost-induced second phase of acclimation is not translocatable.

9. Plants exposed to long days and frost, in time, become fully hardy, while those exposed to short days and high temperatures only reach the first-stage level of acclimation (Fig. 15-9).

These results suggest that the hardiness promoter is either a growth inhibitor or a hormone of some kind. Combined evidence indicates it is a hormone that regulates the flow of cellular information through

The time	Spring	Summer	Early autumn	Late autumn	Winter
The plant	Growing rapidly	Growing slowly — Flowering-fruiting	Growth slows – – – – Photosynthates accumulate; Growth stops – – – – Rest period induced	Leaves drop; Dormant	Dormant
The environment	Lengthening days; Rising temperatures	Long days; Warm temperatures	Shortening days; Spectral changes in sunlight; Warm days — Cool nights	Short days; Frost	Prolonged subfreezing temperature
Acclimation — Biochemical events	(hardiness inhibitor(s) produced in leaves)	(hardiness inhibitor(s) produced in leaves)	Phytochrome P_R ⇄ P_{FR}; (Hardiness promoter(s) synthesized); DNA depression; (Active metabolic changes)	(Synthesis and/or induction of enzymes); Frost induced protein rearrangements exposing new surfaces -disaggregation?-	
Acclimation — Biophysical properties			Altered structural proteins and membranes; (Reduced hydration and increased water permeability)	Cellular components assume stable configurations; Quasi-crystalline water is bound to proteins; (Protoplasm becomes elastic and resistant to dehydration)	Remaining cellular water surrounding macromolecules becomes Highly ordered and tenaciously bound
Status of resistance	Tender — Death at first exotherm (Incapable of acclimation)	Tender — Death at first exotherm (Capable of acclimation)	First stage of acclimation — Death at second exotherm (some acclimation regardless of environment) (inhibition by low temperature)	Second stage of acclimation — Death at third exotherm	Third stage of acclimation — Death at third exotherm (Extremely resistant to dehydration)

Figure 15-10 A model of cold acclimation in hardy woody plants. Numbered arrows indicate the hypothetical sequence of events resulting in the most efficient and complete acclimation. Arrows without numbers identify sequential relationships and alternate acclimation pathways. Parentheses denote events that can be observed experimentally. [After Weiser, 1970]

nucleic acids. Figure 15-10 presents schematically the general hypothesis proposed to explain acclimation in hardy woody plants.

HARDINESS

Flower Bud and Shoot Hardiness

Before 1940, the view was generally held that hardening occurs as growth ceases in the fall in response to shortening days, lowering temperatures, and cultural practices which tend to stop growth—and that the period of greatest hardiness coincides with that of deepest rest. After rest is broken by chilling, hardiness is lost as temperatures rise and growth begins, with plants becoming completely tender about the time of full bloom. It was recognized that in the fall different organs or tissues do not harden at the same rate, nor do they all ultimately reach the same degree of hardiness.

In a recent review, Proebsting (1970) pointed out that during the 1940's it was found that peach-bud hardiness fluctuates in response to winter temperatures, even during rest. Using peach as the test plant, he developed a three-period hardiness model. The first period of hardiness begins at about −21°C and is attained by early November. This level may occur at or before the first frost and is usually quite constant. At this time buds may be hardier than twigs, a condition that later is reversed. If no cold weather (−2.2°C or below) occurs, then the hardiness level remains at −21°C during rest. With cold preconditioning, hardiness may increase to −28°C but will not deharden to above −21°C while buds remain in rest. The second period begins when rest ends and enough warm weather is experienced to initiate growth. Termination of rest is conveniently marked by microspore meiosis in the anthers. As growth advances in response to temperature, the temperature at which 50 percent of the buds are killed (LT_{50} = lethal temperature for 50 percent) gradually moves higher. Hardiness may increase in response to colder temperatures and arrested growth, but if no cold weather occurs, LT_{50} continues to rise slowly until a week before first pink stage, when there is a rapid loss of hardiness. The third period is open-bloom and post-bloom, in which the plant is tender and buds are killed at the first freezing point (Fig. 15-10). However, recent evidence indicates that at this stage some slight hardiness can be acquired by a cool temperature the previous day, and some cultivars are known to be more hardy at anthesis than others.

Proebsting (1970) found that hardening at below 2.2°C occurs at a much slower rate than does dehardening at above this temperature. Continuous cold results in 0.5 to 2 Celsius degrees hardening per day, while mild temperatures can reduce hardiness as much as 0.5 degree per hour. During rest, however, dehardening continues only to −21°C, which is the minimum hardiness level (Fig. 15-11). The foregoing data on peach may conform to the general pattern for fruit species, but peach, with its unmixed solitary flower buds, is a simpler system than species with multiple flowers and—in some species—both flower and leaf primordia in the same bud. In more complex buds—such as those of cherry, apple, pear, and raspberry—the flower bud often is not killed, but the supporting basal tissue is. Other tissues of the whole plant vary in hardiness and usually are killed at temperatures much above that which kills dogwood, previously discussed under acclimation. The hardiness of grafted plants, a much more complex subject, will be considered next.

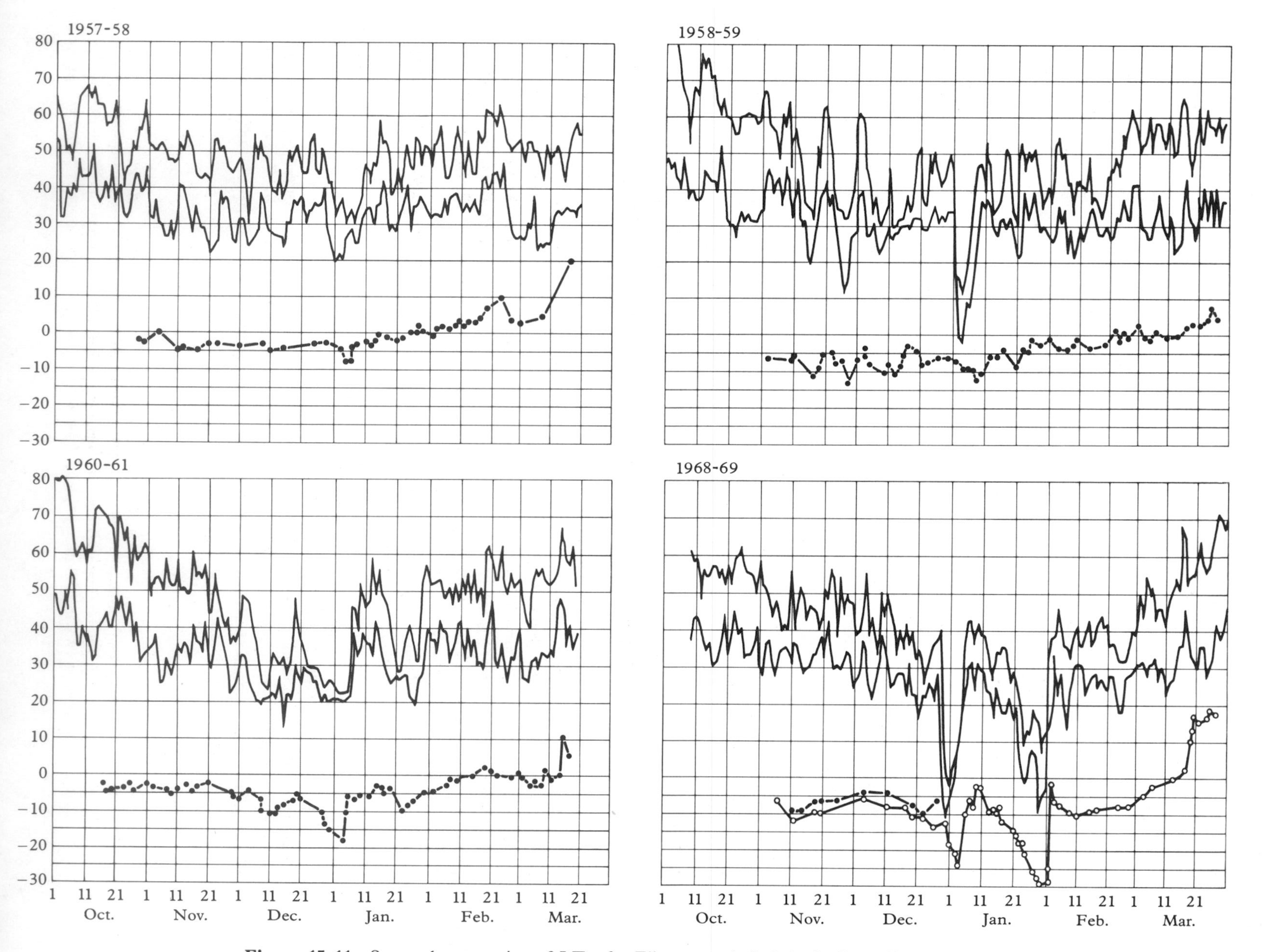

Figure 15-11 Seasonal progression of LT_{50} for Elberta peach fruit buds *(lower line)* and daily maximum and minimum temperatures during four selected winters. The 1968–69 data compare Elberta (*dots in circles*) with Bing cherry (*clear circles*). [After Proebsting, 1970]

Hardiness of Grafted Plants

In fruit production, one of the key limiting factors is freeze damage in the fall or winter to bud, bark, or wood. In part, this is due to the selection of cultivars by criteria other than hardiness and to man's attempt to extend otherwise good cultivars beyond their hardiness range. Damage from a freeze varies between root, trunk, twig, and bud and is further complicated by the presence of at least two and sometimes three genetic systems. Three-piece genetic systems have a root of one type, on which is grafted a trunk or frame of a special type, which in turn is worked to the scion cultivar. These compound genetic systems, physiologically complex are even more complex in hostile environments that result in cold injury to one or more organs or genetic systems.

Also, several other factors influence the extent of cold injury. Such factors as nutrient deficiencies or excesses, diseases and pests, previous crop density, irrigation, tree vigor, pruning, preconditioning temperatures, short-term temperature variations, and the time of season at which the freeze occurs—all affect the extent of injury.

Mechanisms inducing dormancy and rest in plants, discussed previously, are important because growing tissues are nonhardy and are damaged more than dormant ones at any given temperature. While these mechanisms are partially understood for simple plants, they are poorly understood for compound genetic systems.

APPLE One of the most common types of winter injury comes in late fall or early winter. A notable example of such injury occurred in November 1955 in the Northwest states. Temperatures during the first ten days of the month were generally mild. Leaves were still on the trees, and many young trees were still growing. On November 11, temperatures dropped to or slightly below −18°C and remained there for several days. Extensive damage occurred on most single-grafted trees less than 12 years old. The main site of injury was in the lower trunk and crotches. Under these conditions, Yellow Transparent, Hibernal, Haralson, Canada Baldwin, Antonovka, Charlamoff, and Hyslop Crab trunkstocks withstood this type of trunk damage; Spartan trunks were not hardy; East Malling (M) 16 and M 2 roots were quite hardy; M 9 moderately so; M 7 and M 4 were tender. In Iowa, in a similar freeze in November 1940, Duchess, Yellow Transparent, Wealthy, Haralson, and Hawkeye Greening were uninjured.

In late December 1968, temperatures in the apple areas of Washington and British Columbia dropped to as low as −44°C. In the coldest orchards, the cultivars Delicious and Golden Delicious were killed to the hardy stock graft. Injury in one orchard at −32°C was evident only on Hawkeye Greening interstock. No apparent injury was found with interstocks of Yellow Transparent, Astrachan, McIntosh, Beacon, Hibernal, Antonovka, and Ottawa-292. Near Vernon, British Columbia, one orchard of Delicious and Golden Delicious scions was not seriously injured at a temperature of −39°C, apparently because of the heavy sod, which reduced nitrogen levels in late summer, causing better tissue maturity. In a research orchard at Vernon, injury was assessed as follows after −43°C on December 29, 1968:

None: Hever 20, Antonovka, Dr. Bill, Hever 12
Very Slight: Ottawa-271, J. Luke seedling, Hopa crab, Anoka
Slight: Red Astrachan, Minnesota-447
Slight to Moderate: Haralson, McIntosh

Injury, where present, was not severe and normal recovery occurred in 1969.

A review (Lapins, 1963) separates a number of apple rootstocks according to hardiness, as shown in Table 15-1. Roots such as M 9 and M 7 that induce early maturity of the scion tend to protect trees more from fall freezes than mid-winter cold.

Intermediate framestocks are characterized by Lapins (1963) as shown in Table 15-2.

Stuart (1937) tested the hardiness of Malling layered stocks in late winter (March); his results are given in Table 15-3. Note that M 13 stems were hardy, but its roots were tender. Since this test was done in late winter, the results would likely differ from those done in the fall. Fall tests might change the position of the early maturing M 9 relative to the

Table 15-1 Hardiness of apple rootstocks.

Very Tender or Tender	Moderately Hardy to Hardy	Hardy
M 1, 2, 4, 7, 9, 25	M 3, 8, 11, 16	*M. robusta* #5
MM 106, 109	MM104, 111	Maurer's Dab Selections
Commercial seedling	Alnarp 2, N. Spy	Beautiful Arcade seedling
	Antonovka seedling	Charlamoff seedling
	Grahams seedling	*M. baccata* seedling
		M. prunifolia seedling
		M. sargentii seedling
		M. virginiana seedling
		M. zumi seedling

SOURCE: Lapins, 1963.

Table 15-2 Hardiness of intermediate apple framestocks.

Moderately Hardy	Hardy	Very Hardy
Anoka	Antonovka	Anis
Canada Baldwin	Borovinka	Beautiful Arcade
McIntosh	Duchess	Charlamoff
Melba	Haralson	Heyer-12
Astrachan	*M. robusta* #5	Hibernal
Winter St. Lawrence	Robin	*M. prunifolia sikora-1*
	Yellow Transparent	

SOURCE: Lapins, 1963.

Table 15-3 Hardiness of Malling layered stocks.

Tender	Intermediate	Hardy
	Stem tissue	
M 12, 9, 1	M 4, 5, 2	M 3, 13, 7, 16
	Root tissue	
M 1, 13, 2	M 5, 12, 9	M 3, 7, 16, 4

SOURCE: Stuart, 1937.

late maturing M 16. Any factor that increases tissue maturity usually increases hardiness. Factors that delay maturity—such as excess nitrogen, fall nitrogen application, late irrigation and cultivation, early defoliation, heavy cropping, early pruning—all increase freeze injury.

Workers in Ohio (Rollins et al., 1962) explored both environmental and stock-scion effects on hardiness of tender and hardy apples. On November 21, varietal hardiness of tender cultivars was: Franklin > Delicious > Rome Beauty > Baldwin > Staymared. Differences in hardiness between these tender cultivars and the hardy crabs (Hibernal, Columbia, Virginia Crab, *M. robusta* #5) were greater during early dormancy than in February and March. Hibernal was the most hardy all winter; other crabs lost hardiness in late winter. Hibernal interstock increased hardiness of Staymared and Baldwin but not Delicious scions. Terminal twigs of M 2 and 10 were found to be less hardy in late November than M 7, 1, 4, and 5. However, in March, M 2 and M 10 were hardier than the others. Hardiness in early winter conforms to historical observations, but hardiness in late winter may be quite different.

Several studies have shown that either the stock or scion can affect hardiness of another portion of the compound plant. Stuart (1937) showed that while scion cultivars were not influenced by rootstock hardiness, root hardiness was influenced by the scion—an effect not correlated with the hardiness of the scion. In North Dakota, Dolgo Crab scions increased the hardiness of Malling clonal roots. In Maine, Baldwin/Virginia Crab was hardy, but Baldwin/Hibernal interstock was somewhat tender. A reciprocal influence of stock and scion on hardiness was apparent, but again the effects were not predictable. In Poland, root killing during a cold snowless winter was influenced by scion type. Cox's Orange Pippin/M 9 *increased* root hardiness; the hardy Antonovka scion decreased hardiness of M 9 roots. In Oregon, interstocks of Black Twig, Astrachan, Hibernal, Gold Medal, Minnesota-308, and Minnesota-447 increased the hardiness of the lower trunk (Brown et al., 1964).

Hardy interstocks do not harden properly under snow. Haralson trunks covered with snow in November were uncovered later and exposed to −23°C, at which temperature they were killed.

PEAR The review by Lapins (1963) indicates the degrees of hardiness of pear rootstocks shown in Table 15-4.

In Oregon, interstocks of Comice, German Sugar, Orel 15, Vicar, and Flemish Beauty prevented freeze injury to trunks and crotches. Much of the damage

Table 15-4 Hardiness of pear rootstocks.

Tender to Moderately Tender	Moderately Hardy	Hardy
	Quince	
Pillnitz 1,2,3,5 Malling Quince A, C Pfanderquitte	Severnaya	Melitopolskaya
	Pyrus	
Eierbirne seedling Einsiedeln seedling		*P. betulaefolia* Kirchensaller seedling
	Intermediate stocks	
Beurre Hardy	Grune Jagdbirne Neue Poiteau	Bertrams Stammbildner Gute Graue Old Home Sacharnaya

SOURCE: Lapins, 1963.

Table 15-5 Hardiness of *Pyrus* species.

Tender	Intermediate	Hardy
P. calleryana *P. ussuriensis* (one type)	*P. betulaefolia* *P. bretschneideri* *P. phaeocarpa* *P. ussuriensis* (several)	*P. communis* (several) *P. ovoidea* *P. ussuriensis* (several)

sustained was winter sunscald, which was partially prevented by painting the trunks white or by shading to prevent direct solar radiation.

Hardiness studies of *Pyrus* species in Minnesota showed three groups (see Table 15-5).

Field observations after a November freeze in Iowa indicated that domestic pears were injured but not killed. Peaches and plums in the same area were killed. Anjou/Old Home/French seedling in British Columbia came through the 1955 November freeze better than own-stem Anjou. Long-term tests in Oregon showed that rootstocks of *P. serotina* and *P. ussuriensis* induced much greater winter injury to cultivar tops than did *P. communis* roots. This was true regardless of the intermediate stock used. The roots themselves were not injured. Similar responses from Oriental stocks have been observed throughout the Northwest, but no good explanation for this has been advanced. Recently it was shown that both smallpotted Bartlett/*P. calleryana* trees and larger field trees had a lower bud-chilling requirement to break rest than did Bartlett/*P. communis* (Fig. 15-12). With inadequate chilling (850 hours), Bartlett on *P. calleryana* root flowered and forced normally, while flowering and growth were retarded with Bartlett on *P. communis* root. Further studies (Westwood and Chestnut, 1964) showed that buds from Bartlett shoots grown on *P. calleryana* and given 480 hours chilling grew much better than did Bartlett buds with 480 hours chilling grown on *P. communis* root, both bud types of which were placed in a fully chilled host plant (Fig. 15-13).

While a change in chilling requirement of buds does not necessarily reduce hardiness of tissues, it might do so under certain winter conditions. To further explore the rootstock influence in pear, studies winters. It goes dormant in late summer. In its native habitat of constantly cold winters, it easily survives −50°F without injury. Its moderate chilling requirement is completed in early spring as temperatures rise above freezing. This species is ill adapted to the mild winters at Corvallis, Oregon, and it often suffers winter injury. Under Oregon conditions, the trees go dormant in early fall and never suffer damage in early

Figure 15-12 Influence of rootstock on chilling requirement of Bartlett pear buds. *Left.* Forcing of cut shoots from trees on *P. calleryana* rootstock after 850 hours of chilling. *Right.* Similarly treated branches from trees on *P. communis* rootstock. [From Westwood, 1970]

Figure 15-13 Effect of rootstock on growth of Bartlett pear buds following inadequate chilling (480 hours). Two buds (*arrows*) were placed in a fully chilled stock plant and allowed to grow. The bud *on the left* was taken from a 2-year tree on *P. calleryana* rootstock; the bud *on the right* was taken from a 2-year tree on *P. communis* stock. [From Westwood, 1970]

were done to assess rest-period inhibitors of buds as influenced by rootstock (Strausz, 1970). Abscisic acid (ABA) was found to be the principal rest-period inhibitor of *Pyrus,* but the type of rootstock made no consistent difference in ABA content. Thus it was concluded that the lower chilling requirement of trees on *P. calleryana* root was due to a shift in growth promoters rather than to a reduction of ABA. Observations in the Northwest indicate that *P. communis* cultivars on *P. calleryana* stock are less hardy than the same cultivars on *P. communis* root. Strausz (1970) found that winter buds of *P. calleryana, P. serotina,* and *P. ussuriensis* (all shown to reduce hardiness of cultivars grafted to them) contained less ABA than did buds of *P. betulaefolia* and *P. communis,* both reported to be hardy stocks. More work is needed to clarify this complex stock-scion relationship.

P. ussuriensis is a species adapted to cold Siberian

winter. But the mild winter temperatures between 0° and 10°C satisfy its moderate chilling requirement by about January 1, after which it begins to grow in response to even a few days above freezing. In this active state it has been severely injured by moderate late-winter freezes no colder than −7° to −4°C. As this example illustrates, different degrees of hardiness may be reported for a species when it is observed or tested under different conditions.

CHERRY *Prunus mahaleb* rootstocks are generally found to be hardier than *P. avium* (Mazzard). Mahaleb P. I. 193702 and 194098 clones are hardier than mahaleb seedings. Mazzard stocks, including F 12/1 clone, are considered tender. The Hüttner mazzard clones are somewhat less tender; Hüttner 170 is rated relatively hardy. *P. fruticosa* and *P. pennsylvanica* were rated hardier than *P. mahaleb.* In Colorado, sour cherry/mahaleb was hardier and more resistant to chlorosis than sour/mazzard. Likewise, in New York, Montmorency/mahaleb was hardier than mazzard. Stocks infested with nematodes were less hardy. In Pennsylvania, however, Schmidt/mazzard was reported hardier than trees on mahaleb root. Enough variability exists between seed lots of these two species to explain this inconsistency. After the November 1940 freeze in Iowa, Maney (1942) reported that while most peaches and sour cherries were killed, Early Richmond was hardier than Montmorency and English Morello.

Trunkstock hardiness for cherries has been little studied. Mahaleb, *P. pennsylvanica,* and *P. cerasus* (Stockton Morello, Montmorency, and North Star) are known to be hardier than *P. avium* cultivars. There is some evidence in Oregon that the graft unions of high-worked mahaleb are not as hardy as unworked mahaleb.

PLUM AND PEACH Lapins lists Ackermann, Brüssel, and Damascena plums as tender; Brompton, Kroosjes, and Myrobalan as moderately tender; Marianna, Hüttner IV, and St. Julien as moderately hardy; Pershore as hardy; and *P. americana* and *P. besseyi* as very hardy. In the November 1940 freeze, both American and Japanese plums were killed along with peaches, while some sour cherries and pears survived. Root hardiness was ranked from hardy to tender: *P. besseyi* > *P. americana* > Marianna > peach > myrobalan. Also ranked from hardy to tender were *P. americana* > Marianna > *P. davidiana* > myrobalan > southern natural (peach) > Elberta seedling > Florida peach. Stems were hardier than roots; the hardiness of stems and roots from the same plant was related.

Ackerman (1969) found that north Caucasus peach had better bud hardiness than Elberta. A sharp drop from 15° to 20° to −13°C in March killed Elberta buds, whereas more than 25 percent of the buds survived on twenty north Caucasus clones. More than 50 percent of the buds survived on nine of the clones. Hardiness in the bud-swell and bloom stage would be of great value in domestic cultivars. Chili, Greensboro, and Cumberland peach buds resist low winter temperatures well, but high-to-low temperature fluctuations in winter result in less injury to Pallas and Ambergem buds than to those of other cultivars. Although the wood of *P. davidiana* is hardy, the flower buds are not.

Peaches grown in the relatively warm winters in Georgia show injury after sharp fluctuations in temperature following rest (Savage, 1970). Injury to crotch and trunk cambium occurs at −9° to −6°C when preceded by a warm period. In this type of freeze, no root damage occurs. Late winter trunk temperatures during sunny days are as high as 35°C, which is a higher temperature than is reached in summer when trunks are well shaded by foliage. Blake (1935) reported that peach cambium near the ground line remains active into October in some years. Such a condition probably is stimulated by the basipetal flow of auxin and predisposes the trunk and crotches to injury from freezes in the fall and early-winter, when the buds and twigs are more hardy.

Breeding

There is a close relationship in apple between parental and progeny hardiness. Phenotype predicts the genetic potency for hardiness. Tender × tender progeny are mostly tender, but other crosses are variable. Antonovka produces hardy progeny. In Poland, Antonovka × M 9 produced several promising hardy stocks. Lantz and Pickett stated that hardiness is predictable in apple progeny but comes from multiple-factor inheritance. In the November 1940 freeze, most progenies showed the full range from death to no injury. Hardy × tender types resulted in a high percentage of hardy seedlings, but tender × tender crosses resulted in few offspring that were hardier than the parents. Recent progress in breeding for hardiness has been made at the Ottawa Research Station. This work was accelerated by the development of a portable freeze chamber suitable for field use. Both clonal and hybrid seedling apple stocks are being tested; they are not only hardier than Malling and MM stocks but also show a range of size control comparable to that between M 26 and M 2.

Peaches whose cambiums at ground line remain active late in the fall are not hardy in early-winter freezes. White-fleshed seedlings with red at the stone and some red-leafed types were found to be hardy. These can be used as stocks and top-worked 45 centimeters above the ground to avoid ground line injury. Bud hardiness of peach was shown by Mowry (1964) to be quantitatively inherited; an undetermined number of genes is involved. Redskin, Blake, Ranger, Redhaven, and Boone Country as parents produced superior progeny for bud hardiness. Bud hardiness of crosses could be estimated by the average hardiness of the parents. Similar genetic control of blossom hardiness appears to exist with pear. The newly introduced Oregon pear Rogue Red is quite resistant to frost during full bloom. Part of its parentage comes from Seckel, whose blossoms also are resistant to frost.

Nuts

Almonds are not as hardy as peaches and are thus grown commercially in the subtropical latitudes of California and Arizona. Pistachio trees are about as hardy as almond but require more winter chilling to break rest. Pecans are grown mostly in the southern states. Some strains are hardy enough to survive in Michigan, Illinois, and southern Ontario, but the growing season is usually too short for the nuts to mature. The native black walnut, *Juglans nigra*, is hardy to most states, but Persian walnuts *(J. regia)* are not usually hardy—in California, they may grow late in the fall and be killed by moderate freezes below −9°C. In Oregon, the main cultivar, Franquette, was decimated by temperatures near −18°C in November 1955 and again at −24°C in early December 1972 (Fig. 15-14). Some selections from Manchuria and the Carpathian Mountains in Poland are hardy to −40°C. Most of these selections begin growth so early in the spring that they are killed back by spring frosts. Filbert trees are more hardy generally than walnut and peach, but because they bloom in winter, often the limiting factor for production is the freezing of pistillate flowers, which are said to be killed at about −9°C. The butternut, *Juglans cinerea*, is hardier than other nut species and can be grown in the northern U.S. and Canada. The American chestnut, *Castanea dentata*, is hardier than *C. mollissima*, the Japanese chestnut. Some cultivars of *C. mollissima* are somewhat more hardy than peach.

Small Fruits

The strawberry retains its leaves and, with cool preconditioning, is hardy in late fall at −9°C. During winter at −18°C and without snow cover, plants will be damaged; when the soil alternately freezes and thaws, the heaving that results severely damages both roots and crown (Fig. 15-15). Covering with straw or other mulch can protect the plants.

The American grapes *Vitis labrusca* and *V. riparia* are generally hardy in the fruit areas of North America, but European grape *(V. vinifera)* is more tender and requires a longer season to mature the crop. Early winter freezes may kill the new canes or the primary buds on the canes. Muscadine grapes *(V. rotundifolia)* of the southeastern gulf states are not hardy and are thus confined to areas where temperatures are usually

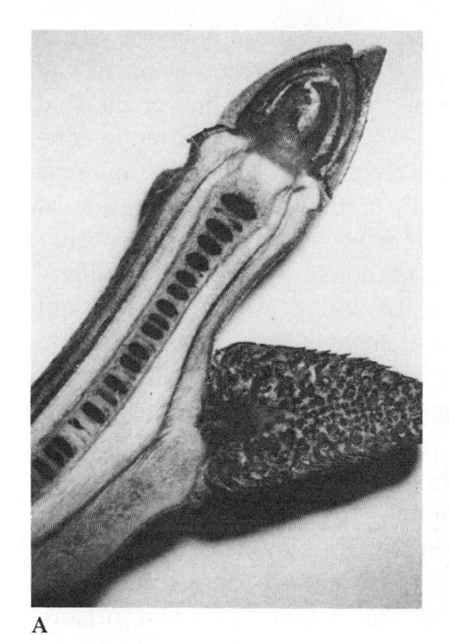
A

B

C

Figure 15-14 Effect of an early winter freeze on Franquette walnut. **(A)** Extensive bud and twig injury. **(B)** Damage to the bark and cambium of the lower trunk. **(C)** Row of Franquette trees (*right*) that died the spring following an early winter freeze. Row on the left is the hardier Spurgeon cultivar. [Courtesy of H. B. Lagerstedt, USDA]

above −12°C. Because of the variability in hardiness and growing requirements of the many cultivars, specific information should be obtained from the local area of interest.

Brambles (*Rubus* species) also vary a great deal in hardiness. The red raspberry is as hardy as most tree species. The variety Chief withstood −46°C in Alberta, but the same freeze killed Taylor plants to the ground. Red raspberry roots are hardy to −19°C, so they are seldom injured. Early winter freezes may kill the tips of new canes if they are still succulent. Premature defoliation by mites or wind also may result in more winter injury. Black raspberries are hardy in northern climates but not as hardy as reds. Erect blackberries are generally hardy in northern areas; western trailing blackberries are less hardy and are thus confined to the coastal valleys of Oregon, Washington, and California.

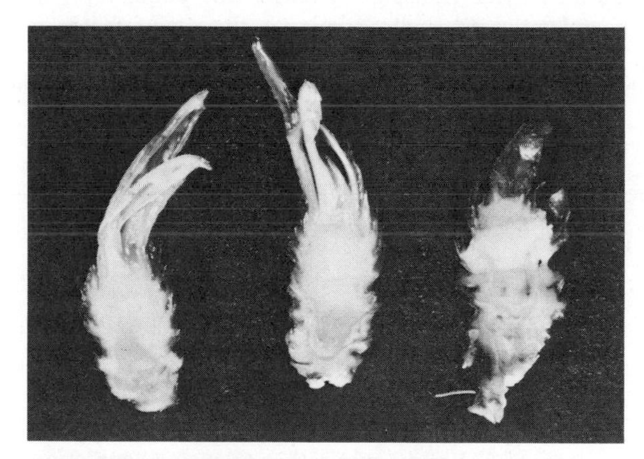

Figure 15-15 Freeze damage to crown and root of strawberry due to frost heaving (*right*). Crowns at left are undamaged. [Courtesy of R. Garren, Jr.]

The highbush blueberry *(Vaccinium australe)* is about as hardy as peach. It may be killed by −29°C but may escape lower temperatures with sufficient protective snow cover. The natural range of the lowbush blueberry *(V. angustifolium)* is from the northeastern United States through southeastern Canada. It is hardy in most fruit-growing areas.

Cranberries survive in northern climates because in winter they are covered with water or snow that protects them from cold. Like the lowbush blueberry, they are native to the most northern fruit areas of North America.

Minor Species

Species with minimal hardiness in temperate latitudes are pomegranate, Chinese gooseberry or kiwi *(Actinidia chinensis)*, fig, mulberry, Oriental persimmon, and Oriental quince *(Chaenomeles lagenaria)*. Those considered moderately hardy are jujube, medlar *(Mespilus germanica)*, North American persimmon, and northern papaw *(Asimina triloba)*. Hardy species include elderberry *(Sambucus* sp.*)*, June or serviceberry *(Amelanchier alnifolia)*, highbush cranberry *(Viburnum trilobum)*, cornelian cherry *(Cornus mas)*, native plums *(Prunus maritima, P. besseyi, P. pumila)*, buffalo berry *(Shepardia argentea)*, and flowering currant *(Ribes odoratum)*.

Summary

The most severe low temperature injury to fruits and nuts usually occurs in late fall or early winter. Under such conditions, tissue maturity is of prime importance. Any condition that prolongs growth delays the physiological "hardening off" process. Some factors that delay or prevent tissue maturity and cold acclimation are high nitrogen nutrition, late cultivation, and irrigation, early defoliation, heavy cropping, and early pruning. Internal (genetic) factors may also delay plant maturity. Some cultivars or species do not cease growth in response to short days as do many temperate zone plants. Also, trees on vigorous rootstocks tend to cease growth later than is desirable.

Hardiness can be affected indirectly in a number of ways. Insects or diseases can cause early defoliation that, in turn, prevents normal cold acclimation. Root pests may devitalize plants or cause an upset in nutrient uptake. Soil conditions may cause nutrient deficiencies or a nutrient imbalance that affects hardiness. Dwarfing rootstocks may enhance early winter hardiness merely by causing shoot growth to cease earlier than would occur with a vigorous stock.

The different organs and/or genetic components of compound trees interact reciprocally; compound trees are thus much more complex than ungrafted plants having a single genetic make-up. Reports on some compound trees state that the influence of the scion upon the stock bears no relationship to the hardiness of the scion. Other studies find predictable effects on hardiness. Some of these apparently conflicting observations result from the use of different methods of assessing hardiness or from testing at different times during the season. Others probably are due to a number of indirect effects, such as differences in balance of IAA, GA, ABA, ethylene, and cytokinins; mineral element balance; pest and disease resistance; or crop load. One or more of these indirect effects may be superimposed upon the direct hardiness interactions (that is, the direct interactions of hardiness promoters and/or inhibitors between root, interstock, and scion).

Auxin balance in the tree plays an important role in the development of hardiness because it influences the amount of cellular activity of plant organs. Growth is negatively correlated with hardiness. Auxin is synthesized mainly in subapical regions of active shoots and in developing seeds within the fruit. Dormant twigs of pear were shown to contain three times as much IAA in trees that had borne seeded fruit as in trees that had borne seedless fruit. Indirect evidence indicates that a fruit-setting factor produced by seed moves downward to the roots and becomes completely systemic the following spring, possibly by upward xylar transport. Because the auxin level is correlated with seedless fruit set, the setting factor is assumed to be auxin. The basipetal movement of auxin through the trunk in the fall could stimulate cambial activity and reduce hardiness of that portion. The genetic make-up of the rootstock would determine (by leakage to the soil solution, enzymatic inactivation, etc.) how much carry-over auxin remained in the system, perhaps to influence hardiness in following years.

Gibberellins also are produced in seed. Gibberellins may decrease fall hardening; they are known to delay cold acclimation. Thus, the observed reduction in hardiness following a large crop could be due to the increase of GA and auxin rather than to a "depletion of reserves," as is so often assumed.

Field observations indicate that trees carrying a

crop retain green foliage later in the fall than do trees with no crop. In November 1955, in Washington, many trees of late cultivars had not been picked at the time of a sudden freeze. The leaves of these were killed and failed to abscise. Freeze damage to such trees was much greater than to trees of the same cultivar that had been picked only a few days before. Detailed observations in 1956 showed that in cultivars that matured after September 1, leaf senescence and fall coloration occurred only after the crop was harvested. Fruit removal quickly and radically altered the physiology of the trees. Fruit on the tree is both a source and a sink: it is a source of auxin and GA and a strong sink for carbohydrates and ABA. After harvest, there must be a concomitant reduction of auxin and GA in the tree and a sharp rise in photosynthate, ABA, and possibly ethylene in the other tissues. These changes seem to favor cold acclimation, because the trees harvested only one week before the 1955 freeze sustained much less injury than unharvested trees. The role of ABA in senescence and initial hardening is unclear, but it probably aids hardening at least by its action as a GA antagonist.

Aside from the possible movement of hardiness promoters and/or inhibitors to and from the organs of a compound tree, the balance of auxin, GA, ethylene, and cytokinins in a three-piece tree must be very complex. The interaction of kinins moving up from the root and auxin and other hormones moving down from the top probably affect early hardiness at least indirectly, if not directly. For example, consider the auxin levels in these compound trees:

1. *Pyrus ussuriensis/P. communis/P. calleryana.*
2. *P. calleryana/P. communis/P. ussuriensis.*

The first has leaves and shoots of a subarctic boreal species, the trunk of a temperate species, and the root of a subtropical species. Growth of this tree in late summer is dominated by the *P. ussuriensis* top. In response to shortening days, it ceases growth in August and is often half-defoliated by October 1. Under these conditions, cambial activity of the *P. communis* trunk ceases early in the fall, and the system is hardy. The second tree system is dominated by the subtropical *P. calleryana.* It does not cease growth in response to short days and will continue shoot growth into November, under favorable conditions. Active cambium permits the budding of *P. calleryana* nursery stock in November. With this species as a top, auxin is produced late in the fall, stimulating late cambial activity of the *P. communis* trunk, thus predisposing it to injury by an early freeze.

Pruning immediately before a freeze greatly increases injury to the tree. The fresh cut apparently stimulates cellular activity and/or creates a strong sink for growth hormones, both of which deharden the tissue, but the true nature of the effect of pruning is not known.

Although much empirical data are available, we know very little about the physiological mechanisms by which one component of a compound plant affects the hardiness of another component. Further experimentation with grafted plants might elucidate the fundamental nature of cold hardiness. For example, all *Pyrus* species are graft compatible, and the genus represents the entire range of hardiness from subarctic species (*P. ussuriensis* is hardy at −50°C) to tropical species (*P. koehnei*, an evergreen, which is killed at −11°C). These extreme types, when grafted together, would facilitate the testing of any general theory of the origin, movement, and fate of hardiness promoters or inhibitors.

SPRING FROST PROTECTION

In 1897, California citrus growers hung wire baskets filled with coal fires in trees to prevent frost. Heating had been used for frost protection even earlier, in the 18th century. Since that time, many kinds of fuels have been used for orchard heating.

Passive control (through selection of climate, site, and cultivar) avoids frosts; active control acts against a local frost. All methods of active control either conserve energy already in the orchard or add heat from a source of energy outside the orchard. The most effective method of control is heating, but heating requires that sufficient fuel be available.

Principles of Control

Zimmerman (1966) has outlined the basic principles of orchard heating. Buds receive heat by radiation, conduction, and convection.

Radiant heat is like light (except that its wavelength is beyond the visible range). Radiant beams travel only in straight lines, and their energy decreases rapidly with distance from their source. In order to receive radiant energy, a bud or other plant part must have no obstruction between it and the source of the radiant beam. The larger the heated surface of the heater, the greater the radiant heat coming from it.

Conduction is the direct transfer of heat between different portions of a body, or different bodies in contact with each other, without any observable motion taking place.

The most important heat-transfer process is that of convection (see Fig. 15-16). Heated air adjacent to the heater expands, becomes lighter, and rises. This up-and-down movement helps transfer heat to the trees. Heaters are most effective where strong inversions occur over the orchard. The heated air rises only until it reaches air of the same temperature in the inversion layer. Thus convection recirculates the warm air into the orchard area.

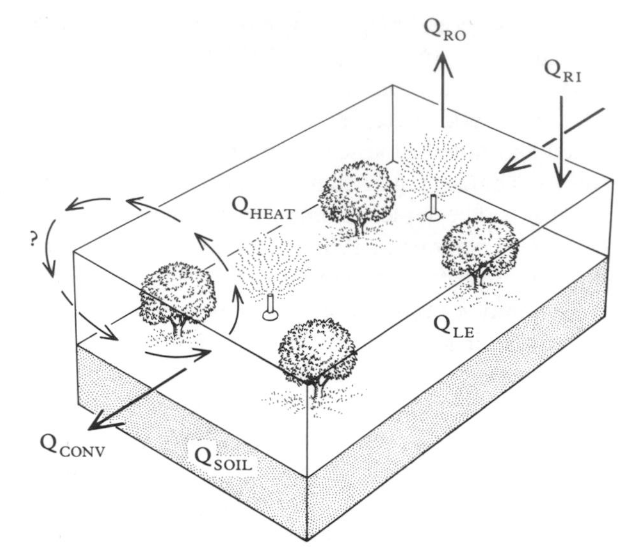

Figure 15-16 An orchard with heaters enclosed in an imaginary volume of air. Various nocturnal heat-transfer components are shown. Q_{RO} is the outgoing radiation. Q_{RI} is the incoming sky radiation. Q_{LE} is the latent transfer of heat by evapotranspiration—dew or frost formation. Q_{conv} is convective and advective transfer. Q_{soil} is the conductive heat transfer from the soil. Q_{heat} is the heat supplied by heaters. If all heat loss is balanced by heat gain, the temperature will remain constant—providing ideal cold protection. The arrows with the question mark are induced flow caused by heating. [After Gerber, 1970]

Normally, temperatures are lower at higher elevations, but inversions occur on still nights when radiation heat loss from the earth's surface cools the layer of air near the ground. This cold, dense air moves down the slopes, filling local valleys and closed basins, displacing the warmer air upward (Fig. 15-17). The degree of temperature inversion near the ground determines the depth of the layer of air that must be warmed to obtain a definite increase in temperature at the ground. Referring to a low or high "ceiling" is another way of stating the amount of temperature inversion. If there is a rapid increase in temperature with increase in elevation (a low ceiling), the surface temperature can be raised several degrees more with the same amount of heating fuel than when the rate of increase is slight. The amount of this temperature inversion varies greatly from night to night, and in different localities. It is mainly determined by the amount of fall in temperature from afternoon to early morning. If the afternoon temperature is high, and it falls to freezing or below on the following morning, the inversion is likely to be great, and orchard heating unusually effective. The most difficult nights, when protection is necessary, are those following cold afternoons when the temperature inversion is slight (high ceiling). Cold arctic advective freezes (with wind) are difficult to protect against because there is a cold air mass with very little inversion. In this situation, heating is the only way of raising the orchard temperature; but often, large amounts of heat are not enough to protect the crop.

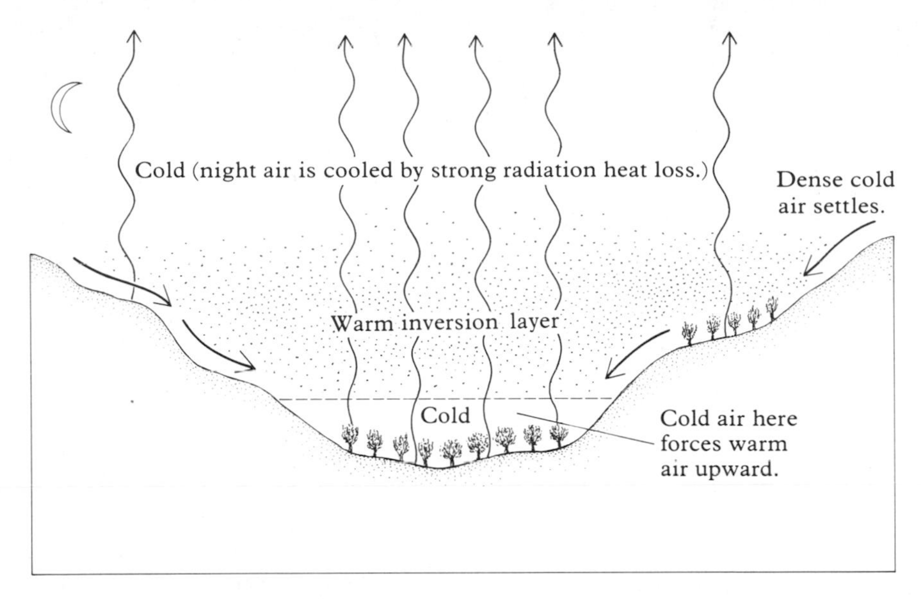
Cold (night air is cooled by strong radiation heat loss.)
Dense cold
air settles.
Warm inversion layer
Cold
Cold air here
forces warm
air upward.

Figure 15-17 A schematic view of a valley with trees planted on the floor and up a slope. On clear, still nights, strong radiation heat loss at the earth's surface cools the air. The dense cold air that is formed settles at the bottom of the valley, forcing warmer air up to a higher level—thus producing a temperature inversion, which is advantageous to the trees on the slope on frosty nights.

Techniques

HEATING DEVICES AND FUELS A wide variety of heaters and fuels have been used. Units may be independent or may be connected to a central fuel supply system.

Individual heaters vary from simple open containers to more complex units designed to burn more efficiently. Several types are shown in Figure 15-18. The return-stack heater is the best of those using oil. It burns 0.3 to 0.6 gallons of oil per hour and is efficient enough that it is relatively smokeless. Fuels used in individual units are usually some form of petroleum oil but may also be solid fuels such as petroleum coke blocks, wax, processed rubber tires, or wood blocks.

Hints on effective heating include pre-season checking of equipment, advance preparation, proper lighting devices, an efficient system for refilling pots, properly tested and placed thermometers, and pretesting of the whole system during daylight. Also, the following should be kept in mind:

1. Smaller units, well spaced, are better than a few large ones.

2. Place a double row of heaters on the upwind side.

3. Frost pockets need more heaters than other areas.

4. Lighting should be done before the critical temperature is reached (see critical bud temperatures below).

5. Oil heaters should be prevented from burning dry.

6. A passing cloud can cause the temperature to rise above the critical level, but heaters should not be

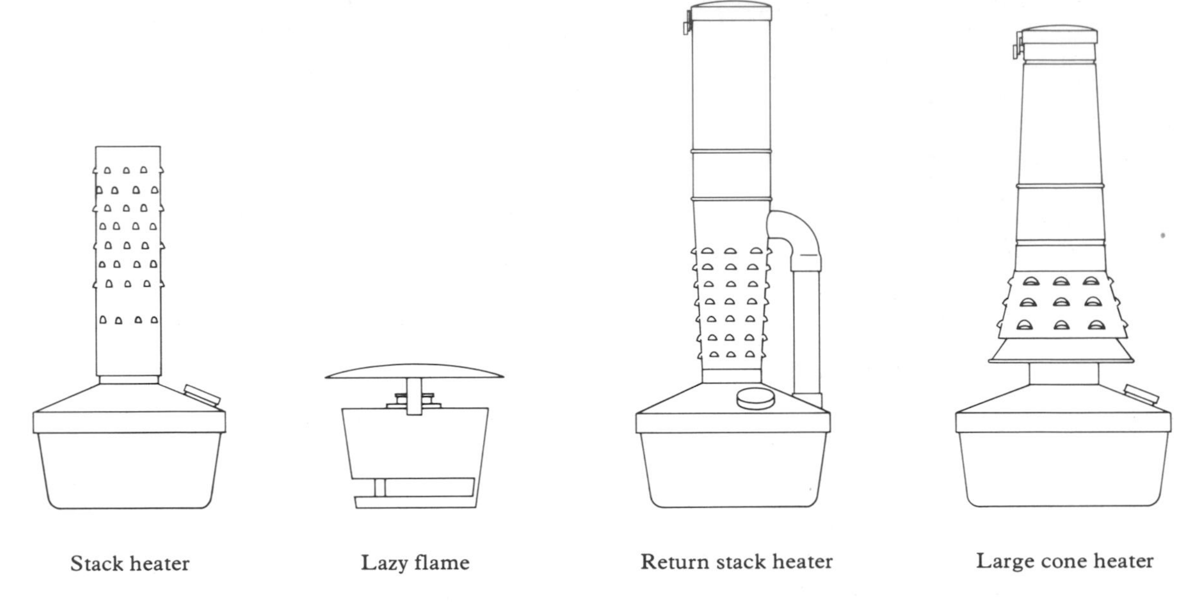

Figure 15-18 Four types of oil-burning orchard heaters.

turned off unless sunrise is near or a general cloud cover has formed over the entire area.

CENTRAL SUPPLY SYSTEMS These systems supply oil, liquid propane gas (LP), or natural gas through underground or surface pipes to the unit heaters throughout the orchard. Figure 15-19 shows several kinds of mantles that cover the combustion chamber. Many types of conventional heaters can be adapted to central supply systems. Pressurized oil systems require special filters, pumps, piping, and nozzles. Natural gas, where available, is the easiest to use because it requires no storage tanks, pumps, or filters, and because it requires minimal heater maintenance. LP gas is the most expensive of the central supply fuels and has a number of special problems. LP boils (becomes a gas) at −42°C. Under pressure in a tank, as LP is released in the lines and vaporizes, the remaining liquid in the tank is cooled by evaporative cooling similar to that of a refrigeration system. When the rate of gas use reaches a certain level, the tank must be heated in order to sustain a high level of vaporization. Hot water jackets or heating mantles are thus installed on the LP tank. Best use of this system requires a 114 cubic meter (30,000 gallon) tank to handle 25 hectares or more of fruit (Ballard and Proebsting, 1972).

The main advantages of central supply systems over self-contained units are savings in labor for refueling and in greater speed of turning them on and off, thus saving fuel during nightly changes in temperature. Performance of various types of heaters is shown in Table 15-6. The current increasing costs of energy will make some types of heating uneconomical and will emphasize the importance of favorable frost-free sites for new plantings.

OVERHEAD SPRINKLERS The principle applied in protection by sprinkling is that as water freezes, it releases 80 kilocalories per liter of water (heat of fusion), which warms the objects in contact with the water and ice. If some free water is maintained on a

A

B

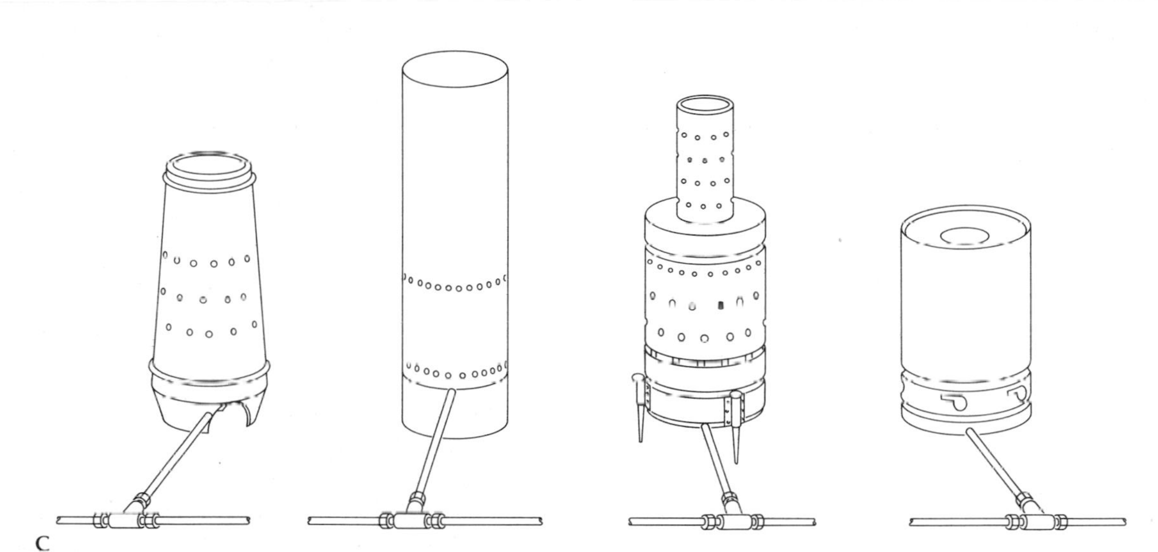

C

Figure 15-19 Central fuel-supply systems. **(A)** A 30,000-gallon tank for liquid propane. **(B)** A homemade mantle for propane-gas system. **(C)** Four different types of burner covers. [After Ballard and Proebsting, 1972]

bud covered with ice, the bud will remain close to 0°C. Sprinklers have been used for low-growing crops such as strawberries and cranberries for a long time. Attempts to control frost in orchards, however, sometimes resulted in excessive limb breakage from ice buildup.

Water protection works best where there are radiation frosts (rather than cold air masses) and only a few degrees protection are needed. Very low humidity might limit the use of water because of increased evaporation, which cools objects from which the water evaporates (heat of vaporization). But con-

Table 15-6 Performance characteristics of orchard heaters.

Heater Type	Percent Radiant Heat	Heat Produced per hour per Unit (k cal)[1]	Number Recommended per ha	Cost per Heater[5]	Air Pollution	Length of Burning	Composition
Tree-Heet	Some	2,015 to 2,150	250 to 500 1.8 kg packages	No heater required	Little	5 hrs.	Petroleum coke fuel blocks
Wax candle	5	6,300 to 25,000	86 to 500	Variable—some do not require heaters	Little	8–10 hrs.	
Propane gas	5 (with open flame)	25,000[2]	86 to 170	50¢ to $20.00	Little	Continuous	
Open pot with oil	5–15	8,800 to 35,000 depending on size	125 to 500	25¢ to $4.00	Some	8–10 hrs. for 10 gal. size	
Rocket heater	24–28	25,000	85	$2.65	Some	4 hrs. per pkg.	Processed chopped scrap tires
Return stack with oil	23–32	12,600 to 530,000 depending on burning rate and size	50 to 135	$8.00 to $9.00	Little	8–18 hrs.	
Natural gas with metal mantle	45–55[3]	25,000	85 to 125	$20.00 to $50.00[4]	Little	Continuous	

SOURCE: *American Fruit Grower,* March 1967.
[1]k cal = kilocalorie, a measure of heat—the quantity required to raise the temperature of one kg of water one degree Celsius (1 BTU = .252 k cal).
[2]At 3.8 liters per hour.
[3]For metal-mantle or tile heaters only. Open flame natural gas will produce only about 5 percent radiant heat.
[4]For metal-mantle and tile heaters.
[5]Costs change rapidly, so listed costs may not represent current prices.

tinuous sprinkling (.38 centimeters/hour) protected the buds on a night that at 7:00 P.M. had a dewpoint of −5° to −9°C at midnight and −4.4°C at 7:00 A.M. Wind also increases evaporation, so sprinkling is best done under calm conditions. In the study cited above, 1.5 Celsius degrees warming occurred with sprinkling on the coldest nights. The advantages of this system in areas where it is effective are lack of air pollution, simplified operation, and a limited requirement for labor. The on-off switch is operated by a thermostat, but the grower should check up to be sure the system is working properly. As in some other systems, summer irrigation can be done with the same pump, lines, and sprinklers (Fig. 15-20). Installation costs are between $1700 and $2100 per hectare. A good water supply is necessary, and screens or filters should be used to prevent clogging of the sprinkler heads (Carpenter, 1966).

WIND MACHINES The successful use of wind depends upon the existence of a warm air layer above the planting. Tower-mounted wind machines require a fairly strong inversion (5 to 8 Celsius degrees) 15 meters above the ground. The warm air is mixed with the cold air of the orchard, raising the temperature. Where needed, heaters can be used with wind machines for more protection (Crawford and Leonard, 1960). However, the horizontal distance the air moves is less with heat because heated air has added buoyancy and rises out of the orchard.

Ground-level mobile wind machines vary in effectiveness, depending upon the thrust force, the strength of the inversion, and whether or not heat is supplied with the wind. This type of machine blows the cold air out of the orchard and permits warmer air from above to come in to replace it. In orchards with dense canopies, this machine can be more effective than the tower-mounted machine. Helicopters are sometimes useful but are good only where a warm inversion layer can be found above the orchard. The helicopter's blades force the warm air downward to displace or mix with colder air at ground level. Rotating fixed fans near the ground are most effective in forcing cold air from frost pockets, thus pulling in warmer air from above (Figs. 15-21 and 15-22).

Figure 15-20 Overhead sprinklers for frost protection. Severe breakage of young trees as a result of the ice load. [Courtesy of P. B. Lombard, Southern Oregon Experiment Station]

Figure 15-21 Wind machine used to protect crops from frost. Successful use of fixed fans is dependent upon the presence of a warm inversion layer above the planting. [Courtesy of J. K. Ballard, Washington Extension Service]

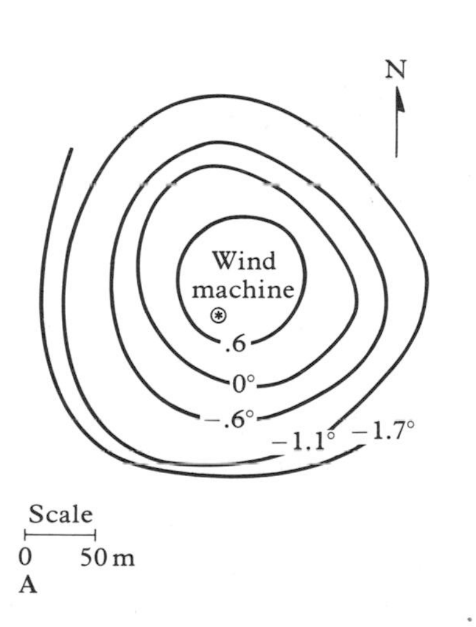

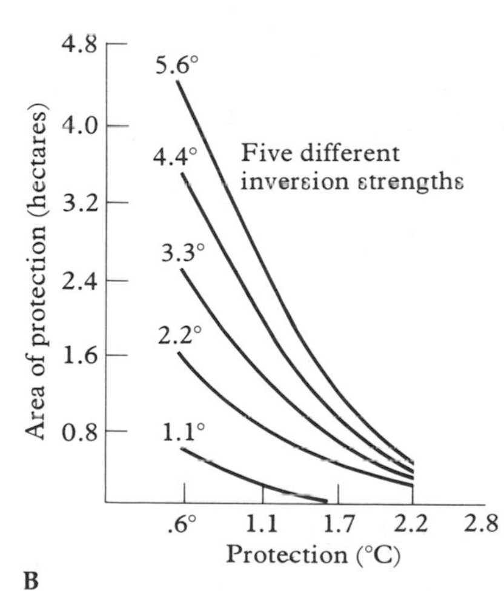

Figure 15-22 **(A)** Pattern of protection around an 85-brake-horsepower wind machine. The isotherm lines cover an area of about 4 hectares; the degree of protection decreases with distance from the machine. **(B)** The area of protection and the amount of protection provided by the wind machine as they are related to inversion strength (i.e., the difference in degrees Celsius between plant-level temperature and that of the inversion layer above it). For example, for an inversion strength of 2.2°, the wind machine should raise the plant-level temperature within an area of 1.6 hectares by .6°C and within an area of .8 hectares by over 1°C. [After Gerber, 1970]

The Relation of Dew Point to Frost

Dew point (DP) is defined as the temperature below which invisible water vapor in the air condenses as dew, fog, or frost. DP is an indication of the total amount of water vapor in the air and is thus more useful than relative humidity (RH) in estimating frosts, because DP does not change with a change in temperature. A high dew point indicates more water in the air than does a low dew point. RH is not useful *per se* in frost prediction because it varies with air temperature. RH is defined as the percent of water vapor in the air relative to the total amount the air could hold at that temperature (warm air will hold more water vapor than cold air). For example, a closed room at 21°C with a DP of 10°C has an RH of 50 percent. Thus, the air has half the water vapor it could hold at saturation. Increase the room temperature to 27°C, and the RH falls to 35 percent, but the DP remains at 10°C. Cool the room to 10°C, and the RH rises to 100 percent, but still the DP remains at 10°C.

Moisture in the air prevents cooling by absorbing radiant heat at 2.7 microns and 5 to 8 micron wavelength, which is being lost from the earth's surface. High fog and clouds are merely visible indicators of vapor in the air and usually indicate that frosts are unlikely. Even if the sky is clear, frosts are unlikely if there is a high DP. As the temperature at night reaches the DP, the condensation of moisture from the air releases a tremendous amount of energy (heat of vaporization), which warms the earth's surface and the atmosphere and tends to prevent frost. When the DP is much below 0°C, plant tissue may be killed before any frost forms. When this occurs it is termed a "black frost."

Growers often want to know the relationship be-

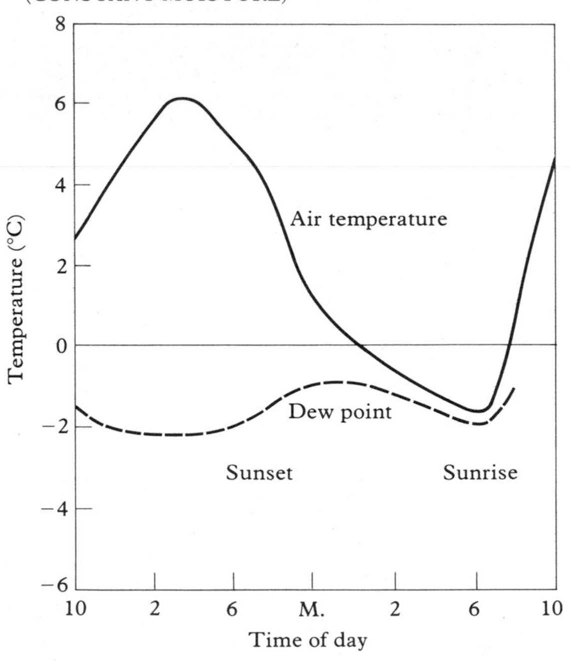

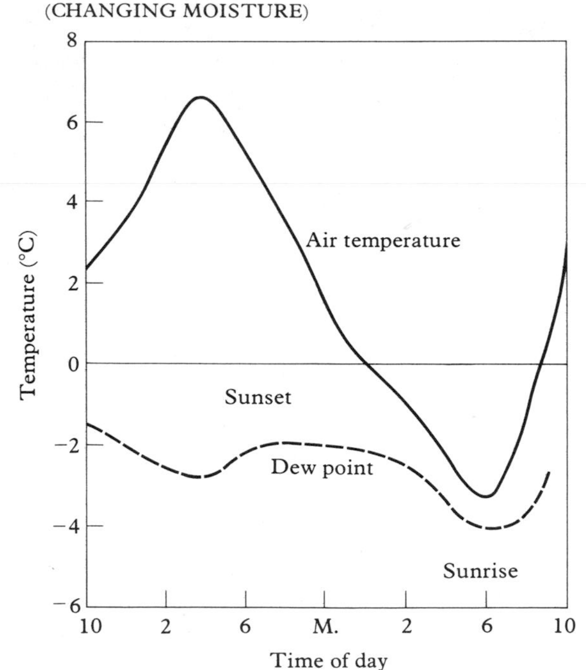

Figure 15-23 The relation of dew point **(A)** to the temperature of air with a constant water-vapor content and **(B)** to the temperature of air with a changing water content. When the moisture content of the air is lowered, nighttime temperature also is lowered. [After Stirm, 1967]

tween DP and freezes. As we have seen, DP directly affects the degree of low temperature. With radiation freezes, the data given above are useful but should be used along with the local weather forecast—there are limitations to using this method. If the DP goes lower (as drier air moves in), the danger of frost will increase (see Fig. 15-23). Air mixing by wind can increase the temperature at tree level, making frost less likely. Also, if clouds move in over the area, the temperature will rise.

Critical Temperatures for Flower Buds

Critical temperatures for fruit-blossom buds at different stages of spring development may be found with the use of Tables 15-7 and 15-8. The temperature at which fruit buds are injured depends primarily on their stage of development. They are most hardy during winter when in dormancy. As they begin to swell and grow, they become less hardy. Not all buds are equally tender. Resistance to frost varies within trees and between orchards, cultivars, and species. Buds that develop slowly tend to be more resistant. To indicate this range, new critical temperatures have been developed that show average temperatures required to kill 10 percent and 90 percent of the buds at different stages (Tables 15-7 and 15-8; Figs. 15-24 and 15-25).

The proportion of buds that is killed by a given temperature is a factor in determining the need for frost control. Orchards with a large number of buds should be able to stand more frost than those with a small number of buds. Consideration should also be given to temperatures preceding cold nights. Prolonged cool weather tends to increase bud hardiness at a given stage. Table 15-7 presents the descriptive stages of buds for the nine numerical categories used in Table 15-8. At the early stages of bud development, there is a wide difference between the temperature required to kill 10 percent of the buds and that required to kill 90 percent. But at full bloom and later, the spread is much narrower. In some cases, at full bloom, a temperature drop of only 1.7 Celsius degrees would make the difference between a 10 percent kill and a 90 percent kill. These new data are important because they tell the grower how important orchard heating will be at any stage of development and how much temperature change is required to save enough buds for a crop.

Accurate temperature readings are important for good frost protection. Properly sheltered thermometers should be placed both inside and outside the protected area. They should be exposed to the circulating air but should be shielded from the sky, direct sun, and from moisture deposits (Fig. 15-26). A common shelter is made by nailing two boards together at right angles and placing them on top of a post 1.4 meters high. They are placed in an open area (never under a tree), facing north. Shelters are often painted white to reduce radiation absorption by the boards.

Table 15-7 Developmental stages of flower buds used in Table 15-8.

Bud Stage	Apple	Pear	Apricot	Cherry	Peach	Prune
1	Silver tip	Scales separate	First swell	First swell	First swell	First swell
2	Green tip	Cluster showing	Tip separate	Side green	Calyx green	Side white
3	Half-inch green	Tight cluster	Red calyx	Green tip	Calyx red	Tip green
4	Tight cluster	First white	First white	Tight cluster	First pink	Tight cluster
5	First pink	Full white	First bloom	Open cluster	First bloom	First white
6	Full pink	First bloom	Full bloom	First white	Full bloom	First bloom
7	First bloom	Full bloom	In shuck	First bloom	Post bloom	Full bloom
8	Full bloom	Post bloom	Green fruit	Full bloom	—	Post bloom
9	Post bloom	—	—	Post bloom	—	—

Table 15-8 Critical temperatures (in degrees Celsius) at which flower buds are killed at various stages of development.

Data	Bud Development Stage								
	1	2	3	4	5	6	7	**FB 8**	9
Apples[4]									
Old standard temp.[1]	−8.9	−8.9	−5.6	−2.8	−2.8	−2.2	−2.2	**−1.7**	−1.7
Ave. temp. for 10% kill[2]	−9.4	−7.8	−5.0	−2.8	−2.2	−2.2	−2.2	**−2.2**	−2.2
Ave. temp. for 90% kill[2]	−17	−12	−9.4	−6.1	−4.4	−3.9	−3.9	**−3.9**	−3.9
Average date (Prosser)[3]	—	3/20	3/27	4/3	4/8	4/11	4/18	**4/25**	—
Pears[5]									
Old standard temp.[1]	−7.8	−5.0	−4.4	−2.2	−1.7	−1.7	**−1.7**	−1.1	
Ave. temp. for 10% kill[2]	−9.4	−6.7	−4.4	−3.9	−3.3	−2.8	**−2.2**	−2.2	
Ave. temp. for 90% kill[2]	−18	−14	−9.4	−7.2	−5.6	−5.0	**−4.4**	−4.4	
Average date (Prosser)[3]	—	3/23	3/31	4/5	4/9	4/14	**4/18**	4/25	
Apricots									
Old standard temp.[1]	—	−5.0	—	−3.9	—	**−2.2**	—	−0.6	
Ave. temp. for 10% kill[2]	−9.4	−6.7	−5.6	−4.4	−3.9	**−2.8**	−2.8	−2.2	
Ave. temp. for 90% kill[2]	—	−18	−13	−10	−7.2	**−5.6**	−4.4	−3.9	
Average date (Prosser)[3]	—	—	3/8	3/16	3/22	**3/28**	4/4	4/18	
Cherries[6]									
Old standard temp.[1]	−5.0	−5.0	−3.9	−2.2	−2.2	−1.7	−1.7	**−1.7**	−1.1
Ave. temp. for 10% kill[2]	−8.3	−5.6	−3.9	−3.3	−2.8	−2.8	−2.2	**−2.2**	−2.2
Ave. temp. for 90% kill[2]	−15	−13	−10	−8.3	−6.1	−4.4	−3.9	**−3.9**	−3.9
Average date (Prosser)[3]	3/5	3/13	3/23	3/27	4/1	4/4	4/8	**4/13**	4/21
Peaches[7]									
Old standard temp.[1]	−5.0	—	—	−3.9	—	**−2.8**	−1.1		
Ave. temp. for 10% kill[2]	−7.8	−6.1	−5.0	−3.9	−3.3	**−2.8**	−2.2		
Ave. temp. for 90% kill[2]	−17	−15	−13	−9.4	−6.1	**−4.4**	−3.9		
Average date (Prosser)[3]	3/7	3/16	3/19	3/29	4/3	**4/11**	4/18		
Prunes[8]									
Old standard temp.[1]	—	—	—	—	−5.0	−2.8	**−2.8**	−1.1	
Ave. temp. for 10% kill[2]	−10	−8.3	−6.7	−4.4	−3.3	−2.8	**−2.2**	−2.2	
Ave. temp. for 90% kill[2]	−18	−16	−14	−8.9	−5.6	−5.0	**−5.0**	−5.0	
Average date (Prosser)[3]	3/13	3/20	3/27	4/3	4/8	4/12	**4/16**	4/23	

SOURCE: Ballard et al., 1971.
NOTE: Full-bloom stage is indicated by **boldface** type.
[1]Critical temperatures as previously published in Wash. State Univ. Extension Memo. 1616.
[2]Average temperatures found by research at the WSU Research and Extension Center, Prosser, to result in 10 percent and 90 percent bud kill.
[3]Average date for this stage at the WSU Research and Extension Center.
[4]For Red Delicious. Rome Beauty, Golden Delicious and Winesap are approximately one degree hardier, except after petal fall, when all cultivars are equally tender.
[5]For Bartlett. Anjou is similar in hardiness but may bloom earlier and therefore may be more tender than Bartlett at the same date.
[6]For Bing. Lambert and Rainier are approximately one-half to one degree hardier through Stage 6.
[7]For Elberta.
[8]For Italian Prunes and Early Italian Prunes.

Figure 15-24 Seven developmental stages of peach buds. These photographs can be used with Table 15-8, which lists critical temperatures for buds at each stage.

1 Silver tip

2 Green tip

3 Half-inch green

4 Tight cluster

5 First pink

6 Full pink

Figure 15-25 Nine developmental stages of apple buds. These photographs can be used with Table 15-8, which lists critical temperatures for buds at each stage.

7 First bloom

8 Full bloom

9 Post bloom

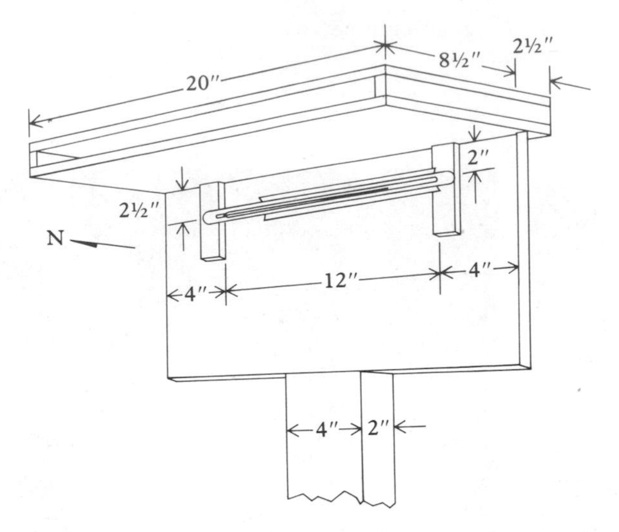

Figure 15-26 Standard agricultural-thermometer shelter, showing the correct placement of the thermometer. For monitoring frost temperature, the straight-tube minimum-registering alcohol thermometer is used. The shelter faces north to prevent direct rays of the sun from striking the thermometer. [After Ballard and Proebsting, 1972]

Cultural and Chemical Protection

Aside from the choice of a site with good air drainage away from the crop, the soil surface itself is important in crop protection and can make a difference of as much as 1.7 Celsius degrees in air temperature. For maximum protection against radiation frosts, the soil should be moist, free of weeds, smooth, and untilled. The energy used to avert frost is stored in the upper 15 centimeters of soil; cover crops, dryness, and tillage tend to insulate and retard the upward flow of heat on frosty nights. Use of taller plants or taller trellis supports for berries and grapes affords some protection with a good temperature inversion (Blanc et al., 1963).

Covering plants can often protect them. Such materials as straw have been used for low-growing strawberries and even dwarf apples and grapes. Artificial stable foams also have been used on strawberries. Cranberries are often covered with water to prevent frost. Low-growing berries or high-density dwarf trees may be protected by muslin or plastic covers. Many plastic films, when dry, do not prevent the loss of heat rays (of 2 to 10 microns wavelength), but when they become coated with a thin layer of condensed moisture, as often occurs, they all are equally effective. Even though coverings alone may not give protection, the addition of a small amount of heat under a cover is effective. Practical use of any of these methods depends upon the cost as related to the risk and the potential gain.

Chemicals have been used with little success until recently. The use of ethephon on stone fruits in the fall has effectively delayed spring bud-break by 4 to 12 days, thus avoiding early frosts. Also, the use of GA_3 on recently frosted young pear fruit has resulted in the development of seedless fruit of some cultivars.

GENERAL REFERENCES

Amen, Ralph D. 1963. The concept of seed dormancy. *Am. Scientist* 51:408–424.

Ballard, J. K.; E. L. Proebsting; R. B. Tukey; and H. Mills. 1971. *Critical temperatures for blossom buds.* Wash. State Agr. Ext. Circ. Nos. 369–374.

Burke, M. J.; L. V. Gusta; H. A. Quamme; C. J. Weiser; and P. H. Li. 1976. Freezing and injury in plants. *Ann. Rev. Plant Physiol.* 27:507–528.

Eagles, C. F., and P. F. Wareing. 1964. The role of growth substances in the regulation of bud dormancy. *Physiologia Plantarum* 17:697–709.

Gerber, J. F. 1970. Crop protection by heating, wind machines, and overhead irrigation (a review). *HortScience* 5:427–431.

Koller, D. 1955. The regulation of germination in seeds. *Bull. Res. Council Israel* 5D:85–108.

Marcus, A. 1969. Seed germination and the capacity for protein synthesis. In *Dormancy and survival.* Academic Press, New York. Pp. 143–160.

Nikolaeva, M. G. 1967. Physiology of deep dormancy in seeds (English translation 1969). *Akad. Nauk USSR.*

Roberts, E. H. 1969. Seed dormancy and oxidation processes. In *Dormancy and survival.* Academic Press, New York. Pp. 161–192.

Vegis, A. 1964. Dormancy in higher plants. *Ann. Rev. Plant. Physiol.* 15:185–224.

Walker, D. R. 1970. Growth substances in dormant buds and seeds (a review). *HortScience* 5:414–417.

Wareing, P. F. 1965. Dormancy in plants. *Sci. Progr.* 53:529–537.

Weiser, C. J. 1970. Cold resistance and acclimation in woody plants (a review). *HortScience* 5:403–408.

16

Pests

This chapter deals with a complex and diverse group of organisms that directly or indirectly influences the health and productivity of plants or the quality of crops. The major groups of pests to be considered are: (1) insects and mites, (2) fungi and bacteria, (3) viruses and mycoplasmas, and (4) herbivorous animals and birds. It is not possible in this book to cover each of these groups in detail; however, the general aspects of pest biology and control will be given for a number of important pests. A knowledge of their life cycles and feeding habits is useful in control procedures. However, because of the rapid changes in the regulations relating to pesticide use and in the number and kind of pesticides available, only general comments on control can be made. Details of pest control for a given crop and area can be obtained from local county extension agents or farm advisors, from local agricultural experiment stations, or pest management consultants.

MITES AND INSECTS

Major Injurious Species

Below are listed some of the major injurious species of spiders and insects. They are grouped by order and family to give the reader an idea of how closely related they are. Following each common name are the Latin binomial, the most important fruit and nut species attacked, and the plant part or parts damaged by the pest.

Class Arachnida
- Order Acarina (ticks and mites)
  - Family Tetranychidae
    - Yellow spider mite, *Eotetranychus carpini borealis* (Ewing), apple and pear leaves
    - European red mite, *Panonychus ulmi* (Koch); fruit tree leaves and fruit
    - Clover mite, *Bryobia praetiosa* Koch; stone and pome fruit leaves
    - Twospotted spider mite, *Tetranychus urticae* Koch; fruit tree leaves
    - Pacific spider mite, *Tetranychus pacificus* McGregor; currant, grape and strawberry
    - McDaniel spider mite, *Tetranychus mcdanieli* McGregor; apple leaves
  - Family Eriophyidae
    - Pear-leaf blister mite, *Eriophyes pyri* (Pagenstecher); pear fruit and leaves
    - Dryberry mite, *Phyllocoptes gracilis* (Nalepa); fruit of Himalaya blackberry
    - Pear rust mite, *Epitrimerus pyri* (Nalepa); pear leaves
    - Apple rust mite, *Aculus schlectendali* (Nalepa); apple leaves, fruit
    - Plum rust mite, *Aculus fockeui* (Nalepa & Trovessart); plum leaves
    - Filbert bud mite, *Pytocoptella avellanae* Nalepa; filbert and hazel buds
    - Fig mite, *Aceria ficus* (Cotte); fig fruit
    - Peach silver mite, *Aculus cornutus* (Banks); peach leaves
    - Grape erineum mite, *Eriophyes vitis* (Pagenstecher); grape leaves
  - Family Tarsonemidae
    - Cyclamen mite, *Steneotarsonemus pallidus* (Banks); strawberry leaf and flower buds

Class Insecta
- Order Thysanoptera
  - Family Thripidae (thrips)
    - Western flower thrips, *Frankliniella occidentalis* (Pergande); fruit species flowers and fruit
    - Pear thrips, *Taeniothrips inconsequens* (Uzel); stone fruit and pear flowers and fruit
- Order Homoptera (planthoppers, aphids, scales)
  - Family Membracidae
    - Buffalo treehopper, *Stictocephala bulbalus* (Fabricius); fruit tree bark and stems
  - Family Cicadellidae
    - Apple leafhopper, *Empoasca maligna* (Walsh); apple leaves
    - Grape leafhopper, *Erythroneura elegantula* Osborn; small fruits and grape leaves
  - Family Psyllidae
    - Pear psylla, *Psylla pyricola* Förster; pear and quince leaves
  - Family Phylloxeridea
    - Grape phylloxera, *Phylloxera vitifoliae* (Fitch); grape roots, leaves
  - Family Aphidae
    - Walnut aphid, *Chromaphis juglandicola* (Kaltenbach); English walnut leaves
    - Leaf curl plum aphid, *Brachycaudis helichrysi* (Kaltenbach); plum leaves and young stems
    - Rosy apple aphid, *Dysaphis plantaginea* (Passerini); apple leaves and fruit
    - Black peach aphid, *Brachycaudus persicae* (Passerini); peach, almond, apricot, plum roots, shoots, and fruit
    - Apple aphid, *Aphis pomi* DeGeer; apple tender tips, leaves, and young fruit
    - Mealy plum aphid, *Hyalopterus pruni* (Geoffroy); plum tender tips and leaves
    - Black cherry aphid, *Myzus cerasi* (Fabricius); sweet cherry leaves
    - Green peach aphid, *Myzus persicae* (Sulzer); stone fruits and walnut leaves
    - Woolly apple aphid, *Eriosoma lanigerum* (Hausmann); apple bark of roots and stems
    - Woolly pear aphid, *Eriosoma pyricola* Baker & Davidson; pear roots and elm leaves
  - Family Pseudococcidae
    - Grape mealybug, *Pseudococcus maritimus* (Ehrhorn); pome fruits, strawberry, walnut, and grape leaves and fruit
  - Family Coccidae
    - European fruit lecanium, *Lecanium corni*

Bouché; pome, stone, small fruits, pecan, and persimmon twigs
Frosted scale, *Lecanium pruinosum* Coquillett; pome, stone, small fruit, and walnut stems
Italian pear scale, *Epidiaspis leperii* (Signoret); pome, stone fruits, and currant stems
Oystershell scale, *Lepidosaphes ulmi* (L.); most fruits and nuts and some stems
San Jose scale, *Quadraspidiotus perniciosus* (Comstock); many fruit species stems and fruit

Order Hemiptera
Family Pentatomidae (bugs)
Consperse stink bug, *Euschistus conspersus* Uhler; blackberry and raspberry fruit

Order Coleoptera (beetles and weevils)
Family Buprestidae
Flatheaded appletree borer, *Crysobothris femorata* Olivier; fruit and nuts, inner bark, and cambium and wood
Family Bostrichidae
Branch and twig borer, *Polycaon confertus* Lec.; deciduous fruits and nut wood
Family Scarabaeidae
Grapevine hoplia, *Hoplia callipyge* Lec.; grape, peach, almond flowers, and young leaves and fruit
Family Carambycidae
Roundheaded appletree borer, *Saperda candida* Fabricius; apple and quince bark and wood
Family Chrysomelidae
Strawberry rootworm, *Paria fragariae* Wilcox; strawberry, raspberry, blackberry roots, and apple and walnut leaves
Family Curculionidae
Fuller rose beetle, *Pantomorus cervinus* (Boheman); pome, stone fruit, persimmon leaves and *Rubus, Ribes,* and strawberry roots and leaves
Strawberry root weevil, *Brachyrhinus ovatus* (L.); pome and stone fruit, grape, strawberry, and raspberry roots and leaves
Pecan weevil, *Balaninus caryae* (Horn); pecan and hickory nuts
Plum curculio, *Conotrachelus nenuphar* (Hbst); pome and stone fruits, blueberry, currant, grape, and young persimmon fruit
Family Scolytidae
Shothole borer, *Scolytus rugulosus* (Ratzeburg); stone and pome fruit, and bark and wood

Order Diptera (flies)
Family Anthomyidae
Raspberry cane borer, *Oberea bimaculata* (Olivier); caneberry shoot tips and canes
Family Trypetidae
Cherry fruit fly, *Rhagoletis cingulata* (Loew); sweet and sour cherry fruit
Western cherry fruit fly, *R. indifferens;* sweet and sour cherry fruit
Walnut husk fly, *R. completa* Cresson; walnut husk and peach fruit at maturity
Apple maggot, *R. pomonella* (Walsh); apple and blueberry fruit
Currant fruit fly, *Epochra canadensis* (Loew); currant and gooseberry fruit

Order Lepidoptera (moths and butterflies)
Family Pyralidae
Navel orangeworm, *Paramyelois transitella* (Walker); mature nuts of regia walnut
Family Gelechiidae
Peach twig borer, *Anarsia lineatella* Zeller; peach, apricot, young twigs and and mature fruit of almond
Family Aegeriidae
Raspberry crown borer, *Bembecia marginata* (Harris); raspberry and blackberry roots and crown
Peachtree borer, *Sanninoidea exitiosa* (Say); peach, apricot, and prune at the crown and lower trunk
Strawberry crown borer, *Tyloderma fragariae* (Riley); strawberry crown
Family Olethreutidae
Eyespotted bud moth, *Spilonota ocellana* (Denis & Schiffermüller); pome and stone fruit buds and leaves
Strawberry leafroller, *Ancylis comptana fragariae* (W. & R.); strawberry, raspberry, and blackberry leaves
Lesser apple worm, *Grapholitha prunivora*

(Walsh); apple, plum, and cherry buds and fruit
Codling moth, *Laspeyresia pomonella* (L.); apple, pear, and walnut fruit
Family Tortricidae
Fruittree leafroller, *Archips argyrospilus* (Walker); pome and stone fruit, walnut, *Rubus*, currant leaves
Obliquebanded leafroller, *Choristoneura rosaceana* (Harris); most tree and small fruit leaves
Apple skin worm, *Tortrix franciscana* (Wlshm.); apple leaves and fruit
Order Hymenoptera (bees, wasps, and ants)
Family Tenthredinidae
Raspberry sawfly, *Monophadnoides geniculatus* (Hartig); raspberry and other *Rubus* leaves
Pearslug, *Caliroa cerasi* (L.); pear leaves

Selected Life Histories

Life histories of pests vary considerably; some pests can cause damage at more than one stage of development. Good pest control depends upon a knowledge of the interactions of many factors. The following life histories indicate these factors for some important mites and insects. Pest control, whether by chemicals, biological means, autocide, or by integrated methods, depends on knowing the details of the life cycle of the pest and the most vulnerable points in that cycle—whether it is when the pests breed or where they overwinter or when they are in some other stage.

SPIDER MITES The European red mite passes the winter on twigs as a small red-orange egg, which hatches in spring, just prior to apple bloom. Young mites crawl to young leaves and suck sap from the undersides. They become full grown after 2 to 4 weeks, depending on the temperature. After mating, the female lays 30 or 35 eggs on the leaves. Four to eight generations per year are produced and soon these generations overlap, resulting in the presence of all stages at one time. Trees attacked are apple, pear, plum, and many other deciduous species. The mite is found in Europe and north of 37° N latitude in North America. (See Fig. 16-1.)

The McDaniel and twospotted mites overwinter as adults in cracks in trunk bark and in trash under the tree. In March and April they move up and infest the leaves, where they multiply rapidly in hot weather. Shortly after migration, each female lays about 50 eggs, and the life cycle may be complete in 2 weeks. Several generations are produced each season. Plants attacked are apple, pear, grape, plum, almond, walnut, and many others (Fig. 16-1). Its distribution is world wide. A closely related species, *T. pacificus* McGregor, is found in the Pacific Coast states.

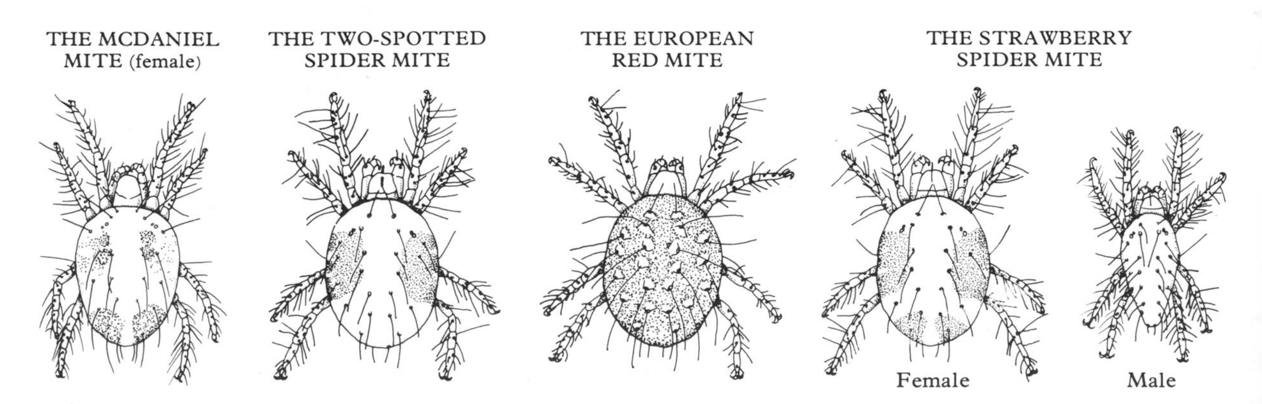

Figure 16-1 A number of common spider mites (greatly enlarged), which are injurious to several fruit species. [After the OSU Extension *Insect Control Handbook*, 1976]

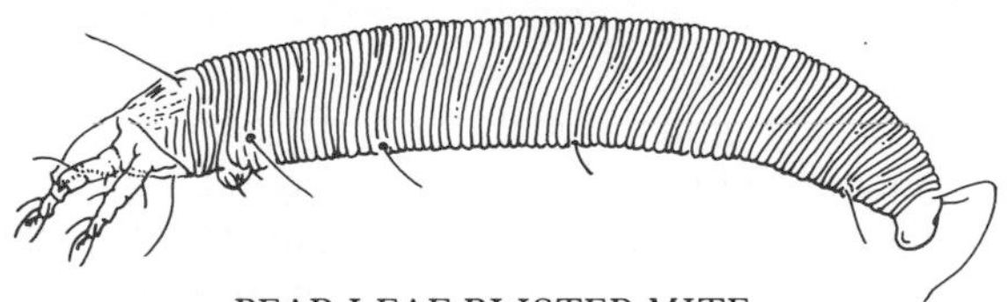

PEAR-LEAF BLISTER MITE

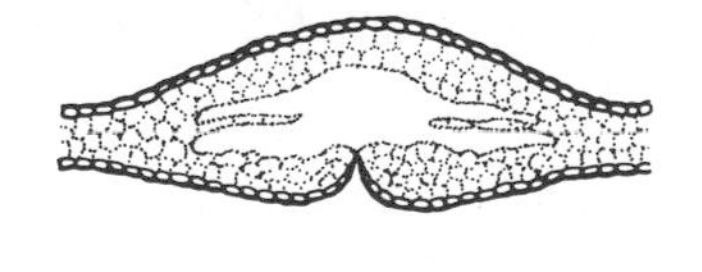

PEAR-LEAF BLISTER

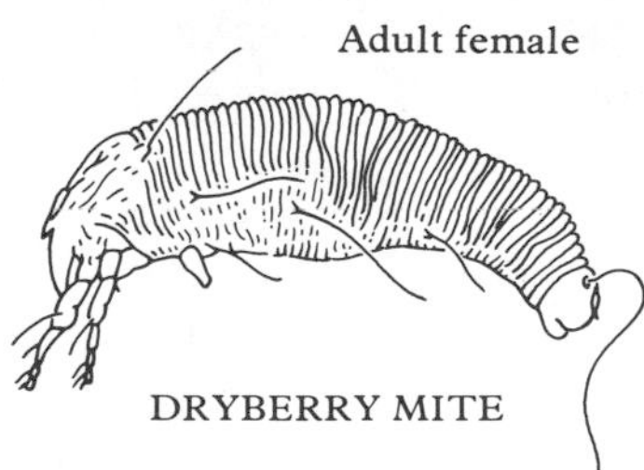

DRYBERRY MITE

Figure 16-2 One microscopic eriophyid mite is the dryberry mite which causes the disorder known as redberry in blackberries. The pear-leaf blister mite is shown with a diagram of a cross-section of a pear leaf showing the internal blister induced by this mite. [After H. H. Keifer, *Bull. of the Calif. Insect Survey,* Vol. 2, No. 1, 1952]

The pear-leaf blister mite, which is less than .02 centimeter in length, overwinters in the bud scales—or, in the Northwest, they winter in the egg stage, hatching in early spring. As the leaves start to come out, mites move to them and feed on the undersides, causing the pale brownish blisters in which eggs are laid later and in which the young develop (Fig. 16-2). There are several generations per year. Trees attacked are pear, apple, and related pome genera. These mites are found generally in all fruit-growing areas.

The dryberry mite is a microscopic mite about .01 centimeters in length that causes red-berry disease of blackberries (Fig. 16-2). The mite overwinters in cane buds as an adult and lays eggs, starting in March, between the bud scales and in the drupelets of berries. Control is by removing old canes early and burning them after harvest is complete. In the spring when buds are about 1.3 centimeters long and again when fruiting arms are 30 centimeters long, lime-sulfur sprays effectively control this mite (Breakey and Brannon, 1946).

GRAPE LEAFHOPPER The grape leafhopper passes the winter in fallen leaves or debris as a pale adult about .3 centimeter long. In spring, when the leaves begin to appear, it flies to them and sucks the sap, attaching to the underside of the leaf. In 2 or 3 weeks, females lay eggs that hatch to pale green nymphs, which feed also on the undersides of leaves. They attain adulthood in 3 to 5 weeks. Two generations per year are produced in most areas. Plants attacked are grape, Virginia creeper, and currant. Other species of leafhopper are *E. tricincta* & Fitch (Midwest) and *E. comes* which is found throughout North America (see Fig. 16-3).

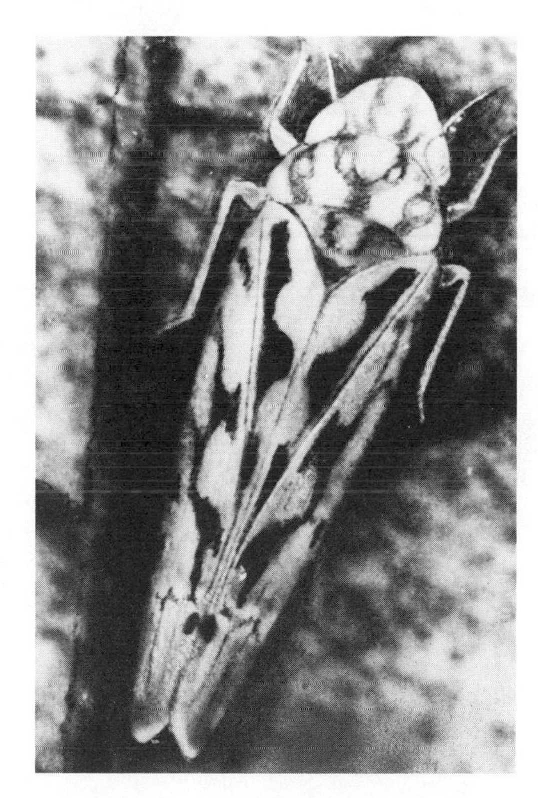

Figure 16-3 Enlarged grape leafhopper feeding on the underside of a grape leaf. [USDA photograph]

PEAR PSYLLA The pear psylla, a small cicada-like insect, is only about .25 centimeter in length and feeds on the leaves of European pears. Heavily infested trees are stunted, lose their leaves early, and the fruit is small in size. This insect also is the vector of pear decline, a disease caused by a mycoplasma. Psyllids secrete honeydew, which makes leaves and fruit sticky. The psylla is found in Europe, Asia Minor, North America, and South America (Fig. 16-4).

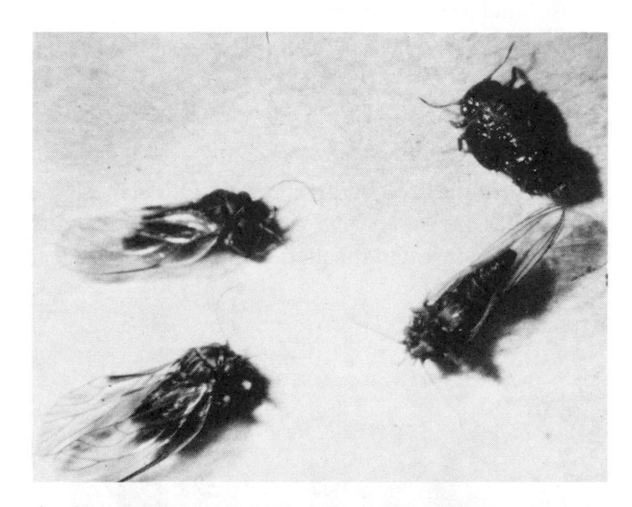

A

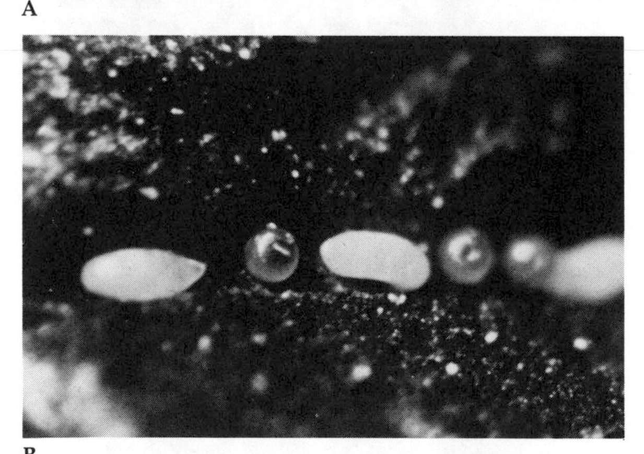

B

Figure 16-4 **(A)** Greatly enlarged adult pear psylla and **(B)** their white oval eggs. This insect, which is only about .25 cm long, feeds on pear leaves and severely debilitates the trees. [A: Courtesy of M. T. AliNiazee; B: USDA photograph]

The adult insect passes the winter behind loose bark or in other sheltered areas both within and outside the orchard. Upon completion of ovarian diapause usually in late winter, the females begin to lay eggs in fissures in bark or around the buds. Dependent upon temperature, the eggs will hatch in 2 to 4 weeks. There are five nymphal stages, the last, referred to as the hardshell, will molt into the adult form. There are three to five generations per year. Some pear species have been found to be resistant to the pear psylla (Westigard et al., 1970).

GRAPE PHYLLOXERA The grape phylloxera is an important insect that has a very complex life cycle, with four distinct adult forms. The winter is passed both as eggs on the canes and as a yellowish aphid on nodules of the grape root. The latter becomes active and feeds on the roots in the spring. The eggs on the canes hatch as growth starts in spring and the aphids move to the leaves and feed. This feeding causes the formation of galls, in which the full-grown insects give birth to live young, which in turn feed and cause more galls to form. Toward fall, those on the roots go above ground and lay eggs on the vines, from which sexual forms of the phylloxera hatch. Each fertilized female lays a single egg on a cane, on which it overwinters. This insect attacks only grape, and affects *Vitis vinifera* to a much greater degree than American species, such as *V. labrusca.* It is found in most areas where grapes are grown (see Fig. 16-5).

GREEN PEACH APHID Green peach aphids, along with several other aphids which infest stone fruits, pass the winter as eggs and hatch in early spring. The new adults give birth to living young during the summer and after several generations on stone fruit trees, winged forms are produced which go to a number of garden plants as summer hosts (see Fig. 16-6).

SAN JOSE SCALE San Jose scale attacks pome fruits, stone fruits, currant, gooseberry, several other bush fruits, and sometimes walnut. It is distributed throughout the fruit-growing regions of the U.S. and Canada as well as many other areas of the world.

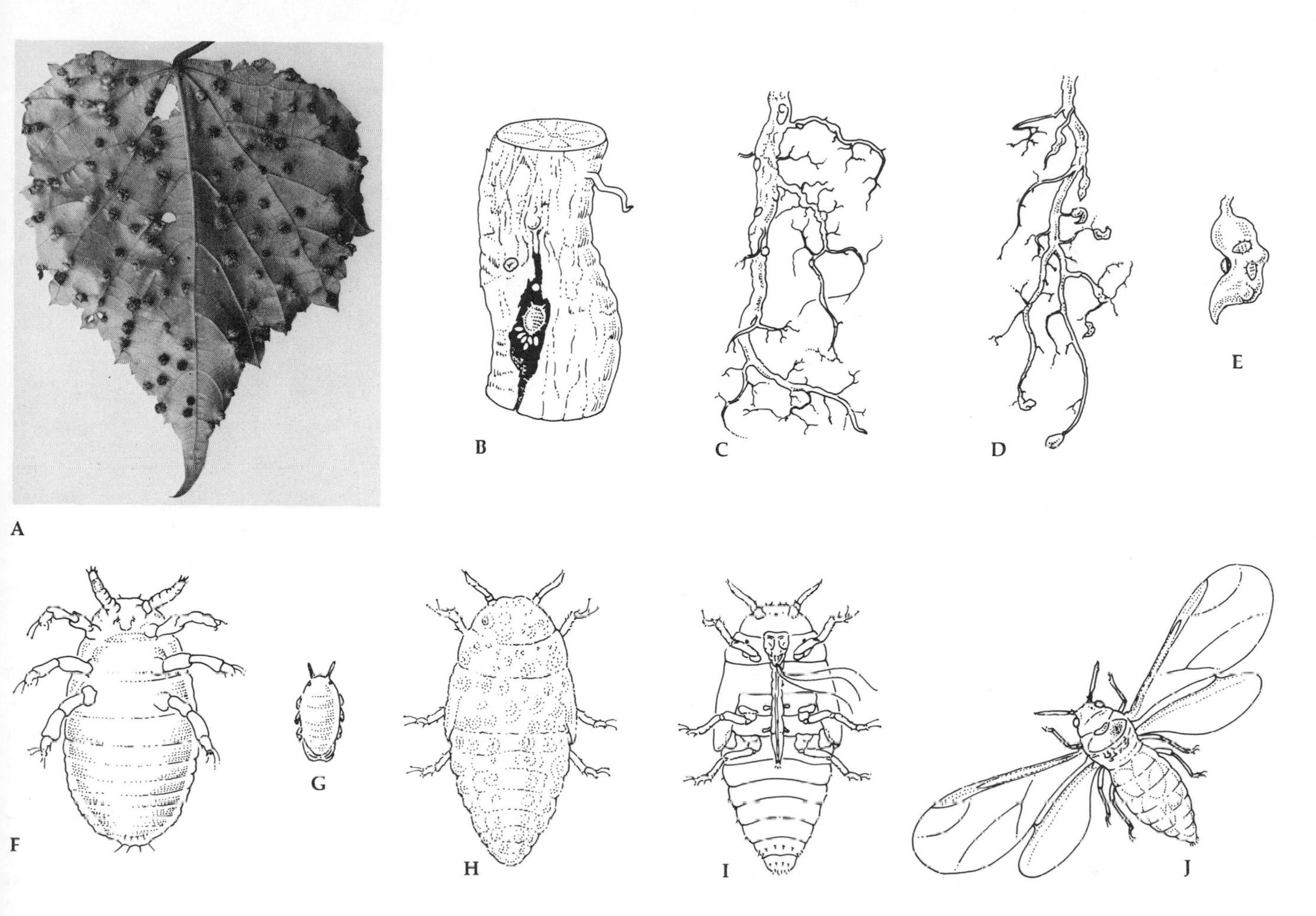

Figure 16-5 **(A)** Galls on a grape leaf infested with grape phylloxera, or root louse. Grape phylloxera is a serious pest of *Vitis vinifera* cultivars except when grafted to resistant rootstocks. [A: USDA photograph; B–J: After W. M. Davidson and R. L. Nougaret, 1921. *The Grape Phylloxera in California,* USDA Bull. 903]

Scales pass the winter as dormant partly grown second instar nymphs. They are fastened to the living bark of the tree and begin growth with the tree in spring, reaching maturity at about full bloom. As shown in Figure 16-7, the larger discs are stationary females, and the smaller ones are the coverings of the mobile males, which emerge and mate with the females. The young, called crawlers, are born alive over a period of weeks; for a short time, they wander over the surface of the branch or fruit. They then insert their thin mouthparts into the living bark or epidermis, after which they lose their legs and secrete a waxy covering that protects them. From two to six generations are produced each year, depending upon the temperature.

FLATHEADED APPLETREE BORER This member of the beetle order passes the winter as a larva, 1.2 to 2.5

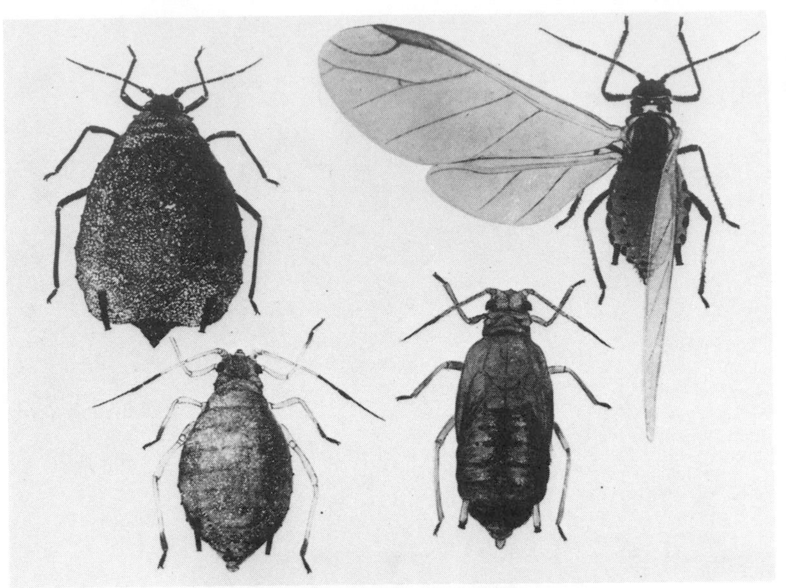

A

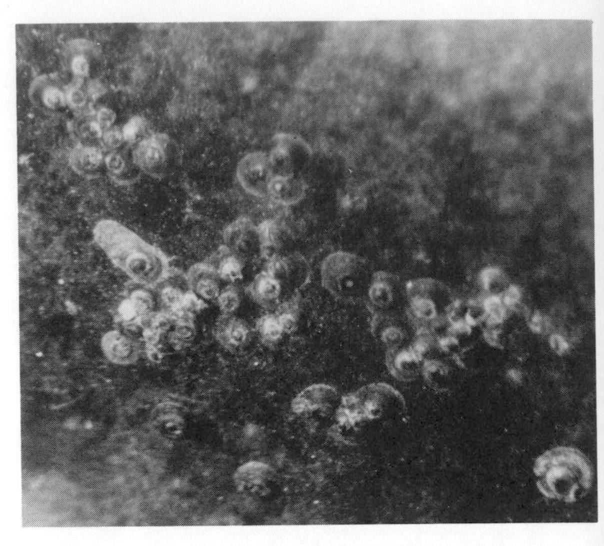

Figure 16-7 The hardshell stage of San Jose scale. The larger objects are females. [USDA photograph]

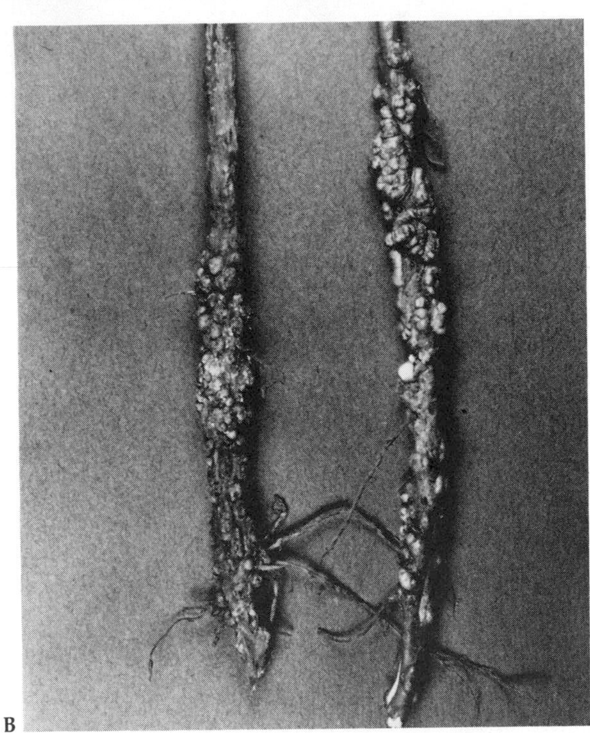

B

Figure 16-6 **(A)** Enlarged rosy apple aphids. **(B)** Root aphids infesting the roots of young nursery trees. [A: USDA photograph; B: Courtesy of M. T. AliNiazee]

centimeters in length, imbedded 2 to 5 centimeters deep in the trunk or branches of the tree (Fig. 16-8). In spring it changes to a yellow pupa and hence to a gray-brown beetle, 1.5 centimeters in length. As with the shothole borer, the females are attracted to injured or unhealthy trees, where they lay their eggs in the cracks of bark. Species attacked are most fruit, nut, woodland, and shade trees. The insect is found in the fruit districts of North America.

STRAWBERRY ROOT WEEVIL The strawberry root weevil passes the winter in surface debris and strawberry crowns, both as a larva and as an adult. The short, dark adult beetles are about .6 centimeter in length; they feed on both berries and foliage (Fig. 16-9). Since only females have been found, they must reproduce parthenogenically. Usually two generations are produced in a season. The round white eggs are laid in the crowns or among the roots of the strawberry plant. In late summer, the eggs hatch into small grubs that overwinter. Plants attacked are strawberry and related small fruits, grape, and many other fruit species. The insect is found mainly in Canada and the northern United States.

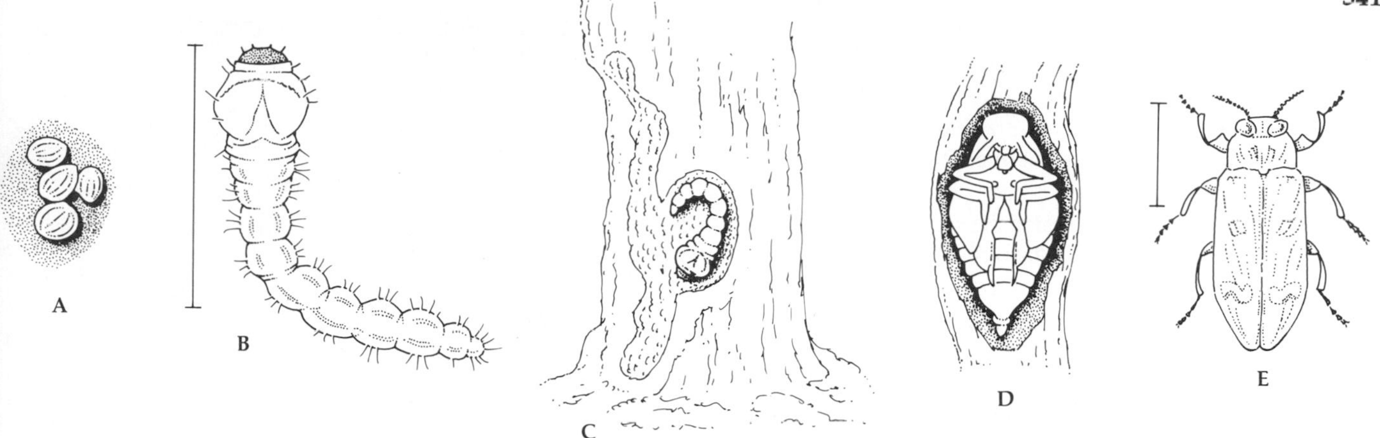

Figure 16-8 The flatheaded appletree borer as **(A)** eggs, **(B, C)** larva, **(D)** pupa, and **(E)** adult. The habitat at the base of a tree trunk is shown at lower left.

Figure 16-9 The strawberry root weevil. **(A)** Larva. **(B)** Pupa. **(C)** Adult. The larvae infest **(D)** the root; the adults feed on **(E)** the leaves. [A, B, C: Courtesy of J. Capizzi; D, E: Courtesy of R. G. Rosenstiel]

PLUM CURCULIO Of a number of snout beetles, the plum curculio is one of the most important pests. The insects pass the winter as dark-colored adult beetles, about .6 centimeter in length, in leaves, rocks, or other debris near orchards. At blossom time in spring, they become active and fly to the trees, feeding on young fruits. They then mate, and the females lay their eggs under the skin of the fruit. Upon hatching in about 10 days, the grubs eat into the flesh of the fruit, where they remain for 2 or 3 weeks. The full-grown larvae then migrate 2.5 to 5 centimeters into the ground and pupate. The adult beetles emerge in about one month (Fig. 16-10). North of 39° N latitude, only one generation per year occurs. Southern states may have two or sometimes three generations per year. Trees attacked are plum, apple, pear, quince, peach, apricot, cherry, grape, persimmon, and other cultivated fruits. The insect's distribution is east of the Rocky Mountains in North America. Related insects are the apple, cherry, and quince curculios.

A

B

Figure 16-10 **(A)** Larval stage of the plum curculio and **(B)** the adult feeding on young fruit. [Courtesy USDA and Federal Extension Service]

CHERRY FRUIT FLY The cherry fruit fly passes the winter in the soil, enclosed in a brown puparium. Adult flies emerge in late spring, and females feed on surfaces of leaves and fruit and lay eggs in the nearly ripe fruit. On hatching, the larvae (maggots) feed on the fruit flesh. When full grown, they drop to the ground and change to pupae in the soil prior to winter (Fig. 16-11). Trees attacked are domestic and wild cherries. The insect is found in the northern U.S. and Canada. A closely related species, the black cherry fruit fly, prefers the sour cherry, *Prunus cerasus,* to the sweet cherry, *Pr. avium.* Another related species is the walnut husk fly.

APPLE MAGGOT The apple maggot is a dark shiny fly .5 centimeter in length, has an orange head and green eyes, white lines on its thorax and abdomen, and dusky yellow legs (Fig. 16-12). The .6-centimeter larvae (maggots) attack apple fruit and sometimes blueberries. The larvae pass the winter as pupae in brown puparia, which are about .6 centimeter in length. They are buried in soil to a depth of 2.5 to 8 centimeters. Adult flies emerge over a period of a month or two in the summer. Females lay their eggs

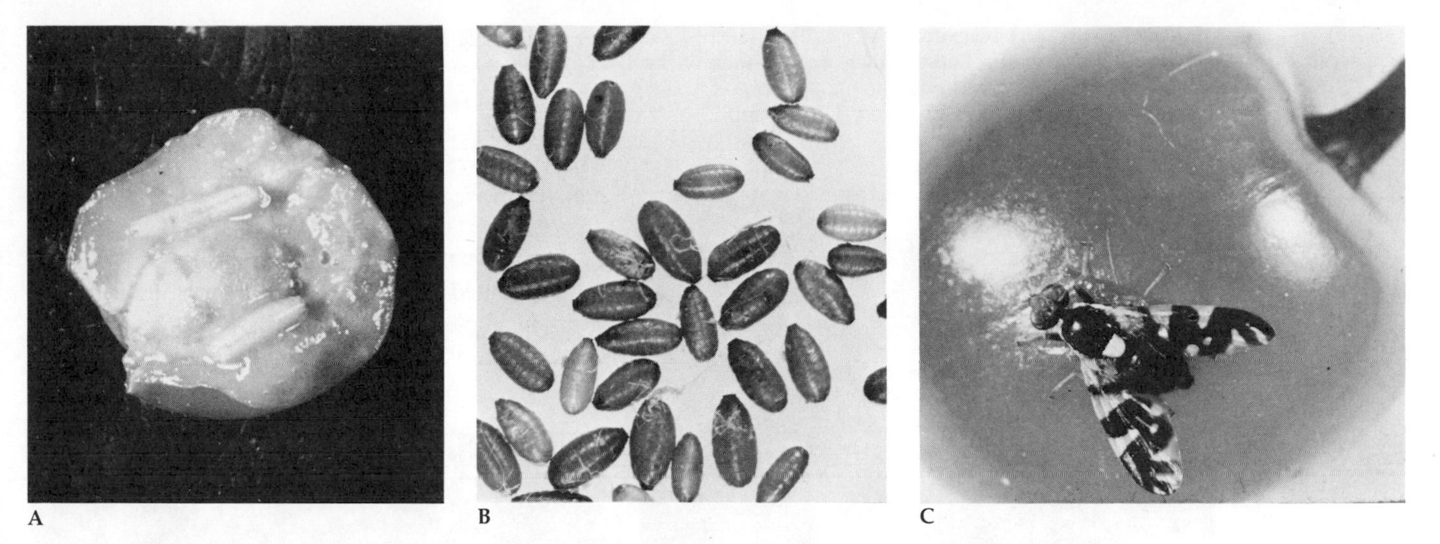

Figure 16-11 **(A)** Larvae of cherry fruit fly feeding on the flesh of a cherry. **(B)** The pupae winter in the soil and **(C)** hatch into adults in late spring. [Courtesy of M. T. AliNiazee]

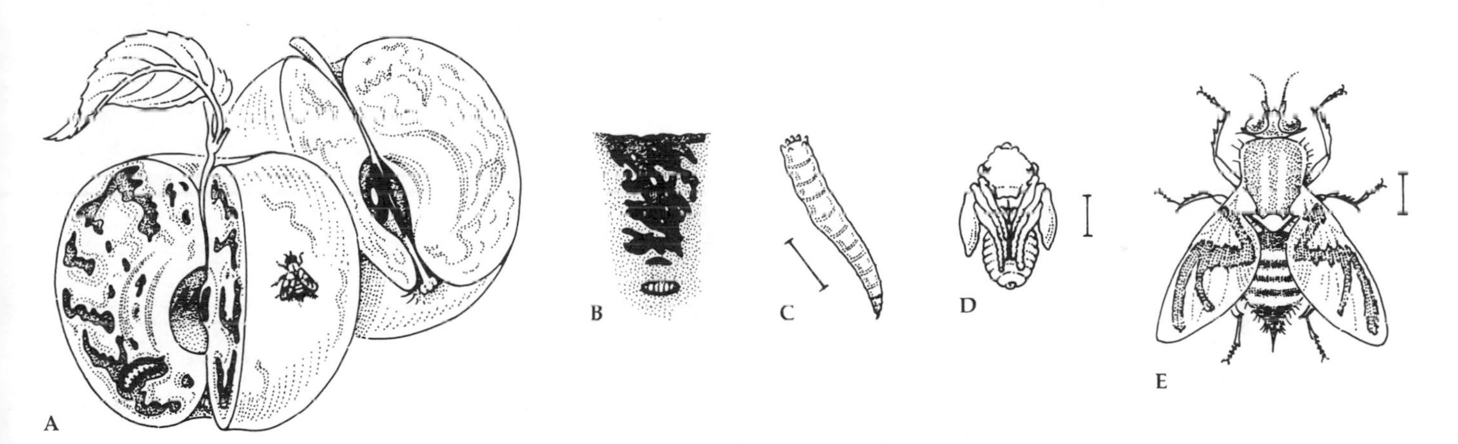

Figure 16-12 The apple maggot. **(A)** Infested fruit. **(B–E)** Life stages of the pest.

singly under the skin of the apple, about 2 to 3 weeks after the flies emerge. Eggs hatch in 5 to 10 days and maggots develop in the fruit, their full growth being attained only after the fruit drops. Larvae then leave the fruit and pupate in the soil. This insect is native to North America in southern Canada and the north central and northeastern regions of the U.S.

PEACHTREE BORER The peachtree borer overwinters as larvae under the bark of trees at the ground line (Fig. 16-13). The worms are up to 1.3 centimeters long in spring and complete their growth to 2.5 centimeters by late May. After pupation, adult moths emerge in July through September; they mate, and the females lay 200 to 800 eggs on the trunks or the

Figure 16-13 **(A)** Peachtree borer as larva and **(B)** as an adult moth. The rectangle on the back of the moth, which appears white in the photograph, is actually reddish in color. [Courtesy USDA and Federal Extension Service]

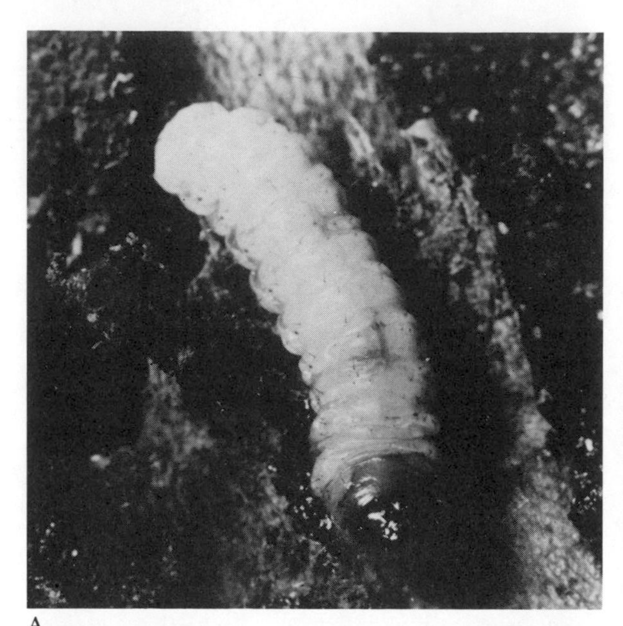

A

B

ground at the base of the trees, where they hatch in about 10 days. Usually, only one generation is produced per year. *Prunus* species attacked are peach, cherry, plum, apricot, and certain ornamental *Prunus*. The insect is found throughout the fruit regions of North America.

CODLING MOTH The codling moth has a pinkish brown-headed larva, 2 centimeters long, that infests fruits of apple, pear, quince, hawthorn, crab, regia walnut, and other fruits and nuts (Fig. 16-14). This

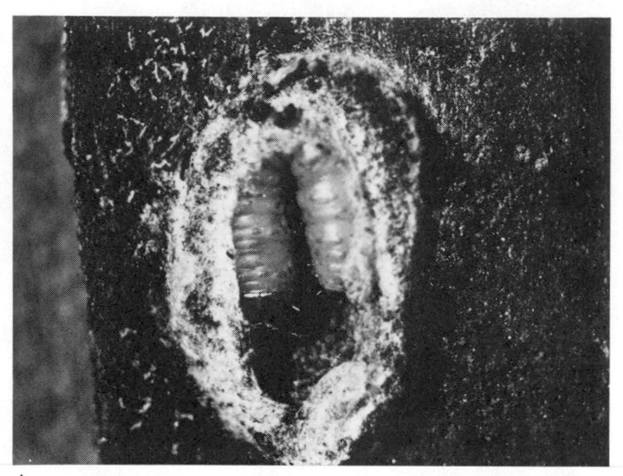

A

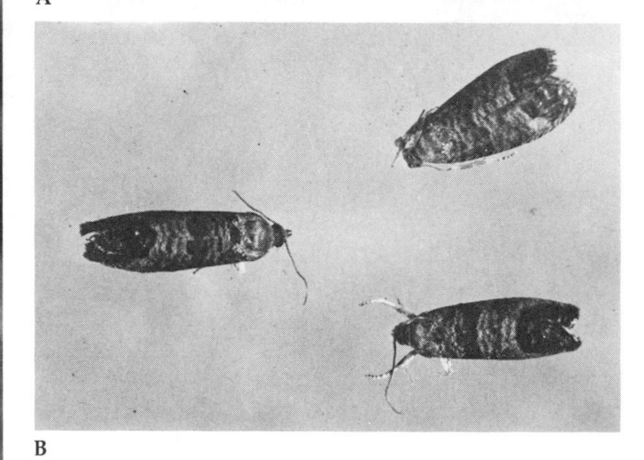

B

Figure 16-14 Codling moth. **(A)** Larvae and **(B)** adults. The full-grown larvae winter in the bark of tree trunks as shown, then mature in spring and lay eggs on young fruit. [Courtesy of J. Capizzi and M. T. AliNiazee]

insect occurs in most apple-growing regions around the world and is one of the most important pests of apple and pear. The insect passes the winter as a full-grown dormant larva in a cocoon spun in the scales of old bark on the trunk or in debris at the base of the tree. It survives cold temperatures, but a temperature of −32°C or lower kills many of them. In midspring, the pupal stage develops; in May, the adult moth, 2 centimeters long, emerges. The insects mate during warm evenings, and eggs are laid singly on leaves, twigs, or fruit. They hatch in one to three weeks, depending on temperature, and make their way into the calyx or side of the fruit. There are from one to four generations per season.

FRUITTREE LEAFROLLER The fruittree leafroller (Fig. 16-15) is 20–25 millimeters long, and its color is fawn or rusty, with irregular spots and markings on the wings. Adults appear during summer and lay eggs on the bark of twigs or trunks of trees. These whitish-gray flat masses of eggs are covered with cement, under which they pass the winter. From March to May, the eggs hatch and the caterpillars feed on young fruit and foliage. Young unfolding leaves are pulled together with webbing to form a hiding place in which they pupate. There is only one brood per year.

RASPBERRY SAWFLY The raspberry sawfly (Fig. 16-16) is about 5.5 millimeters long and is black with yellow and red markings. The pale green larvae reach

Figure 16-15 The fruittree leafroller. **(A)** Larva and **(B)** adult moth, about natural size. [Courtesy of M. T. AliNiazee]

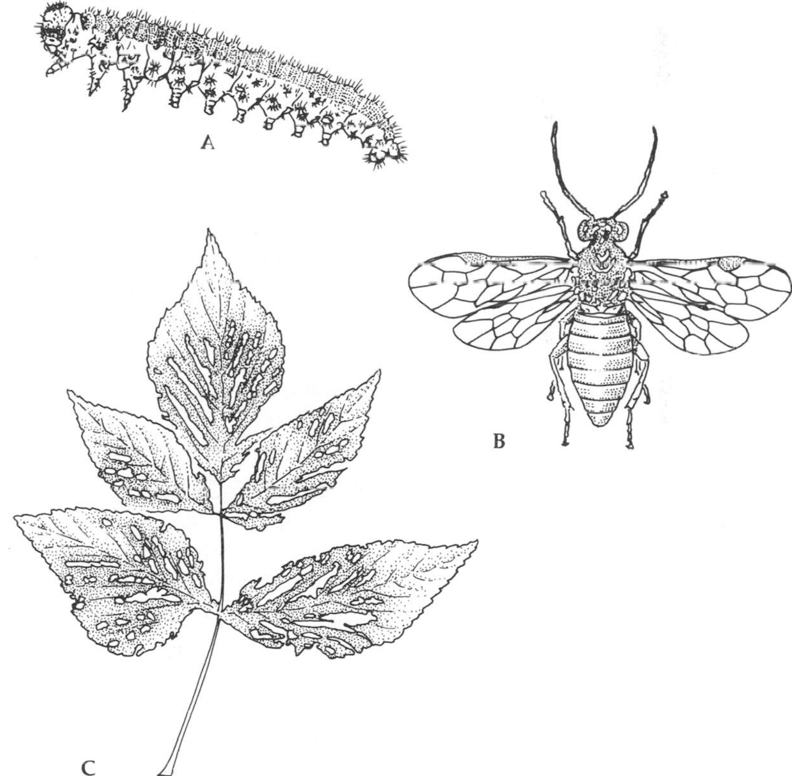

Figure 16-16 Raspberry sawfly. **(A)** Much-enlarged larva and **(B)** adult. **(C)** Injury from larval feeding results in many elongated holes between the veins of the leaves. [After Washington Agr. Exp. Sta., Popular Bull. 164. 1941.]

16 millimeters in length. Adults insert the eggs into leaf tissue in the spring, near blossom time. Larvae feed on the undersides of leaves, mostly of raspberries and loganberries, and eat out many small holes. There are usually two generations per year.

Control. Integrated Pest Management

DEFINITION A wide array of spectacularly effective synthetic organic pesticides became available in the 1940's, and fruit-growers became greatly dependent on chemical control of insects and mites. Over the years, however, the repeated use of chemicals led to pest resistance to chemicals, to the emergence of secondary pests, to the resurgence of pest populations, and to the frequent contamination of the environment. In spite of its drawbacks, however, chemical control has remained the only suitable means of checking specific insects in many situations. Thus, the need has arisen to develop a system of maximizing the advantages of pesticides while minimizing their disadvantages, taking the total environment into account as well as the need for efficient crop production. The term "integrated pest management" has been used to describe this system. It is defined as a pest-management system that, in the context of the associated environment and the population dynamics of the pest species, utilizes all suitable techniques and methods in as compatible a manner as possible and maintains pest populations at levels below those causing economic injury (Glass, 1975).

CONTROL TACTICS The objective of the integrated control system is the optimization of control in terms of the overall economic, environmental, and social needs of mankind. Integrated pest management encompasses the protection of crops from damaging insects, mites, nematodes, plant pathogens, and weeds. This brief discussion will include only insects and mites. The various tactics that may be used in integrated control make use of genetic resistance, cultural factors, biological factors, autocide, pesticide use, attractants and/or repellants, hormonal factors, quarantine, and eradication. Each control tactic will be discussed briefly to indicate ways in which it can be used as part of an integrated pest-management system.

Resistance. Resistance refers to the breeding or selection of crop species or cultivars that are resistant to one or more pests. Examples are: the breeding of mite-tolerant strawberries; the selection of grape rootstocks resistant to phylloxera root louse; the development in England of the Malling-Merton series of clonal apple rootstocks resistant to the woolly apple aphid; the selection of *Pyrus calleryana* and *P. betulaefolia* pear rootstocks resistant to the pear root aphid (Westwood and Westigard, 1969). The tactic of plant resistance to pests is both economically important and environmentally sound; it is compatible with other control tactics, is less expensive, and presents no toxicity or environmental pollution. This tactic could be used much more widely than it now is.

Cultural Control. Cultural practices usually do not result in direct control, but may enhance natural enemies or retard pest populations to an important extent in integrated programs. Untreated abandoned orchards or populations on seedling escapes may be a major source of infestation for commercial plantings. This can cause major problems in the control of insects that attack the fruit itself, such as codling moth, oriental fruit moth, apple maggot, and cherry fruit fly. Sanitation measures within the planting, such as removal or destruction of infested fruit or plant parts, or reduction of overwintering sites help reduce pest populations. Pruning practice may aid in pest control; for example, less severe pruning of peach reduces succulent growth and thus reduces overwintering populations of the oriental fruit moth; also special pruning may permit better spray coverage for control of San Jose scale. Irrigation by overtree sprinkling has helped reduce populations of the McDaniel mite of apple and the pear psylla in pear orchards. Dust on plant leaves causes mites to build up, so cultural practices that reduce dust help to reduce mite populations. Certain ground cover weeds at times are important hosts for beneficial mite and aphid predators. On the other hand, nectar-bearing ground cover species may provide supplementary food for parasites of San Jose scale (Hoyt and Burts, 1974). Even fertilizer applications can affect pest levels. Excess nitrogen, resulting in late succulent growth is conducive to a late season buildup of such insects as pear psylla and the apple aphid. In general, cultural control practices are compatible with other control tactics and thus should play a significant role in integrated pest-management programs.

Biological Control. The regulation of pests by their natural enemies (predators, parasites, and pathogens) is termed "biological control." To realize the full potential of this control tactic it will be necessary to greatly reduce the use of broad-spectrum pesticides, which tend to kill predators as well as plant-feeding species. Some beneficial species are: ladybird beetles, which feed on aphids; lace wings, which feed on aphids and other softbodied insects; ground beetles, which feed on gypsy moth larvae; pirate bugs, which feed on mites; syrphid flies, which feed on aphids; chalcid wasps, which parasitize insect larvae; tachinid flies, which parasitize army worm larvae; predaceous thrips, which feed on mites and scale insects; and predator mites, which feed on phytophagous mites (See Fig. 16-17). In Nova Scotia, integrated control of codling moth by ryania sprays plus other selective chemicals has permitted survival of arthropod predators (MacPhee and Sanford, 1961). Also, biological control has been achieved for such diverse pests as the grape leafhopper, the walnut aphid, and the McDaniel mite of apple. An integrated control for McDaniel mite was reported by Hoyt (1969), who found that, because its mite predator *Typhlodromus occidentalis* has produced strains resistant to organophosphorus pesticides, selective sprays may be used that do not harm the predator. Pre-bloom petroleum oil sprays work well in the integrated program because they kill mites at a time when the least harm can be done to predators. Also, oils have a very low mammalian toxicity. Biological control as part of an integrated system in which selective pesticides do not kill the predators should increase in importance as more is learned about indigenous and introduced natural enemies of plant pests.

Autocidal Control. Autocidal control is achieved by the manipulation (by radiation and other means) of pests, so that they contribute to their own destruction. Autosterilization is one type of autocidal control that has worked well in some situations. Success on a commercial scale has been achieved for codling moth, but at a higher cost than chemical control (Hoyt and Burts, 1974). When the sterile-male technique is used with moths, male moths are reared in large numbers, irradiated with gamma rays to induce sterility, then released in the field to mate with females, which then lay infertile eggs. Success with this tactic depends upon a very low initial field population throughout the complete contiguous area in which the sterile males are released. Autocidal control as part of an integrated system has the advantage of having no deleterious effects on people or the environment.

Chemical Control. The use of chemical control depends upon selective use of pesticides to kill target species and spare beneficial species. Timing is important; for example, during the dormant period, when few parasites and predators are active, oil and other chemicals can be used to kill injurious species. The use of lower rates per hectare and lower volume sprays for some pests has effectively reduced costs and reduced environmental pollution while keeping pest populations at economic levels. This was made possible by the use of new types of equipment and application techniques that provide for better distribution of the pesticide and reduce losses due to drift. One of the main problems associated with intensive pesticide use is that pests become resistant to the pesticide. Usually, when this happens, substitute chemicals have been found to control the resistant species. All alternative means of control should be explored, but it is likely that agriculture must still depend on chemicals for economic control of certain pests in many situations.

Attractants and Repellants. The recent use of attractants (insect pheromones) has led to improved integrated control systems. Major uses have been in (*a*) monitoring for the presence of specific pests, (*b*) establishing population densities for direct control, (*c*) mass trapping, or (*d*) inhibiting mating by

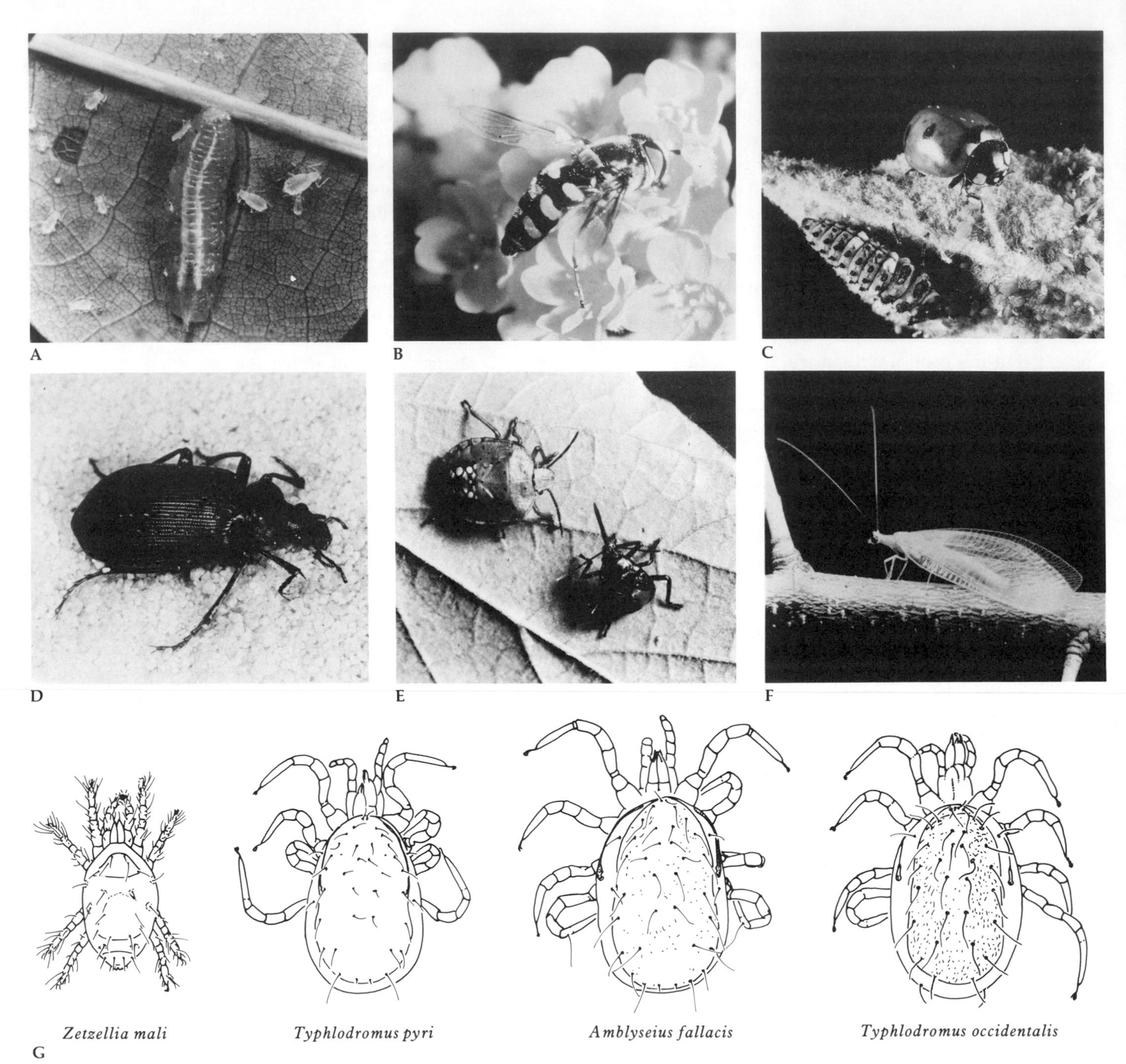

Figure 16-17 Some beneficial insects and mites that feed on harmful species. **(A, B)** Larva and adult syrphid fly. **(C)** Larva and adult ladybird beetle feeding on apple aphids. **(D)** Fiery hunter ground beetle. **(E)** Predaceous stinkbug attacking green stinkbug. **(F)** Lacewing. **(G)** Four species of predaceous mites that feed on phytophagous mites and eggs. [A–F: Courtesy of J. Capizzi, O.S.U., and USDA; G: Courtesy of G. W. Krantz and O.S.U. Extension *Insect Control Handbook*, 1976]

saturating an area with a sex attractant that prevents the males from finding the females. This tactic has many excellent possibilities of enhancing other integrated control measures while offering no environmental pollution, since these attractants are not toxic.

Insect Hormones. Hormones that keep insects in the juvenile state and chemical regulators of insect moulting could be helpful by preventing completion of the pest's life cycle. Although they are not ready for commercial use, juvenile hormones have great potential as a part of integrated pest-management systems, because they are active at low concentrations and are biodegradable.

Quarantine and Eradication. The confinement of a pest to a given area (quarantine) or the elimination of a pest from a given area (eradication) are tactics that are sometimes useful. However, with rapid world travel, quarantine regulations and enforcement are often not sufficient to prevent pests from spreading beyond their present areas. Eradication has not been successful when directed against a widespread pest, but it has sometimes been successful against incipient pests in relatively small areas. At present, the tactics of quarantine and eradication show only limited potential as part of integrated control systems.

SUPPORTIVE TACTICS The best use of integrated pest management depends upon the well-planned and well-executed supportive tactics of reliable population-sampling procedures and monitoring techniques, and the establishing of economic thresholds. The whole approach to integrated pest management is based on the concept of economic thresholds; a population density of a pest below which the cost of applying control measures exceeds the losses incurred by the pest constitutes such a threshold (Glass, 1975). The establishment of economic thresholds is very complicated and must take into account economic alternatives in pest control, relative benefits and hazards, research alternatives, and social considerations. The increasing demand for food will likely bring about more intensive—and more expensive—agriculture and will lead to a frequent adjustment of economic thresholds. For example, many of the pest-control tactics now used are done in part because of the very strict market standards of appearance for fruits. Fruit must have a smooth, blemish-free skin or it becomes a cull to be discarded or sold at a much lower price. The cost of many pest-control practices could be reduced if standards of appearance were changed. Often, surface blemishes do not affect the internal quality or the nutrient value of the crop.

DISEASES

Apple Scab

Apple scab (*Venturia inaequalis* [Cke.] Wint.) is an ascomycete that infects apples throughout the world, except in areas with high temperatures and low relative humidity. Cultivars vary greatly in the degree of susceptibility to the fungus (Fig. 16-18). The fungus

A

B

Figure 16-18 **(A)** Infection of apple by the apple scab fungus. **(B)** A closely related species infects pear. [Courtesy of D. L. Coyier]

overwinters mainly as mycelium in fallen leaves, in which perithecia mature in the spring to produce the ascospores that constitute the primary infective inoculum. Alternate wet and dry conditions and a temperature of 20°C are optimum for ascospore maturation. Under rainy conditions in the spring, ascospores are forcibly ejected into the air and scattered by the wind. Young leaves and fruit are susceptible to primary infection, unless the weather is warm and dry. Once primary infection has occurred, sporulation takes place over a range of weather conditions, making an abundant source of secondary inoculum. Control to prevent primary infection must cover and protect the developing buds before infection occurs and must be repeated at 3- to 8-day intervals to give protection all during the period of infection (green tip to after petal fall). A number of fungicides are used in accordance with local recommendations.

The Powdery Mildews

The powdery mildews (ascomycete; family Erysiphaceae) is a group of disease organisms that affects several fruit and nut species; all powdery mildews are obligate parasites. They are characterized by a grayish-white mycelium that appears on the tips of infected shoots (Fig. 16-19). Many cause most severe damage under dry weather conditions. Powdery mildew of apple (*Podosphaera leucotricha* [Ell. & Ev.] Salm.) overwinters in infected buds, which produce spores to infect other buds in the spring. Mildewcide sprays for control should start at the tight cluster or pink stage and continue at 5- to 7-day intervals through the period of infection. Cold winters and dormant pruning can reduce the amount of mildew overwintering in the buds. Other fruit and nut species afflicted with powdery mildew species are grape (*Uncinula necator* [Schw.] Burr.); gooseberry (*Sphaerotheca mors-uvae* [Schw.] Berk. & Curt.); chestnut (*Phyllactinia corylea* [Pers.] Karst.).

Peach Leaf Curl

Peach leaf curl (*Taphrina deformans* [Berk.] Tul.) is an ascomycete that affects peach, nectarine, and almond throughout the world where these fruits are grown but is not a problem in regions with spring weather that is warm and dry (Fig. 16-20). The organ-

Figure 16-19 **(A)** Pear twig infected with powdery mildew compared with **(B)** normal twig. [Courtesy of D. L. Coyier]

Figure 16-20 Peach leaves infected with peach-leaf curl. [Courtesy of B. J. Moore]

ism overwinters as conidia on the surfaces of the dormant tree. The spores enter the bud between the bud scales in late winter as the buds begin to swell. These conidia penetrate the young leaves in the spring. The disease is most severe in cool (12°–16°C) moist spring conditions. Leaf curl is controlled by a fall or early-spring spray of a suitable fungicide before the bud scales open.

Brown Rot of Stone Fruits

Brown rot of stone fruits (*Sclerotinia* spp.) is an ascomycete that attacks all stone fruits (*Prunus* spp.) wherever they grow throughout Europe, North America, and the rest of the world. Both blossoms and fruit are attacked (Fig. 16-21). The first symptoms are browning and death of blossoms, which persist on the tree. On fruit, it begins as a tiny brown spot and rapidly enlarges to a large, soft, watery, brown area; within a few days the entire fruit is infected. Conidia quickly form on the surface, and spores are

A

B

Figure 16-21 Brown rot of **(A)** plum and **(B)** apricot. Initially infected fruit are soft and later wither and wrinkle. [Courtesy of B. J. Moore]

discharged. If left on the tree, the fruit will dry and mummify and persist indefinitely. In North America, *S. laxa* mainly attacks blossoms and twigs. Both *S. laxa* and *S. fructicola* infect the fruit.

These fungi overwinter in diseased fruit mummies, rotted fruit on the ground, and in twig cankers. *S. laxa* conidia form on the latter under favorable temperature and moisture. Primary inoculum is predominantly from overwintering twig cankers and mummies in the tree; however, fruit on the ground may produce ascospores in spring at 17° to 20°C. After primary infection, secondary spread is by airborne and waterborne conidia at relative humidities of 85 percent and above. The optimum temperature for germination of conidia is about 25°C, and high humidity is also required for germination and infection. Control of blossom infection is achieved by several sprays of organic fungicides before and during anthesis and at petal fall. Control of fruit rot is by one or two fungicide sprays prior to harvest.

Figure 16-22 A pear shoot infected with fire blight. The upper portion—showing wilted and dying leaves and stem exudate—is dead. The arrow shows the downward progress of the active lesion. [Courtesy T. Van der Zwet, USDA]

Fire Blight

Fire blight (*Erwinia amylovora* [Burrill] Winslow et al.) is most severe on pear but may infect apple, quince, hawthorn, and other genera of the subfamily *Pomoideae* (Fig. 16-22). It is the first plant disease known to be caused by a bacterium, reported by T.J. Burrill of the University of Illinois. The disease is indigenous to North America but has now spread to Europe and possibly Asia and other parts of the world. In North America, blight is much less severe in the cool coastal areas and the arid regions of the Pacific slope than in areas east of the Rocky Mountains.

The bacterium attacks mainly twigs and young shoots but may also infect fruit. In severe cases, the entire tree may die (Fig. 16-23). The organism overwinters in blighted twigs and stem cankers. Exudate from the edge of an active lesion in spring contains a great number of bacteria, which may be spread by rain or by insects from flower to flower. Infection takes place through the tender flower tissues, through the epidermis of very young shoots, and through physical injuries on the tree. The organism multiplies rapidly and moves quickly through young tissue of susceptible cultivars and somewhat slower through older woody tissue, depending upon varietal characteristics. Blight can usually be inoculated into succulent shoots of resistant cultivars, but the lesion does not move into the more mature wood of the tree. An exception to this is the cultivar Magness, in which infection occurs on older wood of the trunk. Control is achieved by cutting blighted limbs during summer, taking care to cut below the active lesion in healthy tissue and disinfecting the pruning shears after each cut. Chemical control is by sprays of streptomycin or copper dust during bloom and early post-bloom.

Figure 16-23 Pear tree completely killed by fire blight (*foreground*). Trees in the background are not infected [Courtesy of T. Van der Zwet, USDA]

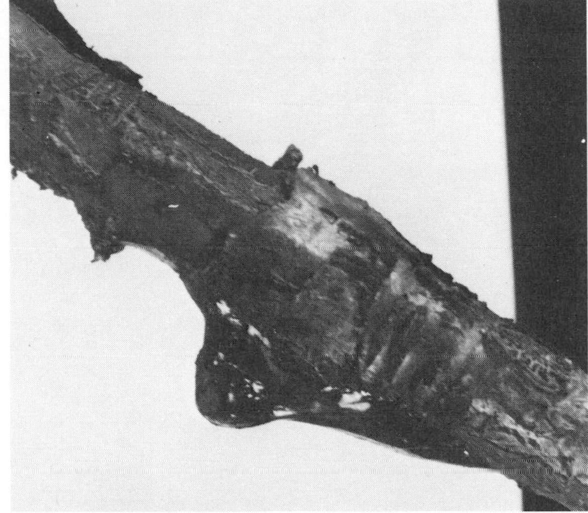

A

B

Figure 16-24 Bacterial canker appears on sweet cherry **(A)** as "gummosis" resulting from tissue injury or **(B)** as "dead bud," in which the spur tissue dies while the tissue of the branch remains healthy. [B: Courtesy of B. J. Moore]

Bacterial Canker

Bacterial canker (*Pseudomonas syringae* van Hall or *P. mors-prunorum* Worm.) is a bacterial disease that occurs on stone fruits, pear, occasionally apple, citrus, as well as lilac, and a number of herbaceous species. It is found in all fruit areas of the world. *P. mors-prunorum* is common in Europe, but its existence in North America has not been confirmed. Of the stone fruits, apricot is most susceptible to bacterial canker, followed closely by sweet cherry and plum. Peach is only moderately susceptible, and almond and sour cherry are seldom affected. The oozing of gum (gummosis) at infection sites (Fig. 16-24) is common with cherry, apricot, and sometimes plum, but is not always found with all cultivars and species.

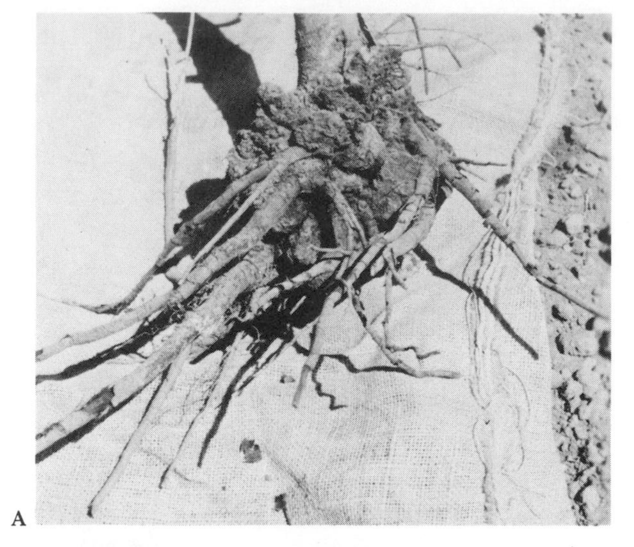

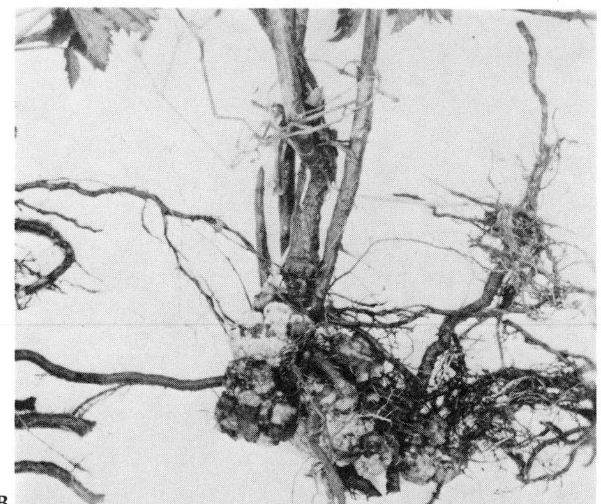

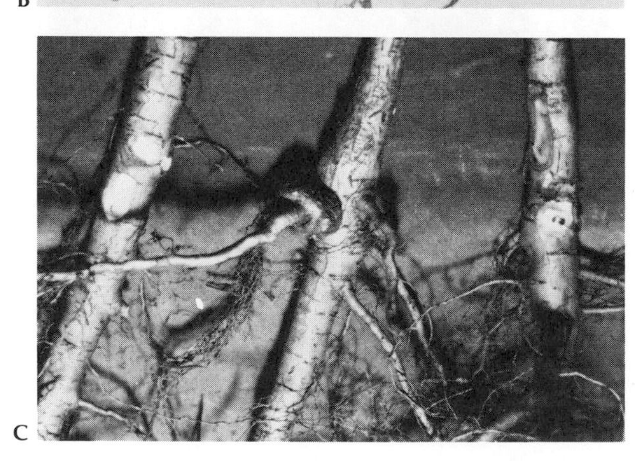

Rain spreads the bacteria about, and they infect the tree through wounds and natural openings, such as through tender tissues at the base of the bud as it begins to grow in the spring. Disease development is most active during the cool, wet period of early spring, in contrast to the warm, moist periods that favor the development of fire blight. Also, frosty weather during bloom and early growth greatly increases the incidence of bacterial canker.

In areas where the "dead bud" phase of the disease is serious, it may be controlled by fall and winter sprays of copper fungicides. These sprays are only surface protectants, however, and are not effective against systemic infection. The use of resistant trunkstocks prevents cankers from becoming established in the lower portion of the tree and reduces infection in the main branches of sweet cherry trees.

Crown Gall

Crown gall (*Agrobacterium tumefaciens* [E. F. Sm. & Towns] Conn.) is a bacterium that infects the crown and roots of apple, pear, peach, apricot, cherry, plum, almond, grape, and caneberries. The cane-gall organism is a different species *(A. rubi)*. Crown-gall organisms are widely distributed in the temperate fruit areas. The organisms live in the roots of many plants and in the soil. Infection takes place through wounds, either man-made or made by subterranean chewing insects or rodents. Deep wounds are more likely to be infected than shallow ones. Galls, stimulated by the presence of bacteria, result from hyperplasia and hypertrophy of cells. The incubation period for crown gall is longer in spring and fall than summer, and the optimum temperature for gall development is lower than that for tree growth (Fig. 16-25). Control is achieved by using gall-free nursery stock, since nothing can be done after the orchard is planted. Where possible, gall-resistant rootstocks should be used to avoid later infection.

Figure 16-25 Crown gall as it appears **(A)** on cherry, **(B)** on boysenberry, and **(C)** on apple. [Courtesy of L. W. Moore]

Root Rot and Collar Rot

Root rot and collar rot (*Phytophthora* spp.) are soil fungi that may cause root rot, as in red stele of strawberry (incited by *P. fragariae* Hickman) or collar rot of fruit trees, in which the bark tissues at the ground line are killed (*P. cactorum* [Leb. & Cohn] Schroet. and other species). This rather large group of phycomycetes attacks roots of a number of fruit species, particularly in wet soil. Control of collar rot is achieved by keeping the surface soil around the trunks and crowns dry and by use of resistant rootstocks. Root rots likewise are controlled by eliminating undue wetness of the soil and by use of rot-resistant rootstocks.

Wilt

Wilt is caused by soil-inhabiting fungi (*Verticillium* spp.) which attack the roots and stems of a large number of plant species, including grape, raspberry, blackberry, strawberry, *Prunus, Malus,* and *Juglans*. The principal means of control is the use of wilt-resistant cultivars. Pre-plant fumigation has sometimes been used to control the fungus.

Viruses

Viruses are too small to be seen with a light microscope, so they are characterized by the symptoms they induce in a host. Particular viruses can only multiply in living cells of the proper host. Transmission of viruses is by (*a*) insect, mite, and nematode vectors and (*b*) by mechanical means, such as sap transmission, budding, and grafting. Viruses of the *Prunus* ring-spot group can be transmitted by infected pollen.

Some common viruses are listed in Table 16-1. Many are latent in some species—that is, they show no visible symptoms. The worst viruses either kill the host or cause markings or injury to the fruit and make it unmarketable. Such is the case with stony pit, star crack, and rough skin viruses. Mosaic and chat fruit cause fruit to be smaller, while vein yellows and rubbery wood cause reduced growth and yield but do not affect fruit size or quality (see Figs. 16-26 and 16-27). Viruses cannot be controlled by chemical treatment, but several weeks of heat treatment of infected plants (at about 37°C) inactivates some viruses, so that propagation of buds or meristem tips can produce a virus-free plant.

Table 16-1 Some common virus diseases of fruits.

Name of Virus	Host	Vector or Natural Mode of Transmission
Mosaic	Apple	
Rubbery wood	Apple	
Star crack	Apple	
Rough skin	Apple	
Stem pitting	Apple	
Chat fruit	Apple	
Stony pit	Pear	
Blister canker	Pear	
Vein yellows	Pear	
Rough bark	Pear	
Ring spot	*Prunus* spp.	Pollen
Necrotic rusty mottle	Sweet cherry	
Sour cherry yellows	Sour cherry	Pollen
Stem pitting	Plum and peach	Nematode
Prune dwarf	Prune	Pollen
Ring pox	Apricot	
Calico	Almond	
Leafroll	Grape	Unknown
Fanleaf	Grape	
Yellows	Strawberry	Strawberry aphid
Crinkles	Strawberry	Strawberry aphid
Witches broom	Strawberry	Strawberry aphid
Green mosaic	Raspberry	Large raspberry aphid
Ring spot	Raspberry	Large raspberry aphid
Mild mottle	Raspberry	Large raspberry aphid
Dwarf	Blackberries	Aphids
Mosaic	*Rubus* spp.	Aphids
Streak	Black raspberry	
Stunt	Highbush blueberry	Leafhoppers
False blossom	Cranberry	Leafhoppers
Bunch	Pecan	Grafting

Until recently, viruses were thought to be the smallest pathogens, but now even smaller pathogens, called viroids, are known to exist.

Mycoplasma

Mycoplasma are micro-organisms that do not have cell walls and can thus pass through filters that trap

Figure 16-26 Some viruses show symptoms only on the fruit. Examples are **(A)** short-stem disease of Bing cherry, and **(B)** stony pit of pear. [A: Courtesy of H. R. Cameron; B: Courtesy of D. L. Coyier]

A

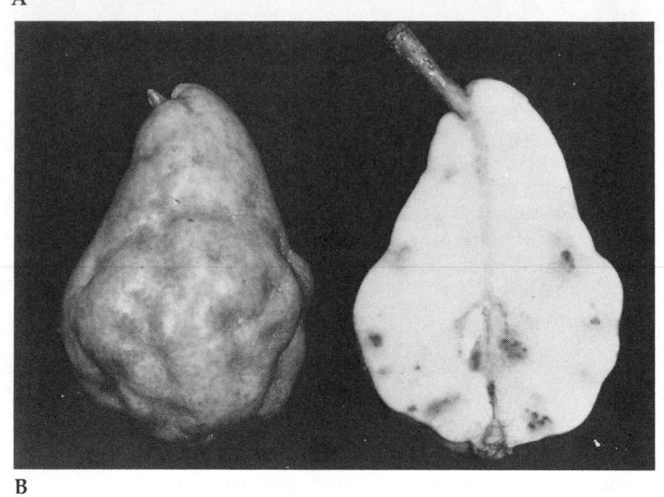

B

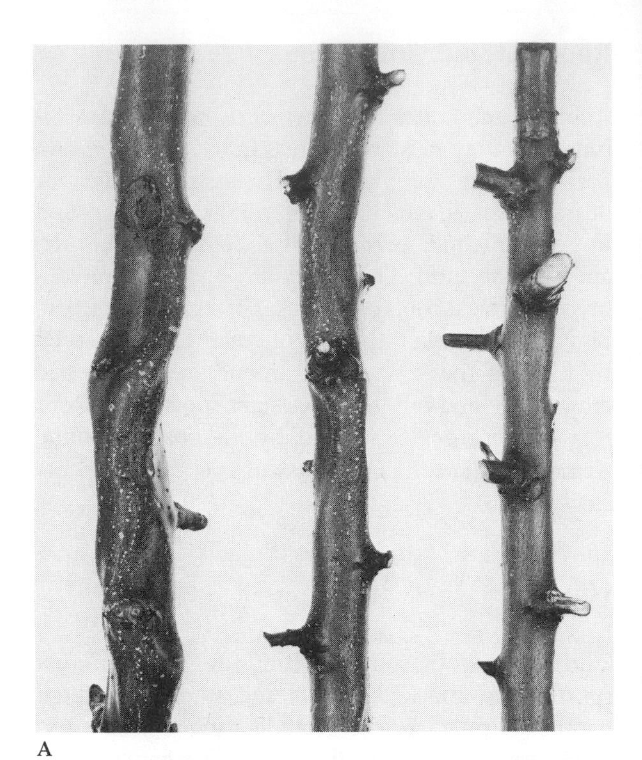

A

B

Figure 16-27 Some viruses show symptoms on stems and leaves rather than on fruit. Some examples are given of stem disorders. **(A)** Flat limb of apple. **(B)** Pear-bark measles. **(C)** Stem pitting of rootstock of Italian prune caused by tomato ring-spot virus. Examples of leaf symptoms are shown also. **(D)** Prune dwarf. **(E)** Mottle leaf of cherry. **(F)** Green-ring yellows of cherry. **(G)** Pear vein yellows (*on right*). **(H)** Red-mottle symptom of vein yellows of some pear cultivars (*on right*). **(I)** Necrotic rusty mottle of sweet cherry. [Courtesy of B. J. Moore and H. R. Cameron]

C
D
E
F
G
H
I

Table 16-2 Some diseases of fruits caused by mycoplasma.

Disease	Host	Vector or Natural Mode of Transmission
Pear decline	*Pyrus* spp.	Pear psylla
X-disease	*Prunus* spp.	A leafhopper
Peach mosaic	Peach	An eriophyid mite
Buckskin	Sweet cherry	
Albino	Sweet cherry	Unknown

bacteria; unlike viruses, mycoplasma can live on inorganic cultures, as do bacteria. In recent years, it was learned that many diseases formerly attributed to viruses are instead caused by mycoplasma (Table 16-2). These organisms vary in size and shape and may produce long filaments that look like fungal hyphae, hence the name—meaning "fungal-form." Elementary bodies ranging in size from 125 to 250 nanometers form from the filaments, and it is these that pass through bacterial filters. Unlike viruses, mycoplasma are lifelike. They do not possess cell walls, as do bacteria, but are surrounded by simple membranes. A number of times, tetracycline has been effective in killing or inhibiting mycoplasma in plants.

RODENTS, DEER, AND BIRDS

Rodents

Mouse damage to crowns and roots occurs mostly between October and spring. Control is by use of poison baits placed in covered runways near the trees. Heavy weeds, grass, or debris around the tree trunks and orchard floor provide nesting sites and protective cover for mice. Thus, it helps to keep the area around the trees clear at all times.

Rabbits damage plants by nipping off young shoots or by gnawing the bark from the trunks or lower stems. Such damage can be serious in dwarf or young trees. Protection may be provided by the use of woven-wire fencing to keep out the rabbits or by the use of certain chemical repellents painted on the bark.

Pocket gophers are serious pests on many fruit species and in plantings of all ages. They dig extensive tunnels, which may cover half a hectare, feeding upon the roots of trees and shrubs. They are controlled by toxic baits, placed in their holes or runways. This is done by hand or by a device that makes an artificial burrow in which it places bait as it is pulled through the soil by a tractor (Fig. 16-28). Baiting is done more readily in the fall, when the rodents are more active near the surface, as is indicated by freshly made mounds of earth. Generally, rodents prefer apple trees to most other species, but they do damage others as well. Cherry trees on *Prunus mahaleb* roots are preferred by gophers over those on mazzard roots. Almost all pear rootstocks are susceptible to gopher damage.

Deer

Deer browsing is especially damaging to young trees and shrubs, although it rarely kills them as does girdling by mice or rabbits. Deer browsing in nurseries also is very damaging because it often makes the plant unsalable. The most positive protection against deer is a woven-wire fence, 2 to 2.5 meters high, around the planting. Some repellents sprayed on the foliage offer partial protection, as does a small bag of blood meal placed in trees at intervals throughout the planting. These treatments, however, must be repeated often to be effective.

Birds

In many areas, bird damage is coming to be a serious problem, both on plants and on ripe fruit. In Europe, buds of apple trees are eaten by bull finches. In the U.S., mature filberts and walnuts are taken by jays and crows. However, most of the damage is done to soft fruits, such as cherries, blueberries, grapes, and caneberries. The starling is perhaps the worst bird pest, but robins, gulls, and others also may be important in local areas. Protection of crops from bird damage has been achieved mainly by keeping the birds out by covering the plants with nets or by devices to scare them away. The latter may be by sounds or

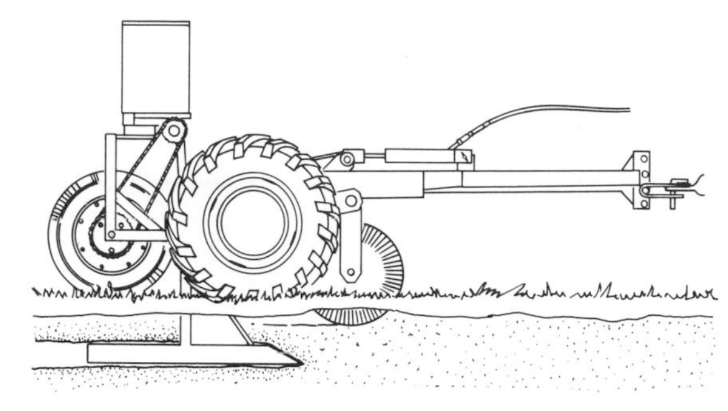

Figure 16-28 Artificial burrow or trail builder. This unit makes a channel 15 centimeters below the surface and deposits poison bait at about 40-cm intervals. Drawing at right shows the position of the trail builder in operation. [Courtesy Rue R. Elston Co., Inc.]

sights, such as reflective, moving strips of foil or intermittent loud blasts from carbide guns. Perhaps the best of such scaring devices is a taped recording of the distress call of the bird in question, played over a public address system in the field. Recent work with chemical repellents indicates that some materials, when sprayed on the crop, can repel birds without leaving a harmful residue. The names of any such chemicals registered for use in local areas can be obtained from the agricultural extension service.

GENERAL REFERENCES

Anderson, D. M. 1975. *Common names of insects.* Entomol. Soc. of Am., Spec. Publ. 75–1.

Anderson, H. W. 1956. *Diseases of fruit crops.* McGraw-Hill, New York.

Brann, J. L.; P. A. Arneson; and G. H. Oberly. 1975. *Tree-fruit production recommendations for commercial growers.* N.Y. Agr. Exp. Sta.

Coop. Extension Service, Wash. State. 1974. *Spray guide for tree fruits in eastern Washington.* Wash. Agr. Ext. Bull. 419.

Glass, E. H. 1975. *Integrated pest management rationale, potential needs, and implementation.* Entomol. Soc. of Am., Spec. Publ. 75–2.

Hoyt, S. C., and E. C. Burts. 1974. Integrated control of fruit pests. *Ann. Rev. Entomology* 19:231–252.

Maramorosch, K.; E. Shikata; and R. R. Granados. 1968. Structures resembling mycoplasma in diseased plants and in insect vectors (trans.) *N.Y. Acad. Sci. Series II.* 30:841–855.

Metcalf, C. L.; W. P. Flint; and R. L. Metcalf. 1951. *Destructive and useful insects.* 3rd ed. McGraw-Hill, New York.

Platt, R. B., and J. F. Griffiths. 1965. *Environmental measurement and interpretation.* Reinhold Pub. Corp., New York.

USDA. 1952. *Insects.* Yearbook of Agr.

USDA. 1953. *Plant diseases.* Yearbook of Agr.

USDA, ARS. 1976. *Virus diseases of stone fruits.* Agr. Handbook 437.

17

Limiting Factors

The foregoing chapters have considered a large number of factors, each of which is important in fruit growing. But in order to be successful, the whole enterprise must be put together so that all of the factors are integrated and in proper balance with the others. Any factor or combination of factors not in good balance can limit the productivity of a planting and hold it to a level far below that possible for an optimized system.

HISTORICAL BACKGROUND

Meyer and Anderson (1939) reported that the early plant scientists recognized that factors such as light, carbon dioxide, soil moisture, and nutrients might limit cropping. J. von Liebig in 1843 proposed his "law of the minimum"—that yield is limited by the factor present in the relatively lowest amount. J. Sachs in 1860 stated that any factor could be too low, at optimum, or too high (the "three cardinal points" concept); he correctly concluded that if a little bit is good, a lot is not necessarily better. F. F. Blackman, working with photosynthesis, in 1905 developed the "principle of limiting factors," asserting that when a process is conditioned by several factors, the rate of the process is limited by the pace of the "slowest" factor. E. A. Mitscherlich in 1909 ramified the old "law of diminishing returns" when he stated that an increase in crop production by a unit increment of a deficient factor is in proportion to the *degree* of deficiency of that factor. In other words, there could be several limiting factors, but a unit increase of one would not result in the same increase as would increases of some other factors, because some are more deficient than others. This concept is more complex than the previous ones but is more nearly in accord with actual experience. It does not go far enough, however, to take into account the fact that sometimes the crop is *reduced* by an increase in only one limiting factor without regard to the others.

In 1936, Paul Macy, working with mineral nutrition, proposed his "critical nutrient concentration" concept, that there is a critical level of each nutrient in a plant, below which the yield decreases, resulting in "poverty adjustment"; "luxury consumption" occurs if the nutrient content is above the critical level, and a large increase in the nutrient produces only a small increase in yield. This idea fits many field situations but still does not take into account simultaneous changes in several nutrients. In 1946, three USDA scientists—C. B. Shear, H. L. Crane, and A. T. Meyers—developed the "nutrient element balance" concept, in which, for a given level of one nutrient, there is an optimum level for all others. If the level of a single nutrient is increased, then the optimum levels for all others changes. Thus, there is no single optimum nutrient level except as it relates to the *balance* of all nutrients in the plant. The idea was carried one step further in 1954 by E. L. Proebsting and A. L. Kenworthy, who found that the optimum nutrient concentration for cherry trees changed when the intensity of light reaching the trees was changed. Thus the proper internal balance of nutrients is intimately related to the total external environment in which the plants are growing. It is important to note that an *excess* of a factor might retard the overall process and thus be limiting, although limiting factors are usually thought of as being deficient.

It now seems logical to generalize the principle of limiting factors to include all factors (physical, chemical, biological, economic, and so forth) relating to fruit growing. The generalized principle can be stated in the following way: *The operation of a reaction, process, system, or organization proceeds at the rate imposed by the most limiting factor(s) essential to the overall process, whether they be deficient or in excess.* When the principle limitation is removed, a new factor then becomes limiting. Methodical elimination of all limiting factors that can be altered results in an optimized system for that particular set of fixed factors.

FIXED FACTORS

There are some major decisions that a fruit grower has only one chance to make. Before a planting is established, one must decide upon the geographic area (climate), the local site, the soil type, the kind of fruit and cultivar to plant, the size of the planting, the spacing, the type of rootstock, and the extent of possible diversification of the whole operation. Each decision regarding one of these factors should be based upon a thorough knowledge of that factor and its relation to the others, because each is more-or-less permanent. A single mistake at this initial stage may doom the enterprise or seriously limit its chance of success. One would not, for example, choose an early-market cultivar for a late-bloom, high-elevation site.

The most important considerations in regard to climate are minimum winter temperatures, length of growing season, temperatures during the growing season, light intensity, and amount and distribution of rainfall. A study of the hardiness of the fruit in question and the frequency of killing temperatures that have occurred during past years at the proposed location are important. The site is chosen mainly on the basis of good soil and drainage and freedom from spring frosts. One grower of stone fruits in central Washington paid 50 percent more per acre for a frost-free site than was being asked for good orchard land low in the valley. Some of his friends questioned the wisdom of paying extra for the high ground, but over the past many years, this grower has saved thousands of dollars in orchard-heating costs and has had good crops of peaches, apricots, plums, and cherries in years when his neighbors at lower elevations had little or no crop.

Either temperature or light intensity can be limiting in some areas. Fruits cannot be grown in central Oregon or in the high plains, because killing frosts occur during the summer. One should always determine the average number of frost-free days in an area before deciding to plant fruits or nuts. Besides this, the maximum temperatures and maximum light intensity determine in part whether a given crop can be successfully grown. Oregon's Willamette Valley, for example, has relatively cool summers with moderate light intensity. Peaches and apricots, which thrive in hot, sunny climates (such as South Carolina and California's Central Valley), do poorly in cool, marine climates. On the other hand, the long 213-day season makes the Willamette Valley a good place to grow filberts, prunes, apples, some kinds of winter pears, and blackberries.

The amount and distribution of annual rainfall may also limit fruit growing. If irrigation is available, this is not so much a problem, but if not, then the kinds of fruit that can be grown are limited. Nonirrigated orchards in areas where most of the moisture comes during winter and spring usually should be planted with cultivars that mature their crop by midsummer, when soil moisture is still adequate for proper sizing of the fruit. Growers in areas with frequent summer rains face much higher costs for disease control because many diseases flourish under such conditions. This is one of the major limiting factors in orcharding in the eastern U.S.

The size of the planting is another factor to consider. A 10-year study indicated that apple and pear orchards of about 12 hectares were more efficient than either larger or smaller ones (Table 17-1). Small orchards (5 hectares) and medium ones (11 hectares) yielded more per hectare than large ones (20 hectares). The cost per box, which is the best overall measure of efficiency, was lowest with the medium-sized orchards. This probably has more to do with management than to farm size, *per se*. It does indicate that planting size influences efficiency. The foregoing examples are given only to indicate the scope and complexity of choosing the site and size of the planting. If these decisions have been sound, then no major limiting factor will be superimposed on the long-term venture.

Table 17-1 Relation of farm size to production costs of apple and pear orchards in Hood River, Oregon.

Farm Size	Hectares in Fruit	Yield (boxes per ha)	Production Costs	
			Per ha	Per Box
Small	5	1013	$1,522	$1.51
Medium	11	983	$1,295	$1.32
Large	20	801	$1,139	$1.42

SOURCE: Mumford, 1962.

Another less obvious but highly important factor that can limit production is rootstock. In a 25-year test it was found that Bartlett pear on *Pyrus calleryana* root bore 50 percent more fruit than those on imported French *(P. communis)* stocks (Westwood et al., 1963). The trees were nearly identical in size and were treated exactly the same, yet the difference in yield over the 25-year period amounted to 25,000 boxes per hectare. A plot of Red Delicious apples on M 1 rootstock yielded 6 boxes more per tree during the first 8 years than those on seedling stocks. The trees were the same size.

Interstock can even be a limiting factor. Work with dwarf pear showed that Bartlett/Old Home/Quince yielded 54 percent more than Bartlett/Hardy/Quince (Table 17-2). These trees grew side by side and received identical treatment, and both interstocks are compatible.

Table 17-2 Effect of interstock on yield of Bartlett pear on Angers Quince Rootstock.

Interstock	Total Yield, First 8 Years (m tons per ha)
Hardy	88
Old Home	136

Tree spacing also may limit production. However, climate, soil, culture, cultivar, pruning method, and rootstock all affect the optimum spacing and thus all are interrelated. Table 17-3 gives an example of yields

Table 17-3 Effect of tree spacing on total yield of pear on Angers Quince root during the first 8 years.

Cultivar	Interstock	Yield (in m tons per ha) for Two Different Tree Spacings	
		6.1 × 2.44 m (667/ha)	4.6 × 1.2 m (1779/ha)
Bartlett	Old Home	37	136
Anjou	None	17	58
Comice	None	45	95

of dwarf pears as related to spacing. Full dwarf trees at 6.1 × 2.44 meters apart did not make efficient use of the land. Each situation is different, so there is no simple rule-of-thumb to ascertain the right spacing. In recent years, however, close rectangular spacing (forming a narrow treewall in one direction) has been found to work well both in filler systems for standard trees and in permanent systems for dwarf trees. Such treewalls lend themselves to mechanical pruning and platform harvesting.

Yield differences related to cultivar are to be expected and, in some cases, a poor choice of cultivar limits the potential success of an operation. Pollinizers and their placement in the orchard are extremely important with some cultivars. Figure 17-1 indicates the results of a 30-year test on Anjou pear using Bartlett as a pollinizer. The best pollinizer placement (solid row, one space from cultivar) resulted in over 14,800 more boxes per hectare during the 30 years than the poorest placement. Obviously, one does not get the best use of fertilizers, pruning, spraying, irrigation, and so forth, where pollination is a limiting factor.

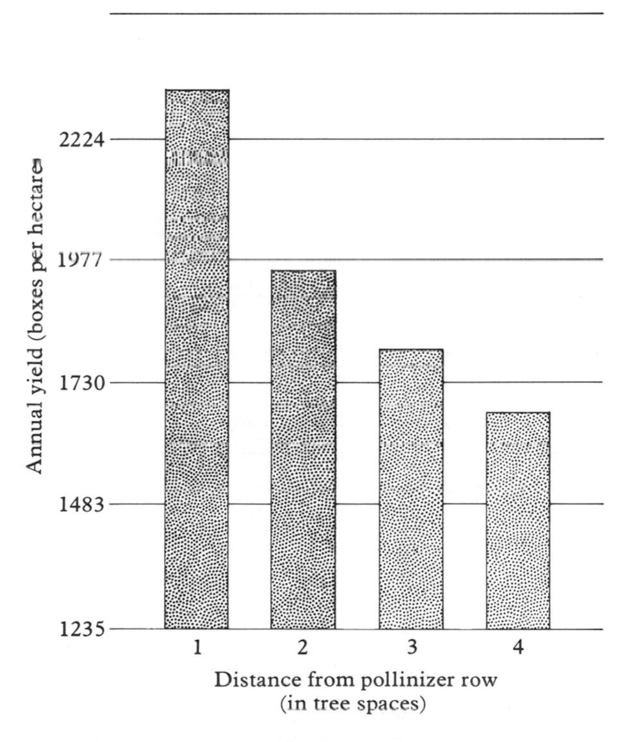

Figure 17-1 Influence of distance from pollinizer row on yield of Anjou pear. [Based on data of Westwood and Grim, 1962]

In regard to these fixed factors, it should be remembered that usually something can be done about a poor decision, even though it may be costly. In England, growers have been advised to replace their dwarf plantings after 15 years in order to take advantage of better cultivars and rootstocks, and to adjust tree spacing if necessary. Also, pollinizers (and major cultivars) can be changed to better ones by top grafting.

NONFIXED LIMITING FACTORS

Of the multitude of changeable limiting factors, a few will suffice to illustrate the importance of single details and timing. In certain cultivars, alternate bearing must be prevented if high production is to be maintained. Chemical thinning is the best method of insuring a good bloom following a heavy crop (see Chapter 9). For example, a Golden Delicious apple plot on M 9 stock bore a heavy crop in 1964. In 1965, trees that were chemically thinned the previous year (Dinitro followed by Sevin) bore 3,469 boxes per hectare, while trees not sprayed bore only 390 boxes per hectare. The yield of the latter trees could not be improved by good cultural practices, because the major limiting factor was the very low bloom density (3 percent on controls versus 39 percent on sprayed).

The level of soil moisture can determine whether another unrelated factor is limiting. Tests were done to learn whether excessive pesticide sprays affected foliage efficiency of apple leaves. The test was set up in an orchard plot already being used in an irrigation study. Table 17-4 shows that Aramite sprays reduced photosynthetic efficiency and fruit size in the ample-moisture plot, but not in the low-moisture plot. In this study, soil moisture was the limiting factor in the one plot; Aramite sprays limited growth in the other. When moisture was limiting, Aramite sprays did not further reduce growth.

Table 17-4 Effects of soil moisture and Aramite sprays on leaf efficiency and fruit growth of apple.

Treatment	Foliage Efficiency (dry wt. gain in mg/100 cm²)	Fruit Volume	
		Golden Delicious (cm³)	Red Delicious (cm³)
Low moisture			
Aramite (8 sprays)	24	172	236
Control	20	177	233
Optimum moisture			
Aramite (8 sprays)	26	208	254
Control	53[1]	227[1]	279[1]

SOURCE: Westwood et al., 1960.
[1] Significantly greater than other means.

Table 17-5 Effects of soil compaction and nitrogen rate on yield of Anjou pear.

Treatment	Yield (m tons per ha)[1]
Low nitrogen, compaction	31
Low nitrogen, no compaction	37
High nitrogen, no compaction	44

SOURCE: Westwood et al., 1964.
[1] Yield in each of the treatments was significantly different.

Sometimes, two factors can be limiting, and an increase of either or both will increase yield. This was true of Anjou pear in a test in which both soil compaction and nitrogen rates were varied (Table 17-5). Compaction alone limited yields as did low nitrogen alone. When compaction was reduced and nitrogen increased, the yield was greater than when only compaction was lessened. Nitrogen status is easily assessed and easily adjusted, but we often overlook less obvious but highly important factors like soil compaction. The increasing use of heavy machinery in orchards with wet soil may increase compaction to the point where it limits potential yield. For this reason, vehicles should be used in the orchard only when there is a good reason.

Factors that enhance fruit color also can be limiting in some areas. The red pigment in apple skin and the blush color in peaches requires direct light but is also affected by such things as pruning, fertilizers, propping, and growing temperature. In very hot climates, particularly where nights are very warm, red apple color is apt to be a major problem, and other practices must be adjusted to improve it. Better-coloring sports can be used, and particular care should be given to nitrogen levels. The best color is found on nitrogen-deficient trees, and each increment of nitrogen added decreases the fruit color somewhat. For this reason, it is better to keep the trees slightly below the optimum nitrogen level rather than in the luxury range. Many growers who have used complete NPK fertilizers have observed that the fruit is better colored than where the same amount of straight nitrogen fertilizer is used. Some are misled into thinking the better color comes from the phosphorus or the potassium in the complete fertilizer, when, in fact, the difference is due to the lower amount of nitrogen applied. For example, a 12-12-12 fertilizer contains 12 percent nitrogen, and ammonium sulfate contains about 21.5 percent nitrogen. To apply similar amounts of actual nitrogen, one must put on nearly twice as many pounds of 12-12-12 as ammonium sulfate, or three times as much as ammonium nitrate. The use of summer sprays of urea after late petal fall should be avoided because later sprays almost completely inhibit color development. Some multiple-element leaf-feed sprays contain urea and so should not be used later than early summer, if at all.

Other practices also may affect fruit color. Severe pruning reduces color both by subsequent shading and by increasing the level of nitrogen in the plant. Adequate fruit thinning is needed because color will not develop properly on overcropping trees. In fact, anything that reduces the effective leaf surface per fruit (such as a heavy mite infestation) will reduce color. Propping in some areas is done primarily to expose the fruit to proper light for better coloring. Thus we see that deficiencies of light and a number of nonfixed factors may limit optimum fruit color.

The whole field of mineral nutrition and fertilizer programs is complex and differs from area to area. Individual nutrients may be limiting by themselves, or they may interact with other practices such as tree spacing, pruning, cover cropping, and so forth, which in turn may limit performance or growth. Nitrogen is the only major fertilizer element needed consistently by tree fruits in the western states. So-called balanced fertilizers are not balanced at all and sometimes cause an imbalance in the soil. Any minor-element deficiency (such as zinc or manganese) usually is handled separately as a spray. Fertilizers containing both major and minor elements are not only expensive but may not contain these elements in the proper form or proportion for best results. Such materials should be used only as suggested by the local agricultural extension service or research stations.

Increasing the level of one element may create a deficiency of another, which then becomes limiting. Figure 17-2 indicates one such example, a test on prunes, in which nitrogen and boron were varied. These data show that increasing nitrogen without adding boron actually decreased yields, but striking increases occurred when both were added. This is an example of a factor in excess being one of the limiting factors. Leaf analyses showed that applications of

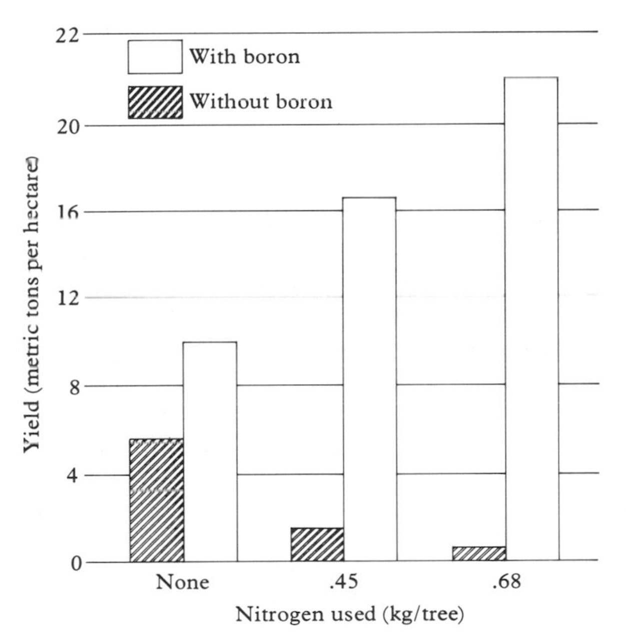

Figure 17-2 Effect of added increments of nitrogen, both with and without added boron (8 pounds per acre), on yield of Italian prune. Nitrogen alone depressed yield, but when boron was added, each increment of nitrogen increased yield. [Based on unpublished data of O. C. Compton]

nitrogen alone caused leaf boron to decline to the deficiency level. This particular effect would not occur everywhere, but other nutrient interactions may be important. For example, calcium tends to make zinc unavailable and, when applied, can cause zinc deficiency.

Another relationship that has been studied is that of nitrogen and phosphorus in prunes. Phosphorus not only failed to increase yields but actually reduced yield when combined with the higher rates of nitrogen. These are but a few examples of the many nutrient interactions that occur. They indicate that the addition of one fertilizer element can alter the uptake of others, and that to maintain the optimum nutrient balance, a knowledge of the effects of environment and mineral interactions on plant development is necessary. Plant-tissue analysis is valuable in assessing these relationships. Although it is not possible to know all of the subtle interactions that take place, it is likely that, at each level of one nutrient, there is an optimum balance for each of the others, and when the level of any nutrient is changed—or if climate changes—then the optimum level for each element is changed.

Lack of pest control can nullify all other production efforts. Failure to control a single insect may be the major limiting factor for a given season. In one instance, failure to control mites in pears not only reduced fruit size in the current year, but drastically reduced yield the following year (Westigard et al., 1966). Once again, production was governed by the weakest link in a long and complex chain.

Not the least of the possible limiting factors is man himself. The training, education, energy, health, ingenuity, and disposition of the fruit grower are each either an asset or a liability. The soundness of one's decisions often outweigh how hard one works at fruit growing. In order to avoid being a limiting factor, the operator should not work so hard at routine chores that he or she lacks time to attend to the important managerial tasks that no one else can (or should) do.

The net result of a well-balanced fruit operation is a high net income per hectare. A study of apple and pear production costs showed that orchards producing 2,500 boxes of apples per hectare paid out only

$.70 per box in costs, while those producing only 600 boxes per hectare paid out $1.91 per box. Likewise, Bartlett pear orchards that yielded 370 boxes per hectare cost $3.20 per box while those yielding 1,730 boxes per hectare cost only $1.02 per box to produce. These figures would not fit precisely each situation, but the principle of "high yield, low cost" is to be found wherever fruits are grown. Keeping this fact in mind, it is much more profitable to do an excellent job on 20 hectares than a mediocre job on 40 hectares. The general relationship of production costs to hectare yields for pears is shown in Figure 17-3. Clearly, "average" yields are not good enough to warrant staying in business.

Every facet should be examined and evaluated as a possible limiting factor, including the choice of a location, site, soil, cultivars, spacing, rootstock, production practices, farm size, bookkeeping practice, method of farm management, and marketing procedures. The key to using the principle of limiting factors in fruit growing is for the grower to be aware of the possible limiting factors and to recognize which ones can be altered. Season by season, and week by week, he or she should think through the *present situation* and find the weakest points that can be upgraded. It is just as important to know what *not* to do as what to do. The cost of an unnecessary operation can quickly erase the profit. Successful growers take time to visit other growers from whom they borrow new ideas that can be adapted and used to increase efficiency.

The examples used here are not exhaustive and are not intended as specific suggestions for growing fruit in every area but rather are offered as illustrations of the generalized principle of limiting factors. The only ones who can really help with specific problems are local experiment-station and extension workers, trained and experienced in the problems peculiar to the area.

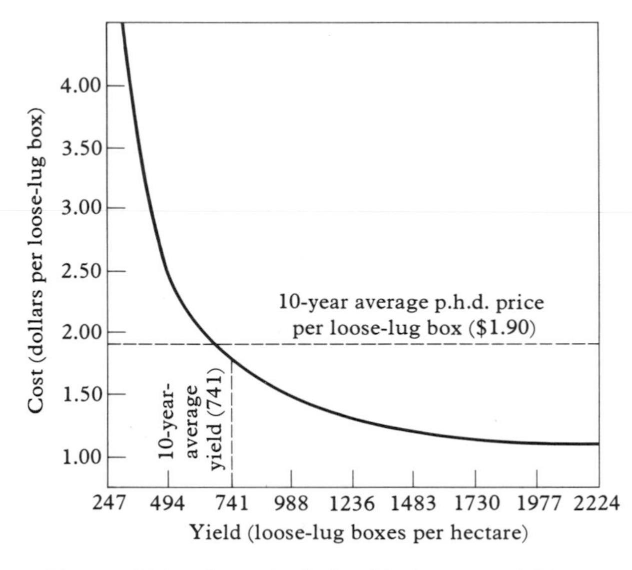

Figure 17-3 General relationship between yield per hectare and cost of Bartlett-pear production. Although costs and yields vary, this general relationship exists for all fruits. (The abbreviation "p.h.d." stands for "packing house door" and refers to the price paid to a grower for fruit as delivered to the packing house.) [After Green, Irish, and Mumford, 1960]

KINDS OF DECISIONS

A careful review of the five fixed factors listed below will tell whether serious limiting factors exist as "built in" features of an orchard. (If so, agricultural research and extension staff can be asked to help determine major corrective measures. Records should be such that it is known which blocks are unproductive. These blocks should be removed and replanted.)

- Geographic location
  - Winter climate
  - Growing climate
- Local site
  - Air drainage and frost hazard
  - Soil type, depth, and drainage
  - Orchard size
- Kinds, cultivars, and strains to plant
  - Proportion of different fruits and cultivars
  - Kind and placement of pollinizers
- Rootstocks
  - Growth control

- Efficiency in cropping
- Disease resistance

Plant spacing
- For maximum cropping
- For orderly filler removal

The following nonfixed factors must be reviewed periodically:

Chemical sprays
- Pest control
- Thinning
- Growth regulators and special sprays

Cultural practices
- Weed control and cover cropping
- Irrigation
- Fertilizer program and mineral deficiencies
- Pruning, thinning, and propping

Frost control

Pollinating insects

Economics and management
- Bookkeeping and record taking
- Efficient use of equipment and buildings
- Efficient use of labor (foremen and workers)
- Packaging and marketing
- Long-term planning
- Financial structure of the business

Even though accurate information can be obtained from many sources regarding a planting, it is a mistake for a grower to ask outsiders to make decisions for him (for example, what cultivar or rootstock to use). The well-informed grower is in the best position to make such decisions.

THE OPTIMUM SYSTEM

When any possible limitations of the foregoing factors have been eliminated so that all factors are in balance at maximum intensity, an optimum system has been achieved. Although an optimum system, in the absolute sense, is not usually attainable, it does provide a model or goal to strive for. Periodic comparison of one's own farm with this ideal provides a sound way of finding and eliminating limiting factors.

GENERAL REFERENCES

Chaplin, M. H.; M. N. Westwood; and R. L. Stebbins. 1973. Estimates of tree size, bearing surface, and yield of tree fruits and nuts. *Proc. Oregon Hort. Soc.* 64:94-98.

Meyer, B. S., and D. B. Anderson. 1939. *Plant physiology,* D. van Nostrand, New York. Chapters 21 and 33.

Shear, C. B.; H. L. Crane; and A. T. Myers. 1946. Nutrient-element balance: A fundamental concept in plant nutrition. *Proc. Am. Soc. Hort. Sci.* 47:239-248.

Glossary

Abscisic acid (ABA). A complex natural growth inhibitor, thought to be the principal inhibitor in the dormant (resting) stage of buds.

Abscise. To cut off or to fall off. "Abscission" is the noun.

Absorb. To take up water, mineral nutrients, or applied chemicals by roots or other plant organs.

Acaulescent. Apparently without a stem.

Achene. A dry, hard, one-seeded indehiscent fruit in which the seed is attached to the ovary wall at one point.

Acuminate. Pointed; tapering to a point.

Acute. Sharp at the end (as opposed to obtuse or blunt).

Adjuvant. A substance added to a chemical spray to enhance the action of the main ingredient.

Adnate. United with.

Adsorb. The adhesion of gases, liquids, or dissolved substances to the surface of a solid.

Adventitious. Arising new from an unusual place, as a bud or a sprout arising from a root or internode.

Aggregate fruit. Made up of two or more carpels from a single flower, plus the stem axis; e.g., blackberry, strawberry.

Air layerage. A method of stimulating rooting of an attached stem by excluding light, under moist conditions, from a portion of the stem.

Alkaline. Basic rather than acid; i.e., having a pH above 7.0. Alkaline soils contain a considerable amount of lime or sodium salts.

Allopolyploid. A polyploid having dissimilar sets of chromosomes, usually differentiated in pairs.

Alternate leaves. Leaves placed singly on the stem (rather than opposite in pairs), or rather than whorled (if more than two).

Androecium. The stamens of a flower collectively.

Anion. Negatively charged ion; e.g., Cl^-, NO_3^-, SO_4^{--}.

Antagonism of salts. A mutual counteraction of the movement of salts through cell membranes.

Anthesis. The expanded or full-bloom stage of flowers.

Anthocyanin. Any of a class of water-soluble pigments, including most of those imparting red or blue color to fruits or flowers.

Apomixis. Asexual seed production by reproduction budding in the ovary without fertilization.

Articulate. Jointed.

Ascorbic acid. $C_6H_8O_6$, vitamin C, essential to the diet of man, found in citrus, strawberry, black currant, mustard species, and many other plants.

Asexual. Without sex. Asexual propagation is the production of a new plant by vegetative means: budding, grafting, rooting of cuttings, or division.

Auxin. The natural plant-growth hormone indole-3-acetic acid.

Available moisture. That portion of soil water available for uptake by plant roots.

Axil. The angle between a petiole and the stem it is attached to.

Bacterial canker. An area of injury, often necrotic, of stems by bacteria; e.g., *Pseudomonas* sp.

Basic number. The number of chromosomes found in gametes of a diploid plant or a diploid ancestor of a polyploid plant.

Berry. A simple fruit derived from one flower, in which the parts remain succulent. The true berry is derived from an ovary (as is the grape); the false berry from an ovary plus receptacle tissue (as is the blueberry).

Biennial. Occurring once every two years. A plant that completes its life cycle in two years and then dies.

Bifid. Two-cleft.

Bitter pit. A physiological disorder of apples in which discolored corky areas form beneath the skin in the flesh near the calyx end, caused by calcium deficiency resulting from moisture stress.

Black end (hard end). A physiological disorder of the calyx end of pear fruit caused by certain rootstocks (e.g., *Pyrus pyrifolia, P. ussuriensis*).

Blackheart. Darkening of inner areas of woody stems by deposits that prevent conduction or food storage. Caused by death of cells from winter freeze damage.

Boreal. Of or pertaining to the North; of northern latitudes.

Bract. A specialized leaf or leaflike part, usually at the base of a flower or inflorescence.

Breba. The first or spring crop of figs.

British thermal unit (BTU). The amount of heat required to raise the temperature of one pound of water one degree Fahrenheit.

Bud sport. A new strain or clone arising from a bud in which a mutation occurred in the bud initial.

Callose. A carbohydrate found in the sieve plates of sieve tubes. It normally builds up and blocks phloem translocation during winter dormancy.

Callus. Wound-healing tissue, arising first as undifferentiated parenchyma tissue at the edges of a bud, graft, or other wound.

Calorie. The amount of energy required to raise the temperature of one gram of water one degree Celsius. (1 food calorie = 1000 calories.)

Campanulate. Bell-shaped.

Cane. Stems of such fruits as raspberry, blackberry, and grape.

Capillary movement. The elevation or depression of the surface of a liquid in very small spaces (usually at a rate of less than 0.1 mm per day), induced by the relative attraction of the molecules of the liquid for each other and for those of the solid.

Caprification. The process of pollination of the fig flower with pollen from the caprifig by the blastophaga wasp.

Caprifig. The wild or uncultivated fig, used to pollinate the domestic fig.

Carotinoid. Any of a number of orange-red or yellow pigments similar to carotene found in leaves, fruits and flowers.

Carpel. A modified leaf forming the structure enclosing the seed.

Cation. A positive ion (e.g., Na^+, Ca^{++}, K^+, $NH_4{}^+$).

Catkin. A scaly cylindric floral spike.

Caulescent. Having a conspicuous stem.

Cellulose. An inert carbohydrate (glucose polymer), the chief constituent of cell walls.

Chalaza. The point of an ovule or seed where the integuments are united to the nucellus.

Chemotropism. The bending of a stem by growth in response to a chemical stimulus.

Chloroplast. A plastid (in a leaf cell or other green organ) containing chlorophyll.

Chlorosis. The yellowing of leaves; may be due to lime-induced iron deficiency.

Chromosome. Rod-like structures in the cell nucleus that contain the genes, the units of heredity.

Cleistogamy. Self-pollination without the flower opening.

Clone. A group of plants originating as parts of the same individual, from buds or cuttings or division.

Coleoptile. A hollow living cylinder enclosing the epicotyl of a newly germinated grain seedling; it is attached to the seedling at the first node. Wheat and oat coleoptiles are used to study growth regulators and hormones.

Colloid. A particle larger than most inorganic molecules but which remains suspended indefinitely in water and has a large specific surface.

Compatible. (1) Of sex cells, the ability to unite and form a viable embryo. (2) Of grafting, the formation of a successful, long-lived stock/scion union.

Conduplicate. Folded upon itself lengthwise.

Connate organs. Organs congenitally more or less united.

Cordate. Heartshaped.

Cork spot (drought spot). A physiological disorder of pear fruits similar to bitter pit in apple.

Corymb. Inflorescence with a short rachis and longer lower pedicels, forming a flat or convex cluster—the outermost flowers being the first to open, as in pear.

Cotyledons. The primary leaves of the embryo; seed leaves.

Coulure. Failure of grape blossoms to set, resulting in premature drop.

Crenate. Leaf margins with rounded scallops.

Cuneate. Leaves that are wedge shaped at the base.

Cyme. An inflorescence whose single terminal flower opens first, followed by those of secondary and tertiary axes (as in apple).

Cytokinins. A class of plant hormones that stimulate cell divisions and delay senescence; e.g., benzyl adenine, zeatin, kinetin.

Deciduous. Falling off, as the calyx of some fruits and leaves in autumn.

Degree-day. In pomology, a heat unit representing one degree of temperature above a given mean daily temperature on a given day. Since crops require a specific amount of energy in order to mature, the number of accumulated degree-days indicates how close a crop is to maturity.

Dehiscence. The natural bursting of anthers or other organs to release the contents.

Dentate. Leaf margins with coarse, pointed teeth.

Determinate. With the terminal (center) flower of an inflorescence opening first, thus stopping terminal growth. The lateral flowers open somewhat later.

Dichogamy. The condition in which the male and female flowers of a plant do not mature at the same time.

Dicliny. The condition in which male and female organs are separate, in different flowers.

Differentiation, floral. Flower or inflorescence development after initiation.

Dimorphic. Have two forms, as the leaves of juvenile and adult plants of many species.

Dioecious. Staminate and pistillate flowers on different plants.

Diploid. A plant with two sets of chromosomes, as found normally in somatic cells of the sporophyte phase.

Dormancy. The condition of a bud or seed characterized by lack of visible growth. Deciduous plants not in leaf are said to be dormant.

Drought spot. *See* cork spot.

Drupe. A fruit derived entirely from an ovary, one-seeded, with a pericarp, fleshy mesocarp, and stony endocarp (as peach, cherry, and plum).

Drupelet. A small drupe; the unit, many of which make up the fruit of such species as raspberry and blackberry.

Effective pollination period (EPP). The embryo-sac longevity minus the time required to complete pollination and pollen-tube growth.

Emarginate. Notched at the summit.

Emasculation. Removal of the anthers of a flower prior to dehiscence and the shedding of pollen.

Embryo. The rudimentary juvenile plant usually contained in the seed.

Embryo sac. The cell in the ovule that produces the egg and in which the embryo is formed following fertilization.

Emulsion. Any colloidal suspension of one liquid in another.

Endocarp. The ligneous inner ovary wall of the drupe; the stony part of the pit containing the seed(s) of stone fruits.

Endodermis. A primary tissue between the cortex and the pericycle in stems and roots.

Endogenous. Originating within the plant.

Endosperm. Tissue surrounding the embryo in seeds. It is triploid tissue derived from triple fusion of the second sperm nucleus and the two polar nuclei of the embryo sac.

Entire leaf margin. Smooth, with no marginal teeth or pattern.

Epidermis. The thin layer of cells forming the outer covering of plants.

Epicotyl. The stem above the cotyledons of a young seedling.

Epigynous. Flower type in which the floral parts arise conjointly from above the ovary (as in apple and pear).

Epinasty. Abnormal growth of leaves or stems, often twisted or misshapen. Caused by exposure to certain growth regulators.

Etiolation. The condition of spindly white growth caused by excluding light.

Exine. The outer coat of a pollen grain.

Exocarp. The outer layer of fruits derived from an ovary, as with the skin of a peach.

Exocortis. The shedding of the outer bark of trees.

Exogenous. Arising or coming from an external source (as does an applied chemical).

Eye, grape. The compound bud, consisting of a primary bud and several secondary buds that usually do not develop.

Fasciation. The condition of several stems fused together in a single plane, causing abnormal and deformed growth.

Fascicle. Condensed cluster.

Fecundity. The ability of flowers to produce viable seed.

Feeder roots. Small lateral roots arising from the main roots of trees.

Fertile. (1) Of a flower: capable of producing seed. (2) Of soil: rich in essential mineral nutrients.

Fertilization. (1) The union of the male cell with the egg. (2) The application of nutrient fertilizers to the soil.

Filament. The stalk on which the anther is attached.

Filiform. Thread-like.

Flaccid. Soft and limber; non-turgid.

Floret. One of the closely clustered small flowers making up a compound flower (such as the mulberry).

Foot-candle. The light intensity of one standard candle at a distance of one foot from a light source.

Foxy. The characteristic flavor of some American grapes, or of wine or juice made from them.

Fruit set. The persistence and development of an ovary or adjacent tissue following anthesis.

Fulvous. Reddish-yellow, tawny.

Fusiform. Spindle-shaped; rounded and tapering from the middle to both ends as in some roots.

Gametes. Specialized haploid cells of each sex, which must fuse in fertilization to form a diploid zygote and embryo.

Gene. The unit of heredity contained in the chromosome and is passed to the next generation via sexual reproduction.

Genotype. The entire genetic constitution of the organism.

Genus. The second division of classification up from species and below family; the first name of a binomial (e.g., *Pyrus* in *Pyrus fauriei*).

Gibberellic acid (GA_3). Gibberellin A_3, one of a related group of plant hormones found in fungi and higher plants.

Glabrous. Smooth, hairless.

Glaucous. Covered with a whitish bloom, as a plum.

Glucoside. One of a large group of natural compounds that yield glucose and some other substance on hydrolysis.

Glycoside. One of a group of compounds yielding a sugar and another substance on hydrolysis.

Graft. A form of asexual propagation in which a scion or bud of one plant is placed in intimate contact with another until they grow together.

Gram calorie. *See* Calorie.

Growth regulator. Any of several classes of natural or synthetic chemicals that in some way regulate plant growth. (Growth regulators are listed on pages 283–284.)

Gummosis. A general (nonspecific) disorder, particularly of stone fruits, in which exudation and deposit of gum occurs.

Gum spot. A disorder, particularly of stone fruits, in which there are small local gum exudations from the tissue, usually the fruit.

Guttation. The exudation of water from the ends of the vascular system at the margins of leaves under humid conditions.

Gynaecium. The pistil or pistils of a flower.

Haploid. Having only one set of chromosomes, as the gametophyte generation of plants.

Hard end. *See* Black end.

Hardpan. A hard impervious layer in the soil, usually in the subsoil.

Heat of fusion. The heat released when a liquid freezes. For water, the heat of fusion is 80 calories per gram.

Heat of vaporization. The heat energy required to vaporize a liquid. For water, it is 540 calories per gram (at 100°C).

Hermaphrodite. A flower with both stamens and pistil.

Hesperidium. A berrylike fruit with a leathery exocarp (e.g., citrus fruits).

Heterostyly. The presence of styles of two or more forms or lengths on a flower.

Hirsute. Covered with long, stiff hairs.

Homogamous. Having stamens and pistils that mature at the same time.

Hormone. A substance produced in minute amounts in one part of the plant and transported to another part where it evokes a response.

Hydrolysis. Decomposition of a compound into other compounds by the chemical addition of the elements of water (H^+ and OH^-).

Hydrophyte. A plant that grows naturally in the water.

Hygroscopic. The ability of a substance to attract and hold moisture from the air.

Hygroscopic coefficient. The percentage of soil water retained in contact with a saturated atmosphere in the absence of any other source of water.

Hypanthium. The cup-shaped receptacle of a flower.

Hyphae. The threadlike elements of the mycelium of a fungus.

Hypocotyl. The stem below the cotyledons and above the root of a young seedling.

Hypogenous. Growing beneath or on the undersurface, as of leaves.

Imbibition. The initial taking up of water; particularly used to describe the action of seeds prior to germination.

Imbricate. Curved and overlapping like roof tiles (e.g., bud scales of a flower bud; scales of a fir cone).

Imperfect. Of a flower, containing only one sex.

Incompatible. *See* compatible.

Indeterminate. Continuing to grow. With inflorescences, the opening of the lower (lateral) flowers first, and the terminal ones opening later.

Induction, floral. The condition required for floral initiation.

Initiation, floral. The first discernible change from a vegetative bud to a floral primordium.

Integuments. The maternal layers (usually two) that grow around the ovule, eventually forming the seedcoat.

Interfertile. Two or more varieties that are able to cross-pollinate each other and produce seed.

Interfruitful. The ability of one variety to cross-pollinate another, resulting in either seeded or seedless fruit.

Internode. The area of a stem between two nodes.

Intersterile. Two or more varieties that cannot successfully cross-pollinate each other and produce seeds.

Intine. The inner coat of the pollen grain.

Intrasterile. The condition of a group of varieties in which none are cross compatible with each other.

Inversion temperature. The condition in which cold air settles beneath warm air on a clear night.

Involucre. A collection or rosette of bracts subtending a flower cluster, umbel, etc.

Involute. Rolled inwards from the edges.

June drop. The final post-bloom shedding of fruits, often occurring in late May or June.

Juvenile. The stage of a seedling plant in which floral initiation is not possible, often accompanied by distinct leaf or stem morphology.

Kilocalorie. One thousand calories, a food calorie.

Lamellate. Having a thin plate or layer.

Lanceolate. Lance-shaped.

Langley. A measure of radiant energy equal to one gram calorie per square centimeter.

Latent bud. A bud, usually concealed, more than one year old, which remains dormant for an indefinite period. Under certain conditions, such as severe pruning, a latent bud may grow.

Leaching. The process of removing soluble minerals from the soil by their downward movement to the water table.

Leaf area index (LAI). The total leaf area of a plant or plants divided by the land area covered by the plants.

Leaflet. One of the separate blades of a compound leaf.

Ligneous. Woody.

Locule. The cell or cavity of an ovary (or anther).

Long day (LD) plant. One that is stimulated by long days to initiate some process, usually flowering.

Lumen. (1) The light emitted by one candle in a unit solid angle (*see* Lux). (2) The cavity enclosed by cell walls.

Lux. The illumination received from one candle on one square meter at a distance of one meter; one lumen per square meter; 0.0929 foot-candle. (*See* Foot candle.)

Mamme. The first crop of the "male" fig or caprifig.

Measles. A physiological disorder of apple trees in which the bark of twigs and young branches becomes rough and scaly. May be caused by boron deficiency, manganese toxicity, or a virus.

Meristem. Actively dividing cells of undifferentiated tissue, as cambium or the tips of a growing shoot or root.

Meristematic. Of or pertaining to the qualities of a meristem.

Mesophyll. The cells of the middle portion of the leaf, containing chloroplasts.

Mesophyte. A plant species adapted to moderate moisture rather than to very wet or very dry conditions.

Metaxenia. The physiological effects of foreign pollen on the maternal tissue of the fruit.

Millerandage. A condition in grape, in which the ovary persists but the seeds remain small or do not attain the usual size.

Monoecious. Separate male and female flowers borne on the same plant (as in filbert, walnut, pecan, chestnut, for example).

Mulch. Material, usually organic, used to cover the soil surface around plants.

Multiple fruit. Formed by fusion of carpels from many flowers, plus stem axis and accessory tissues.

Mutation. The spontaneous change in a hereditary unit, commonly called a bud sport.

Mycorrhiza. A fungus on the surface or within the tissues, usually roots, of higher plants in a symbiotic relationship.

Necrotic. Dead or dying tissues. "Necrosis" is the noun.

Nectary. An organ or part of a plant that secretes nectar.

Nematocide. A chemical that kills nematodes.

Node. The point on a stem from which buds and leaves arise.

Non-ionic. Undissociated state of a compound or molecule; not separating into charged ions in solution.

Nucellus. The tissue of the ovule in which the embryo sac develops. In mature seed, it is a thin remnant between the seedcoat and the endosperm layer.

Nucleic acids. Organic compounds of high molecular weight found in the nucleus and cytoplasm of cells, containing the genetic information of the organism.

Nut. A hard, indehiscent, usually one-seeded fruit, derived from carpels plus stem axis plus accessory parts.

Oblique. Neither perpendicular nor parallel to a given line or surface; slanting; sloping.

Obovate. Inversely egg-shaped, with the narrow end at the base.

Obtuse. Blunt in form, not sharp or acute; of a leaf or petal rounded at the extremity.

Osmosis. The diffusion of fluids through a semipermeable membrane.

Ovary. The enlarged lower part of the pistil, enclosing the ovules or young seeds.

Ovate. Or ovoid. Egg-shaped.

Ovicide. A chemical that kills the eggs of mite and insect pests.

Ovule. The portion of the ovary that contains the embryo sac and the egg cell, and which, after fertilization, develops into a seed.

Palmate. Leaflets of a compound leaf borne at the end of the petiole.

Panicle. A branching raceme.

Parthenocarpy. Development of a fruit without fertilization and seed.

Parthenogenesis. Development of an egg without fertilization.

Pedate. Palmately parted leaf, or a leaf divided with lateral lobes cleft or divided.

Pedicel. The stalk bearing a single flower of an inflorescence.

Peduncle. The stalk supporting either an entire inflorescence or a solitary flower.

Perfect flower. Having both male and female parts.

Perianth. The outer floral parts considered together.

Perigynous. A flower in which petals and stamens are attached to the margin of the hypanthium.

pH. The symbol for the logarithm of the reciprocal of the hydrogen ion concentration in gram atoms per liter (e.g., pH 5 = $H^+ 10^{-5}$).

Phenotype. The observable hereditary traits of a plant, which are the interactions of genotype and environment.

Phloem. The food-conducting tissue of plants composed of sieve elements, parenchyma cells, fibers, and sclereids. Commonly called "bark."

Photoperiodism. The response of a plant to day length.

Photosynthesis. The synthesis, in the presence of light, of carbohydrates and other organic compounds from simple molecules through the aid of chlorophyll.

Phototropism. Directional growth in response to light.

Phyllotaxy. The arrangement or pattern of leaves on a stem.

Phytotoxin. A substance toxic to plants.

Pilose. Covered with soft slender hairs.

Pinnate. Of a compound leaf, having leaflets arranged on either side of a common petiole.

Pith. The central cylinder of parenchyma tissue in the stems of dicots.

Plumose. Feathery.

Plumule. The bud of the ascending axis of a plant while still in the embryo.

Pollen tube. The growth extension in the style following the germination of the pollen grain (the male gametophyte is a part of the pollen tube).

Pollination. The transfer of pollen to the female stigma.

Pollinator. The agent of pollen transfer, usually bees.

Pollinizer. The producer of pollen; the variety used as a source of pollen for cross-pollination.

Polyembryony. The production of two or more embryos from a single ovule.

Polygamous. Bearing both unisexual and perfect flowers on the same or different plants of the same species.

Polyploid. Having more than twice the haploid number of chromosomes.

Polysaccharide. A carbohydrate containing three or more monosaccharide units (such as starch, cellulose, inulin).

Pome. The fruit type derived from the fusion of the ovaries, calyx cup, and floral tube, produced by the apple, pear, quince, and other members of the subfamily Pomoideae.

Primordium. The first recognizable but undifferentiated stage of a developing organ.

Procumbent. Lying along the ground but not putting out roots.

Profichi. The second crop of the caprifig.

Proliferation. A rapid and repeated production of new parts or tissues.

Protandry. The shedding of pollen before the stigmas are receptive.

Protogyny. The stigmas being receptive before pollen is shed.

Pseudohermaphrodite. Functional unisexualism in a plant that bears perfect flowers.

Pubescent. Covered with down or fine, short hairs.

Pyriform. Pear-shaped.

Q_{10}. Temperature coefficient. The rate of change in a process or reaction with each change of ten Celsius degrees in temperature.

Quiescence. Dormancy of buds imposed by external conditions.

Raceme. A simple indeterminate inflorescence in which the flowers are borne on short stalks along a common axis.

Rachis. The axis of a floral spike.

Radical. The rudimentary root of a germinating seed.

Receptacle. The modified or expanded portion of an axis that bears the organs of a single flower or the florets of a flower head.

Reflexed. Bent or turned back.

Relative humidity (RH). The amount of water vapor in the air relative to the greatest amount that could be held at a given temperature. For example 50 percent RH at 20° C is half the amount of water possible at that temperature.

Reniform. Kidney-shaped.

Respiration. The process by which oxygen and carbohydrates are assimilated into an organism and oxidation products, CO_2 and water are given off.

Respiratory coefficient. The ratio of CO_2 given off to O_2 used in respiration.

Rest. Dormancy of seeds and buds imposed by internal physiological blocks, which are removed by chilling.

Reticulate. Netted.

Revolute. Rolled backwards.

Rhizome. A rootlike underground stem, usually horizontal, with roots below and shoots from the upper surface.

Ringing. The girdling of the bark of trunks or branches of trees.

Root cap. The specialized structure at the tip of roots.

Root hair. The elongated tubular extension of an epidermal cell of the root, serving to absorb water and minerals.

Rugose. Rough and wrinkled.

Sapwood. The layer of wood (xylem) just inside the cambium; wood used to conduct xylar sap upward in the plant.

Scape. A leafless peduncle arising from the ground.

Scarification. The act of scratching or wearing away, as the scarifying of hard seedcoats to aid germination.

Scarious. Thin, dry, and membranous.

Scion. A detached stem, usually dormant, used in asexual propagation by grafting.

Self-fertilization. The fertilization of the egg of a variety or flower type by its own pollen.

Self-pollination. Pollination by pollen from the same flower or tree.

Sepals. The individual lobes of the calyx.

Serrate. Saw-toothed, with teeth pointing toward apex.

Sessile. A plant leaf, flower, or part attached directly at the base without an intervening stalk.

Soil fumigant. A chemical injected or incorporated into the soil to control soil-borne diseases or pests.

Somatic. In plants, the diploid body cells; vegetative rather than reproductive cells.

Sporogenous. Producing spores.

Sporophyte. The diploid generation of a plant, arising from the fertilized egg (zygote).

Sport. A variety or strain arising from a bud mutation.

Stamen. A male portion (of a flower) that produces pollen; a stamen consists of an anther and a filament.

Staminate flower. A flower producing pollen but not pistils.

Stele. The central cylinder of vascular tissue, etc., in the root or stem of a plant.

Stellate. Star-shaped.

Sterility. The inability to produce viable seeds.

Stigma. The terminus of the style, on which the pollen is placed in pollination.

Stipule. One of a pair of lateral appendages, often leaf-like, at the base of the petiole of many plants.

Stolon. A slender branch, as a runner or prostrate stem, which takes root at the tip, developing a new plant.

Stomata. Small openings in leaves or stems through which gas exchange takes place.

Stratification, seed. Moist chilling to remove rest, so that germination can occur.

Suborbicular. Nearly round.

Sucker. An undesirable sprout from the roots or crown of a tree.

Sulcate. Grooved longitudinally.

Supercooling. Cooling hydrated tissue or a liquid below its freezing point without solidification occurring.

Superior ovary. Flower in which the ovary is attached above the other parts.

Surfactant. A detergent, adjuvant, or other surface-active agent that enhances penetration and action of applied chemicals.

Suture. The connecting line on one side of stone fruits from tip to base.

Syconium. The fruit of the fig, in which the inflorescence is borne on the inside of a balloon-like receptacle; a type of multiple fruit.

Syncarp. The fruit of the mulberry, composed of many druplets from many flowers plus the fleshy inflorescence axis; a type of multiple fruit.

Synergism. A mutual interaction of two substances that results in an effect beyond the additive effects of the two when used separately.

Systemic. The state of being distributed throughout the plant system.

Tap root. The central root of a plant which grows down rather than laterally.

Temperature coefficient. *See* Q_{10}.

Temperature inversion. See *Inversion.*

Tendril. A filiform leafless organ of vines, often attaching to by twining around another body. In grape, a flowerless inflorescence.

Ternate. In threes.

Testa. Seedcoat.

Tetrad. A group of four pollen grains derived from a single microspore mother cell.

Tetraploid. An individual with four sets of chromosomes.

Torus. (1) The receptacle of a flower; part of the axis on which the flower parts are inserted. (2) A thickening in the membrane of a bordered pit.

Translocation. The movement of water and other dissolved substances through the conducting system of the plant.

Transpiration. The loss of water vapor from plant surfaces, chiefly through the stomata of leaves.

Triploid. An individual with three sets of chromosomes.

Tropism. Movement by growth of stems or other plant parts.

Turgor. The normal distention of plant cells resulting from internal pressure exerted against the cell walls, as when water is absorbed.

Umbel. An umbrella-shaped inflorescence.

Urceolate. Pitcher-shaped; swelling out like the body of a pitcher and contracting at the orifice.

Vacuole. The cavity within the cell containing the cell sap.

Variety. A clone, sort, or cultivar that is propagated asexually.

Vernation. The arrangement of leaves in the bud.

Villose. Having long soft hairs.

Viscosity. The thickness of a liquid, measured by its resistance to flow.

Volatile. Evaporating readily and passing off as a gas.

Water core. A condition of apple fruits in which the inner tissues become watersoaked; a condition of overmaturity.

Waterlogged soil. An improperly drained soil in which water content remains too high for normal plant growth.

Watersprout. A vigorous shoot arising from the trunk or main branch of a tree.

Wilting coefficient. The percentage of moisture in the soil when permanent wilting of plants takes place.

Xenia. The physiological effect of foreign pollen on maternal fruit tissue. (This term formerly was applied to pollen effects on the embryo and endosperm).

Xerophyte. Desert plant.

Xylem. The woody portion of trees and shrubs, interior to the cambium.

Yield efficiency. Crop yield per unit of plant size or per unit of land area (e.g., kilograms of yield per square centimeter trunk, or yield per hectare.)

Zygote. The diploid cell formed by the union of the male nucleus and the egg.

Appendixes

Appendix A Chemical symbols used in the text.

Symbol	Element
B	Boron
C	Carbon
Ca	Calcium
Cl	Chlorine
Cu	Copper
H	Hydrogen
Fe	Iron
K	Potassium
Mg	Magnesium
Mn	Manganese
Mo	Molybdenum
N	Nitrogen
O	Oxygen
P	Phosphorus
S	Sulfur
Zn	Zinc

Appendix B International System of Units (metric system).

Basic Units	Supplementary Units
Meter for length (m)	Radian, a plane angle (ra)
Kilogram for mass (kg)	Steradian, a solid angle (ster)
Second for time (sec)	
Ampere for electric current (amp)	
Kelvin scale for temperature (°K)	
Mole for amount of a substance (M)	
Candela for luminous intensity (can)	

Prefixes for the Basic Units

Large Quantities	Small Quantities
deka 10^{1}	deci 10^{-1}
hecto 10^{2}	centi 10^{-2}
kilo 10^{3}	milli 10^{-3}
mega 10^{6}	micro 10^{-6}
giga 10^{9}	nano 10^{-9}
tera 10^{12}	pico 10^{-12}
peta 10^{15}	femto 10^{-15}
exa 10^{18}	atto 10^{-18}

Combinations

Luminous flux (the lumen; candela-steradian)
Illuminance (the lux; lumens per square meter)
Frequency (the hertz; cycles per sec)
Force (the newton; m-kg per sec per sec)
Pressure or stress (the pascal; newton per square meter)
Heat or work (the joule; newton-meter)
Radiant flux or power (the watt; joule per sec)
Electric charge or quantity (the coulomb; ampere-sec)
Electric potential, electromotive force, or e.m.f. (the volt; watts per ampere)
Inductance (the henry; webers per ampere)
Radioactivity (becquerel; number of nuclear transitions per sec)
Absorbed radiation dose (the gray; joules per kg)
Capacitance (the farad; coulombs per volt)
Electrical resistance (the ohm; volts per ampere)
Density of magnetic flux (the weber; volt-sec)

Non-I.S. units widely used with I.S. units

Celsius (centigrade) temperature scale (0°C = 273.18° Kelvin)
Liter (10^{-3} cubic meters)
Metric ton (megagram)
Minute, hour, day, year

$\pi = 3.1416$
d = diameter
r = radius
L = length
Circumference of a circle $= \pi d = 3.1416d$
Area of a circle $= \pi r^2 = .7854d^2$
Surface area of a sphere $= 4\pi r^2 = \pi d^2 = 12.57r^2$
Volume of a sphere $= \frac{4}{3}\pi r^3 = .5236d^3$
Volume of a cylinder = cross-sectional area $\times$ L $= .7854d^2 \times L$
Spheroids (trees, fruits, cells).
Where $a = \frac{1}{2}$ major axis, $b = \frac{1}{2}$ minor (shortest) axis:
Prolate spheroid (taller than wide) Volume $= \frac{4}{3}\pi ab^2 = 4.189ab^2$
Oblate spheroid (wider than tall) Volume $= \frac{4}{3}\pi a^2b = 4.189a^2b$

$$\text{Specific gravity (density)} = \frac{\text{g weight}}{\text{cm}^3 \text{ volume}}$$

Appendix D Useful values.

Heat Quantity small calorie (cal) = heat necessary to raise 1 gram of water from 3.5°C to 4.5°C large calorie = 1000 cal (1 kcal) (1 food cal) British thermal unit (BTU) = heat necessary to raise 1 pound of water from 39° to 40°F = 252 cal
Work erg = 1 dyne acting through a distance of 1 cm joule = 10^7 ergs
Atmospheric Pressure atmosphere (atmos) = pressure exerted by 76 cm of Hg (13.595 g/cm^3) = 1.0133 bars = 1033.2 g/cm^2 = 2116.2 lbs./ft.^{2} bar = 0.98692 atmos = 1.01971 $\times$ 10^4 kg/m^2 = 14.504 lbs./in^2
Illuminance, Light Intensity lumen = unit of luminous flux (1 candle through unit distance) lux = 1 lumen per m^2 (1 candle-meter) foot-candle = 1 lumen per ft^2
Concentration part per million (ppm) = 1 mg/liter = 1 g/1000 liters = .3785 g/100 gallons

Appendix E Fahrenheit/Celsius (centigrade) temperature conversions.

Formulae: $°C = \frac{5}{9}(°F - 32)$ $°F = (\frac{9}{5} \times °C) + 32$

°F	°C	°F	°C	°F	°C	°F	°C	°F	°C
-50	-45.6	-9	-22.8	32	0	73	22.8	114	45.6
-49	-45.0	-8	-22.2	33	.6	74	23.3	115	46.1
-48	-44.4	-7	-21.7	34	1.1	75	23.9	116	46.7
-47	-43.9	-6	-21.1	35	1.7	76	24.4	117	47.2
-46	-43.3	-5	-20.6	36	2.2	77	25.0	118	47.8
-45	-42.8	-4	-20.0	37	2.8	78	25.6	119	48.3
-44	-42.2	-3	-19.4	38	3.3	79	26.1	120	48.9
-43	-41.7	-2	-18.9	39	3.9	80	26.7	121	49.4
-42	-41.1	-1	-18.3	40	4.4	81	27.2	122	50.0
-41	-40.6	0	-17.8	41	5.0	82	27.8	123	50.6
-40	-40.0	1	-17.2	42	5.6	83	28.3	124	51.1
-39	-39.4	2	-16.7	43	6.1	84	28.9	125	51.7
-38	-38.9	3	-16.1	44	6.7	85	29.4	126	52.2
-37	-38.3	4	-15.6	45	7.2	86	30.0	127	52.8
-36	-37.8	5	-15.0	46	7.8	87	30.6	128	53.3
-35	-37.2	6	-14.4	47	8.3	88	31.1	129	53.9
-34	-36.7	7	-13.9	48	8.9	89	31.7	130	54.4
-33	-36.1	8	-13.3	49	9.4	90	32.2	131	55.0
-32	-35.6	9	-12.8	50	10.0	91	32.8	132	55.6
-31	-35.0	10	-12.2	51	10.6	92	33.3	133	56.1
-30	-34.4	11	-11.7	52	11.1	93	33.9	134	56.7
-29	-33.9	12	-11.1	53	11.7	94	34.4	135	57.2
-28	-33.3	13	-10.6	54	12.2	95	35.0	136	57.8
-27	-32.8	14	-10.0	55	12.8	96	35.6	137	58.3
-26	-32.2	15	-9.4	56	13.3	97	36.1	138	58.9
-25	-31.7	16	-8.9	57	13.9	98	36.7	139	59.4
-24	-31.1	17	-8.3	58	14.4	99	37.2	140	60.0
-23	-30.6	18	-7.8	59	15.0	100	37.8	141	60.6
-22	-30.0	19	-7.2	60	15.6	101	38.3	142	61.1
-21	-29.4	20	-6.7	61	16.1	102	38.9	143	61.7
-20	-28.9	21	-6.1	62	16.7	103	39.4	144	62.2
-19	-28.3	22	-5.6	63	17.2	104	40.0	145	62.8
-18	-27.8	23	-5.0	64	17.8	105	40.6	146	63.3
-17	-27.2	24	-4.4	65	18.3	106	41.1	147	63.9
-16	-26.7	25	-3.9	66	18.9	107	41.7	148	64.4
-15	-26.1	26	-3.3	67	19.4	108	42.2	149	65.0
-14	-25.6	27	-2.8	68	20.0	109	42.8	150	65.6
-13	-25.0	28	-2.2	69	20.6	110	43.3	151	66.1
-12	-24.4	29	-1.7	70	21.1	111	43.9	152	66.7
-11	-23.9	30	-1.1	71	21.7	112	44.4	153	67.2
-10	-23.3	31	-.6	72	22.2	113	45.0	154	67.8
								155	68.3

Appendix F Foot/meter conversions.

1 foot = .305 meters				1 meter = 3.281 feet			
Even Feet	Meters	Even Feet	Meters	Even Meters	Feet	Even Meters	Feet
1	.305	26	7.93	1	3.28	11	36.09
2	.61	27	8.24	1.5	4.92	11.5	37.73
3	.92	28	8.54	2	6.56	12	39.37
4	1.22	29	8.84	2.5	8.20	12.5	41.01
5	1.52	30	9.15	3	9.84	13	42.65
6	1.83	31	9.46	3.5	11.48	13.5	44.29
7	2.14	32	9.76	4	13.12	14	45.93
8	2.44	33	10.06	4.5	14.76	14.5	47.57
9	2.74	34	10.37	5	16.40	15	49.21
10	3.05	35	10.68	5.5	18.04	15.5	50.85
11	3.36	36	10.98	6	19.68	16	52.49
12	3.66	37	11.28	6.5	21.32	16.5	54.13
13	3.96	38	11.59	7	22.97	17	55.77
14	4.27	39	11.90	7.5	24.61	17.5	57.41
15	4.58	40	12.20	8	26.25	18	59.05
16	4.88	41	12.50	8.5	27.89	18.5	60.69
17	5.18	42	12.81	9	29.53	19	62.33
18	5.49	43	13.11	9.5	31.17	19.5	63.97
19	5.80	44	13.42	10	32.81	20	65.62
20	6.10	45	13.72	10.5	34.45		
21	6.40	46	14.03				
22	6.71	47	14.33				
23	7.02	48	14.64				
24	7.32	49	14.94				
25	7.62	50	15.25				

Appendix G Acre/hectare conversions.

acres × .40468 = *ha*				ha × 2.471 = *acres*			
Even Acres	Hectares	Even Acres	Hectares	Even Hectares	Acres	Even Hectares	Acres
.1	.04	2.4	.97	.1	.25	2.4	5.93
.2	.08	2.5	1.01	.2	.49	2.5	6.18
.3	.12	2.6	1.05	.3	.74	2.6	6.42
.4	.16	2.7	1.09	.4	.99	2.7	6.67
.5	.20	2.8	1.13	.5	1.24	2.8	6.92
.6	.24	2.9	1.17	.6	1.48	2.9	7.17
.7	.28	3.0	1.21	.7	1.73	3.0	7.41
.8	.32	3.1	1.25	.8	1.98	3.1	7.66
.9	.36	3.2	1.29	.9	2.22	3.2	7.91
1.0	.40	3.3	1.34	1.0	2.47	3.3	8.15
1.1	.45	3.4	1.38	1.1	2.72	3.4	8.40
1.2	.49	3.5	1.42	1.2	2.97	3.5	8.65
1.3	.53	3.6	1.46	1.3	3.21	3.6	8.90
1.4	.57	3.7	1.50	1.4	3.46	3.7	9.14
1.5	.61	3.8	1.54	1.5	3.71	3.8	9.39
1.6	.65	3.9	1.58	1.6	3.95	3.9	9.64
1.7	.69	4.0	1.62	1.7	4.20	4.0	9.88
1.8	.73	5.0	2.02	1.8	4.45	5.0	12.36
1.9	.77	6.0	2.43	1.9	4.69	6.0	14.83
2.0	.81	7.0	2.83	2.0	4.94	7.0	17.30
2.1	.85	8.0	3.24	2.1	5.19	8.0	19.77
2.2	.89	9.0	3.64	2.2	5.44	9.0	22.23
2.3	.93	10.0	4.05	2.3	5.68	10.0	24.71

Appendix H Metric/imperial conversions.

AREA		
Metric to Imperial		
1 square centimeter (cm^2)	=	0.155 sq inch
	=	100 sq millimeters
1 square meter (m^2)	=	1,550 sq inches
	=	10.764 sq feet
	=	1.196 sq yards
	=	10,000 sq centimeters
1 square kilometer (km^2)	=	0.3861 sq mile
	=	1,000,000 sq meters
1 hectare (ha)	=	2.471 acres
	=	10,000 sq meters
Imperial to Metric		
1 square inch (in^2)	=	6.452 sq centimeters
	=	1/144 sq foot
	=	1/1296 sq yard
1 square foot (ft^2)	=	929.088 sq centimeters
	=	0.0929 sq meter
1 square yard (yd^2)	=	8,361.3 sq centimeters
	=	0.8361 sq meter
	=	1,296 sq inches
	=	9 sq feet
1 square mile (mi^2)	=	2.59 kilometers
	=	640 acres
1 acre (ac)	=	0.4047 hectare
	=	43,560 sq feet
	=	4,840 sq yards
	=	4,046.87 sq meters

LENGTH		
Metric to Imperial		
1 millimicron (nanometer, nm)	=	0.001 micron
1 micron (micrometer, μ)	=	0.001 millimeter
1 millimeter (mm)	=	0.001 meter
	=	0.0394 inch
1 centimeter (cm)	=	10 millimeters
	=	0.3937 inch
	=	0.01 meter
1 meter (m)	=	39.37 inches
	=	3.281 feet
	=	1,000 millimeters
	=	100 centimeters
	=	1.2 varas
1 kilometer (km)	=	3,281 feet
	=	1,094 yards
	=	0.621 mile
	=	1,000 meters

Imperial to Metric		
1 inch (in.)	=	25.4 millimeters
	=	2.54 centimeters
1 foot (ft)	=	30.48 centimeters
	=	0.3048 meter
	=	12 inches
1 yard (yd)	=	0.9144 meter
	=	91.44 centimeters
	=	3 feet
1 mile (mi)	=	1,609.347 meters
	=	1.609 kilometers
	=	5,280 feet
	=	1,760 yards

MASS (WEIGHT)

Metric to Imperial		
1 milligram (mg)	=	0.001 gram
	=	0.0154 grain
1 centigram (cg)	=	0.01 gram
	=	0.1543 grain
1 gram (g)	=	0.0353 avoirdupois ounce
	=	15.4324 grains
1 kilogram (kg)	=	1,000 grams
	=	353 avoirdupois ounces
	=	2.2046 avoirdupois pounds
1 metric ton (mega gram)	=	1,000 kilograms
	=	2,204.6 pounds
	=	1.102 short tons
	=	0.984 long ton

Imperial to Metric		
1 grain (gr)	=	1/7000 avoirdupois pound
	=	0.064799 gram
1 ounce (avoirdupois) (oz)	=	28.3496 grams
	=	437.5 grains
	=	1/16 pound
1 pound (avoirdupois) (lb)	=	453.593 grams
	=	0.45369 kilograms
	=	16 ounces
1 short ton (ton)	=	907.184 kilograms
	=	0.9072 metric ton
	=	2,000 pounds

YIELD

Metric to Imperial		
1 kilogram per hectare	=	0.89 pound per acre
1 cubic meter per hectare	=	14.2916 cubic feet per acre

(continued)

Appendix H (continued)

Imperial to Metric		
1 pound per acre	=	1.121 kilograms per hectare
1 ton (2,000 lb) per acre	=	2.242 metric tons per hectare
1 cubic foot per acre	=	0.0699 cubic meter per hectare
1 bushel (50 lb) per acre	=	56.1 kilograms per hectare
VOLUME		
Metric to Imperial		
1 liter (*l*)	=	1.057 U.S. quarts liquid
	=	0.9081 quart, dry
	=	0.2642 U.S. gallon
	=	0.221 Imperial gallon
	=	1,000 milliliters or cc
	=	0.0353 cubic foot
	=	61.02 cubic inches
	=	0.001 cubic meter
1 cubic meter (m^3)	=	61,023.38 cubic inches
	=	35.314 cubic feet
	=	1.308 cubic yards
	=	264.17 U.S. gallons
	=	1,000 liters
	=	28.38 U.S. bushels
	=	1,000,000 cubic centimeters
	=	1,000,000,000 cubic millimeters
Imperial to Metric		
1 fluid ounce (oz)	=	1/128 gallon
	=	29.57 cubic centimeters
	=	29.562 milliliters
	=	1.805 cubic inches
	=	0.0625 U.S. pint (liquid)
1 U.S. quart liquid (qt)	=	946.3 milliliters
	=	57.75 cubic inches
	=	32 fluid ounces

1 U.S. quart liquid (qt)	=	4 cups
	=	1/4 gallon
	=	2 U.S. pints (liquid)
	=	0.946 liter
1 quart dry	=	1.1012 liters
	=	67.20 cubic inches
	=	2 pints (dry)
	=	0.125 peck
	=	1/32 bushel
1 cubic inch (in^3)	=	16.387 cubic centimeters
1 cubic foot (ft^3)	=	28,317 cubic centimeters
	=	0.0283 cubic meter
	=	28.316 liters
	=	7.481 U.S. gallons
	=	1,728 cubic inches
1 U.S. gallon (gal)	=	16 cups
	=	3.785 liters
	=	231 cubic inches
	=	4 U.S. quarts liquid
	=	8 U.S. pints liquid
	=	8.3453 pounds of water
	=	128 fluid ounces
	=	0.8327 British Imperial gallon
1 British Imperial gallon	=	4.546 liters
	=	1.201 U.S. gallons
	=	277.42 cubic inches
1 U.S. bushel (bu)	=	35.24 liters
	=	2,150.42 cubic inches
	=	1.2444 cubic feet
	=	0.03524 cubic meter
	=	2 pecks
	=	32 quarts (dry)
	=	64 pints (dry)

SOURCE: After Janick et al., 1974.

Appendix I Diameter/circumference/cross-sectional area/spherical-volume conversions.

Trunk, Branch, or Fruit			Trunk or Branch Cross-Sectional Area (cm^2)	Volume of Fruit as Sphere (cm^3)	Trunk, Branch, or Fruit			Trunk or Branch Cross-Sectional Area (cm^2)	Volume of Fruit as Sphere (cm^3)
Diameter (cm)	Diameter (inches)	Circumference (cm)			Diameter (cm)	Diameter (inches)	Circumference (cm)		
					5.0	1.97	15.7	19.63	65.45
0.1	.039	.314	.0079	.00052	.1	2.01	16.0	20.43	69.48
.2	.079	.628	.0314	.00419	.2	2.05	16.3	21.24	73.62
.3	.118	.942	.0707	.01414	.3	2.09	16.6	22.06	77.96
.4	.157	1.26	.1257	.0335	.4	2.13	17.0	22.90	82.47
.5	.197	1.57	.1963	.0654	.5	2.16	17.3	23.76	87.13
.6	.236	1.88	.2827	.1131	.6	2.20	17.6	24.63	91.94
.7	.276	2.20	.3848	.1796	.7	2.24	17.9	25.52	96.97
.8	.315	2.51	.5027	.2681	.8	2.28	18.2	26.42	102.2
.9	.354	2.82	.6362	.3817	.9	2.32	18.5	27.34	107.5
1.0	.394	3.14	.7854	.5236	6.0	2.36	18.8	28.27	113.1
.1	.433	3.46	.9503	.6969	.1	2.40	19.2	29.22	118.9
.2	.472	3.77	1.131	.9048	.2	2.44	19.5	30.19	124.8
.3	.512	4.09	1.327	1.150	.3	2.48	19.8	31.17	130.9
.4	.551	4.40	1.539	1.437	.4	2.52	20.1	32.17	137.2
.5	.591	4.71	1.767	1.767	.5	2.56	20.4	33.18	143.8
.6	.630	5.03	2.011	2.145	.6	2.60	20.7	34.21	150.5
.7	.669	5.34	2.270	2.572	.7	2.64	21.0	35.26	157.5
.8	.709	5.65	2.545	3.054	.8	2.68	21.4	36.32	164.6
.9	.748	5.96	2.835	3.591	.9	2.72	21.7	37.39	172.0
2.0	.787	6.28	3.142	4.189	7.0	2.76	22.0	38.48	179.6
.1	.827	6.60	3.464	4.849	.1	2.80	22.3	39.59	187.4
.2	.866	6.90	3.801	5.576	.2	2.83	22.6	40.72	195.4
.3	.906	7.24	4.155	6.372	.3	2.87	22.9	41.85	203.7
.4	.945	7.55	4.524	7.236	.4	2.91	23.2	43.01	212.2
.5	.984	7.85	4.909	8.179	.5	2.95	23.6	44.18	220.9
.6	1.024	8.16	5.309	9.205	.6	2.99	23.9	45.36	229.9
.7	1.063	8.49	5.726	10.30	.7	3.03	24.2	46.57	239.0
.8	1.102	8.80	6.158	11.49	.8	3.07	24.5	47.78	248.5
.9	1.142	9.10	6.605	12.77	.9	3.11	24.8	49.02	258.1
3.0	1.18	9.42	7.07	14.14	8.0	3.15	25.1	50.27	268.1
.1	1.22	9.75	7.55	15.60	.1	3.19	25.4	51.53	278.2
.2	1.26	10.05	8.04	17.16	.2	3.23	25.7	52.81	288.7
.3	1.30	10.37	8.55	18.82	.3	3.27	26.1	54.11	299.4
.4	1.34	10.68	9.08	20.58	.4	3.31	26.4	55.42	310.3
.5	1.38	11.00	9.62	22.45	.5	3.35	26.7	56.75	321.5
.6	1.42	11.30	10.18	24.43	.6	3.39	27.0	58.09	333.1
.7	1.46	11.61	10.75	26.52	.7	3.43	27.3	59.45	344.8
.8	1.50	11.91	11.34	28.73	.8	3.46	27.6	60.82	356.8
.9	1.54	12.25	11.95	31.06	.9	3.50	27.9	62.21	369.1
4.0	1.57	12.6	12.57	33.51	9.0	3.54	28.2	63.62	381.7
.1	1.61	12.9	13.20	36.09	.1	3.58	28.6	65.04	394.6
.2	1.65	13.2	13.85	38.79	.2	3.62	28.9	66.48	407.7
.3	1.69	13.5	14.52	41.63	.3	3.66	29.2	67.93	421.2
.4	1.73	13.8	15.21	44.60	.4	3.70	29.5	69.40	434.9
.5	1.77	14.1	15.90	47.71	.5	3.74	29.8	70.88	448.9
.6	1.81	14.4	16.62	50.97	.6	3.78	30.1	72.38	463.2
.7	1.85	14.8	17.35	54.35	.7	3.82	30.5	73.90	477.9
.8	1.89	15.1	18.10	57.91	.8	3.86	30.8	75.43	492.8
.9	1.93	15.4	18.86	61.58	.9	3.90	31.1	76.98	508.0

Trunk, Branch, or Fruit			Trunk or Branch Cross-Sectional Area (cm²)	Volume of Fruit as Sphere (cm³)
Diameter (cm)	Diameter (inches)	Circumference (cm)		
10.0	3.94	31.4	78.5	524
.1	3.98	31.7	80.1	539
.2	4.02	32.0	81.7	556
.3	4.06	32.4	83.3	572
.4	4.10	32.7	84.9	589
.5	4.13	33.0	86.6	606
.6	4.17	33.3	88.2	624
.7	4.21	33.6	89.9	641
.8	4.25	33.9	91.6	660
.9	4.29	34.2	93.3	678
11.0	4.33	34.5	95.0	697
.1	4.37	34.9	96.8	716
.2	4.41	35.2	98.5	736
.3	4.45	35.5	100.3	756
.4	4.49	35.8	102.1	776
.5	4.53	36.1	103.9	796
.6	4.57	36.4	105.7	817
.7	4.61	36.8	107.5	839
.8	4.65	37.1	109.4	860
.9	4.68	37.4	111.2	882
12.0	4.72	37.7	113.1	905
.1	4.76	38.0	115.0	928
.2	4.80	38.3	116.9	951
.3	4.84	38.6	118.8	974
.4	4.88	39.0	120.8	998
.5	4.92	39.3	122.7	1023
.6	4.96	39.6	124.7	1047
.7	5.00	39.9	126.7	1072
.8	5.04	40.2	128.7	1098
.9	5.08	40.5	130.7	1124
13.0	5.12	40.8	132.7	1150
.1	5.16	41.2	134.8	1177
.2	5.20	41.5	136.8	1204
.3	5.24	41.8	138.9	1232
.4	5.28	42.2	141.0	1260
.5	5.31	42.5	143.1	1288
.6	5.35	42.8	145.3	1317
.7	5.39	43.1	147.4	1346
.8	5.43	43.4	149.6	1376
.9	5.47	43.7	151.7	1406
14.0	5.51	44.0	153.9	1437
.1	5.55	44.3	156.1	1468
.2	5.59	44.6	158.4	1499
.3	5.63	44.9	160.6	1531
.4	5.67	45.2	162.9	1563
.5	5.71	45.6	165.1	1596
.6	5.75	45.9	167.4	1629
.7	5.79	46.2	169.7	1663
.8	5.83	46.5	172.0	1698
.9	5.87	46.8	174.4	1732
15.0	5.91	47.1	176.7	1767
.1	5.94	47.4	179.1	1803
.2	5.98	47.7	181.5	1839
.3	6.02	48.1	183.9	1876
.4	6.06	48.4	186.3	1912
.5	6.10	48.7	188.7	1950
.6	6.14	49.0	191.1	1988
.7	6.18	49.3	193.6	2026
.8	6.22	49.6	196.1	2065
.9	6.26	49.9	198.6	2105
16.0	6.30	50.2	201.1	2145
.1	6.34	50.6	203.6	2185
.2	6.38	50.9	206.1	2226
.3	6.42	51.2	208.7	2268
.4	6.46	51.5	211.2	2310
.5	6.50	51.8	213.8	2352
.6	6.54	52.1	216.4	2395
.7	6.57	52.5	219.0	2438
.8	6.61	52.8	221.7	2483
.9	6.65	53.1	224.3	2527
17.0	6.69	53.4	227.0	2572
.1	6.73	53.7	229.7	2618
.2	6.77	54.0	232.4	2664
.3	6.81	54.3	235.1	2711
.4	6.85	54.7	237.8	2758
.5	6.89	55.0	240.5	2806
.6	6.93	55.3	243.3	2855
.7	6.97	55.6	246.0	2903
.8	7.01	55.9	248.8	2953
.9	7.05	56.2	251.6	3003
18.0	7.09	56.5	254.5	3054
.1	7.13	56.9	257.3	3105
.2	7.17	57.2	260.2	3157
.3	7.21	57.5	263.0	3209
.4	7.24	57.8	265.9	3262
.5	7.28	58.1	268.8	3315
.6	7.32	58.4	271.7	3369
.7	7.36	58.7	274.6	3424
.8	7.40	59.1	277.6	3479
.9	7.44	59.4	280.6	3535
19.0	7.48	59.7	283.5	3591
.1	7.52	60.0	286.5	3648
.2	7.56	60.3	289.5	3706
.3	7.60	60.6	292.6	3764
.4	7.64	60.9	295.6	3823
.5	7.68	61.3	298.6	3882
.6	7.72	61.6	301.7	3943
.7	7.76	61.9	304.8	4003
.8	7.80	62.2	307.9	4064
.9	7.83	62.5	311.0	4126

(continued)

Appendix I (continued)

Trunk, Branch, or Fruit			Trunk or Branch Cross-Sectional Area (cm^2)	Trunk, Branch, or Fruit			Trunk or Branch Cross-Sectional Area (cm^2)
Diameter (cm)	Diameter (inches)	Circumference (cm)		Diameter (cm)	Diameter (inches)	Circumference (cm)	
20.0	7.87	62.8	314	25.0	9.84	78.5	491
.1	7.91	63.1	317	.1	9.88	78.9	495
.2	7.95	63.5	320	.2	9.92	79.2	499
.3	7.99	63.8	324	.3	9.96	79.5	503
.4	8.03	64.1	327	.4	10.00	79.8	507
.5	8.07	64.4	330	.5	10.04	80.1	511
.6	8.11	64.7	333	.6	10.08	80.4	515
.7	8.15	65.0	337	.7	10.12	80.7	519
.8	8.19	65.3	340	.8	10.16	81.1	523
.9	8.23	65.7	343	.9	10.20	81.4	527
21.0	8.27	66.0	346	26.0	10.24	81.7	531
.1	8.31	66.3	350	.1	10.28	82.0	535
.2	8.35	66.6	353	.2	10.31	82.3	539
.3	8.39	66.9	356	.3	10.35	82.6	543
.4	8.43	67.2	360	.4	10.39	82.9	547
.5	8.46	67.5	363	.5	10.43	83.3	552
.6	8.50	67.9	366	.6	10.47	83.6	556
.7	8.54	68.2	370	.7	10.51	83.9	560
.8	8.58	68.5	373	.8	10.55	84.2	564
.9	8.62	68.8	377	.9	10.59	84.5	568
22.0	8.66	69.1	380	27.0	10.63	84.8	573
.1	8.70	69.4	384	.1	10.67	85.1	577
.2	8.74	69.7	387	.2	10.71	85.5	581
.3	8.78	70.1	391	.3	10.75	85.8	585
.4	8.82	70.4	394	.4	10.79	86.1	590
.5	8.86	70.7	398	.5	10.83	86.4	594
.6	8.90	71.0	401	.6	10.87	86.7	598
.7	8.94	71.3	405	.7	10.91	87.0	603
.8	8.98	71.6	408	.8	10.94	87.3	607
.9	9.02	71.9	412	.9	10.98	87.7	611
23.0	9.06	72.3	415	28.0	11.02	88.0	616
.1	9.09	72.6	419	.1	11.06	88.3	620
.2	9.13	72.9	423	.2	11.10	88.6	625
.3	9.17	73.2	426	.3	11.14	88.9	629
.4	9.21	73.5	430	.4	11.18	89.2	633
.5	9.25	73.8	434	.5	11.22	89.5	638
.6	9.29	74.1	437	.6	11.26	89.8	642
.7	9.33	74.5	441	.7	11.30	90.2	647
.8	9.37	74.8	445	.8	11.34	90.5	651
.9	9.41	75.1	449	.9	11.38	90.8	656
24.0	9.45	75.4	452	29.0	11.42	91.1	661
.1	9.49	75.7	456	.1	11.46	91.4	665
.2	9.53	76.0	460	.2	11.50	91.7	670
.3	9.57	76.3	464	.3	11.54	92.0	674
.4	9.61	76.7	468	.4	11.57	92.4	679
.5	9.65	77.0	471	.5	11.61	92.7	683
.6	9.68	77.3	475	.6	11.65	93.0	688
.7	9.72	77.6	479	.7	11.69	93.3	693
.8	9.76	77.9	483	.8	11.73	93.6	697
.9	9.80	78.2	487	.9	11.77	93.9	702

Trunk, Branch, or Fruit			Trunk or Branch Cross-Sectional Area (cm²)
Diameter (cm)	Diameter (inches)	Circumference (cm)	
30.0	11.81	94.2	707
.1	11.85	94.6	712
.2	11.89	94.9	716
.3	11.93	95.2	721
.4	11.97	95.5	726
.5	12.01	95.8	731
.6	12.05	96.1	735
.7	12.09	96.4	740
.8	12.13	96.8	745
.9	12.17	97.1	750
31.0	12.20	97.4	755
.1	12.24	97.7	760
.2	12.28	98.0	765
.3	12.32	98.3	769
.4	12.36	98.6	774
.5	12.40	99.0	779
.6	12.44	99.3	784
.7	12.48	99.6	789
.8	12.52	99.9	794
.9	12.56	100.2	799
32.0	12.60	100.5	804
.1	12.64	100.8	809
.2	12.68	101.2	814
.3	12.72	101.5	819
.4	12.76	101.8	824
.5	12.80	102.1	830
.6	12.83	102.4	835
.7	12.87	102.7	840
.8	12.91	103.0	845
.9	12.95	103.4	850
33.0	12.99	103.7	855
.1	13.03	104.0	860
.2	13.07	104.3	866
.3	13.11	104.6	871
.4	13.15	104.9	876
.5	13.19	105.2	881
.6	13.23	105.6	887
.7	13.27	105.9	892
.8	13.31	106.2	897
.9	13.35	106.5	903
34.0	13.38	106.8	908
.1	13.42	107.1	913
.2	13.46	107.4	919
.3	13.50	107.8	924
.4	13.54	108.1	929
.5	13.58	108.4	935
.6	13.62	108.7	940
.7	13.66	109.0	946
.8	13.70	109.3	951
.9	13.74	109.6	957
35.0	13.78	110.0	962
.1	13.82	110.3	968
.2	13.86	110.6	973
.3	13.90	110.9	979
.4	13.94	111.2	984
.5	13.98	111.5	990
.6	14.02	111.8	995
.7	14.06	112.2	1001
.8	14.09	112.5	1007
.9	14.13	112.8	1012
36.0	14.17	113.1	1018
.1	14.21	113.4	1024
.2	14.25	113.7	1029
.3	14.29	114.0	1035
.4	14.33	114.4	1041
.5	14.37	114.7	1046
.6	14.41	115.0	1052
.7	14.45	115.3	1058
.8	14.49	115.6	1064
.9	14.53	115.9	1069
37.0	14.57	116.2	1075
.1	14.61	116.6	1081
.2	14.65	116.9	1087
.3	14.69	117.2	1093
.4	14.72	117.5	1099
.5	14.76	117.8	1104
.6	14.80	118.1	1110
.7	14.84	118.4	1116
.8	14.88	118.8	1122
.9	14.92	119.1	1128
38.0	14.96	119.4	1134
.1	15.00	119.7	1140
.2	15.04	120.0	1146
.3	15.08	120.3	1152
.4	15.12	120.6	1158
.5	15.16	121.0	1164
.6	15.20	121.3	1170
.7	15.24	121.6	1176
.8	15.28	121.9	1182
.9	15.31	122.2	1188
39.0	15.35	122.5	1195
.1	15.39	122.8	1201
.2	15.43	123.2	1207
.3	15.47	123.5	1213
.4	15.51	123.8	1219
.5	15.55	124.1	1225
.6	15.59	124.4	1232
.7	15.63	124.7	1238
.8	15.67	125.0	1244
.9	15.71	125.3	1250

(continued)

Appendix I (continued)

Trunk, Branch, or Fruit			Trunk or Branch	Trunk, Branch, or Fruit			Trunk or Branch
Diameter (cm)	Diameter (inches)	Circumference (cm)	Cross-Sectional Area (cm^2)	Diameter (cm)	Diameter (inches)	Circumference (cm)	Cross-Sectional Area (cm^2)
40.0	15.75	125.7	1257	45.0	17.72	141.4	1590
.1	15.79	126.0	1263	.1	17.76	141.7	1598
.2	15.83	126.3	1269	.2	17.80	142.0	1605
.3	15.87	126.6	1276	.3	17.83	142.3	1612
.4	15.91	126.9	1282	.4	17.87	142.6	1619
.5	15.94	127.2	1288	.5	17.91	142.9	1626
.6	15.98	127.5	1295	.6	17.95	143.3	1633
.7	16.02	127.9	1301	.7	17.99	143.6	1640
.8	16.06	128.2	1307	.8	18.03	143.9	1647
.9	16.10	128.5	1314	.9	18.07	144.2	1655
41.0	16.14	128.8	1320	46.0	18.11	144.5	1662
.1	16.18	129.1	1327	.1	18.15	144.8	1669
.2	16.22	129.4	1333	.2	18.19	145.1	1676
.3	16.26	129.7	1340	.3	18.23	145.5	1684
.4	16.30	130.1	1346	.4	18.27	145.8	1691
.5	16.34	130.4	1353	.5	18.31	146.1	1698
.6	16.38	130.7	1359	.6	18.35	146.4	1706
.7	16.42	131.0	1366	.7	18.39	146.7	1713
.8	16.46	131.3	1372	.8	18.43	147.0	1720
.9	16.50	131.6	1379	.9	18.46	147.3	1728
42.0	16.54	131.9	1385	47.0	18.50	147.7	1735
.1	16.57	132.3	1392	.1	18.54	148.0	1742
.2	16.61	132.6	1399	.2	18.58	148.3	1750
.3	16.65	132.9	1405	.3	18.62	148.6	1757
.4	16.69	133.2	1412	.4	18.66	148.9	1765
.5	16.73	133.5	1419	.5	18.70	149.2	1772
.6	16.77	133.8	1425	.6	18.74	149.5	1780
.7	16.81	134.1	1432	.7	18.78	149.9	1787
.8	16.85	134.5	1439	.8	18.82	150.2	1794
.9	16.89	134.8	1445	.9	18.86	150.5	1802
43.0	16.93	135.1	1452	48.0	18.90	150.8	1810
.1	16.97	135.4	1459	.1	18.94	151.1	1817
.2	17.01	135.7	1466	.2	18.98	151.4	1825
.3	17.05	136.0	1472	.3	19.02	151.7	1832
.4	17.09	136.3	1479	.4	19.06	152.1	1840
.5	17.13	136.7	1486	.5	19.09	152.4	1847
.6	17.17	137.0	1493	.6	19.13	152.7	1855
.7	17.20	137.3	1500	.7	19.17	153.0	1863
.8	17.24	137.6	1507	.8	19.21	153.3	1870
.9	17.28	137.9	1514	.9	19.25	153.6	1878
44.0	17.32	138.2	1521	49.0	19.29	153.9	1886
.1	17.36	138.5	1527	.1	19.33	154.3	1893
.2	17.40	138.9	1534	.2	19.37	154.6	1901
.3	17.44	139.2	1541	.3	19.41	154.9	1909
.4	17.48	139.5	1548	.4	19.45	155.2	1917
.5	17.52	139.8	1555	.5	19.49	155.5	1924
.6	17.56	140.1	1562	.6	19.53	155.8	1932
.7	17.60	140.4	1569	.7	19.57	156.1	1940
.8	17.64	140.7	1576	.8	19.61	156.5	1948
.9	17.68	141.1	1583	.9	19.65	156.8	1956
				50.0	19.68	157.1	1963

Bibliography

Abeles, F. B. 1973. *Ethylene in plant biology.* Academic Press, New York.

Ackerman, W. L. 1969. Fruit bud hardiness of North Caucasus seedlings and other foreign peach introductions. *Fruit Vars. and Hort. Digest* 23:14–16.

Ackley, W. B. 1962. Question box. *Proc. Wash. State Hort. Assoc.* 58:215.

Ackley, W. B.; P. C. Crandall; and T. S. Russell. 1958. The use of linear measurements in estimating leaf areas. *Proc. Am. Soc. Hort. Sci.* 72:326–330.

Alban, L. A. 1958. *Interpretation of soil test.* Oregon Agr. Exp. Sta. Circ. 587.

Aldrich, W. W. 1936. Relative efficiency of spur and shoot leaves for fruit growth of pears. *Proc. Am. Soc. Hort. Sci.* 34:227–232.

Aldrich, W. W.; M. R. Lewis; R. A. Work; A. Lloyd Ryall; and F. C. Reimer. 1940. *Anjou pear responses to irrigation in a clay adobe soil.* Oregon Agr. Exp. Sta. Bull. 374.

Aldrich, W. W.; R. A. Work; and M. R. Lewis. 1935. Pear root concentration in relation to soil moisture extraction in heavy clay soil. *J. Agr. Res.* 50: 975–988.

Ali, C. N., and M. N. Westwood. 1968. Juvenility as related to chemical content and rooting of stem cuttings of *Pyrus* species. *Proc. Am. Soc. Hort. Sci.* 93:77–82.

Allen, F. W. 1932. Physical and chemical changes in the ripening of deciduous fruits. *Hilgardia* 6:381–441.

Amen, Ralph D. 1963. The concept of seed dormancy. *Am. Scientist* 51:408–424.

Am. Fruit Grower. 1967. Here's the latest in frost control. 87:11–16.

Am. Fruit Grower. 1969. Where we stand on mechanical harvesting of strawberries. 89:13–51.

Anderson, D. M. 1975. *Common names of insects.* Entomol. Soc. of Am., Spec. Publ. 75–1.

Anderson, H. W. 1956. *Diseases of fruit crops.* McGraw-Hill, New York.

Argles, G. K. 1937. *A review of literature on stock-scion incompatibility of fruit trees with particular reference to pome and stone fruits.* Imperial Bureau Hort. and Plantation Crops Tech. Communication 9.

Ashton, C. D.; S. W. Edgecombe; R. K. Gerber; and O. Kirk. 1950. *How to prune peach trees in Utah.* Utah State Agr. Ext. Bull. 199.

Awad, M., and A. L. Kenworthy. 1963. Clonal rootstock scion variety and time of sampling influences in apple leaf composition. *Proc. Am. Soc. Hort. Sci.* 83:68–73.

Bain, J. M. 1961. Some morphological, anatomical, and physiological changes in the pear fruit (*Pyrus communis* var. Williams) during development and following harvest. *Australian J. Botany* 9:99–123.

Bain, J. M., and R. N. Robertson. 1951. The physiology of growth of apple fruits I. Cell size, cell number, and fruit development. *Australian J. Sci. Res.* 4:75–91.

Baker, G. A., and R. M. Brooks. 1944. Climate in relation to deciduous fruit production in California. III. Effect of temperature on number of days from full bloom to harvest of apricot and prune fruits. *Proc. Am. Soc. Hort. Sci.* 45:95–104.

Ballard, J. K., and E. L. Proebsting. 1972. *Frost and frost control in Washington orchards.* Wash. Agr. Ext. Bull. 634.

Ballard, J. K.; E. L. Proebsting; R. B. Tukey; and H. Mills. 1971. *Critical temperatures for blossom buds.* Wash. Agr. Ext. Circ. Nos. 369–374.

Barlow, H. W. B. 1959. Root/shoot relationships in fruit trees. *Scientific Hort.* 14:35–41.

Baron, L. C., and R. L. Stebbins. 1972. *Growing filberts in Oregon.* Oregon Ext. Bull. 628.

Batjer, L. P. 1954. Nutrient utilization—apple and peach. *Proc. Wash. State Hort. Assoc.* 50:68–72.

Batjer, L. P. 1961. Scoring to increase fruit set—chemical thinning to reduce it. *Proc. Wash. State Hort. Assoc.* 57:89–91.

Batjer, L. P., and H. D. Billingsley. 1964. *Apple thinning with chemical sprays.* Wash. Agr. Exp. Sta. Bull. 651.

Batjer, L. P.; H. D. Billingsley; M. N. Westwood; and B. L. Rogers. 1957. Predicting harvest size of apples at different times during the growing season. *Proc. Am. Soc. Hort. Sci.* 70:46–57.

Batjer, L. P., and G. C. Martin. 1965. The influence of night temperature on growth and development of Early Redhaven peaches. *Proc. Am. Soc. Hort. Sci.* 87:139–144.

Batjer, L. P.; H. A. Schomer; E. J. Newcomer; and D. L. Coyier. 1967. *Commercial pear growing.* USDA Agr. Handbook 330.

sion of effect of naphthaleneacetic acid on apple drop as determined by localized applications. *Proc. Am. Soc. Hort. Sci.* 51:77–84.

Batjer, L. P., and M. N. Westwood. 1958a. Seasonal trend of several nutrient elements in leaves and fruits of Elberta peach. *Proc. Am. Soc. Hort. Sci.* 71:116–126.

Batjer, L. P., and M. N. Westwood. 1958b. Size of Elberta and J. H. Hale peaches during the thinning period as related to size at harvest. *Proc. Am. Soc. Hort. Sci.* 72:102–105.

Batjer, L. P., and M. N. Westwood. 1960. 1-Naphthyl *N*-methylcarbamate, a new chemical for thinning apples. *Proc. Am. Soc. Hort. Sci.* 75:1–4.

Batjer, L. P., and M. N. Westwood. 1963. Effects of pruning, nitrogen, and scoring on growth and bearing characteristics of young Delicious apple trees. *Proc. Am. Soc. Hort. Sci.* 82:5–10.

Benson, N. R. 1972. Questions and answers dealing with planting and fertilizing fruit trees and controlling weeds. *Proc. Wash. State Hort. Assoc.* 68:62–66.

Benson, N. R.; R. M. Bullock; I. C. Chmelir; and E. S. Degman. 1957. Effects of levels of nitrogen and pruning on Starking and Golden Delicious apples. *Proc. Am. Soc. Hort. Sci.* 70:27–39.

Blake, M. A. 1935. Types of varietal hardiness in the peach. *Proc. Am. Soc. Hort. Sci.* 33:240–244.

Blanc, M. L.; H. Geslin; I. A. Holzberg; and B. Mason. 1963. *Protection against frost damage.* World Meteorol. Org., Tech. Note No. 51.

Blanpied, G. D. 1964. The relationship between growing season temperature and bloom dates and length of the growing season of Red Delicious apples in North America. *Proc. Am. Soc. Hort. Sci.* 34:72–81.

Bolkhovskikh, Z.; V. Grif.; T. Matvejeva; and O. Zakharyeva. 1969. *Chromosome numbers of flowering plants.* Acad. of Sci., USSR.

Boyton, D. 1939. Soil atmosphere and the production of new rootlets by apple tree root systems. *Proc. Am. Soc. Hort. Sci.* 37:19–26.

Bradt, O. A.; A. Hutchinson; C. L. Ricketson; and G. Tehrani. 1974. *Fruit varieties.* Ontario Ministry of Agr. and Food, Publ. 430.

Brann, J. L.; P. A. Arneson; and G. H. Oberly. 1975. *Tree-fruit production recommendations for commercial growers.* N.Y. Agr. Exp. Sta.

Breakey, E. P., and D. H. Brannon. 1946. *The control of the blackberry mite.* Wash. Agr. Ext. Bull. 346.

Brison, F. R. 1974. *Pecan culture.* Capital Printing, Austin, Texas.

Brown, D. S. 1952. Climate in relation to deciduous fruit

production in California. V. The use of temperature records to predict the time of harvest of apricots. *Proc. Am. Soc. Hort. Sci.* 60:197–203.

Brown, G. G.; W. M. Mellenthin; and L. Childs. 1964. *Observations on winter injury to apple and pear trees in the Hood River Valley.* Oregon Agr. Exp. Sta. Bull. 595.

Bukovac, M. J. 1963. Induction of parthenocarpic growth of apple fruits with gibberellins A_3 and A_4. *Botan. Gaz.* 124:191–195.

Bukovac, M. J., and S. Nakagawa. 1968. Gibberellin-induced asymmetric growth of apple fruits. *Hort. Science* 3:172–174.

Bullock, R. M. 1952. A study of some inorganic compounds and growth promoting chemicals in relation to fruit cracking of Bing cherries at maturity. *Proc. Am. Soc. Hort. Sci.* 59:243–253.

Burke, M. J.; L. V. Gusta; H. A. Quamme; C. J. Weiser; and P. H. Li. 1976. Freezing and injury in plants. *Ann. Rev. Plant Physiol.* 27:507–528.

Burkhart, D. J., and M. N. Westwood. 1964. Inducing young trees to bear fruit. *Am. Fruit Grower* 84:24.

Cain, J. C. 1970. Optimum tree density for apple orchards. *HortScience* 5:232–234.

Cajlachjan, M. Ch. 1937. Concerning the hormonal nature of plant development processes. *Comptes Rendus (Doklady) de l'Academie des Sciences de l'USSR* 16:227–230.

Callan, N., and M. M. Thompson, 1976. Progress on fruit set of Italian prune. *Proc. Oregon Hort. Soc.* 67:21–23.

Cameron, H. R. 1971. Effect of root or trunk stock on susceptibility of orchard trees to *Pseudomona syringae. Plant Disease Reporter* 55:421–423.

Cameron, H. R., and M. N. Westwood. 1968. Solution pocket: A physiological problem of brined sweet cherries. *HortScience* 3:141–143.

Cameron, H. R.; M. N. Westwood; and P. B. Lombard. 1969. Resistance of *Pyrus* species and cultivars to *Erwinia amylovora* (blight). *Phytopathology* 59:1813–1815.

Carpenter, Dunbar. 1966. Frost control by overhead sprinkling. *Proc Oregon. Hort. Soc.* 58:148–150.

Chan, B., and J. C. Cain. 1967. Effect of seed formation on subsequent flowering of apple. *Proc. Am. Soc. Hort. Sci.* 91:63–68.

Chandler, W. H. 1957. *Deciduous orchards.* 3rd ed. Lea & Febiger, Philadelphia.

Chandler, W. H.; M. H. Kimball; G. L. Philp; W. P. Tufts; and G. P. Weldon. 1937. *Chilling requirements for opening of buds on deciduous orchard trees and some other plants in California.* Calif. Agr. Exp. Sta. Bull. 611.

Chaplin, M. H.; and M. N. Westwood; and A. N. Roberts. 1972. Effects of rootstock on leaf element content of Italian prune (*Prunus domestica* L.). *J. Am. Soc. Hort. Sci.* 97:641–644.

Chaplin, M. H.; M. N. Westwood; and R. L. Stebbins. 1973. Estimates of tree size, bearing surface, and yield of tree fruits and nuts. *Ann. Proc. Oregon Hort. Soc.* 64:94–98.

Claypool, L. L.; W. H. Dempsey; P. Esau; and M. W. Miller. 1962. Physical and chemical changes in French prunes during maturation in coastal valleys. *Hilgardia* 33:311–318.

Childers, N. F., Ed. 1966. *Fruit nutrition.* 2nd ed. Horticultural Publications, New Brunswick, New Jersey.

Coe, F. M. 1945. *Cherry rootstocks.* Utah Agr. Exp. Sta. Bull. 319.

Cooke, J. R. 1970. Mathematical determination of surface area and volume for developing apple, lemon, and peach fruits. *Proc. Ann. Meet., Am. Soc. Agr. Eng.* 70–338: 1–36.

Coop. Extension Service, Wash. State. 1974. *Spray guide for tree fruits in eastern Washington.* Wash. Agr. Ext. Bull. 419.

Couey, M. 1971. Researcher outlines three types of cherry pitting. *Goodfruit Grower,* December, p. 12.

Crafts, A. S.; H. B. Currier; and C. R. Stocking. 1949. *Water in the physiology of plants. Chronica Botanica,* Waltham Mass.

Crandall, P. C., and J. E. George. 1967. Here's a harvest of mechanical caneberry harvesters. *Am. Fruit Grower* 87:20–38.

Crane, J. C. 1969. The role of hormones in fruit set and development. *HortScience* 4:108–111.

Crane, J. C. 1971. The unusual mechanism of alternate bearing in the Pistachio. *HortScience* 6:489–490.

Crane, J. C. 1973. Abscission of pistachio inflorescence buds as affected by leaf area and number of nuts. *J. Am. Soc. Hort. Sci.* 98:591–592.

Crane, M. B., and W. J. C. Lawrence. 1938. *The genetics of garden plants.* Macmillan, London.

Crawford, T. V., and A. S. Leonard. 1960. Wind machine–orchard heater system for frost protection. *Calif. Agr.* 14:10–13.

Dana, M. N., and G. C. Clingbeil. 1966. *Cranberry growing in Wisconsin.* Wisconsin Agr. Circ. 654.

Davis, L. D., and M. M. Davis. 1948. Size in canning peaches. The relation between the diameter of cling peaches early in the season and at harvest. *Proc. Am. Soc. Hort. Sci.* 51:225–230.

Day, L. H. 1953. *Rootstocks for stone fruits.* Calif. Agr. Ext. Bull. 736.

Dean, A. 1970. *Vote for vitality and good health—Elect vegetables and fruits.* Michigan Agr. Ext. Bull. E-686.

Dedolph, R. R.; J. A. Stevens; R. N. Monfort; and H. B. Tukey. 1961. The effect of apples on the general health of college students. *Michigan Quart. Bull.* 44:230–234.

Degman, E. S. 1961. Filler-tree management. *Proc. Wash. State Hort. Assoc.* 57:63.

Denne, M. P. 1960. The growth of apple fruitlets and the effect of early thinning on fruit development. *Ann. Botany* [N. S.] 24:397–406.

Dennis, F. G., Jr. 1967. The role of seeds in apple fruit development. *Farm Research* 33:12–13.

Dennis, F. G., Jr. 1973. Physiological control of fruit set and development with growth regulators. *Acta Horticulturae* 34:251–257.

Dennis, F. G., Jr., and L. J. Edgerton. 1966. Effects of gibberellins and ringing upon apple fruit development and flower bud formation. *Proc. Am. Soc. Hort. Sci.* 88:14–24.

Devlin, R. M. 1967. *Plant physiology.* Reinhold Publishing Corp., New York.

Dewey, D. H., and M. Uota. 1953. Post-harvest applications of chemicals for ripening cannery Bartlett pears. *Proc. Am. Soc. Hort. Sci.* 61:246–250.

Dilley, D. R. 1969. Hormonal control of fruit ripening. *HortScience* 4:111–114.

Doughty, C. C.; Max E. Patterson; and A. Y. Shawa. 1967. Storage longevity of McFarlin cranberry as influenced by certain growth retardants and stage of maturity. *Proc. Am. Soc. Hort. Sci.* 91:192–204.

Eagles, C. F., and P. F. Wareing. 1964. The role of growth substances in the regulation of bud dormancy. *Physiologia Plantarum* 17:697–709.

Epstein, E. 1973. Roots. *Scientific American,* May, pp. 48–58.

Esau, K. 1953. *Plant anatomy,* Chapters 19 and 20. Wiley, New York.

Evans, H. C., and R. S. Marsh. 1957. *Apple color—its development and sales appeal.* West Va. Agr. Exp. Sta. Bull. 396.

Eyring, C. F. 1948. *Essentials of Physics,* Chapters 12 and 17. Prentice-Hall, New York.

Ferguson, J. H. A. 1960. A comparison of two planting systems in orchards as regards the amount of radiation intercepted by the trees. *Neth. J. Agr. Sci.* 8:271–280.

Fisher, D. V. 1940. A three year study of maturity indices for harvesting Italian prunes. *Proc. Am. Soc. Hort. Sci.* 37:183–186.

Fisher, D. V. 1962. Heat units and number of days required to mature some pome and stone fruits in various areas of North America. *Proc. Am. Soc. Hort. Sci.* 80:114–124.

Fisher, H.; P. Griminger; E. R. Sostman; M. K. Brush. 1965. Dietary pectin and blood cholesterol. *J. of Nutrition* 86:113.

Fisher, H.; P. Griminger, H. S. Weiss; and W. G. Siller. 1964. Avian atherosclerosis: Retardation by pectin. *Science* 146:1063–1064.

Fogle, H. W. et. al. 1955. A comparison of several Italian-type plum varieties. *Proc. Wash. State Hort. Assoc.* 51:109–113.

Ford, Harry W. 1960. A hand instrument for estimating height and width of citrus trees. *Proc. Am. Soc. Hort. Sci.* 76:245–247.

Forde, H. I. 1975. Walnuts. In *Advances in fruit breeding,* edited by J. Janick and J. Moore. Purdue Univ. Press, Lafayette, Indiana.

Forde, H. I., and W. H. Griggs. 1975. Pollination and blooming habits of walnuts. *Calif. Agr. Ext. Serv. Leaf.* 2753.

Fridley, R. B. 1974. Engineering view of high-density planting as related to mechanical harvesting. *Proc. Oregon Hort. Soc.* 65:67–73.

Fridley, R. B., and P. A. Adrian. 1966. *Mechanical harvesting equipment for deciduous tree fruits.* Calif. Agr. Exp. Sta. Bull. 825.

Fuchigami, L. H.; D. R. Evert; and C. J. Weiser. 1971. A translocatable cold-hardiness promoter. *Plant Physiol.* 47:164–167.

Galletta, G. J. 1975. Blueberries and cranberries. In *Advances in fruit breeding,* edited by J. Janick and J. Moore. Purdue Univ. Press, Lafayette, Indiana.

Galston, A. W., and P. J. Davies. 1969. Hormonal regulation in higher plants. *Science* 163:1288–1297.

Gardner, F. E.; C. P. Marth; and L. P. Batjer. 1939. Spraying with plant-growth substances for control of the pre-harvest drop of apples. *Proc. Am. Soc. Hort. Sci.* 37:415–428.

Gardner, W. R. 1960. Dynamic aspects of water availability to plants. *Soil Sci.* 89:63–73.

Garren, R., and J. F. Sprowls. 1975. *Growing red raspberries in Oregon.* Ore. Agr. Ext. Circ. 764.

Gates, D. M. 1965. Heat transfer in plants. *Scientific American,* December, pp. 76–84.

Gauch, H. G. 1972. *Inorganic plant nutrition.* Dowden, Hutchinson & Ross, Inc., Stroudburg, Pennsylvania.

Geiger, R. 1957. *The climate near the ground.* Harvard Univ. Press, Cambridge, Mass.

Gerber, J. F. 1970. Crop protection by heating, wind machines, and overhead irrigation (a review). *HortScience* 5:426–431.

Gerhardt, F., and D. F. Allmendinger. 1946. The influence of naphthaleneacetic acid spray on the maturity and storage physiology of apples, pears, and sweet cherries. *J. Agr. Res.* 73:189–206.

Gerhardt, F., and H. English. 1945. Ripening of the Italian prune as related to maturity and storage. *Proc. Am. Soc. Hort. Sci.* 62:205–209.

Gerhardt, F.; H. English; C. P. Harley; and E. Smith. 1943. The influence of maturity and storage temperature on the ripening behavior and dessert quality of the Italian prune. *Proc. Am. Soc. Hort. Sci.* 42:247–252.

Gerhardt, F., and H. A. Schomer. 1955. Possible maturity standards for picking early strains of Italian prunes. *Wash. State Hort. Assoc.* 51:185–188.

Glass, E. H. 1975. *Integrated pest management rationale, potential needs, and implementation.* Entomol. Soc. of Am., Spec. Pub. 75–2.

Green, W. J.; A. E. Irish; and C. D. Mumford. 1960. *Cost of producing apples and pears in the Hood River Valley.* Oregon Agr. Exp. Sta. Bull. 573.

Griggs, W. H. 1953. *Pollination requirements of fruits and nuts.* Calif. Agr. Ext. Serv. Circ. 424.

Haller, M. H., 1941. *Fruit pressure testers and their practical application.* USDA Circ. 627.

Handbook of chemistry and physics. Chem. Rubber Co. Press, Cleveland, Ohio. (Published annually.)

Hansen, E. 1946. Effect of 2,4-dichlorophenoxyacetic acid on the ripening of Bartlett pears. *Plant Physiol.* 21:588–592.

Hansen, E. 1959. Maturity study for Italian prunes intended for drying. *Proc. Oregon Hort. Soc.* 51:73–75.

Hansen, E. 1961. Climate in relation to post-harvest physiological disorders of apples and pears. *Proc. Oregon Hort. Soc.* 53:54–58.

Hansen, E. 1966. Post-harvest physiology of fruits. *Ann. Rev. Plant Physiol.* 17:459–480.

Harris, G. H. 1926. The activity of apple and filbert roots, especially during the winter months. *Proc. Am. Soc. Hort. Sci.* 23:414–422.

Hartman, H. 1926. *Studies relating to the harvesting of Italian prunes for canning and fresh fruit shipment.* Oregon Agr. Exp. Sta. Circ. 75.

Hartmann, H. T., and D. E. Kester. 1975. *Plant propagation.* 3rd ed. Prentice-Hall, Englewood Cliffs, New Jersey.

Hedden, S. L.; H. P. Gaston; and J. H. Levin. 1959. Harvesting blueberries mechanically. *Michigan Quart. Bull.* 42:24.

Hedrick, U. P. 1925. *The small fruits of New York.* N.Y. Agr. Exp. Sta., Geneva, N.Y.

Heinicke, A. J. 1921. Some relations between circumference and weight, and between root and top growth of young apple trees. *Proc. Am. Soc. Hort. Sci.* 18:222–227.

Heinicke, D. R. 1964. The micro-climate of fruit trees. III. The effect of tree size on light penetration and leaf area in Red Delicious apple trees. *Proc. Am. Soc. Hort. Sci.* 85:33–41.

Heinicke, D. R. 1966. Characteristics of McIntosh and Red Delicious apples as influenced by exposure to sunlight during the growing season. *Proc. Am. Soc. Hort. Sci.* 89:10–13.

Heinicke, D. R. 1975. *High-density apple orchards—planting, training, and pruning.* USDA Agr. Handbook No. 458.

Helgeson, J. P. 1968. The cytokinins. *Science* 161:974–981.

Hess, C. E. 1961. The physiology of root initiation in easy and difficult-to-root cuttings. *The Hormolog* 3:3–6.

Hodgeman, C. D. 1947. *Handbook of chemistry and physics.* Chem. Rubber Co. Press, Cleveland, Ohio. (Published annually.)

Hoyt, S. C. 1969. Integrated chemical control of insects and biological control of mites on apple in Washington. *J. Econ. Entomol.* 62:74–86.

Hoyt, S. C., and E. C. Burts. 1974. Integrated control of fruit pests. *Ann. Rev. Entomol.* 19:231–252.

Jackson, D. I., and B. G. Coombe. 1966. Gibberellin-like substances in the developing apricot fruit. *Science* 154:277–278.

Jackson, D. I., and G. B. Sweet. 1972. Flower initiation in temperate woody plants (a review). *Hort. Abstracts* 42:9–25.

Janick, J. 1972. *Horticultural science.* 2d ed. W. H. Freeman and Company, San Francisco.

Janick, J., and J. N. Moore. 1975. *Advances in fruit breeding.* Purdue Univ. Press, Lafayette.

Janick, J.; R. W. Schery; F. W. Woods; and V. W. Ruttan. 1970. *Plant agriculture.* W. H. Freeman and Company, San Francisco.

Janick, J.; R. W. Schery; F. W. Woods; and V. W. Ruttan. 1974. *Plant science: An introduction to world crops.* 2nd ed. W. H. Freeman and Company, San Francisco.

Jankiewicz, L. S. 1966. Crotch angle formation in young apple trees grown in reversed position. *Acta Agrobotanica* 18:19–27.

Kenworthy, A. L. 1972. *Trickle irrigation—The concept and guidelines for use.* Michigan Agr. Exp. Sta. Res. Rep. 165.

Khan, A. A. 1966. "Morphactins" destroy plant responses to gravity and light. *Farm Res.* 32:2–3.

Kidd, F. D., and C. West, 1932. Effects of ethylene and of apple vapors on the ripening of fruits. *Ann. Rep. Food Invest. Board* (London). Pp. 55–58.

Kochan, W. J. 1962. *Iron chelate control of chlorosis in peach trees.* Idaho Agr. Exp. Sta. Res. Bull. 384.

Kochan, W. J.; L. Verner; A. Kamal; and R. Braun. 1962. *Control of fruit-dropping in Italian prunes by foliar sprays of 2,4,5-TP.* Idaho Agr. Exp. Sta. Res. Bull. 378.

Koller, D. 1955. *The regulation of germination in seeds.* Bull. Res. Council Israel 5D:85–108.

Kramer, P. J. 1937. The relation between rate of transpiration and rate of absorption of water in plants. *Am. J. Bot.* 24:10–15.

Kramer, P. J. 1969. *Plant and soil water relationships: A modern synthesis.* McGraw-Hill, New York.

Kraus, E. J. 1913. *The pollination of the pomaceous fruits. I. Gross morphology of the apple.* Oregon Agr. Exp. Sta., Res. Bull. 1 (part 1).

Krueger, R. R. 1972. The geography of the orchard industry in Canada. In *Readings in Canadian geography,* edited by R. M. Irving. Holt, Rinehart, & Winston, Ltd., Toronto. Pp. 216–241.

Lagerstedt, H. B. 1972a. Ethephon, a filbert harvest aid. *Proc. Nut Growers Soc. Oregon and Wash.* 57:65–69.

Lagerstedt, H. B. 1972b. Nut research progress report for 1971. *Proc. Nut Growers Soc. Oregon and Wash.* 57:37–41.

Lagerstedt, H. B. 1975. Filberts. In *Advances in fruit breeding,* edited by J. Janick and J. N. Moore. Purdue Univ. Press, Lafayette, Indiana. Pp. 456–489.

Lang, A. 1970. Gibberellins: Structure and metabolism. *Ann. Rev. Plant Physiol.* 21:537–570.

Lapins, K. 1963. Cold hardiness of rootstocks and framebuilders for tree fruits. *Can. Dept. of Agr. Bull.,* SP32.

LaRue, J. H., and M. Gerdts. 1973. *Growing plums in California.* Calif. Agr. Exp. Sta. Circ. 563.

Lavee, S. 1973. Dormancy and bud break in warm climates: Considerations of growth-regulator involvement. *Acta Horticulturae* 34:225–234.

Leopold, A. C., and P. E. Kriedemann. 1975. *Plant growth and development.* 2nd ed. McGraw-Hill, New York.

Letham, D. S. 1961. Influence of fertilizer treatment on apple fruit composition and physiology. *Australian J. Agr. Res.* 12:600–611.

Letham, D. S. 1969. Cytokinins and their relation to other phytohormones. *BioScience* 19:309–316.

Leuty, S. J., and M. J. Bukovac. 1968. Effect of naphthaleneacetic acid on abscission of peach fruits in relation to endosperm development. *Proc. Am. Soc. Hort. Sci.* 92:124–134.

Levin, J. H. 1969. Mechanical harvesting of food. *Science* 166:968–974.

Lombard, P. B.; C. B. Cordy; and E. Hansen. 1971. Relation of post-bloom temperature to Bartlett pear maturation. *J. Am. Soc. Hort. Sci.* 96:799–801.

Lombard, P. B., and A. E. Mitchell. 1962. Anatomical and hormonal development of Redhaven peach seeds as related to the timing of naphthaleneacetic acid for fruit thinning. *Proc. Am. Soc. Hort. Sci.* 80:163–171.

Lombard, P. B., and M. N. Westwood. 1975. Effect of hedgerow orientation on solar radiation and cropping within a pear orchard. *HortScience* 10:312.

Lombard, P. B.; M. Westwood; and M. Thompson. 1971. Effective pollination—The facts of life behind uniform pear cropping. *Proc. Oregon Hort. Soc.* 62:31–36.

Looney, N. E. 1968. Light regimes within standard-size apple trees as determined spectrophotometrically. *Proc. Am. Soc. Hort. Sci.* 93:1–6.

Luckwill, L. C. 1953. Studies of fruit development in relation to plant hormones I. Hormone production by the developing apple seed in relation to fruit drop. *J. Hort. Sci.* 28:14–24.

Luckwill, L. C., and P. Whyte. 1968. *Hormones in the xylem sap of apple trees.* Soc. Chem. Industry Monograph 31:87–101.

Lutz, J. M., and R. E. Hardenburg. 1968. *The commercial storage of fruits, vegetables, and florist and nursery crops.* USDA Agr. Handbook 66.

MacDaniels, L. H. 1940. *The morphology of the apple and other pome fruits.* N.Y. Agr. Exp. Sta. Mem. 230.

MacDaniels, L. H., and A. J. Heinicke. 1929. *Pollination and other factors affecting set of fruit with special reference to the apple.* N.Y. Agr. Exp. Sta. Bull. 497.

MacPhee, A. W., and K. H. Sanford. 1961. The influence of spray programs on the fauna of apple orchards in Nova Scotia XII—Second supplement to VII. Effects of beneficial arthopods. *Can. Entomologist* 93:671–673.

Madden, G. D., and H. L. Malstrom. 1975. Pecans and hickories. In *Advances in fruit breeding,* edited by J. Janick and J. N. Moore. Purdue Univ. Press, Lafayette, Indiana.

Magness, J. R. 1928. Observations on color development in apples. *Proc. Am. Soc. Hort. Sci.* 25:289–292.

Magness, J. R., and F. L. Overley. 1929. Relation of leaf area to size and quality of apples and pears. *Proc. Am. Soc. Hort. Sci.* 26:160–162.

Magness, J. R., and G. F. Taylor. 1925. *An improved type of pressure tester for the determination of fruit maturity.* USDA Circ. 350.

Maney, T. J. 1942. Fruit tree injury resulting from the Midwest blizzard of November 1940. *Proc. Am. Soc. Hort. Sci.* 40:215–219.

Maramorosch, K.; E. Shikata; and R. R. Granados. 1968. Structures resembling mycoplasma in diseased plants and in insect vectors (trans.). *N.Y. Acad. Sci.* (Series II) 30:841–855.

Marcus, A. 1969. Seed germination and the capacity for protein synthesis. In *Dormancy and survival.* Academic Press, New York. Pp. 143–160.

Martin, G. C.; D. S. Brown; and M. M. Nelson. 1970. Apple shape changing possible with cytokinin and gibberellin sprays. *Calif. Agr.*, April, p. 14.

Martin, J. T., and D. J. Fisher. 1965. Surface structure of plant roots. *Ann. Rep. Long Ashton Res. Sta.*, pp. 251–254.

Martin, D. and T. L. Lewis. 1952. Physiology of growth in apple fruits. III. Cell characteristics and respiratory activity of light and heavy crop trees. *Australian J. Sci. Res.* 5:315–327.

Martin, D.; T. L. Lewis; and J. Cerny. 1954. The physiology of growth in apple fruits. VIII. Between tree variation in cell physiology in relation to disorder incidence. *Australian J. Biol. Sci.* 7:211–220.

Martin, D.; T. L. Lewis; and J. Cerny. 1964. Apple fruit cell numbers in relation to cropping alternation and certain treatments. *Australian J. Agr. Res.* 15:905–919.

Martin, G. C.; H. I. Forde; and E. F. Serr. 1969. Yield performance of (*Juglans regia*) 'Payne' on seedlings of Northern California black walnut (*Juglans hindsii*) rootstock at three planting distances. *HortScience* 4:130–131.

Maxie, E. C., and L. L. Claypool. 1957. Heat injury in prunes. *Proc. Am. Soc. Hort. Sci.* 69:116–121.

Maxie, E. C., and J. C. Crane. 1968. Effect of ethylene on growth and maturation of the fig, *Ficus carica* L., fruit. *Proc. Am. Soc. Hort. Sci.* 92:255–267.

Mellenthin, W. M. 1966. Effect of climatic factors on fruit maturity and quality of pears. *Proc. Wash. State Hort. Assoc.* 62:67–69.

Melnick, V. L. M.; L. Holm; and E. Struckmeyer. 1964. Physiological studies on fruit development by means of ovule transplantation *in vivo. Science* 145:609–611.

Metcalf, C. L.; W. P. Flint; and R. L. Metcalf. 1951. *Destructive and useful insects.* 3rd ed. McGraw-Hill, New York.

Meyer, B. S., and D. B. Anderson. 1939. *Plant physiology,* Chapters 21 and 33. D. van Nostrand, New York.

Meyers, T., and M. Weidenhamer. 1966. *Homemaker's use of and opinions about selected fruits and fruit products.* USDA Stat. Reporting Serv. Rep. 765.

Millier, W. F.; G. E. Rehkugler; R. A. Pellerin; J. A. Throop; and R. B. Bradley. 1975. Pamper your apples with multilevel harvester. *Western Fruit Grower* 95:19–22.

Mitchell, F. G.; R. Guillou; and R. A. Parsons. 1972. *Commercial cooling of fruits and vegetables.* Calif. Agr. Exp. Sta. Ext. Man. 43.

Mitchell, J. W. 1966. Present status and future of plant regulating substances. *Agr. Sci. Rev.* 4:27–36.

Mowry, J. B. 1964. Inheritance of cold hardiness of dormant peach flower buds. *Proc. Am. Soc. Hort. Sci.* 85:128–133.

Mumford, D. C. 1962. *A minimum-sized economic orchard unit, Hood River Valley, Oregon.* Oregon Agr. Exp. Sta. Circ. 611.

Murray, C. D. 1927. A relationship between circumference and weight in trees and its bearing on branching angles. *J. Gen. Physiol.* 10:725–729.

Nikolaeva, M. G. 1967. Physiology of deep dormancy in seeds (English translation 1969). *Akad. Nauk USSR.*

Nobel, P. S. 1974. *Introduction to biophysical plant physiology.* W. H. Freeman and Company, San Francisco.

Norton, R. A.; C. J. Hansen; H. J. O'Reilly; and W. H. Hart. 1963a. *Rootstocks for apricots in California.* Calif. Agr. Exp. Sta. Ext. Leaf. 156.

Norton, R. A.; C. J. Hansen; H. J. O'Reilley; and W. H. Hart. 1963b. *Rootstocks for peaches and nectarines in California.* Calif. Agr. Exp. Sta. Ext. Leaf. 157.

Norton, R. A.; C. J. Hansen; H. J. O'Reilly; and W. H. Hart. 1963c. *Rootstocks for plums and prunes in California.* Calif. Agr. Exp. Sta. Ext. Leaf. 158.

Norton, R. A.; C. J. Hansen; H. J. O'Reilly; and W. H. Hart. 1963d. *Rootstocks for sweet cherries in California.* Calif. Agr. Exp. Sta. Ext. Leaf. 159.

Olmstead, C. W. 1956. American orchard and vineyard regions. *Econ. Geography* 32:189–236.

Olsen, K. L., and H. A. Schomer. 1964. *Oxygen and carbon dioxide levels for controlled atmosphere storage of Starking and Golden Delicious apples.* USDA Market Res. Rep. 653.

Oskamp, J. 1935. *Soils in relation to fruit growing in New York.* Part VI. N.Y. Agr. Exp. Sta. Bull. 626.

Ourecky, D. K. 1974. *Minor fruits in New York State.* N.Y. Agr. Ext. Inf. Bull. 11.

Ourecky, D. K., and J. P. Tomkins. 1974. *Raspberry growing in New York State.* N.Y. Agr. Exp. Sta. Ext. Bull. 1170.

Overbeek, J. van. 1962. *Endogenous regulators of fruit growth.* Campbell Soup Co., Plant Science Symposium. Pp. 37–56.

Overcash, J. P., and J. A. Campbell. 1955. The effects of intermittent warm and cold periods on breaking the rest period of peach leaf buds. *Proc. Am. Soc. Hort. Sci.* 66:87–92.

Overholser, E. L.; A. J. Winkler; and H. E. Jacob. 1923. *Factors influencing the development of internal browning of the Yellow Newtown apple.* Calif. Agr. Exp. Sta. Bull. 370.

Painter, J. H.; P. W. Miller; and I. C. MacSwan. 1958. *Shell perforation and poor sealing of walnuts.* Oregon Agr. Ext. Circ. 647.

Pearce, S. C. 1952. Studies in the measurement of apple trees I. The use of trunk girths to estimate tree size. In *Ann. Rep. 1951, East Malling Res. Sta.,* pp. 101–104.

Pieniazek, S. A. 1973. China—16 years after. *International Fruit World* (Basle, Switzerland).

Plass, G. N. 1959. Carbon dioxide and climate. *Scientific American,* July, pp. 41–47.

Platt, R. B., and J. F. Griffiths. 1965. *Environmental measurement and interpretation.* Reinhold Pub. Corp., New York.

Posnette, A. F. 1963. *Virus diseases of apples and pears.* Commonwealth Bureau Hort. and Plantation Crops Tech. Communication 30. (East Malling, England.)

Powell, L. E., and C. Pratt. 1960. Hormones and peach development. *Farm Res. Bull.,* Cornell, 27:11.

Proebsting, E. L. 1936. Kelsey spot of plums in California. *Proc. Am. Soc. Hort. Sci.* 34:272–274.

Proebsting, E. L. 1970. Relation of fall and winter temperatures to flower bud behavior and wood hardiness of deciduous fruit trees (a review). *HortScience* 5:422–424.

Proebsting, E. L., Jr., and H. W. Fogle. 1957. Prune leaf curl and gum spot as influenced by crop load. *Proc. Wash. State Hort. Assoc.* 53:86.

Proebsting, E. L., Jr., and A. L. Kenworthy. 1954. Growth and leaf analysis of Montmorency cherry trees as influenced by solar radiation and intensity of nutrition. *Proc. Am. Soc. Hort. Sci.* 63:41–48.

Proebsting, E. L., Jr., and H. H. Mills. 1961. Response of Richards Early Italian prunes to 2,4,5-TP with varied time of application. *Proc. Wash. State Hort. Assoc.* 57:164–165.

Rehder, A. 1947. *Manual of cultivated trees and shrubs.* Macmillan, New York.

Reimer, F. C. 1925. *Blight resistance in pears and characteristics of pear species and stocks.* Ore. Agr. Exp. Sta. Bull. 214.

Reimer, F. C. 1950. *Development of blight resistant French pear rootstocks.* Ore. Agr. Exp. Sta. Bull. 485.

Richardson, D. G.; D. Kirk; and R. Cain. 1975. Brining cherries mechanical harvesting experiments. *Proc. Oregon Hort. Soc.* 66:18–22.

Roberts, A. N. 1962. Cherry rootstocks. *Proc. Oregon Hort. Soc.* 54:95–98.

Roberts, A. N., and W. M. Mellenthin. 1957. Propagating clonal rootstocks. Oregon Agr. Circ. 578.

Roberts, A. N., and W. M. Mellenthin. 1964. Unusual ratio measures efficiency and vigor. *Western Fruit Grower* 18:26–27.

Roberts, E. H. 1969. Seed dormancy and oxidation processes. In *Dormancy and survival.* Academic Press, New York. Pp. 161–192.

Roberts, K. O. 1964. Maturity of French prunes in relation to time of harvest. Master's Thesis, Oregon State Univ.

Robertson, R. N., and J. F. Turner. 1951. The physiology of growth of apple fruits. II. Respiratory and other metabolic activities as functions of cell number and cell size in fruit development. *Australian J. Sci. Res.* 4:92–107.

Rogers, B. L., and L. P. Batjer. 1954. Seasonal trends of six nutrient elements in the flesh of Winesap and Delicious apple fruits. *Proc. Am. Soc. Hort. Sci.* 63:67–73.

Rogers, B. L.; L. P. Batjer; and A. H. Thompson. 1953. Seasonal trend of several nutrient elements in Delicious apple leaves expressed on a percent and unit area basis. *Proc. Am. Soc. Hort. Sci.* 61:1–5.

Rollins, H. A., Jr., F. S. Howlett, and F. H. Emmert. 1962. Factors affecting apple hardiness and methods of measuring resistance of tissue to low temperature injury. *Ohio Agr. Exp. Sta. Bull.* 901.

Rood, Paul. 1957. Development and evaluation of objective maturity indices for California freestone peaches. *Proc. Am. Soc. Hort. Sci.* 70:104–112.

Rose, D. H., and R. C. Wright. 1952. Commodity storage requirements. In USDA *Refrigerating data book,* 4th ed., Chapter 19.

Rowe, R. N., and P. B. Catlin. 1971. Differential sensitivity to waterlogging and cyanogenesis by peach, apricot, and plum roots. *J. Am. Soc. Hort. Sci.* 96:305–308.

Ryall, A. L., and W. T. Pentzer. 1974. *Handling, transportation, and storage of fruits and vegetables.* The Avi Publishing Co., Westport, Connecticut.

Ryall, A. L.; E. Smith; and W. T. Pentzer. 1941. The elapsed period from full bloom as an index of harvest maturity of pears. *Proc. Am. Soc. Hort. Sci.* 38:273–281.

Ryugo, K., and W. Micke. 1975. Vladimir, a promising dwarfing rootstock for sweet cherry. *HortScience* 10:585.

Sando, C. E. 1937. Coloring matters of Grimes Golden, Jonathan, Stayman, and Winesap apples. *J. Biol. Chem.* 117:45–46.

Sastry, K. K. S., and R. M. Muir. 1963. Gibberellins: Effect on diffusible auxin in fruit development. *Science* 140:494–495.

Savage, E. F. 1970. Cold injury as related to cultural management and possible protective devices for dormant peach trees (a review). *HortScience* 5:425–427.

Schrader, A. L., and P. C. Marth. 1931. Light intensity as a factor in development of apple color and size. *Proc. Am. Soc. Hort. Sci.* 28:552–555.

Schuster, C. E., and R. E. Stephenson. 1940. *Soil moisture, root distribution, and aeration as factors in nut production in Western Oregon.* Ore. Agr. Exp. Sta. Bull. 372.

Scott, D. H.; G. M. Darrow; and F. J. Lawrence. 1972. *Strawberry varieties in the United States.* USDA Farmers' Bull. 1043.

Scott, D. H.; and A. D. Draper; and G. M. Darrow. 1973. *Commercial blueberry growing.* USDA Farmers' Bull. 2254.

Serr, E. F., and H. I. Forde. 1968. Ten new walnut varieties released. *Calif. Agr.* 22:8–10.

Shaw, J. K. 1914. *A study of variation in apples.* Mass. Agr. Exp. Sta. Bull. 149.

Shear, C. B.; H. L. Crane; and A. T. Myers. 1946. Nutrient-element balance: A fundamental concept in plant nutrition. *Proc. Am. Soc. Hort. Sci.* 47:239–248.

Shearer, M. N.; P. B. Lombard; W. M. Mellenthin; and L. W. Martin. 1974. *Drip irrigation research in Oregon.* Oregon Agr. Exp. Sta. Spec. Rep. 412.

Sheets, W. A.; R. M. Bullock; and R. Garren, Jr. 1972. Effects of plant density, training, and pruning on blackberry yield. *J. Am. Soc. Hort. Sci.* 97:262–264.

Siegelman, H. W., and S. B. Hendricks. 1958. Photocontrol of anthocyanin synthesis in apple skin. *Plant Physiol.* 33:185–190.

Skene, D. S. 1966. The distribution of growth and cell division in the fruit of Cox's Orange Pippin. *Ann. Botany* 30:493–512.

Slayter, R. O., and I. C. McIlroy. 1961. *Practical microclimatology.* UNESCO, Paris.

Snyder, J. C., and R. D. Bartram. 1970. *Grafting fruit trees.* Pacific Northwest Bull. 62.

Snyder, J. C., and W. A. Luce. 1950. *Pruning apple and pear trees.* Wash. Agr. Ext. Bull. 381.

Stebbins, R. L. 1967. Goodfruit growing in Oregon—mechanical harvesting of cherries. *The Goodfruit Grower* 17:18–19.

Stebbins, R. L. 1971. *Growing prunes.* Oregon Agr. Ext. Circ. 773.

Stebbins, R. L. 1976. *Training and pruning apple and pear trees.* Pac. Northwest Ext. Pub., Pacific Northwest Bull. 156.

Stebbins, R. L., and W. L. Bluhm. 1975. *When to pick apples and pears.* Oregon Ext. Fact Sheet 147.

Stebbins, R. L.; D. H. Dewey; and V. E. Shull. 1972. Calcium crystals in apple stem, petiole, and fruit tissue. *HortScience* 7:492–493.

Stebbins, R. L.; H. G. Lagerstedt; W. Roberts; and L. Baron. 1967. *Walnut varieties for Oregon.* Oregon Agr. Ext. Fact Sheet 122.

Stebbins, R. L.; M. N. Westwood; and P. B. Lombard. 1972. *Pear rootstocks for Oregon.* Oregon Agr. Ext. Fact Sheet 61.

Steward, F. C. 1968. *Growth and organization in plants.* Addison-Wesley, Reading, Massachusetts.

Stirm, W. L. 1967. What dewpoint means to the fruit grower. *Am. Fruit Grower* 87:18.

Stuart, N. W. 1937. Cold hardiness of some apple understocks and the reciprocal effect of stock and scion on hardiness. *Proc. Am. Soc. Hort. Sci.* 35:386–389.

Strausz, S. D. 1969. A study of the physiology of dormancy in the genus *Pyrus.* Doctoral thesis, Oregon State Univ.

Strydom, D. K. 1975. South Africa pear and apple production. Present and future. *Proc. Wash. State Hort. Assoc.* 71:140–150.

Sudds, R. H., and R. D. Anthony. 1928. The correlation of trunk measurements with tree performance in apples. *Proc. Am. Soc. Hort. Sci.* 25:244–246.

Swingle, D. B. 1946. *The textbook of systematic botany.* McGraw-Hill, New York.

Tavernetti, J. R. 1948. *Construction of farm refrigerators and freezers.* Calif. Agr. Exp. Sta. Circ. 387.

Thompson, A. H., and L. P. Batjer. 1950. Effect of various soil treatments for correcting arsenic injury of peach trees. *Soil Sci.* 69:281–290.

Thompson, M. M. 1967. Role of pollination in nut development. *Proc. Nut Growers Soc. Oregon and Wash.* 53:31–36.

Tucker, L. R. 1934. *A varietal study of the susceptibility of sweet cherries to cracking.* Idaho Agr. Exp. Sta. Bull. 211.

Tucker, L. R., and L. Verner. 1932. *Prune maturity and storage.* Idaho Agr. Exp. Sta. Bull. 196.

Tufts, W. P., and E. B. Morrow. 1925. Fruit bud differentiation in deciduous fruits. *Hilgardia* 1:3–14.

Tukey, H. B. 1933. Growth of peach embryo in relation to growth of fruit and season of ripening. *Proc. Am. Soc. Hort. Sci.* 30:209–218.

Tukey, H. B. 1942. Time interval between full bloom and fruit maturity for several varieties of pears, apples, peaches, and cherries. *Proc. Am. Soc. Hort. Sci.* 40:133–140.

Tukey, H. B. 1964. *Dwarfed fruit trees.* Macmillan, New York.

Tukey, L. D. 1959. *Periodicity in growth of fruits of apples, peaches, and sour cherries with some factors influencing this development.* Penn. Agr. Exp. Sta. Bull. 661.

Tukey, L. D. 1964. A linear electronic device for continuous measurement and recording of fruit enlargement and contraction. *Proc. Am. Soc. Hort. Sci.* 84:653–660.

Unrath, C. R. 1973. Interaction of irrigation, evaporative cooling, and ethephon on fruit quality of Delicious apple. *HortScience* 8:318–319.

Upshall, W. H., ed. 1976. *History of fruit growing and handling in the United States of America and Canada.* 1860–1972. The American Pomological Society, Univ. Park, Pennsylvania.

USDA. 1948. *Woody-plant seed manual.* Misc. Pub. 654.

USDA. 1952. *Insects.* Yearbook of Agr.

USDA. 1953. *Plant diseases.* Yearbook of Agr.

USDA. 1961. *Seeds.* Yearbook of Agr.

USDA. 1970. *Growing raspberries.* Farmers' Bull. 2165.

USDA. 1974. *Crop production, 1974 annual summary.* Stat. Reporting Service.

USDA. 1974. *Seeds of woody plants in the United States.* Agr. Handbook 450.

USDA, ARS. 1976. *Virus diseases of stone fruits.* Agr. Handbook 437.

Vegis, A. 1964. Dormancy in higher plants. *Ann. Rev. Plant Physiol.* 15:185–224.

Verner, L. 1946. *Vegetative propagation of plants.* Idaho Agr. Ext. Circ. 95.

Verner, L. 1955. *Hormone relations in the growth and training of apple trees.* Idaho Agr. Exp. Sta. Res. Bull. 28.

Verner, L. 1962. *A new kind of dendrometer.* Idaho Agr. Exp. Sta. Bull. 389.

Verner, L., and E. C. Blodgett. 1931. *Physiological studies of the cracking of sweet cherries.* Idaho Agr. Exp. Sta. Bull. 184.

Verner, L., and D. F. Franklin. 1960. *Training and pruning Italian prune trees.* Idaho Agr. Exp. Sta. Bull. 335.

Verner, L.; W. J. Kochan; C. E. Loney; D. C. Moore; and A. L. Kamal. 1962. *Internal browning of fresh Italian prunes.* Idaho Agr. Exp. Sta. Res. Bull. 56.

Verner, L.; W. J. Kochan; D. O. Ketchie; A. Kamal; R. W. Braun; J. W. Berry, Jr.; and M. E. Johnson. 1962. *Trunk growth as a guide in orchard irrigation.* Idaho Agr. Exp. Sta. Res. Bull. 52.

Walker, D. R. 1970. Growth substances in dormant buds and seeds (a review). *HortScience* 5:414–417.

Walker, D. R., and S. D. Seeley. 1973. The rest mechanism in decidious fruit trees as influenced by plant growth substances. *Acta Horticulturae* 34:235–239.

Wang, C. Y.; W. M. Mellenthin; and E. Hansen. 1971. Effect of temperature on development of premature ripening in Bartlett pears. *J. Am. Soc. Hort. Sci.* 96:122–126.

Wang, C. Y.; W. M. Mellenthin; and E. Hansen. 1972. Maturation of Anjou pears in relation to chemical composition and reaction of ethylene. *J. Am. Soc. Hort. Sci.* 97:9–12.

Wareing, P. F. 1965. Dormancy in plants. *Sci. Progr.* 53:529–537.

Waring, J. H. 1920. The probable value of trunk circumference as an adjunct to fruit yield in interpreting apple orchard experiments. *Proc. Am. Soc. Hort. Sci.* 17:179–185.

Watt, B. K., and A. L. Merrill. 1963. *Composition of foods.* USDA Agr. Handbook 8.

Weaver, R. J. 1972. *Plant growth substances in agriculture.* W. H. Freeman and Company, San Francisco.

Weiser, C. J. 1970. Cold resistance and acclimation in woody plants. (a review). *HortScience* 5:403–408.

Wellensiek, S. J. 1972. Growth regulators in fruit production. *Acta Horticulturae* 34:1–507.

Went, F. W. 1957. Climate and agriculture. *Scientific American,* June, pp. 82–94.

Westigard, P. H.; P. B. Lombard; and J. H. Grim. 1966. Preliminary investigations of the effect of feeding of various levels of two-spotted spider mite on its Anjou pear host. *Proc. Am. Soc. Hort. Sci.* 89:117–122.

Westigard, P. H.; M. N. Westwood; and P. B. Lombard. 1970. Host preference and resistance of *Pyrus* species to the pear psylla, *Psylla pyricola* Foerster. *J. Am. Soc. Hort. Sci.* 95:34–36.

Westwood, M. N. 1962. Seasonal changes in specific gravity and shape of apple, pear, and peach fruits. *Proc. Am. Soc. Hort. Sci.* 80:90–96.

Westwood, M. N. 1965. A cyclic carbonate and three new carbamates as chemical thinners for apple. *Proc. Am. Soc. Hort. Sci.* 86:37–40.

Westwood, M. N. 1968. Some factors affecting abscission of stone fruits. *Proc. Oregon Hort. Soc.* 60:51–54.

Westwood, M. N. 1969. Tree size control as it relates to high-density orchard systems. *Proc. Wash. State Hort. Assoc.* 65:92–94.

Westwood, M. N. 1970. Rootstock-scion relationships in hardiness of deciduous fruit trees (a review). *HortScience* 5:418–421.

Westwood, M. N. 1972. The role of growth regulators in rooting. *Acta Horticulturae* 34:89–92.

Westwood, M. N., and L. P. Batjer. 1960. Effect of environment and chemical additives on absorption of naphthaleneacetic acid by apple leaves. *Proc. Am. Soc. Hort. Sci.* 76:16–29.

Westwood, M. N.; L. P. Batjer; and H. D. Billingsley. 1960. Effects of several organic spray materials on the fruit growth and foliage efficiency of apple and pear. *Proc. Am. Soc. Hort. Sci.* 76:59–67.

Westwood, M. N.; L. P. Batjer; and H. D. Billingsley. 1967. Cell size, cell number, and fruit density of apples as related to fruit size, position in cluster and thinning method. *Proc. Am. Soc. Hort. Sci.* 91:51–62.

Westwood, M. N., and H. O. Bjornstad. 1968a. Chilling requirements of dormant seeds of fourteen pear species as related to their climatic adaptation. *Proc. Am. Soc. Hort. Sci.* 92:141–149.

Westwood, M. N., and H. O. Bjornstad. 1968b. Effects of gibberellin A_3 on fruit shape and subsequent seed dormancy of apple. *HortScience* 3:19–20.

Westwood, M. N., and H. O. Bjornstad. 1970a. Cherry rootstocks for Oregon. *Proc. Oregon Hort. Soc.* 61:76–79.

Westwood, M. N., and H. O. Bjornstad. 1970b. Some factors affecting rain cracking of sweet cherries. *Proc. Oregon Hort. Soc.* 61:70–75.

Westwood, M. N., and L. T. Blaney. 1963. Non-climatic factors affecting the shape of apple fruits. *Nature* 200:802–803.

Westwood, M. N., and D. J. Burkhart. 1968. Climate influences shape of Delicious. *Am. Fruit Grower* 88:26.

Westwood, M. N.; H. R. Cameron; P. B. Lombard; and C. B. Cordy. 1971. Effects of trunk and rootstock on decline, growth and performance of pear. *J. Am. Soc. Hort. Sci.* 96:147–150.

Westwood, M. N.; M. H. Chaplin; and A. N. Roberts. 1973. Effects of rootstock on growth, bloom, yield, maturity, and fruit quality of prune (*Prunus domestica* L.) *J. Am. Soc. Hort. Sci.* 98:352–357.

Westwood, M. N., and N. E. Chestnut. 1964. Rest-period chilling requirement of Bartlett pear as related to *Pyrus calleryana* and *P. communis* rootstocks. *Proc. Am. Soc. Hort. Sci.* 84:82–87.

Westwood, M. N.; E. S. Degman; and J. H. Grim. 1964. *Effects of long-term fertilizer and management practices on growth and yield of pears grown in a clay adobe soil.* Oregon Agr. Exp. Sta. Tech. Bull. 82.

Westwood, M. N., and R. K. Gerber. 1958. Seasonal light-intensity and fruit-quality factors as related to the method of pruning peaches. *Proc. Am. Soc. Hort. Sci.* 72:85–91.

Westwood, M. N., and J. H. Grim. 1962. Effect of pollinizer placement on long-term yield of Anjou, Bartlett, and Bosc pears. *Proc. Am. Soc. Hort. Sci.* 81:103–107.

Westwood, M. N., and P. B. Lombard. 1966. Pear rootstocks. *Ann. Rep. Oregon State Hort. Soc.* 58:61–68.

Westwood, M. N.; P. B. Lombard; and H. O. Bjornstad. 1976. Performance of Bartlett pear on standard and Old Home × Farmingdale clonal rootstocks. *J. Am. Soc. Hort. Sci.* 101:161–164.

Westwood, M. N.; P. B. Lombard; and L. D. Brannock. 1968. Effect of seeded fruits and foliar appled auxin on seedless fruit set of pear the following year. *HortScience* 3:168–169.

Westwood, M. N.; F. C. Reimer; and V. L. Quackenbush. 1963. Long-term yield as related to ultimate tree size of three pear varieties grown on rootstocks of five *Pyrus* species. *Proc. Am. Soc. Hort. Sci.* 82:103–108.

Westwood, M. N., and A. N. Roberts. 1970. The relationship between trunk cross-sectional area and weight of apple trees. *J. Am. Soc. Hort. Sci.* 95:28–30.

Westwood, M. N.; A. N. Roberts; and H. O. Bjornstad. 1976. Comparison of Mazzard, Mahaleb, and hybrid rootstocks for Montmorency cherry (*Prunus cerasus* L.) *J. Am. Soc. Hort. Sci.* 101:268–269.

Westwood, M. N., and P. H. Westigard. 1969. Degree of resistance among pear species to the woolly pear aphid, *Eriosoma pyricola*. *J. Am. Soc. Hort. Sci.* 94:91–93.

Wilhelm, S. 1974. The garden strawberry: A study of its origin. *Am. Scientist* 62:264–271.

Williams, M. W., and L. P. Batjer. 1964. Site and mode of action of 1-Naphthyl *N*-methylcarbamate (Sevin) in thinning apple. *Proc. Am. Soc. Hort. Sci.* 85:1-10.

Williams, M. W.; H. D. Billingsley; and L. P. Batjer. 1969.

Early season harvest size prediction of Bartlett pears. *J. Am. Soc. Hort. Sci.* 94:596-598.

Williams, M. W., and E. A. Stahly. 1968. Effect of cytokinins on apple shoot development from axillary buds. *HortScience* 3:68-69.

Williams, M. W., and E. A. Stahly. 1969. Effect of cytokinins and gibberellins on shape of Delicious apple fruits. *J. Am. Soc. Hort. Sci.* 94:17-18.

Williams, R. R. 1965. The effect of summer nitrogen applications on the quality of apple blossom. *J. Hort. Sci.* 40:31-41.

Wilson, W. C. 1966. The anatomy and physiology of citrus fruit abscission induced by iodoacetic acid. Doctoral thesis, Univ. of Florida.

Wittenbach, V. A., and M. J. Bukovac. 1974. Cherry fruit abscission: Evidence for time of initiation and the involvement of ethylene. *Plant Physiology* 54:494–498.

Wittenbach, V. A., and M. J. Bukovac. 1975. Cherry fruit abscission: Peroxidase activity in the abscission zone in relation to separation. *J. Am. Soc. Hort. Sci.* 100:387-391.

Wittwer, S. H. 1971. Growth regulants in agriculture. *Outlook in Agr.* 6:205-217.

Woodroof, J. G. 1967. *Tree nuts.* Two vols. Avi. Pub. Co., Westport, Connecticut.

Work, R. A. 1939. Soil moisture control by irrigation. *Agr. Eng.* 20:3-12.

Zielinski, Q. B. 1964. Resistance of sweet cherry varieties to fruit cracking in relation to fruit and pit size and fruit color. *Proc. Am. Soc. Hort. Sci.* 84:98-102.

Zielinski, Q. B.; W. A. Sistrunk; and W. M. Mellenthin. 1959. *Sweet cherries for Oregon.* Oregon Agr. Exp. Sta. Bull. 570.

Zimmerman, L. 1966. New and old ways of frost control. *Proc. Oregon Hort. Soc.* 58:136-138.

Zimmermann, M. H. 1963. How sap moves in plants. *Scientific American,* March, pp. 132–142.

Zimmerman, P. W., and A. E. Hitchcock. 1933. Initiation and stimulation of adventitious roots caused by unsaturated hydrocarbon gases. *Contribution Boyce Thompson Institute* 5:351-369.

Zimmerman, R. H. 1972. Juvenility and flowering in wood plants (a review). *HortScience* 7:447-455.

Index of Scientific Names

Subject Index

*Asterisked words are defined in the Glossary.